Dynamic Spin Fluctuation Theory of Metallic Magnetism

Nikolai B. Melnikov • Boris I. Reser

Dynamic Spin Fluctuation Theory of Metallic Magnetism

Springer

Nikolai B. Melnikov
Moscow State University
Moscow, Russia

Boris I. Reser
Institute of Metal Physics
Russian Academy of Sciences
Ekaterinburg, Russia

ISBN 978-3-319-92972-9 ISBN 978-3-319-92974-3 (eBook)
https://doi.org/10.1007/978-3-319-92974-3

Library of Congress Control Number: 2018944841

Printed on acid-free paper

This Springer imprint is published by the registered company Springer International Publishing AG part of Springer Nature.
The registered company address is: Gewerbestrasse 11, 6330 Cham, Switzerland

To Our Parents

Preface

Accurate description of itinerant electron magnetism at finite temperatures remains an important and to a large extent an unresolved problem. This book presents recent theoretical developments and their applications to ferromagnetic metals.

The objective is twofold: firstly, we introduce the dynamic spin fluctuation theory that takes into account both local and long-wave spin fluctuations. We explain the fundamental role of quantum spin fluctuations in the mechanism of metallic magnetism and illustrate the theory by applying it to real metals and alloys.

Secondly, we provide an accurate and self-contained presentation of the many-body techniques such as Green functions and functional integral method by giving a number of worked-out examples. Most of the many-body textbooks view superconductivity as a key application domain and do not consider magnetism in detail. Our book fills this gap in the literature and could be useful to a wide range of physicists working in solid-state physics, both theoreticians and experimentalists. The introductory chapters are accessible to graduate students.

Our purpose is not only to present the results but also to explain how to obtain them. "Brevity is a sister of talent" but a stepmother of an ordinary man. Therefore, we derive most of the formulae in such a detail that the reader could reproduce them. On the other hand, we limit theoretical methods to those that are essential for developing and explaining our approach.

The discussion in the book always refers to single-crystal one-domain samples. We are interested in bulk magnetic properties of metals; therefore translational invariance is always assumed. We consider an ideal crystal without impurities and neglect the anisotropy effects, i.e. consider cubic ferromagnets. Where possible, we follow the notation of Purcell, Kittel, Raimes, White and Kim.[1] In the introductory chapters we retain $\hbar$, g, μ_B, etc., to give the correct dimensions.

One of us (B.R.) started doing physics as a graduate student of S.V. Vonsovskii and is grateful to him for the interest in the metallic magnetism. We are grateful to V.I. Grebennikov for a long-term, friendly and fruitful cooperation. It is a pleasure to thank N.M. Plakida, M. Probert and L.M. Sandratskii for useful discussions. We would like to thank the editor S.K. Heukerott for her kind support and assistance during the writing process. Last but not least, we thank G.V. Paradezhenko, who read the manuscript thoroughly and spotted some inaccuracies and typos. However, the final responsibility for the content of this book and all remaining typos lies solely with us.

Corrections and suggestions will be gratefully received and may be addressed by email to melnikov@cs.msu.ru.

Moscow, Russia
Ekaterinburg, Russia
March 2018

Nikolai B. Melnikov
Boris I. Reser

[1] E.M. Purcell, *Electricity and Magnetism*, 2nd edn. (McGraw-Hill, New York, 1985); C. Kittel, *Introduction to Solid State Physics*, 8th edn. (Wiley, New York, 2005); S. Raimes, *The Wave Mechanics of Electrons in Metals* (North-Holland, Amsterdam, 1970); R.M. White, *Quantum Theory of Magnetism*, 3rd edn. (Springer, Berlin, 2007); D.J. Kim, *New Perspectives in Magnetism of Metals* (Kluwer/Plenum, New York, 1999).

Contents

List of Symbols

As is customary in physics, we often omit the infinite limits of integration; the symbol $\propto$ denotes proportionality, and 0^+ stands for a positive infinitesimal. Any physical quantity is denoted by the same symbol independently of the number of arguments. In particular, we denote the function and its Fourier transform by the same symbol with different arguments. Bold letters denote vectors; calligraphic letters are used for operators. We omit the tensor product notation, which is explained in Appendix A.1.5. We use n and s_z for the mean values instead of $\bar{n}$ and $\bar{s}_z$, respectively, where it does not lead to confusion with the operators. In the DSFT, all susceptibilities are given in units of $g^2\mu_B^2/2$.

Symbol	Meaning	Definition
$\{A, B\}$	Anticommutator $\{A, B\} = AB + BA$	
$[A, B]$	Commutator $[A, B] = AB - BA$	
$\langle A \rangle$ or $\bar{A}$	Average of a quantity A	
$A^{\mathrm{a}}_{\mathbf{k}\sigma}(\omega)$	Advanced spectral function	Section 6.1
$A^{\mathrm{r}}_{\mathbf{k}\sigma}(\omega)$	Retarded spectral function	Section 6.1
$A^{\alpha\beta}_{jj'}(\tau, \tau')$	Matrix of the quadratic form $F^{(2)}(V)$	Section A.3.3
$A^{\alpha\beta}_{\mathbf{q}m}$	Fourier transform of $A^{\alpha\beta}_{jj'}(\tau, \tau')$	Section A.3.3
$A^{\alpha\beta}_{jj'}(t)$	Oscillating part of $F^{\alpha\beta}_{jj'}(t)$	Section 13.1
$A(t)$	Oscillating part of $F(t)$	Section 13.1
$A(\omega)$	Fourier transform of $A(t)$	Section 13.1
$A(\Omega)$	Scattering amplitude	Section F.2
$\mathbf{A}$	Magnetic vector potential	Section 15.1
$a^\dagger$	Creation operator for an electron	Section 3.3
a	Annihilation operator for an electron	Section 3.3
a	Lattice constant	Section 12.1.2
B	Bulk modulus	Section 11.2.3
$B(\varepsilon)$	Bose function	Section 5.2.2
$b^\dagger$	Creation operator for a phonon	Section G
b	Annihilation operator for a phonon	Section G
$C(r, T)$	Normalized correlation function	Section 15.3
c	Velocity of light	
D	Spin-wave stiffness constant	Equation (5.37)
D_0	Spin-wave stiffness constant at $T = 0$	Section 12.1.2
D_0	Magneto-volume coupling constant for $q = 0$	Section 11.2.3
E_k	Energy of a neutron	Section 14.1
E_λ	Energy of the crystal	Section 14.1

e	Elementary positive charge	
$\mathbf{e}_s$	Polarization vector of the sth normal mode	Section 14.3
$\mathrm{e}^{-2W(\kappa)}$	Debye-Waller factor	Section 14.3
$F, \mathcal{F}$	(Helmholtz) free energy	Section 2.2.1
$F_0(V)$	Energy of the fluctuating field V	Section 8.2
$F_1(V)$	Free energy of electrons in the field V	Section 8.2
$F(V)$	$F(V) = F_0(V) + F_1(V)$	Section 8.3
$F^{(2)}(V)$	Quadratic approximation of $F(V)$	Equation (9.6)
$F(\mathbf{q}, \mathbf{k})$, $F(q)$	Magnetic form-factor	Section 3.3
$F(\mathbf{q}, \omega)$	Lindhard function	Section 3.4
$F_{-+}(\mathbf{q}, \omega)$	Transverse Lindhard function	Equation (5.21)
$F_{+-}(\mathbf{q}, \omega)$	Transverse Lindhard function	Equation (5.23)
$F^{\alpha\beta}_{jj'}(t)$	Dynamic spin correlation function	Section 13.1
$F(t)$	Temporal correlation function	Section 13.1
$F(\omega)$	Fourier transform of $F(t)$	Section 13.1
$f(\varepsilon)$	Fermi function	Section 3.4
$G^{\mathrm{a}}_{jj'\sigma\sigma'}(t, t')$	Advanced real-time Green function	Equation (6.14)
$G^{\mathrm{r}}_{jj'\sigma\sigma'}(t, t')$	Retarded real-time Green function	Equation (6.1)
$G^{\mathrm{a}}_{\mathbf{k}\sigma}(\omega)$	Advanced Green function in Fourier space	Section 6.1
$G^{\mathrm{r}}_{\mathbf{k}\sigma}(\omega)$	Retarded Green function in Fourier space	Section 6.1
$G^{\mathrm{a}0}_{\mathbf{k}}(\omega)$	Zeroth advanced Green function in Fourier space	Section 6.1
$G^{\mathrm{r}0}_{\mathbf{k}}(\omega)$	Zeroth retarded Green function in Fourier space	Section 6.1
$G_{jj'\sigma\sigma'}(\tau, \tau')$	Temperature Green function	Equation (6.23)
$G_{\mathbf{k}\sigma}(\mathrm{i}\omega_n)$	Temperature Green function in Fourier space	Section 6.1.2
$G^{0}_{\mathbf{k}}(\mathrm{i}\omega_n)$	Zeroth temperature Green function in Fourier space	Equation (6.37)
$G, G(V)$	Green function in the field V	Section 8.3
$\bar{G}, \langle G(V) \rangle$	Mean Green function	Section 8.3
$g_\sigma(\varepsilon)$	Single-site Green function	Section 10.1
$\mathbf{H}$	Magnetic field	Section 2.1.1
$H_\pm$	Circular components of magnetic field	Section 2.1.2
$\mathbf{H}(\mathbf{r}, t)$	Magnetic field in real space	Section 2.1.1
$\mathbf{H}(\mathbf{q}, \omega)$	Magnetic field in Fourier space	Section 2.1.1
$\mathcal{H}$	Interacting electrons Hamiltonian	
$\mathcal{H}'$	$\mathcal{H}' = \mathcal{H} - \mu\mathcal{N}_\mathrm{e}$	Section 2.2.1
$\mathcal{H}_0$	One-electron part of $\mathcal{H}$	
$\mathcal{H}_\mathrm{I}$	Interaction part of $\mathcal{H}$	
$\mathcal{H}(V)$	Hamiltonian of electrons in the field V	Section 8.2
J	Exchange interaction constant	
j	Site index	

$\mathbf{k}$	Wavevector	
k_B	Boltzmann constant	
$\mathbf{M}$	Magnetization	Section 2.1.1
M	Atomic mass	Section 14.3
$M_\pm$	Circular components of magnetization	Section 2.1.2
$\mathbf{M}(\mathbf{r}, t)$	Magnetization in real space	Section 2.1.1
$\mathbf{M}(\mathbf{q}, \omega)$	Magnetization in Fourier space	Section 2.1.1
$M(q)$	q-Dependent effective moment	Section 2.2.2
m_{eff}	Curie-Weiss effective moment, $m_{\text{eff}} = M(0)$	Section 2.2.2
$\boldsymbol{\mathcal{M}}$	Magnetic moment operator	Section 2.2.1
$\boldsymbol{\mathcal{M}}(\mathbf{r})$	Magnetic moment in real space	Section 2.2.2
$\boldsymbol{\mathcal{M}}(\mathbf{q})$	Magnetic moment in Fourier space	Section 2.2.2
m_e	Mass of an electron	Section 3.1
m	Mass of a neutron	Section 14.1
$\mathbf{m}$	Magnetic moment operator of an electron	Section 2.2.1
m_0	Magnetic moment at $T = 0$	Section 10.5
m_L	Local magnetic moment	Section 5.2.2
N	Number of atoms (lattice sites in the crystal)	
N_e	Number of electrons in the crystal	
$\mathcal{N}_e$	Total number of electrons operator	Section 3.3
N_b	Number of energy bands	Chapter 13
N_d	Number of d bands	
N_σ	Total number of electrons with the spin σ	Section 4.2
$N(\varepsilon)$	Number of electron states	Section 13.3
n	Band index	
n_e	Number of electrons per atom $n_e = N_e/N$	
$n(\mathbf{r})$	Charge density operator	Section 3.3
n_j	Single-site (atomic) charge	Section 4.1
$n(E_k)$	Neutron density of states	Section 14.1
$n(\omega)$	Phonon density of states	Section 14.3
$n(\varepsilon)$	Electron nonmagnetic density of states	Section 13.3.2
$n_\sigma(\varepsilon)$	Electron spin-polarized density of states	Section 13.3.2
$n_{\nu\sigma}(\varepsilon)$	Electron spin-polarized density of states in the ν-th subband	Section 13.3.2
$\mathbf{P}$	Neutron beam polarization vector	Section 15.1
$P_{\mathbf{k}\to\mathbf{k}'}$	Transition probability from $\mathbf{k}$ to $\mathbf{k}'$	Section 14.2
$p(V)$	Probability density function	Chapter 9
$p^{(2)}(V)$	Gaussian probability density function	Chapter 9
$\mathbf{q}$	Wavevector	
q_B	Radius of a sphere with the same volume as the Brillouin zone (for electrons)	
q_D	Debye radius (same as q_B but for phonons)	Section G.3
$\mathbf{R}_j$	Lattice site	
r_S	Radius of the Slater sphere	Section 13.3

r_c	Correlation radius	Equation (15.32)
$r_{1/2}$	Halfwidth of the correlation function $C(r, T)$	Section 15.3
$\mathbf{S}$	Neutron's spin	Section 15.1
$\bar{S}_z$	Total spin of the crystal	Section 4.2
$S^{\alpha\beta}(\kappa, \omega)$	Scattering function	Equation (15.5)
$\mathbf{s} = (s_x, s_y, s_z)$	Spin operator	Section 3.1
$s = (\mathbf{q}, i)$	Normal mode corresponding to the wavevector $\mathbf{q}$ and polarization $i = 1, 2, 3$	Section 14.3.1
$\mathbf{s}(\mathbf{r})$	Spin density operator in real space	Section 3.3
$\mathbf{s}(\mathbf{q})$	Spin density operator in Fourier space	Section 3.3
$\tilde{s}_\kappa$	Component of the vector $\mathbf{s}_\kappa$ perpendicular to the direction of the scattering vector $\boldsymbol{\kappa}$	Section 15.1
$\mathbf{s}_j$	Single-site (atomic) spin	Equation (3.30)
s_L	Local spin	Section 5.2.2
$s_L(\omega)$	Local spin in the frequency interval $[-\omega, \omega]$	Section 13.1
$\bar{s}_z$	$\bar{s}_z = \bar{S}_z/N$	
T	Temperature	
T	Scattering T-matrix	Section 9.4.3
T_C	Curie temperature	
T_τ	"Time"-ordering operator	Section 6.1.2
T_1^{-1}	Longitudinal nuclear spin relaxation rate	Section 13.4.2
T_2^{-1}	Transverse nuclear spin relaxation rate	Section 13.4.2
t	(Dynamic) time variable	
t_0	Single-site spin relaxation time	Section 13.1
U	Coulomb interaction constant	
$\tilde{U}$	Fourier transform of U	Equation (4.6)
u	(Effective) interaction constant	Section 8.2.2
$\tilde{u}$	Fourier transform of u	Section 9.2
$\mathbf{u}_j$	Lattice site displacement from the equilibrium position $\mathbf{R}_j$	Section 14.3
$V, \mathcal{V}$	Volume of the crystal	
$V(\mathbf{r})$	Periodic potential of the crystal lattice	Equation (3.2)
$V(\mathbf{r})$	Scattering potential in real space	Section 14.1
V_κ	Scattering potential in Fourier space	Section 14.1
$V_j(\tau)$	Fluctuating field in real space	Section 8.2.2
$V_{\mathbf{q}m}$	Fluctuating field in Fourier space	Section 8.2.2
v	Velocity of sound	
W	Bandwidth	
$w_{nj\sigma}(\mathbf{r})$	Wannier function	Section 3.1
$w_{\mathbf{k}\to\mathbf{k}'}$	Transition probability per unit time from the state $\mathbf{k}$ to the state $\mathbf{k}'$	Section 14.1
Z	Canonical partition function	
z	Complex variable	

$\beta = 1/T$		Section B.1
Γ, γ	Lorentzian function halfwidth	Section A.2.3
$\delta(t), \delta(\mathbf{r})$	Dirac delta function in time and space	Section A.2.1
ε	Energy variable	
$\varepsilon_{\mathbf{k}}$	Energy spectrum of $\mathcal{H}_0$	Equation (3.38)
$\varepsilon_{\mathbf{k}\sigma}$	Spin-polarized energy spectrum	Section 4.2
ε_{F}	Fermi level	Section 4.2
$\zeta(x)$	Riemann zeta function	
ζ^{α}	Mean-square field fluctuation $\zeta^{\alpha} = \langle \Delta V_{\alpha}^2 \rangle'$	Section 11.1
η	Small positive parameter	
η	Corrections coefficient in the extended DSFT	Section 11.2
Θ_{C}	Paramagnetic Curie temperature	Equation (2.59)
Θ_{D}	Debye temperature	Section 14.3
$\theta(t)$	Step function	Section A.2.1
$\boldsymbol{\kappa}$	Scattering vector $\boldsymbol{\kappa} = \mathbf{k} - \mathbf{k}'$	Section 14.1
λ	State of the crystal	Chapter 14
μ	Chemical potential	
$\boldsymbol{\mu}$	Neutron's magnetic moment	Section 15.1
μ_{B}	Bohr magneton	Equation (3.41)
ν	Band index	
$\nu(\varepsilon)$	Electron nonmagnetic density of states	Section 3.4
$\nu_{\sigma}(\varepsilon)$	Electron spin-polarized density of states	Section 3.4
Ξ	Grand canonical partition function	
$\xi_{\mathbf{k}}$	$\xi_{\mathbf{k}} = \varepsilon_{\mathbf{k}} - \mu$	
ρ	Canonical density matrix	Equation (2.24)
ρ'	Grand canonical density matrix	Equation (2.25)
ρ_j	Local density matrix	Section 6.1.2
$\Sigma_{\mathbf{k}\sigma}(\omega)$	Self-energy	Section 6.1
$\Sigma_{\sigma}(\varepsilon)$	Coherent potential	Section 10.1
σ	Spin index	
$\mathrm{d}^2\sigma/\mathrm{d}\Omega'\mathrm{d}E_{k'}$	Differential scattering cross-section	Section 14.1
τ	(Thermodynamic) "time" variable	Section 6.1.2
$\Phi(\mathbf{r}_1, \ldots, \mathbf{r}_{N_e})$	Determinantal function (Slater determinant)	Section 3.2
Φ	Flux density of the incident neutrons	Section 14.1
$\varphi(\mathbf{r})$	Wave function of an electron	Section 3.1
$\varphi_{n\mathbf{k}\sigma}(\mathbf{r})$	Bloch function	Equation (3.2)
$\chi_{\alpha\beta}$	Generalized susceptibility	Section 2.1.2
χ_{+}, χ_{-}	Transverse generalized susceptibilities	Equation (2.16)
$\chi_{\uparrow}, \chi_{\downarrow}$	Spin states of an electron	Section 3.1
χ_{CW}	Curie-Weiss susceptibility	Equation (2.59)
χ_{P}	Pauli susceptibility	Equation (3.58)
χ_{S}	Stoner susceptibility	Equation (4.32)
$\chi_{\alpha\beta}(\mathbf{r}, t)$	Dynamic susceptibility in real space	Section 2.1.1
$\chi_{\alpha\beta}(\mathbf{q}, \omega)$	Dynamic susceptibility in Fourier space	Section 2.1.1

$\chi_{-+}(\mathbf{q}, \omega)$	Transverse dynamic susceptibility in Fourier space	Section 2.2.2
$\chi_{+-}(\mathbf{q}, \omega)$	Transverse dynamic susceptibility in Fourier space	Section 2.2.2
$\chi^0_{zz}(\mathbf{q}, \omega)$	Zeroth dynamic susceptibility in Fourier space	Section 3.4
$\chi_{\alpha\beta}(\mathbf{q}, \mathrm{i}\omega_m)$	Thermodynamic susceptibility in Fourier space	Section 6.2.2
$\chi^0_{zz}(\mathbf{q}, \mathrm{i}\omega_m)$	Zeroth thermodynamic susceptibility in Fourier space	Section 6.2.2
$\chi^{\alpha\beta}_{\mathbf{q}m}$	DSFT enhanced susceptibility	Section 10.4
$\chi^{0\alpha\beta}_{\mathbf{q}m}$	DSFT unenhanced susceptibility	Section 9.2
$\chi^{0\alpha}_{\mathrm{L}}(\varepsilon)$	DSFT local unenhanced susceptibility	Section 10.3
$\chi^{+-}_{\mathbf{q}m}$, $\chi^{-+}_{\mathbf{q}m}$	DSFT transverse susceptibilities	Section 12.1
$\chi^{0+-}_{\mathbf{q}m}$, $\chi^{0-+}_{\mathbf{q}m}$	DSFT unenhanced transverse susceptibilities	Section 12.1
$\chi^{0-+}_{\mathrm{L}}(\varepsilon)$	DSFT local unenhanced transverse susceptibility	Section 12.1
$\Psi(\mathbf{r}_1, \ldots, \mathbf{r}_{N_e})$	Many-electron wave function	Section 3.2
$\psi(\mathbf{r})$, $\psi^\dagger(\mathbf{r})$	Field operators	Section 3.3
$\psi_\mathbf{k}$	Wave function of a neutron	Section 14.1
Ω	Thermodynamic potential	Section 2.2.1
Ω	Scattering solid angle	Section 14.1
$\Omega_1(V)$	Thermodynamic potential in the field V	Section 8.3
Ω_{BZ}	Volume of the Brillouin zone	Section 5.2.2
Ω_{WS}	Volume of the Wigner-Seitz cell	Section 5.2.2
Ω_j	Wigner-Seitz cell at the jth lattice site	Section 3.3
ω	Angular frequency	
ω_0	Nuclear magnetic resonance frequency	Section 13.4.2
ω_n	Even "frequency" $\omega_n = (2n + 1)\pi T/\hbar$	Section 6.1.2
ω_m	Odd "frequency" $\omega_m = 2\pi m T/\hbar$	Section 6.2.2
$\omega_m(T)$	Magnetic volume change	Section 11.2.3
ω_s	Vibration frequency of the sth normal mode	Section 14.3
ω_D	Debye frequency	Section 14.3

1 Introduction

Here we give a historical overview of major theoretical developments in metallic magnetism, comment on selected bibliography on magnetism [1–7] and describe the organization of the book.

The first theory of ferromagnetism in metals was developed by *Slater* [8] and *Stoner* [9].[1] This theory is now called the Stoner model. The model replaces the pair interaction of electrons by an interaction of each electron with a spin-dependent mean field that is calculated self-consistently at each temperature. The input data of the model are the electron density of states (DOS) and the so-called Stoner parameter of a particular metal. The mean-field theory of itinerant magnets gives correct qualitative temperature dependence of magnetization, but has its shortcomings: the calculated Curie temperature T_C is too high, the uniform static susceptibility does not follow the Curie-Weiss law, etc. This is not surprising since the theory does not take correlation effects into account.

Description of exchange and correlation effects was improved by the density-functional theory [11–14]. The local spin-density approximation (LSDA) of the density-functional theory gives good quantitative agreement for magnetic properties at *zero* temperature [15–20]. However, attempts to describe *temperature* dependence of magnetic properties within the LSDA do not lead to satisfactory results.

The Stoner theory considers only the spin-flip excitations and thus the temperature dependence of magnetization is only due to the thermal smearing of the Fermi level. The spin-wave excitations were taken into account within the random-phase approximation (RPA) in the single-band Hubbard model by *Izuyama*, *Kim* and *Kubo* [21]. Band calculations for real metals were carried out by several authors [22, 23] and results were found in good agreement with neutron scattering experiments at low temperatures ($T \ll T_C$). At finite temperatures, results of the RPA do not agree with experiment, because the RPA neglects the feedback of spin waves on the thermal equilibrium state.

Considerable progress in explaining temperature dependence of magnetic characteristics was made with the advent of spin fluctuation theory (SFT). The physical picture is as follows. The itinerant electron system is treated as a system of single-site spins, where each spin is the integral of the spin density over the Wigner-Seitz cell centred at the lattice site. At $T = 0$ all spins are aligned along the spontaneous magnetization. At $T > 0$ the spins start fluctuating in direction and modulus. These spin fluctuations are not entirely chaotic because of the exchange interaction. Therefore, the mean spin is nonzero but decreases as the amplitude of the spin fluctuations increase with temperature. At the Curie temperature T_C, the mean spin vanishes but the local spin moment does not. Moreover, the spin directions at neighbouring sites are correlated even above T_C. Information about the local spin moment, spin relaxation time and short-range order can be obtained from temporal and spatial correlation functions. The correlation functions cannot be observed explicitly, but their Fourier transforms can be probed with various nuclear magnetic resonance and neutron scattering techniques. Comparison of experiments and SFT helps to determine the relevance of the above picture.

The development of SFT followed two different directions. The first one is the *phenomenological* SFT, which is based on expansions of the Ginzburg-Landau type of the free energy (for a review see, e.g. [5, 6, 24]). This approach can be justified in the weak ferromagnets and, to some extent, in paramagnetic metals, where long-wave fluctuations of small frequencies dominate. However, it is not a priori clear how well the long-wave approximation is justified in ferromagnetic metals for temperatures that are not close to T_C.

[1] For discussion of earlier works, see [10].

N. B. Melnikov, B. I. Reser, *Dynamic Spin Fluctuation Theory of Metallic Magnetism*,
https://doi.org/10.1007/978-3-319-92974-3_1

The second approach to SFT is the *microscopic* treatment of the spin fluctuations, based on the functional integral method [25, 26]. The static single-site SFT [27–33] uses the coherent-potential approximation (CPA). Originally the CPA was applied for averaging the Green function[2] in binary alloys (see, e.g. [35]). In the SFT the averaging is carried out over an infinite set of fluctuating field configurations. Moreover, the probability density of the fluctuating field is calculated self-consistently. The static single-site approximation of SFT allowed to explain the Curie-Weiss susceptibility and to obtain a reasonable estimate of the Curie temperature, but it gives a small effective moment in the Curie-Weiss law, too rapid decrease ($\propto T$) of magnetization at low temperatures, etc. The main reason of these shortcomings is that the static approximation neglects the quantum nature of the thermal spin fluctuations.

In the series of papers [36–38], the disordered local moment approach [27, 28] was used to generalize the realistic band structure calculations to finite temperatures. The directional fluctuations of the local magnetic moments were taken into account in the CPA of the Korringa-Kohn-Rostoker (KKR) method, which allowed to calculate various magnetic characteristics of iron and nickel in the paramagnetic state (a relativistic extension of this method was introduced by Deák et al. [39]). Another single-site approach [40] is based on the dynamical CPA in the functional integral method. As it was shown in [7], the dynamical CPA is equivalent to the dynamical mean-field theory [41].

To go beyond the single-site approximation in SFT a self-consistent Gaussian approximation was suggested in [42, 43]. Practical use of this method required further development. The long-wave approximation, used by Hertz and Klenin [42,43], was extended in [44] by taking spatial correlations into account. However, both [42,43] and [44] described paramagnets and thus the feedback of spin fluctuations on magnetization was missing. The nonlocal approximation applicable in the whole temperatures range was developed in [45, 46].

The dynamic SFT [45,46] takes into account both single-site and nonlocal interactions. The spin fluctuations are treated by the functional integral method without mapping of the itinerant electron system onto an effective Hamiltonian with classical spins (see, e.g. [47–50]). Application of the dynamic SFT to transition metals [51–54] and Fe-Ni Invar alloys [55,56] showed good quantitative agreement over a wide range of temperatures (for a review, see [57]).

In the last two decades, the magnetism of metals has been discussed in a number of monographs; we would like to comment on seven of them [1–7].

The book by *White* [1] is a unique textbook on magnetism that covers a wide range of topics and uses linear response theory as a basis for understanding a variety of magnetic phenomena. Though the author describes his book as a "poor man's theory of magnetic phenomena", it has been and still remains one of the most influential textbooks in the field. *White* confines the scope of the book in such a way that he does not use many-body techniques. Needless to say that books including many-body techniques become more specific.

Yosida [2] covers a range of topics, including the magnetic properties of itinerant electron systems. In particular, the problem of strong electron correlation is discussed in relation to 3d electrons in iron group metals.

The book by *Kim* [3] is an excellent introduction to the magnetism of metals in the broad sense. This book provides an accessible and self-contained presentation and served a specimen for us in the way the material should be explained. The emphasis of *Kim*'s book is on the role of phonons and electron–phonon interaction. The electrons are considered as the electron gas with the interaction treated mainly within the RPA.

In the books by *Mohn* [4] and *Kübler* [5] thermal spin fluctuations are treated by means of phenomenological models of the Ginzburg–Landau type. As we have already mentioned, this approach applies to the weak ferromagnets and paramagnetic metals rather than to the ferromagnetic metals. The phenomenological spin fluctuation theory (SFT) of *Takahashi* [6] is based on the assumption that the local moment is nearly constant with temperature. This assumption can be justified only for the weak ferromagnets as well.

Kakehashi [7] presented a microscopic treatment of magnetism that is based on the coherent-potential approximation to the functional integral method. This approach uses local approximation, just as the dynamical mean-field theory. The dynamics is taken into account through a limited number of frequencies used on top of the static approximation. As a result the theory has difficulties in describing nonlocal spin excitations such as spin waves and in predicting the magnetic short-range order. As the author admits: "Non-local theory of dynamical spin fluctuations which goes beyond the dynamical CPA is left as a problem of future concern".

Our book presents a dynamic spin fluctuation theory (DSFT) that takes into account both local and long-wave spin fluctuations with all frequencies. This way we are able to describe magnetic properties of metals and alloys in a wide temperature interval including room temperatures. We also show that our theory produces correct low- and high-temperature

[2]The name *Green's function* is perhaps more common, but the omission of the possessive is consistent with the use of the names Schrödinger equation, Fermi surface, Stoner theory, etc. (see, e.g. [34]).

asymptotic behaviour. The use of electronic density of states of the real metal as an input data, allows us to reduce the gap between the spin fluctuation theory and the band theoretical approach and thus to investigate the relationship between metallic magnetism and electronic structure.

The main text can be divided into *two parts*. Chapters 2–6 are introductory; they explain the mechanism of metallic magnetism and present necessary many-body techniques. Chapters 7–15 form the core of the book; they present the DSFT and its applications. In the Appendices, we give proofs of some relations used in the main text and collect supplementary material that can be of interest on its own. The book is organized as follows.

In the *introductory part*, Chap. 2 gives a clear and readable presentation of basic theoretical concepts, which makes the prerequisites minimal. We introduce magnetic susceptibility and derive its general properties. The microscopic treatment is based on the quantum-statistical theory of linear response. Chapter 3 summarizes necessary facts from quantum mechanics and statistical mechanics of electrons in a periodic crystal lattice, including the second-quantization. In Chap. 4 we derive the Stoner mean-field theory from the Hartree-Fock approximation and discuss results of the band calculations for real metals. In Chap. 5 we calculate magnetic susceptibility in the RPA and discuss the magnetic excitations in metals. For simplicity, we confine the presentation in Chaps. 4 and 5 to the single-band Hubbard model.

Stoner theory and RPA employ only elementary methods. To proceed further, we discuss the Green functions at finite temperatures (Chap. 6). First, we consider the fermion-type Green functions that describe energy spectrum and electron correlations. We introduce both the real-time and thermodynamic Green functions and establish the relation between them. Next, we consider the boson-type Green functions, i.e. the real-time and thermodynamic susceptibilities. We obtain the RPA result one more time to show the direct method of calculating the real-time susceptibility from the equation of motion. To go beyond the RPA, either through diagram technique or functional integral method, one should consider the thermodynamic susceptibility. The relation between the dynamic and thermodynamic susceptibilities is derived in a general case and explicitly illustrated in the example of noninteracting electrons.

The *main part* of the book is devoted to the DSFT. In Chap. 7 we illustrate the key idea of replacing the pair interaction by the interaction with a fluctuating field in the example of the Ising model, the simplest model that exhibits magnetic phase transition. We discuss different effects of spin fluctuations on temperature dependence of magnetization and phase transition.

Chapter 8 gives an introduction to the functional integral method in SFT. We begin by constructing a multiband Hubbard Hamiltonian and deriving Hund's rule for metals (the well-known Hund's rule refers to a single atom). The development of the functional integral method itself requires a special form of the model Hamiltonian. We express the multiband Hubbard Hamiltonian in terms of the atomic charge and spin. In the functional integral formalism, we derive expression for the free energy, mean and local spin and spin correlator.

Chapter 9 describes the Gaussian approximation of the fluctuating field in the functional integral method. First, we describe the simplest saddle-point approximation that leads to the Stoner mean-field equations and RPA dynamic susceptibility. The *optimal* Gaussian approximation in the DSFT utilizes a quadratic approximation of the free energy based on a variational principle, which we describe in a rather general form here. The optimal Gaussian approximation allows to take both quantum nature (dynamics) and spatial correlation (nonlocality) of thermal fluctuations of the electron spin density.

The next three chapters present the DSFT and its possible modifications at low and high temperatures. In Chap. 10 we describe the DSFT, show the calculation results for basic magnetic characteristics in different approximations of the theory and compare them with experiment. In Chap. 11 we consider problems of the temperature dependence that appear in the DSFT if the spin fluctuations become large. The temperature dependence in the DSFT can become unstable at high temperatures, well below the Curie temperature T_C.[3] We explain a possible solution to this problem and present the results of the *extended* DSFT in metals and alloys. In Chap. 12 we consider low-temperature versions of the theory, compare its results with the RPA and discuss the temperature region where one should go beyond the spin-wave approximation in describing magnetic and neutron-scattering measurements.

Chapter 13 is devoted to studying the spin-correlation effects in metals. We begin with qualitative estimates of the correlation effects on the magnitude and relaxation time of a single-site spin. Then we apply the DSFT to calculate temperature dependency of the local susceptibility, dynamic spin correlation function, local magnetic moment and nuclear spin-relaxation rates.

[3] In ferromagnetic metals, this first-order-like behaviour happens too far from T_C to be interpreted in the framework of the critical phenomena, which is not considered in the book (for a discussion of the *critical region* see, e.g. [1] and refs. therein).

One of the main experimental methods in studying magnetic properties of metals and alloys is the neutron scattering. Chapter 14 introduces to neutron scattering theory for *itinerant electron* magnets. We obtain an expression for the neutron scattering cross-section and estimate the effect of lattice vibrations. Chapter 15 studies the spatial spin correlator in the DSFT and compares its Fourier transform (effective moment) with the polarized neutron scattering experiment.

Appendices are structured as follows. Appendix A gives a quick introduction to various mathematical methods. Most of them are used in many-body systems far beyond spin fluctuation theory. In explaining mathematical methods we tried to be as much down to earth as possible and always illustrate general concepts with concrete examples from the main text. Appendix B introduces some less familiar mathematical tools: the ordered exponential, functional derivative, Stratonovich-Hubbard transformation and optimal Gaussian approximation. In Appendix C we derive and summarize necessary formulae related to the Fourier transformations. Appendices D and E give detailed proofs of specific results in RPA and DSFT. In Appendices F and G we collected necessary material on the scattering theory and phonons. Appendix H presents calculation methods for various integrals such as the Fermi integrals over energies and integrals over the Brillouin zone. Finally, in Appendix I we give a short overview of the Fortran code we developed and used for the DSFT calculations.

References

1. R.M. White, *Quantum Theory of Magnetism*, 3rd edn. (Springer, Berlin, 2007)
2. K. Yosida, *Theory of Magnetism*, 2nd edn. (Springer, Berlin, 1998)
3. D.J. Kim, *New Perspectives in Magnetism of Metals* (Kluwer/Plenum, New York, 1999)
4. P. Mohn, *Magnetism in the Solid State*, 2nd edn. (Springer, Berlin, 2006)
5. J. Kübler, *Theory of Itinerant Electron Magnetism*, 2nd edn. (Oxford University Press, Oxford, 2009)
6. Y. Takahashi, *Spin Fluctuation Theory of Itinerant Electron Magnetism* (Springer, Berlin, 2013)
7. Y. Kakehashi, *Modern Theory of Magnetism in Metals and Alloys* (Springer, Berlin, 2012)
8. J.C. Slater, Phys. Rev. **49**, 537 (1936)
9. E.C. Stoner, Proc. R. Soc. Lond. A **165**, 372 (1938)
10. A. Sommerfeld, H. Bethe, in *Handbuch der Physik*, vol. 24, 2nd edn. (Springer, Berlin, 1933)
11. U. von Barth, L. Hedin, J. Phys. C Solid State Phys. **5**, 1629 (1972)
12. O. Gunnarsson, B.I. Lundqvist, Phys. Rev. B **13**, 4274 (1976)
13. W. Kohn, P. Vashishta, in *Theory of the Inhomogeneous Electron Gas*, ed. by S. Lundqvist, N.H. March (Plenum, New York, 1983), pp. 79–147
14. R.O. Jones, O. Gunnarsson, Rev. Mod. Phys. **61**, 689 (1989)
15. O. Gunnarsson, J. Phys. F: Met. Phys. **6**, 587 (1976)
16. O. Gunnarsson, Physica **91B**, 329 (1977)
17. U.K. Poulsen, J. Kollår, O.K. Andersen, J. Phys. F: Met. Phys. **6**, L241 (1976)
18. O.K. Andersen, J. Madsen, U.K. Poulsen, O. Jepsen, J. Kollår, Physica **86–88B**, 249 (1977)
19. J.F. Janak, Phys. Rev. B **16**, 255 (1977)
20. V.L. Moruzzi, J.F. Janak, A.R. Williams, *Calculated Electronic Properties of Metals* (Pergamon, New York, 1978)
21. T. Izuyama, D.J. Kim, R. Kubo, J. Phys. Soc. Jpn. **18**(7), 1025 (1963)
22. J.F. Cooke, J.W. Lynn, H.L. Davis, Phys. Rev. B **21**, 4118 (1980)
23. J. Callaway, A.K. Chatterjee, S.P. Singhal, A. Ziegler, Phys. Rev. B **28**, 3818 (1983)
24. T. Moriya, *Spin Fluctuations in Itinerant Electron Magnetism* (Springer, Berlin, 1985)
25. R.L. Stratonovich, Sov. Phys. Dokl. **2**, 416 (1958)
26. J. Hubbard, Phys. Rev. Lett. **3**(2), 77 (1959)
27. J. Hubbard, Phys. Rev. B **19**, 2626 (1979)
28. J. Hubbard, Phys. Rev. B **20**, 4584 (1979)
29. H. Hasegawa, J. Phys. Soc. Jpn. **46**, 1504 (1979)
30. H. Hasegawa, J. Phys. Soc. Jpn. **49**, 178 (1980)
31. H. Hasegawa, J. Phys. Soc. Jpn. **49**, 963 (1980)
32. V.I. Grebennikov, Y.I. Prokopjev, O.B. Sokolov, E.A. Turov, Phys. Met. Metallogr. **52**, 1 (1981)
33. B.I. Reser, V.I. Grebennikov, Phys. Met. Metallogr. **83**, 127 (1997)
34. S. Raimes, *Many-Electron Theory* (North-Holland, Amsterdam, 1972)
35. H. Ehrenreich, L.M. Schwartz, in *Solid State Physics: Advances in Research and Applications*, vol. 31, ed. by H. Ehrenreich, F.S.D. Turnbull (Academic, New York, 1976), p. 149
36. B.L. Gyorffy, A.J. Pindor, J. Staunton, G.M. Stocks, H. Winter, J. Phys. F: Met. Phys. **15**, 1337 (1985)
37. J. Staunton, B.L. Gyorffy, A.J. Pindor, G.M. Stocks, H. Winter, J. Phys. F: Met. Phys. **15**, 1387 (1985)
38. J. Staunton, B.L. Gyorffy, G. Stocks, J. Wadsworth, J. Phys. F: Met. Phys. **16**, 1761 (1986)
39. A. Deák, E. Simon, L. Balogh, L. Szunyogh, M. dos Santos Dias, J.B. Staunton, Phys. Rev. B **89**, 224401 (2014)
40. Y. Kakehashi, Phys. Rev. B **65**, 184420 (2002)
41. G. Kotliar, S.Y. Savrasov, K. Haule, V.S. Oudovenko, O. Parcollet, C.A. Marianetti, Rev. Mod. Phys. **78**, 865 (2006)
42. J.A. Hertz, M.A. Klenin, Phys. Rev. B **10**(3), 1084 (1974)
43. J.A. Hertz, M.A. Klenin, Physica B&C **91B**, 49 (1977)

44. V.I. Grebennikov, J. Magn. Magn. Mater. **84**, 59 (1990)
45. V.I. Grebennikov, Phys. Solid State **40**, 79 (1998)
46. B.I. Reser, V.I. Grebennikov, Phys. Met. Metallogr. **85**, 20 (1998)
47. L.M. Sandratskii, Adv. Phys. **47**, 91 (1998)
48. A.I. Lichtenstein, M.I. Katsnelson, G. Kotliar, Phys. Rev. Lett. **87**, 067205 (2001)
49. A. Ruban, S. Khmelevskyi, P. Mohn, B. Johansson, Phys. Rev. B **75**, 054402 (2007)
50. P.-W. Ma, S.L. Dudarev, Phys. Rev. B **86**, 054416 (2012)
51. V.I. Grebennikov, S.A. Gudin, Phys. Met. Metallogr. **85**, 258 (1998)
52. B.I. Reser, J. Phys. Condens. Matter **11**, 4871 (1999)
53. B.I. Reser, J. Phys. Condens. Matter **12**, 9323 (2000)
54. B.I. Reser, J. Phys. Condens. Matter **14**, 1285 (2002)
55. B.I. Reser, J. Phys. Condens. Matter **16**, 361 (2004)
56. B.I. Reser, Phys. Met. Metallogr. **103**, 373 (2007)
57. N.B. Melnikov, B.I. Reser, Phys. Met. Metallogr. **117**(13), 1328 (2016)

2 Basics of Metallic Magnetism

In nearly every theory there exist steps that are omitted in the theoretical papers and not treated in the textbooks. These steps are obviously designed to keep the experimental physicists in their place. (H. Frauenfelder, The Mössbauer Effect, *Benjamin, New York, 1962)*

This chapter gives a brief introduction to magnetism of metals and is designed in such a way as to make the book self-contained. We begin with basic topics and elements of the phenomenological theory. We introduce the notion of magnetic susceptibility, which is central in metallic magnetism, and derive some of its most general properties that will be used in the following chapters. Next, we describe the magnetization that originates from the spin of the electrons and their correlated motion. We present the quantum-statistical theory of linear response and some of its applications. Our discussion of microscopic properties is limited to those that do not rely on a specific model Hamiltonian. For a more extensive introduction to magnetism, see, for instance, Refs. [1–7].

2.1 Magnetic Susceptibility: Macroscopic Approach

2.1.1 Generalized Magnetic Susceptibility

Any system in an applied field may be characterized by a response function. If the magnetic field **H** acts as an "input" and the magnetization **M** is the "output", the response function χ is the magnetic susceptibility: $\mathbf{M} = \chi \mathbf{H}$. In general, the applied field may depend on space and time. The resulting magnetization will also vary in space and time.

Let us consider the magnetization $\mathbf{M}(\mathbf{r}, t)$ associated with a particular magnetic field $\mathbf{H}(\mathbf{r}, t)$. These quantities are related to their Fourier components by

$$\mathbf{H}(\mathbf{r}, t) = \frac{1}{V} \sum_{\mathbf{q}} \frac{1}{2\pi} \int \mathbf{H}(\mathbf{q}, \omega)\, \mathrm{e}^{\mathrm{i}(\mathbf{qr}-\omega t)}\, \mathrm{d}\omega, \tag{2.1}$$

$$\mathbf{M}(\mathbf{r}, t) = \frac{1}{V} \sum_{\mathbf{q}} \frac{1}{2\pi} \int \mathbf{M}(\mathbf{q}, \omega)\, \mathrm{e}^{\mathrm{i}(\mathbf{qr}-\omega t)}\, \mathrm{d}\omega, \tag{2.2}$$

where $\mathbf{q}$ is the wavevector and ω is the frequency (V is the volume of the magnet). The generalized wavevector- and frequency-dependent susceptibility is defined by

$$M_\alpha(\mathbf{q}, \omega) = \sum_{\mathbf{q}'} \int \sum_{\beta} \chi_{\alpha\beta}(\mathbf{q}, \mathbf{q}', \omega, \omega') H_\beta(\mathbf{q}', \omega')\, \mathrm{d}\omega',$$

where α, β are Cartesian coordinates x, y, z, or shortly,

$$\mathbf{M}(\mathbf{q}, \omega) = \sum_{\mathbf{q}'} \int \chi(\mathbf{q}, \mathbf{q}', \omega, \omega') \mathbf{H}(\mathbf{q}', \omega')\, \mathrm{d}\omega', \tag{2.3}$$

N. B. Melnikov, B. I. Reser, *Dynamic Spin Fluctuation Theory of Metallic Magnetism*,
https://doi.org/10.1007/978-3-319-92974-3_2

where $\chi(\mathbf{q}, \mathbf{q}', \omega, \omega')$ is the susceptibility tensor. Substitution of (2.3) in (2.2) gives

$$\mathbf{M}(\mathbf{r}, t) = \frac{1}{V}\sum_{\mathbf{q}\mathbf{q}'}\frac{1}{2\pi}\iint \chi(\mathbf{q}, \mathbf{q}', \omega, \omega')\mathrm{e}^{\mathrm{i}(\mathbf{q}\mathbf{r}-\omega t)}\mathbf{H}(\mathbf{q}', \omega')\,\mathrm{d}\omega\,\mathrm{d}\omega'.$$

Using the Fourier transform

$$\mathbf{H}(\mathbf{q}', \omega') = \iint \mathbf{H}(\mathbf{r}', t')\,\mathrm{e}^{-\mathrm{i}(\mathbf{q}'\mathbf{r}'-\omega' t')}\,\mathrm{d}\mathbf{r}'\mathrm{d}t',$$

we obtain

$$\mathbf{M}(\mathbf{r}, t) = \iint \chi(\mathbf{r}, \mathbf{r}', t, t')\,\mathbf{H}(\mathbf{r}', t')\,\mathrm{d}\mathbf{r}'\mathrm{d}t', \tag{2.4}$$

where

$$\chi(\mathbf{r}, \mathbf{r}', t, t') = \frac{1}{V}\sum_{\mathbf{q}\mathbf{q}'}\frac{1}{2\pi}\iint \chi(\mathbf{q}, \mathbf{q}', \omega, \omega')\,\mathrm{e}^{\mathrm{i}(\mathbf{q}\mathbf{r}-\omega t)}\mathrm{e}^{-\mathrm{i}(\mathbf{q}'\mathbf{r}'-\omega' t')}\,\mathrm{d}\omega\,\mathrm{d}\omega'.$$

We consider bulk magnetism in an ideal crystal with cyclic boundary conditions. Due to translational invariance, the susceptibility must be a function only of the relative coordinate $\mathbf{r} - \mathbf{r}'$. This implies that in the wavevector-dependent susceptibility $\mathbf{q} = \mathbf{q}'$. Furthermore, if the medium is stationary, the temporal dependence is $t - t'$, which implies a response at the same frequency $\omega = \omega'$. Therefore, the susceptibility takes the form

$$\chi(\mathbf{q}, \mathbf{q}', \omega, \omega') = \chi(\mathbf{q}, \omega)\delta_{\mathbf{q}\mathbf{q}'}\delta_{\omega\omega'}$$

and relation (2.3) can be written as

$$\mathbf{M}(\mathbf{q}, \omega) = \chi(\mathbf{q}, \omega)\mathbf{H}(\mathbf{q}, \omega). \tag{2.5}$$

Relation between original quantities (2.4) takes the form

$$\mathbf{M}(\mathbf{r}, t) = \iint \chi(\mathbf{r} - \mathbf{r}', t - t')\,\mathbf{H}(\mathbf{r}', t')\,\mathrm{d}\mathbf{r}'\mathrm{d}t',$$

where

$$\chi(\mathbf{r}, t) = \frac{1}{V}\sum_{\mathbf{q}}\frac{1}{2\pi}\int \chi(\mathbf{q}, \omega)\,\mathrm{e}^{\mathrm{i}(\mathbf{q}\mathbf{r}-\omega t)}\,\mathrm{d}\omega$$

and its Fourier transform is

$$\chi(\mathbf{q}, \omega) = \iint \chi(\mathbf{r}, t)\,\mathrm{e}^{-\mathrm{i}(\mathbf{q}\mathbf{r}-\omega t)}\,\mathrm{d}\mathbf{r}\,\mathrm{d}t. \tag{2.6}$$

Since $\chi(\mathbf{r}, t)$ is real, its Fourier transform satisfies the following relation:

$$\left(\chi_{\alpha\beta}(\mathbf{q}, \omega)\right)^* = \chi_{\alpha\beta}(-\mathbf{q}, -\omega), \tag{2.7}$$

where the asterisk stands for the complex conjugate.

If the crystal has the inversion symmetry, which is the case for the majority of transition metals and alloys, we have

$$\chi_{\alpha\beta}(\mathbf{q}, \omega) = \chi_{\alpha\beta}(-\mathbf{q}, \omega). \tag{2.8}$$

From (2.7) we obtain

$$\chi_{\alpha\beta}(\mathbf{q},\omega) = \left(\chi_{\alpha\beta}(\mathbf{q},-\omega)\right)^*.$$

Therefore, the real part is an even function and the imaginary part is an odd function of ω:

$$\mathrm{Re}\chi_{\alpha\beta}(\mathbf{q},\omega) = \mathrm{Re}\chi_{\alpha\beta}(\mathbf{q},-\omega), \tag{2.9}$$

$$\mathrm{Im}\chi_{\alpha\beta}(\mathbf{q},\omega) = -\mathrm{Im}\chi_{\alpha\beta}(\mathbf{q},-\omega). \tag{2.10}$$

In general $\chi(\mathbf{r},t)$ depends on $\mathbf{H}(\mathbf{r},t)$. We consider the case of small magnetic fields, so that the dependence between $\mathbf{M}(\mathbf{r},t)$ and $\mathbf{H}(\mathbf{r},t)$ is linear.

2.1.2 Symmetry Relations

In the most general case the susceptibility tensor $\chi(\mathbf{r},t)$ has the form

$$\chi = \begin{bmatrix} \chi_{xx} & \chi_{xy} & \chi_{xz} \\ \chi_{yx} & \chi_{yy} & \chi_{yz} \\ \chi_{zx} & \chi_{zy} & \chi_{zz} \end{bmatrix}$$

(for brevity, we temporarily omit the $\mathbf{r}$ and t dependence).

In the ferromagnetic state, the crystal has a preferential direction of the spontaneous magnetization $\mathbf{M}$, which we chose to align the z-axis. Then all components of the tensor χ must be invariant under any rotation about the z-axis. Under an arbitrary rotation, the tensor components change as (see, e.g. [8])

$$\chi'_{\alpha'\beta'} = \sum_{\alpha\beta} a_{\alpha\alpha'} a_{\beta\beta'} \chi_{\alpha\beta}, \tag{2.11}$$

where $a_{\alpha\alpha'}$ and $a_{\beta\beta'}$ are the cosines between the old axis α, β and the new ones α', β'. For the rotation by the angle φ about the z-axis, we have

$$\begin{aligned} &a_{xx'} = a_{yy'} = \cos\varphi, \quad a_{zz'} = 1, \\ &a_{yx'} = -a_{xy'} = \sin\varphi, \\ &a_{zx'} = a_{zy'} = a_{xz'} = a_{yz'} = 0. \end{aligned} \tag{2.12}$$

It suffices to take into account the invariance of three components, for instance χ_{xx}, χ_{xz} and χ_{zx}. Substituting the coefficients (2.12) to the formula (2.11) and using $\chi'_{\alpha'\beta'} = \chi_{\alpha\beta}$, we obtain

$$\begin{aligned} \sin^2\varphi\,(\chi_{yy} - \chi_{xx}) + \sin\varphi\cos\varphi\,(\chi_{xy} + \chi_{yx}) &= 0, \\ (\cos\varphi - 1)\chi_{xz} + \sin\varphi\,\chi_{yz} &= 0, \\ (\cos\varphi - 1)\chi_{zx} - \sin\varphi\,\chi_{zy} &= 0. \end{aligned}$$

Since the angle φ is arbitrary, we have

$$\chi_{xx} = \chi_{yy}, \qquad \chi_{xy} = -\chi_{yx}, \qquad \chi_{xz} = \chi_{yz} = \chi_{zx} = \chi_{zy} = 0.$$

Thus, the susceptibility tensor of a system with the axial symmetry is written in the form

$$\chi = \begin{bmatrix} \chi_{xx} & \chi_{xy} & 0 \\ -\chi_{xy} & \chi_{xx} & 0 \\ 0 & 0 & \chi_{zz} \end{bmatrix}. \tag{2.13}$$

Note that in the paramagnetic state, the susceptibility tensor is diagonal, with equal components, so that any direction can be taken as the symmetry axis. Thus,

$$\chi = \begin{bmatrix} \chi_{xx} & 0 & 0 \\ 0 & \chi_{xx} & 0 \\ 0 & 0 & \chi_{xx} \end{bmatrix}.$$

In a system with axial symmetry it is useful to introduce the circular components of the magnetic field and magnetization,

$$H_\pm = H_x \pm \mathrm{i}H_y, \qquad M_\pm = M_x \pm \mathrm{i}M_y. \tag{2.14}$$

If the field **H** has only H_+ component it corresponds to a transverse vector that is circularly polarized with *right-hand* rotation (for details, see, e.g. [9, 10]). Similarly, the field **H** with only H_- component corresponds to a transverse vector that is circularly polarized with *left-hand* rotation. Using (2.13) and (2.14), it is easy to verify that H_+ produces only M_+ component and H_- produces only M_- component,[1]

$$M_\pm = \frac{1}{2}\chi_\pm H_\pm, \tag{2.15}$$

where

$$\frac{1}{2}\chi_+ = \chi_{xx} - \mathrm{i}\chi_{xy}, \qquad \frac{1}{2}\chi_- = \chi_{xx} + \mathrm{i}\chi_{xy}. \tag{2.16}$$

From (2.13) we also have

$$M_z = \chi_{zz} H_z.$$

The components $\chi_\pm$ are called *transverse susceptibilities* and χ_{zz} is called *longitudinal susceptibility*.

2.1.3 Dispersion Relations

Here we establish a general relation between the real and imaginary parts of the susceptibility

$$\chi_{\alpha\beta}(\mathbf{q},\omega) = \mathrm{Re}\chi_{\alpha\beta}(\mathbf{q},\omega) + \mathrm{i}\mathrm{Im}\chi_{\alpha\beta}(\mathbf{q},\omega).$$

If the system obeys the principle of causality, then $\chi(\mathbf{r}-\mathbf{r}', t-t') = 0$ for $t < t'$. Hence the time integral in (2.6) runs only from 0 to ∞; that is,

$$\chi(\mathbf{q},\omega) = \int_0^\infty \chi(\mathbf{q},t)\,\mathrm{e}^{\mathrm{i}\omega t}\,\mathrm{d}t. \tag{2.17}$$

Then $\chi(\mathbf{q},\omega)$ can be analytically continued into the upper half of the complex plane $z = \omega + \mathrm{i}\eta$,

$$\chi(\mathbf{q},z) = \int_0^\infty \chi(\mathbf{q},t)\,\mathrm{e}^{\mathrm{i}\omega t}\mathrm{e}^{-\eta t}\,\mathrm{d}t. \tag{2.18}$$

[1] The factor $1/2$ in the definition of $\chi_\pm$ is introduced for consistency with further notation.

Indeed, if the integral in (2.17) is bounded, then so is the integral in (2.18), since the factor $e^{-\eta t}$ with $\eta > 0$ can only improve the convergence of the integral.

From the physical argument, it follows that $\mathrm{Im}\chi_{\alpha\beta}(\mathbf{q}, \omega)$ must vanish as $\omega \to \infty$, but $\mathrm{Re}\chi_{\alpha\beta}(\mathbf{q}, \omega)$, in general, tends to a nonzero value (for details, see, e.g. [7, 9]). If we define

$$\lim_{\omega\to\infty} \mathrm{Re}\chi_{\alpha\beta}(\mathbf{q}, \omega) = \chi_{\alpha\beta}(\mathbf{q}, \infty),$$

then $\chi_{\alpha\beta}(\mathbf{q}, z) - \chi_{\alpha\beta}(\mathbf{q}, \infty)$ is an analytic function in the upper half-plane which vanishes as $|z| \to \infty$. We consider a closed contour C that runs from $-\infty$ to ∞ along the real axis and closes in the upper half plane. The residue theorem then says

$$\oint_C \frac{\chi_{\alpha\beta}(\mathbf{q}, z) - \chi_{\alpha\beta}(\mathbf{q}, \infty)}{z - \omega + \mathrm{i}0^+} \mathrm{d}z = 0,$$

because the pole of the integrand at $z = \omega - \mathrm{i}0^+$ lies in the lower half plane, outside the contour. The integral over the arc in the upper half-plane is equal to zero. By applying the Sokhotsky formula (A.44), we write the integral over the real axis as

$$\mathcal{P} \int \frac{\chi_{\alpha\beta}(\mathbf{q}, \omega') - \chi_{\alpha\beta}(\mathbf{q}, \infty)}{\omega' - \omega} \mathrm{d}\omega' - \mathrm{i}\pi \left(\chi_{\alpha\beta}(\mathbf{q}, \omega) - \chi_{\alpha\beta}(\mathbf{q}, \infty)\right) = 0,$$

where $\mathcal{P}$ denotes the Cauchy principle value of the integral following it. Taking the real and imaginary parts, we obtain the result

$$\mathrm{Re}\big[\chi_{\alpha\beta}(\mathbf{q}, \omega) - \chi_{\alpha\beta}(\mathbf{q}, \infty)\big] = \frac{1}{\pi} \mathcal{P} \int \frac{\mathrm{Im}\chi_{\alpha\beta}(\mathbf{q}, \omega')}{\omega' - \omega} \mathrm{d}\omega', \tag{2.19}$$

$$\mathrm{Im}\chi_{\alpha\beta}(\mathbf{q}, \omega) = -\frac{1}{\pi} \mathcal{P} \int \frac{\mathrm{Re}\big[\chi_{\alpha\beta}(\mathbf{q}, \omega') - \chi_{\alpha\beta}(\mathbf{q}, \infty)\big]}{\omega' - \omega} \mathrm{d}\omega', \tag{2.20}$$

where ω' stands for $\omega' + \mathrm{i}0^+$. Relations of this type are frequently termed the *Kramers-Kronig relations* (see, e.g. [5–7, 9]). They show that the real and imaginary parts of the susceptibility are not independent but related to each other (for numerical calculation of the integral (2.19), see Appendix H.1).

2.2 Magnetic Susceptibility: Microscopic Approach

2.2.1 Magnetization and Spin

Since the spin of *electrons* is principally responsible for magnetism in a metal, we consider the system of interacting electrons. Quantum mechanically, this system is described by a Hamiltonian operator $\mathcal{H}$. To find the magnetization, we must take the expectation value of the magnetic moment operator $\boldsymbol{\mathcal{M}} = (\mathcal{M}_x, \mathcal{M}_y, \mathcal{M}_z)$,

$$(\Psi, \mathcal{M}_\alpha \Psi) = \int \Psi^* \mathcal{M}_\alpha \Psi \, \mathrm{d}\mathbf{r}_1 \ldots \mathrm{d}\mathbf{r}_{N_e},$$

where N_e is the total number of electrons. The wave function $\Psi(\mathbf{r}_1, \ldots, \mathbf{r}_{N_e})$ can be represented as a superposition of the eigenfunctions $\Psi_k(\mathbf{r}_1, \ldots, \mathbf{r}_{N_e})$ of the Hamiltonian $\mathcal{H}$,

$$\Psi(\mathbf{r}_1, \ldots, \mathbf{r}_{N_e}) = \sum_{k'} c_{k'} \Psi_{k'}(\mathbf{r}_1, \ldots, \mathbf{r}_{N_e}).$$

The expectation value then becomes

$$(\Psi, \mathcal{M}_\alpha \Psi) = \sum_{kk'} c_k^* c_{k'} \mathcal{M}^\alpha_{kk'},$$

where $\mathcal{M}^{\alpha}_{kk'} = (\Psi_k, \mathcal{M}_\alpha \Psi_{k'})$ is the matrix element. The fact that we are describing the system at a temperature T implies that the system is in equilibrium with some heat bath, i.e. it cannot be described by a unique wave function Ψ (for details see, e.g. [11, 12]). Taking the statistical average over different states Ψ (the ensemble average), we come to

$$\langle \mathcal{M}_\alpha \rangle = \sum_{kk'} \langle c_k^* c_{k'} \rangle \mathcal{M}^{\alpha}_{kk'}. \tag{2.21}$$

Introducing the density matrix ρ with the elements

$$\rho_{kk'} = \langle c_{k'}^* c_k \rangle,$$

we rewrite the quantum-statistical average (2.21) in the form

$$\langle \mathcal{M}_\alpha \rangle = \sum_{kk'} \rho_{k'k} \mathcal{M}^{\alpha}_{kk'} = \mathrm{Tr}(\rho \mathcal{M}_\alpha).$$

Notice that we derived the average of the magnetic moment over the entire system. If we are interested in the magnetization at the point $\mathbf{r}$, this behaviour can be obtained by

$$\boldsymbol{\mathcal{M}}(\mathbf{r}) = \sum_{i=1}^{N_e} \mathbf{m}_i \delta(\mathbf{r} - \mathbf{r}_i). \tag{2.22}$$

Since the delta function $\delta(\mathbf{r} - \mathbf{r}_i)$ has dimensions of a reciprocal volume, $\boldsymbol{\mathcal{M}}(\mathbf{r})$ is the magnetic moment operator per unit volume. Here $\mathbf{m}_i$ is the magnetic moment operator associated with the spin of ith electron. Therefore, the magnetization $\mathbf{M}(\mathbf{r}) = \langle \boldsymbol{\mathcal{M}}(\mathbf{r}) \rangle$ becomes

$$\mathbf{M}(\mathbf{r}) = \mathrm{Tr}(\rho \boldsymbol{\mathcal{M}}(\mathbf{r})). \tag{2.23}$$

We recall that when the system is in equilibrium through energy exchange with a heat bath (canonical ensemble), the density matrix ρ is determined by

$$\rho = \frac{1}{Z} \mathrm{e}^{-\mathcal{H}/T}, \qquad Z = \mathrm{Tr} \mathrm{e}^{-\mathcal{H}/T}, \tag{2.24}$$

where Z is the partition function and T is temperature (in energy units). In studying a many-body system such as interacting electrons, it is convenient to assume that the system is in equilibrium with a large bath not only through energy exchange but through particle exchange as well (the grand canonical ensemble). In this case, the density matrix is given by

$$\rho' = \frac{1}{\Xi} \mathrm{e}^{-\mathcal{H}'/T}, \qquad \Xi = \mathrm{Tr} \mathrm{e}^{-\mathcal{H}'/T}. \tag{2.25}$$

Here Ξ is the grand canonical partition function and $\mathcal{H}' = \mathcal{H} - \mu \mathcal{N}_e$, where $\mathcal{N}_e$ is the total number of electrons operator and μ is the chemical potential.

The thermodynamic potential Ω and the (Helmholtz) free energy F are related to the partition functions Ξ and Z as

$$\Omega = -T \ln \Xi, \qquad F = -T \ln Z.$$

The functions $\Omega(V, T, \mu)$ and $F(V, T, N_e)$, where V is the volume of the crystal and $N_e = \langle \mathcal{N}_e \rangle$ is the average number of electrons in the system, satisfy the relation

$$F = \Omega + \mu N_e. \tag{2.26}$$

To calculate F, it is convenient first to obtain Ω, then use (2.26).

2.2.2 Linear Response Theory

If the system is under a time-dependent external perturbation, such as magnetic field, we have to solve the time-dependent Schrödinger equation

$$\mathrm{i}\hbar\frac{\partial}{\partial t}\Psi(\mathbf{r}_1,\ldots,\mathbf{r}_{N_\mathrm{e}},t)=\mathcal{H}_\mathrm{tot}(t)\Psi(\mathbf{r}_1,\ldots,\mathbf{r}_{N_\mathrm{e}},t), \tag{2.27}$$

where the total Hamiltonian contains the interaction with the external perturbation:

$$\mathcal{H}_\mathrm{tot}(t)=\mathcal{H}+\mathcal{H}_\mathrm{ext}(t). \tag{2.28}$$

To describe the dynamics of the corresponding density matrix

$$\rho_\mathrm{tot}(t)=\rho+\rho_\mathrm{ext}(t),$$

we expand the wave function Ψ in the eigenfunctions Ψ_k of the unperturbed Hamiltonian $\mathcal{H}$,

$$\Psi(\mathbf{r}_1,\ldots,\mathbf{r}_{N_\mathrm{e}},t)=\sum_{k'}c_{k'}(t)\Psi_{k'}(\mathbf{r}_1,\ldots,\mathbf{r}_{N_\mathrm{e}}),$$

substitute in the Schrödinger equation (2.27), multiply both sides by Ψ_k^* on the left and integrate over $\mathbf{r}_1,\ldots,\mathbf{r}_{N_\mathrm{e}}$. Then, taking into account the orthogonality of the eigenfunctions, we obtain

$$\mathrm{i}\hbar\frac{\partial c_k(t)}{\partial t}=\sum_{k'}\mathcal{H}^\mathrm{tot}_{kk'}(t)\,c_{k'}(t), \tag{2.29}$$

where $\mathcal{H}^\mathrm{tot}_{kk'}(t)=(\Psi_k,\mathcal{H}_\mathrm{tot}(t)\Psi_{k'})$. From Eq. (2.29) and its complex conjugate, for $\rho^\mathrm{tot}_{kk'}(t)=\langle c^*_{k'}(t)c_k(t)\rangle$ we derive

$$\mathrm{i}\hbar\frac{\partial\rho^\mathrm{tot}_{kk'}(t)}{\partial t}=\sum_{k''}\left(\mathcal{H}^\mathrm{tot}_{kk''}(t)\rho^\mathrm{tot}_{k''k'}(t)-\rho^\mathrm{tot}_{kk''}(t)\mathcal{H}^\mathrm{tot}_{k''k'}(t)\right)=[\mathcal{H}_\mathrm{tot}(t),\rho_\mathrm{tot}(t)]_{kk'}.$$

Thus, the equation of motion for the density matrix is[2]

$$\mathrm{i}\hbar\frac{\partial\rho_\mathrm{tot}(t)}{\partial t}=[\mathcal{H}_\mathrm{tot}(t),\rho_\mathrm{tot}(t)]. \tag{2.30}$$

It is now convenient to introduce the *interaction* representation of the density matrix,

$$\rho^\mathrm{I}_\mathrm{tot}(t)=\mathrm{e}^{\mathrm{i}\mathcal{H}t/\hbar}\rho_\mathrm{tot}(t)\,\mathrm{e}^{-\mathrm{i}\mathcal{H}t/\hbar}. \tag{2.31}$$

Differentiating (2.31) and using (2.30) and (2.28), we obtain

$$\begin{aligned}\frac{\mathrm{d}\rho^\mathrm{I}_\mathrm{tot}(t)}{\mathrm{d}t}&=\frac{\mathrm{i}}{\hbar}\mathrm{e}^{\mathrm{i}\mathcal{H}t/\hbar}[\mathcal{H},\rho_\mathrm{tot}(t)]\,\mathrm{e}^{-\mathrm{i}\mathcal{H}t/\hbar}+\mathrm{e}^{\mathrm{i}\mathcal{H}t/\hbar}\frac{\partial\rho_\mathrm{tot}(t)}{\partial t}\mathrm{e}^{-\mathrm{i}\mathcal{H}t/\hbar}\\&=-\frac{\mathrm{i}}{\hbar}\mathrm{e}^{\mathrm{i}\mathcal{H}t/\hbar}[\mathcal{H}_\mathrm{ext}(t),\rho_\mathrm{tot}(t)]\,\mathrm{e}^{-\mathrm{i}\mathcal{H}t/\hbar}.\end{aligned}$$

[2]This equation is sometimes called the Liouville, or von Neumann equation (see, e.g. [5]).

Direct integration gives

$$\rho^{\mathrm{I}}_{\mathrm{tot}}(t) = \rho^{\mathrm{I}}_{\mathrm{tot}}(-\infty) + \frac{\mathrm{i}}{\hbar}\int_{-\infty}^{t} \mathrm{e}^{\mathrm{i}\mathcal{H}t'/\hbar}\,[\rho_{\mathrm{tot}}(t'), \mathcal{H}_{\mathrm{ext}}(t')]\,\mathrm{e}^{-\mathrm{i}\mathcal{H}t'/\hbar}\,\mathrm{d}t'. \tag{2.32}$$

If the perturbation $\mathcal{H}_{\mathrm{ext}}(t)$ is turned on adiabatically, then $\rho_{\mathrm{tot}}(-\infty) = \rho$. In the expression $[\rho_{\mathrm{tot}}(t), \mathcal{H}_{\mathrm{ext}}(t)] = [\rho + \rho_{\mathrm{ext}}(t), \mathcal{H}_{\mathrm{ext}}(t)]$, we neglect the second-order perturbation term $[\rho_{\mathrm{ext}}(t), \mathcal{H}_{\mathrm{ext}}(t)]$. Thus, substituting (2.31) in (2.32), we arrive at

$$\rho_{\mathrm{tot}}(t) = \rho + \frac{\mathrm{i}}{\hbar}\int_{-\infty}^{t} \mathrm{e}^{\mathrm{i}\mathcal{H}(t'-t)/\hbar}\,[\rho, \mathcal{H}_{\mathrm{ext}}(t')]\,\mathrm{e}^{-\mathrm{i}\mathcal{H}(t'-t)/\hbar}\,\mathrm{d}t'.$$

The magnetization is given by

$$\mathbf{M}(\mathbf{r}, t) = \mathrm{Tr}(\rho_{\mathrm{tot}}(t)\boldsymbol{\mathcal{M}}(\mathbf{r})). \tag{2.33}$$

If the system is ordered in the absence of the applied field, then $\mathbf{M}(\mathbf{r}, -\infty) = \mathrm{Tr}(\rho\boldsymbol{\mathcal{M}}(\mathbf{r}))$ is nonzero. The response of such a system is defined by the difference $\mathbf{M}(\mathbf{r}, t) - \mathbf{M}(\mathbf{r}, -\infty)$ resulting from the applied field. However, in the following, for simplicity, we shall understand $\mathbf{M}(\mathbf{r}, t)$ to be the response to the applied field. Then, commuting the integral with the trace, we have

$$\mathbf{M}(\mathbf{r}, t) = \frac{\mathrm{i}}{\hbar}\int_{-\infty}^{t} \mathrm{Tr}\big(\boldsymbol{\mathcal{M}}(\mathbf{r})\,\mathrm{e}^{\mathrm{i}\mathcal{H}(t'-t)/\hbar}\,[\rho, \mathcal{H}_{\mathrm{ext}}(t')]\,\mathrm{e}^{-\mathrm{i}\mathcal{H}(t'-t)/\hbar}\big)\,\mathrm{d}t'.$$

Using the cyclic property of trace, we write

$$\mathrm{Tr}\big(\boldsymbol{\mathcal{M}}(\mathbf{r})\,\mathrm{e}^{\mathrm{i}\mathcal{H}(t'-t)/\hbar}\,[\rho, \mathcal{H}_{\mathrm{ext}}(t')]\,\mathrm{e}^{-\mathrm{i}\mathcal{H}(t'-t)/\hbar}\big) = \mathrm{Tr}\big(\boldsymbol{\mathcal{M}}(\mathbf{r}, t-t')[\rho, \mathcal{H}_{\mathrm{ext}}(t')]\big),$$

where

$$\boldsymbol{\mathcal{M}}(\mathbf{r}, t) = \mathrm{e}^{\mathrm{i}\mathcal{H}t/\hbar}\boldsymbol{\mathcal{M}}(\mathbf{r})\,\mathrm{e}^{-\mathrm{i}\mathcal{H}t/\hbar}.$$

Therefore,

$$\mathbf{M}(\mathbf{r}, t) = \frac{\mathrm{i}}{\hbar}\int_{-\infty}^{t} \mathrm{Tr}\big(\boldsymbol{\mathcal{M}}(\mathbf{r}, t-t')\,[\rho, \mathcal{H}_{\mathrm{ext}}(t')]\big)\,\mathrm{d}t'. \tag{2.34}$$

By applying the cyclic property of trace once again, we have

$$\mathrm{Tr}\big(\boldsymbol{\mathcal{M}}(\mathbf{r}, t-t')[\rho, \mathcal{H}_{\mathrm{ext}}(t')]\big) = -\mathrm{Tr}\big(\rho[\boldsymbol{\mathcal{M}}(\mathbf{r}, t-t'), \mathcal{H}_{\mathrm{ext}}(t')]\big).$$

Making use of the unperturbed average (2.23), we write (2.34) in the form

$$\mathbf{M}(\mathbf{r}, t) = -\frac{\mathrm{i}}{\hbar}\int_{-\infty}^{t} \big\langle[\boldsymbol{\mathcal{M}}(\mathbf{r}, t-t'), \mathcal{H}_{\mathrm{ext}}(t')]\big\rangle\,\mathrm{d}t'. \tag{2.35}$$

If the perturbation $\mathcal{H}_{\mathrm{ext}}(t)$ is generated by the space- and time-dependent magnetic field $\mathbf{H}(\mathbf{r}, t)$, we have

$$\mathcal{H}_{\mathrm{ext}}(t) = -\int \sum_{\beta} \mathcal{M}_{\beta}(\mathbf{r}) H_{\beta}(\mathbf{r}, t)\,\mathrm{d}\mathbf{r}. \tag{2.36}$$

We assume that the magnetic field $\mathbf{H}(\mathbf{r}, t)$ changes slowly, so that the perturbed system moves from one equilibrium to another. Then, using (2.35), we write the response as

$$M_{\alpha}(\mathbf{r}, t) = \frac{\mathrm{i}}{\hbar}\iint_{-\infty}^{t} \sum_{\beta}\big\langle[\mathcal{M}_{\alpha}(\mathbf{r}, t-t'), \mathcal{M}_{\beta}(\mathbf{r}')]\big\rangle\, H_{\beta}(\mathbf{r}', t')\,\mathrm{d}\mathbf{r}'\,\mathrm{d}t'. \tag{2.37}$$

Comparing (2.37) with (2.4), we obtain the following expression for the susceptibility:

$$\chi_{\alpha\beta}(\mathbf{r}, \mathbf{r}', t - t') = \frac{\mathrm{i}}{\hbar}\langle[\mathcal{M}_\alpha(\mathbf{r}, t - t'), \mathcal{M}_\beta(\mathbf{r}')]\rangle\theta(t - t'),$$

where $\theta(t) = 0$ for $t < 0$ and $\theta(t) = 1$ for $t > 0$. Passing to the *spatial* Fourier transforms and taking into account the translational invariance, we come to

$$\chi_{\alpha\beta}(\mathbf{q}, t) = \frac{\mathrm{i}}{\hbar}\langle[\mathcal{M}_\alpha(\mathbf{q}, t), \mathcal{M}_\beta(-\mathbf{q})]\rangle\theta(t).$$

Making the *time* Fourier transformation, we obtain[3]

$$\chi_{\alpha\beta}(\mathbf{q}, \omega) = \frac{\mathrm{i}}{\hbar}\int_0^\infty \langle[\mathcal{M}_\alpha(\mathbf{q}, t), \mathcal{M}_\beta(-\mathbf{q})]\rangle\, \mathrm{e}^{\mathrm{i}\omega t}\, \mathrm{d}t, \tag{2.38}$$

where ω stands for $\omega + \mathrm{i}0^+$ (for details, see Appendix C.5).

Of particular interest is the longitudinal susceptibility $\chi_{zz}(0, 0)$. At the point where the longitudinal susceptibility diverges, $\chi_{zz}(0, 0) = \infty$, even an infinitesimal magnetic field produces a finite magnetization. The pole of $\chi_{zz}(0, 0)$ therefore describes the phase transition from para- to ferromagnetism.

In Sect. 2.1.2, for a system with axial symmetry, we defined the transverse susceptibility as the response to the circular magnetic field. Using the circular components (2.14), we can write the magnetic energy (2.36) as

$$\mathcal{H}_{\mathrm{ext}}(t) = -\int\left[\frac{1}{2}\Big(\mathcal{M}_+(\mathbf{r})H_-(\mathbf{r}, t) + \mathcal{M}_-(\mathbf{r})H_+(\mathbf{r}, t)\Big) + \mathcal{M}_z(\mathbf{r})H_z(\mathbf{r}, t)\right]\mathrm{d}\mathbf{r}. \tag{2.39}$$

As we discussed, H_+ produces only M_+ and H_- produces only M_-. Keeping only the first term in $\mathcal{H}_{\mathrm{ext}}(t)$, from (2.35) we obtain

$$M_-(\mathbf{r}, t) = \frac{1}{2}\frac{\mathrm{i}}{\hbar}\iint_{-\infty}^{t} \langle[\mathcal{M}_-(\mathbf{r}, t - t'), \mathcal{M}_+(\mathbf{r}')]\rangle\, H_-(\mathbf{r}', t')\, \mathrm{d}\mathbf{r}'\, \mathrm{d}t'.$$

Making the Fourier transform, we come to

$$M_-(\mathbf{q}, \omega) = \frac{1}{2}\chi_{-+}(\mathbf{q}, \omega)H_-(\mathbf{q}, \omega) \tag{2.40}$$

(compare this with expression (2.15) for χ_-), where the transverse susceptibility is given by

$$\chi_{-+}(\mathbf{q}, \omega) = \frac{\mathrm{i}}{\hbar}\int_0^\infty \langle[\mathcal{M}_-(\mathbf{q}, t), \mathcal{M}_+(-\mathbf{q})]\rangle\, \mathrm{e}^{\mathrm{i}\omega t}\, \mathrm{d}t. \tag{2.41}$$

This formula explains the notation $\chi_{-+}(\mathbf{q}, \omega)$ for the transverse component in the linear response theory, which is standard within the metallic magnetism literature (see, e.g. [5, 7]) and will be used from now on. Similarly, we define the other transverse susceptibility,

$$\chi_{+-}(\mathbf{q}, \omega) = \frac{\mathrm{i}}{\hbar}\int_0^\infty \langle[\mathcal{M}_+(\mathbf{q}, t), \mathcal{M}_-(-\mathbf{q})]\rangle\, \mathrm{e}^{\mathrm{i}\omega t}\, \mathrm{d}t. \tag{2.42}$$

[3] This is also called Kubo formula for the magnetic susceptibility (see, e.g. [5]).

The two transverse susceptibilities are related by

$$\left(\chi_{-+}(\mathbf{q},\omega)\right)^* = \chi_{+-}(-\mathbf{q},-\omega)$$

or, equivalently,

$$\mathrm{Re}\chi_{-+}(\mathbf{q},\omega) = \mathrm{Re}\chi_{+-}(-\mathbf{q},-\omega), \tag{2.43}$$

$$\mathrm{Im}\chi_{-+}(\mathbf{q},\omega) = -\mathrm{Im}\chi_{+-}(-\mathbf{q},-\omega). \tag{2.44}$$

Transverse susceptibility has two types of singularities that determine different magnetic excitations: isolated poles determine low-energy spectrum of spin waves and branch cuts determine higher energy spin-flip excitations (for details, see Sect. 5.2).

2.2.3 Fluctuation-Dissipation Theorem

The linear response theory allows to establish a relation between the susceptibility and ordinary correlator. Let us consider the function

$$\begin{aligned} f_{\alpha\beta}(\mathbf{q},\omega) &= \frac{\mathrm{i}}{\hbar}\int_{-\infty}^{\infty}\Big\langle\Big[\Delta\mathcal{M}_\alpha(\mathbf{q},t),\Delta\mathcal{M}_\beta(-\mathbf{q})\Big]\Big\rangle \mathrm{e}^{\mathrm{i}\omega t}\,\mathrm{d}t \\ &= \frac{\mathrm{i}}{\hbar}\int_{-\infty}^{0}\Big\langle\Big[\Delta\mathcal{M}_\alpha(\mathbf{q},t),\Delta\mathcal{M}_\beta(-\mathbf{q})\Big]\Big\rangle \mathrm{e}^{\mathrm{i}\omega t}\,\mathrm{d}t \\ &\quad + \frac{\mathrm{i}}{\hbar}\int_{0}^{\infty}\Big\langle\Big[\Delta\mathcal{M}_\alpha(\mathbf{q},t),\Delta\mathcal{M}_\beta(-\mathbf{q})\Big]\Big\rangle \mathrm{e}^{\mathrm{i}\omega t}\,\mathrm{d}t, \end{aligned} \tag{2.45}$$

where $\Delta\mathcal{M} = \mathcal{M} - \langle\mathcal{M}\rangle$. First, we relate this function to the susceptibility (2.38). Using the commutator property (see Appendix A.1.2)

$$\Big\langle\Big[\mathcal{M}_\alpha(\mathbf{q},t),\mathcal{M}_\beta(-\mathbf{q})\Big]\Big\rangle = \Big\langle\Big[\Delta\mathcal{M}_\alpha(\mathbf{q},t),\Delta\mathcal{M}_\beta(-\mathbf{q})\Big]\Big\rangle,$$

we write

$$\chi_{\alpha\beta}(\mathbf{q},\omega) = \frac{\mathrm{i}}{\hbar}\int_{0}^{\infty}\langle[\Delta\mathcal{M}_\alpha(\mathbf{q},t),\Delta\mathcal{M}_\beta(-\mathbf{q})]\rangle \mathrm{e}^{\mathrm{i}\omega t}\,\mathrm{d}t. \tag{2.46}$$

In the first integral on the right-hand side of (2.45), by applying the cyclic property of trace we move the operators $\mathrm{e}^{\pm\mathrm{i}\mathcal{H}t/\hbar}$ that enclose $\Delta\mathcal{M}_\alpha(\mathbf{q})$ so that they enclose $\Delta\mathcal{M}_\beta(-\mathbf{q})$ instead. Then making the change of variable $t \to -t$ in the integral, we have

$$\begin{aligned} &\frac{\mathrm{i}}{\hbar}\int_{-\infty}^{0}\Big\langle\Big[\Delta\mathcal{M}_\alpha(\mathbf{q},t),\Delta\mathcal{M}_\beta(-\mathbf{q})\Big]\Big\rangle \mathrm{e}^{\mathrm{i}\omega t}\,\mathrm{d}t \\ &\quad = \frac{\mathrm{i}}{\hbar}\int_{0}^{\infty}\Big\langle\Big[\Delta\mathcal{M}_\alpha(\mathbf{q}),\Delta\mathcal{M}_\beta(-\mathbf{q},t)\Big]\Big\rangle \mathrm{e}^{-\mathrm{i}\omega t}\,\mathrm{d}t. \end{aligned}$$

Substituting this in (2.45) and recalling (2.46), we obtain

$$f_{\alpha\beta}(\mathbf{q},\omega) = \chi_{\alpha\beta}(\mathbf{q},\omega) - \chi_{\beta\alpha}(-\mathbf{q},-\omega).$$

Using relation (2.7), we have

$$f_{\alpha\beta}(\mathbf{q},\omega) = \chi_{\alpha\beta}(\mathbf{q},\omega) - \left(\chi_{\beta\alpha}(\mathbf{q},\omega)\right)^*. \tag{2.47}$$

Now we relate the function $f_{\alpha\beta}(\mathbf{q},\omega)$ to the correlator. Using the cyclic property of trace, we obtain

$$\int_{-\infty}^{\infty}\langle\Delta\mathcal{M}_\alpha(\mathbf{q},t)\Delta\mathcal{M}_\beta(-\mathbf{q})\rangle\,\mathrm{e}^{\mathrm{i}\omega t}\,\mathrm{d}t=\int_{-\infty}^{\infty}\langle\Delta\mathcal{M}_\beta(-\mathbf{q})\Delta\mathcal{M}_\alpha(\mathbf{q},t+\mathrm{i}\hbar/T)\rangle\,\mathrm{e}^{\mathrm{i}\omega t}\,\mathrm{d}t.$$

Making the transformation $t\to t-\mathrm{i}\hbar/T$ in the integral on the right-hand side, we rewrite this relation in the form

$$\int_{-\infty}^{\infty}\langle\Delta\mathcal{M}_\alpha(\mathbf{q},t)\Delta\mathcal{M}_\beta(-\mathbf{q})\rangle\,\mathrm{e}^{\mathrm{i}\omega t}\,\mathrm{d}t=\mathrm{e}^{\hbar\omega/T}\int_{-\infty}^{\infty}\langle\Delta\mathcal{M}_\beta(-\mathbf{q})\Delta\mathcal{M}_\alpha(\mathbf{q},t)\rangle\,\mathrm{e}^{\mathrm{i}\omega t}\,\mathrm{d}t. \tag{2.48}$$

Using (2.48), we write the first line of (2.45) as

$$f_{\alpha\beta}(\mathbf{q},\omega)=\left(1-\mathrm{e}^{-\hbar\omega/T}\right)\frac{\mathrm{i}}{\hbar}\int_{-\infty}^{\infty}\langle\Delta\mathcal{M}_\alpha(\mathbf{q},t)\Delta\mathcal{M}_\beta(-\mathbf{q})\rangle\,\mathrm{e}^{\mathrm{i}\omega t}\,\mathrm{d}t. \tag{2.49}$$

From (2.47) and (2.49), we finally obtain

$$\begin{aligned}&\int_{-\infty}^{\infty}\langle\Delta\mathcal{M}_\alpha(\mathbf{q},t)\Delta\mathcal{M}_\beta(-\mathbf{q})\rangle\,\mathrm{e}^{\mathrm{i}\omega t}\,\mathrm{d}t\\&\quad=\frac{\mathrm{i}\hbar}{\mathrm{e}^{-\hbar\omega/T}-1}\left[\chi_{\alpha\beta}(\mathbf{q},\omega)-\left(\chi_{\beta\alpha}(\mathbf{q},\omega)\right)^*\right].\end{aligned} \tag{2.50}$$

By taking the inverse Fourier transform this can be written as

$$\begin{aligned}&\langle\Delta\mathcal{M}_\alpha(\mathbf{q},t)\Delta\mathcal{M}_\beta(-\mathbf{q})\rangle\\&\quad=\frac{1}{2\pi}\int\frac{\mathrm{i}\hbar}{\mathrm{e}^{-\hbar\omega/T}-1}\left[\chi_{\alpha\beta}(\mathbf{q},\omega)-\left(\chi_{\beta\alpha}(\mathbf{q},\omega)\right)^*\right]\mathrm{e}^{-\mathrm{i}\omega t}\,\mathrm{d}\omega.\end{aligned} \tag{2.51}$$

Relation (2.51) is called the *fluctuation-dissipation theorem* (see, e.g. [5,7]). In particular, for $\alpha=\beta$, we can rewrite (2.50) as

$$\int_{-\infty}^{\infty}\langle\Delta\mathcal{M}_\alpha(\mathbf{q},t)\Delta\mathcal{M}_\alpha(-\mathbf{q})\rangle\,\mathrm{e}^{\mathrm{i}\omega t}\,\mathrm{d}t=\frac{-\hbar}{\mathrm{e}^{-\hbar\omega/T}-1}\,2\mathrm{Im}\chi_{\alpha\alpha}(\mathbf{q},\omega)$$

and rewrite (2.51) as

$$\langle\Delta\mathcal{M}_\alpha(\mathbf{q},t)\Delta\mathcal{M}_\alpha(-\mathbf{q})\rangle=-\frac{1}{\pi}\int\frac{\hbar}{\mathrm{e}^{-\hbar\omega/T}-1}\,\mathrm{Im}\chi_{\alpha\alpha}(\mathbf{q},\omega)\,\mathrm{e}^{-\mathrm{i}\omega t}\,\mathrm{d}\omega. \tag{2.52}$$

Components of the magnetic moment operator do not commute with each other. Therefore, it is sometimes convenient to define the correlator as the average of the symmetrized product

$$\frac{1}{2}\left\langle\left\{\Delta\mathcal{M}_\alpha(\mathbf{q},t),\,\Delta\mathcal{M}_\beta(-\mathbf{q})\right\}\right\rangle,$$

where

$$\left\{\Delta\mathcal{M}_\alpha(\mathbf{q},t),\,\Delta\mathcal{M}_\beta(-\mathbf{q})\right\}=\Delta\mathcal{M}_\alpha(\mathbf{q},t)\Delta\mathcal{M}_\beta(-\mathbf{q})+\Delta\mathcal{M}_\beta(-\mathbf{q})\Delta\mathcal{M}_\alpha(\mathbf{q},t)$$

is the anticommutator. To relate this correlator to the imaginary part of the susceptibility, similar to (2.45) we introduce the function

$$g_{\alpha\beta}(\mathbf{q},\omega)=\frac{\mathrm{i}}{\hbar}\int_{-\infty}^{\infty}\left\langle\left\{\Delta\mathcal{M}_\alpha(\mathbf{q},t),\,\Delta\mathcal{M}_\beta(-\mathbf{q})\right\}\right\rangle\mathrm{e}^{\mathrm{i}\omega t}\,\mathrm{d}t.$$

Using (2.48), we have

$$g_{\alpha\beta}(\mathbf{q},\omega) = \left(1+\mathrm{e}^{-\hbar\omega/T}\right)\frac{\mathrm{i}}{\hbar}\int_{-\infty}^{\infty}\left\langle \Delta\mathcal{M}_\alpha(\mathbf{q},t),\,\Delta\mathcal{M}_\beta(-\mathbf{q})\right\rangle \mathrm{e}^{\mathrm{i}\omega t}\,\mathrm{d}t.$$

Then from (2.49) we obtain

$$g_{\alpha\beta}(\mathbf{q},\omega) = \coth\left(\frac{\hbar\omega}{2T}\right) f_{\alpha\beta}(\mathbf{q},\omega)$$

or, equivalently,

$$\int_{-\infty}^{\infty}\langle\{\Delta\mathcal{M}_\alpha(\mathbf{q},t),\,\Delta\mathcal{M}_\beta(-\mathbf{q})\}\rangle\,\mathrm{e}^{\mathrm{i}\omega t}\,\mathrm{d}t$$
$$= -\mathrm{i}\hbar\coth\left(\frac{\hbar\omega}{2T}\right)\left[\chi_{\alpha\beta}(\mathbf{q},\omega)-\left(\chi_{\beta\alpha}(\mathbf{q},\omega)\right)^*\right]. \tag{2.53}$$

The inverse Fourier transform is written as

$$\langle\{\Delta\mathcal{M}_\alpha(\mathbf{q},t),\,\Delta\mathcal{M}_\beta(-\mathbf{q})\}\rangle$$
$$= -\frac{\mathrm{i}\hbar}{2\pi}\int\coth\left(\frac{\hbar\omega}{2T}\right)\left[\chi_{\alpha\beta}(\mathbf{q},\omega)-\left(\chi_{\beta\alpha}(\mathbf{q},\omega)\right)^*\right]\mathrm{e}^{-\mathrm{i}\omega t}\,\mathrm{d}t. \tag{2.54}$$

In particular, for $\alpha=\beta$, we rewrite (2.53) as

$$\int_{-\infty}^{\infty}\langle\{\Delta\mathcal{M}_\alpha(\mathbf{q},t),\,\Delta\mathcal{M}_\alpha(-\mathbf{q})\}\rangle\,\mathrm{e}^{\mathrm{i}\omega t}\,\mathrm{d}t = 2\hbar\coth\left(\frac{\hbar\omega}{2T}\right)\mathrm{Im}\chi_{\alpha\alpha}(\mathbf{q},\omega) \tag{2.55}$$

and (2.54) takes the form

$$\langle\{\Delta\mathcal{M}_\alpha(\mathbf{q},t),\,\Delta\mathcal{M}_\alpha(-\mathbf{q})\}\rangle = \frac{\hbar}{\pi}\int\coth\left(\frac{\hbar\omega}{2T}\right)\mathrm{Im}\chi_{\alpha\alpha}(\mathbf{q},\omega)\,\mathrm{e}^{-\mathrm{i}\omega t}\mathrm{d}\omega. \tag{2.56}$$

At high temperatures $T\gg T_\mathrm{C}$, both formulae (2.52) and (2.56) give the same result for $t=0$. Indeed, using $\mathrm{e}^{\hbar\omega/T}\approx 1+\hbar\omega/T$, we write (2.52) as

$$\langle\mathcal{M}_\alpha(\mathbf{q})\mathcal{M}_\alpha(-\mathbf{q})\rangle = \frac{T}{\pi}\int\frac{\mathrm{Im}\chi_{\alpha\alpha}(\mathbf{q},\omega)}{\omega}\,\mathrm{d}\omega.$$

Taking into account the Kramers-Kronig relation

$$\mathrm{Re}\chi_{\alpha\alpha}(\mathbf{q},0) = \frac{1}{\pi}\int\frac{\mathrm{Im}\chi_{\alpha\alpha}(\mathbf{q},\omega)}{\omega}\,\mathrm{d}\omega$$

and the fact that $\mathrm{Im}\chi_{\alpha\alpha}(\mathbf{q},\omega)$ is an odd function of ω, we come to

$$\langle\mathcal{M}_\alpha(\mathbf{q})\mathcal{M}_\alpha(-\mathbf{q})\rangle = T\,\chi_{\alpha\alpha}(\mathbf{q},0), \tag{2.57}$$

where $\chi_{\alpha\alpha}(\mathbf{q},0)$ is the static susceptibility. Similarly, using

$$\coth\left(\frac{\hbar\omega}{2T}\right)\approx\frac{2T}{\hbar\omega},$$

we write (2.56) as

$$\frac{1}{2}\langle\{\mathcal{M}_\alpha(\mathbf{q}),\,\mathcal{M}_\alpha(-\mathbf{q})\}\rangle = \frac{T}{\pi}\int\frac{\mathrm{Im}\chi_{\alpha\alpha}(\mathbf{q},\omega)}{\omega}\,\mathrm{d}\omega$$

and thus,

$$\frac{1}{2}\langle\{\mathcal{M}_\alpha(\mathbf{q}), \mathcal{M}_\alpha(-\mathbf{q})\}\rangle = T\,\chi_{\alpha\alpha}(\mathbf{q}, 0). \tag{2.58}$$

This is the *high-temperature* form of the fluctuation-dissipation theorem. Note that the right-hand sides of (2.57) and (2.58) are the same. Equating the left-hand sides of (2.57) and (2.58), we obtain

$$\langle\mathcal{M}_\alpha(\mathbf{q})\mathcal{M}_\alpha(-\mathbf{q})\rangle = \langle\mathcal{M}_\alpha(-\mathbf{q})\mathcal{M}_\alpha(\mathbf{q})\rangle.$$

i.e. on the average $\mathcal{M}_\alpha(\mathbf{q})$ and $\mathcal{M}_\alpha(-\mathbf{q})$ commute. Therefore, we can also interpret (2.57) or (2.58) as the *classical* form of the fluctuation-dissipation theorem.

In the paramagnetic state, the uniform static susceptibility $\chi_{\alpha\alpha}(0, 0)$ in metals follows the *Curie-Weiss law* (see, e.g. [4])

$$\chi^\alpha_{\mathrm{CW}} = \frac{m^2_{\mathrm{eff}}}{3(T - \Theta_{\mathrm{C}})}, \qquad T > \Theta_{\mathrm{C}}, \tag{2.59}$$

where m_{eff} is the effective magnetic moment and Θ_{C} is the paramagnetic Curie temperature. The high-temperature version of the fluctuation-dissipation theorem (2.57) yields

$$\frac{1}{N}\chi_{\alpha\alpha}(0, 0) = \frac{M^2(0)}{3T}, \tag{2.60}$$

where

$$M^2(q) = \frac{1}{N}\langle\boldsymbol{\mathcal{M}}(\mathbf{q})\boldsymbol{\mathcal{M}}(-\mathbf{q})\rangle \tag{2.61}$$

is the square of the *q-dependent effective moment.* Comparing formulae (2.59) and (2.60), for $T \gg \Theta_{\mathrm{C}}$ we obtain

$$\chi^\alpha_{\mathrm{CW}} = N^{-1}\chi_{\alpha\alpha}(0, 0), \qquad M(0) = m_{\mathrm{eff}}. \tag{2.62}$$

References

1. R.M. Bozorth, *Ferromagnetism* (Van Nostrand, Princeton, 1951)
2. S.V. Tyablikov, *Methods in the Quantum Theory of Magnetism* (Springer, New York, 1967)
3. S.V. Vonsovskii, *Magnetism* (Wiley, New York, 1974)
4. S. Chikazumi, *Physics of Ferromagnetism*, 2nd edn. (Clarendon, Oxford, 1997)
5. D.J. Kim, *New Perspectives in Magnetism of Metals* (Kluwer/Plenum, New York, 1999)
6. C. Kittel, *Introduction to Solid State Physics*, 8th edn. (Wiley, New York, 2005)
7. R.M. White, *Quantum Theory of Magnetism*, 3rd edn. (Springer, Berlin, 2007)
8. J.F. Nye, *Physical Properties of Crystals*, 2nd edn. (Clarendon, Oxford, 1985)
9. C.P. Slichter, *Principles of Magnetic Resonance*, 3rd edn. (Springer, Berlin, 1990)
10. A.G. Gurevich, G.A. Melkov, *Magnetization Oscillations and Waves* (CRC, Boca Raton, 1996)
11. L.D. Landau, E.M. Lifshitz, *Statistical Physics, Course on Theoretical Physics*, vol. 5, 3rd edn. (Pergamon, Oxford, 1985)
12. K. Huang, *Statistical Mechanics*, 2nd edn. (Wiley, New York, 1987)

Many-Electron Problem

3

...we can make the one possible combination which is antisymmetric, and it will both satisfy the exclusion principle, and will be an approximate solution of Schrodinger's equation. This combination is conveniently written as a determinant... (J.C. Slater, Phys. Rev. ***34****, 1293 (1929))*

In this chapter we summarize the essentials of quantum mechanics and statistical mechanics that are used later in the book. First, we recall the necessary facts about one-electron problem in a periodic crystal lattice (for details see, e.g. [1–3]). Then the second quantization method is described and illustrated by concrete examples (see also [4–7]). In particular, we calculate the paramagnetic susceptibility of noninteracting electrons in the field of the crystal lattice. As an immediate application of this result we briefly discuss the RKKY oscillation.

3.1 One-Electron States

An electron in a crystal can be described by the wave function $\varphi(\mathbf{r})$, which satisfies the following Schrödinger equation:

$$\left(-\frac{\hbar^2}{2m_{\mathrm{e}}}\nabla^2 + V(\mathbf{r})\right)\varphi(\mathbf{r}) = \varepsilon\varphi(\mathbf{r}). \tag{3.1}$$

Here ∇^2 is the Laplace operator, ε is the energy and m_{e} is the mass of an electron. The potential energy $V(\mathbf{r})$ is periodic with the period of the lattice:

$$V(\mathbf{r}+\mathbf{R}_j) = V(\mathbf{r}), \qquad \mathbf{R}_j = j_1\mathbf{a}_1 + j_2\mathbf{a}_2 + j_3\mathbf{a}_3, \tag{3.2}$$

where $\mathbf{a}_1, \mathbf{a}_2, \mathbf{a}_3$ are the primitive vectors of the lattice and j_1, j_2, j_3 are integers. We assume the cyclic boundary conditions

$$\varphi(\mathbf{r}+L\mathbf{a}_1) = \varphi(\mathbf{r}+L\mathbf{a}_2) = \varphi(\mathbf{r}+L\mathbf{a}_3) = \varphi(\mathbf{r}),$$

where L is a large integer. Then, according to *Bloch's theorem*, a solution to (3.1) is given by

$$\varphi_{\mathbf{k}}(\mathbf{r}) = \mathrm{e}^{\mathrm{i}\mathbf{kr}}u_{\mathbf{k}}(\mathbf{r}), \tag{3.3}$$

where $\mathbf{k}$ is the *wavevector* and $u_{\mathbf{k}}(\mathbf{r})$ is a periodic function:

$$u_{\mathbf{k}}(\mathbf{r}+\mathbf{R}) = u_{\mathbf{k}}(\mathbf{r})$$

($\mathbf{R}$ is a lattice vector).

It is easily seen that $\mathbf{k}$ in Bloch's theorem is determined up to a reciprocal lattice vector $\mathbf{K}$:

$$\varphi_{\mathbf{k}}(\mathbf{r}) = \varphi_{\mathbf{k}+\mathbf{K}}(\mathbf{r}).$$

N. B. Melnikov, B. I. Reser, *Dynamic Spin Fluctuation Theory of Metallic Magnetism*,
https://doi.org/10.1007/978-3-319-92974-3_3

Indeed,

$$\varphi_{\mathbf{k}}(\mathbf{r}) = e^{i(\mathbf{k}+\mathbf{K})\mathbf{r}}[e^{-i\mathbf{K}\mathbf{r}}u_{\mathbf{k}}(\mathbf{r})],$$

where $e^{-i\mathbf{K}\mathbf{r}}u_{\mathbf{k}}(\mathbf{r})$ is a periodic function, because $e^{i\mathbf{K}\mathbf{R}} = 1$. Thus, the energies are also periodic

$$\varepsilon_{\mathbf{k}} = \varepsilon_{\mathbf{k}+\mathbf{K}}.$$

In order to determine the wavevectors **k** uniquely, we restrict it to a primitive cell of the reciprocal lattice. Then solutions to the Schrödinger equation form an infinite number of states for each **k** in the primitive cell and we label them by the band index n. The wave function

$$\varphi_{n\mathbf{k}}(\mathbf{r}) = e^{i\mathbf{k}\mathbf{r}}u_{n\mathbf{k}}(\mathbf{r}) \tag{3.4}$$

corresponding to the energy $\varepsilon_{n\mathbf{k}}$ is called the *Bloch function*. Throughout the book we assume that **k** belongs to the reciprocal-lattice primitive cell centred at the origin, which is called the *Brillouin zone*.

The spin of an electron is given by the operator $\mathbf{s} = (s_x, s_y, s_z)$. The components s_α act in a two-dimensional complex space and satisfy the same commutation relations as the ones of the angular momentum operator:

$$[s_x, s_y] \equiv s_x s_y - s_y s_x = i s_z, \qquad \text{etc.}$$

Denoting the eigenvectors of the operator s_z by $\chi_\uparrow$ and $\chi_\downarrow$, we have

$$s_z\chi_\uparrow = \frac{1}{2}\chi_\uparrow, \qquad s_z\chi_\downarrow = -\frac{1}{2}\chi_\downarrow.$$

The spin-flip operators $s_\pm = s_x \pm i s_y$ satisfy the relations

$$\begin{aligned} s_-\chi_\uparrow &= \chi_\downarrow, & s_-\chi_\downarrow &= 0, \\ s_+\chi_\uparrow &= 0, & s_+\chi_\downarrow &= \chi_\uparrow. \end{aligned}$$

Since s_α do not commute, they cannot be diagonalized simultaneously. We choose a coordinate system such that s_z is diagonal with the eigenvalues $\pm 1/2$ (in units of $\hbar$). Then the spin operator can be represented as $\mathbf{s} = \frac{1}{2}\boldsymbol{\sigma}$, where $\boldsymbol{\sigma} = (\sigma_x, \sigma_y, \sigma_z)$ is the vector of *Pauli matrices*

$$\sigma_x = \begin{bmatrix} 0 & 1 \\ 1 & 0 \end{bmatrix}, \quad \sigma_y = \begin{bmatrix} 0 & -i \\ i & 0 \end{bmatrix}, \quad \sigma_z = \begin{bmatrix} 1 & 0 \\ 0 & -1 \end{bmatrix}. \tag{3.5}$$

A state of an electron can now be characterized by the product of the orbital wave function and the spin eigenfunction:

$$\varphi_\sigma(\mathbf{r}) = \varphi(\mathbf{r})\,\chi_\sigma,$$

where $\sigma = \uparrow\downarrow$ is the spin index. Thus, the Bloch function (3.4) transforms to $\varphi_{n\mathbf{k}\sigma}(\mathbf{r}) = \varphi_{n\mathbf{k}}(\mathbf{r})\chi_\sigma$.

For our purposes it is more convenient to use the *Wannier functions* $w_{nj\sigma}(\mathbf{r}) = w_{nj}(\mathbf{r})\chi_\sigma$, which are related to the Bloch functions by the following relations:

$$w_{nj}(\mathbf{r}) = \frac{1}{\sqrt{N}}\sum_{\mathbf{k}}^{\text{BZ}} e^{-i\mathbf{k}\mathbf{R}_j}\varphi_{n\mathbf{k}}(\mathbf{r}). \qquad \varphi_{n\mathbf{k}}(\mathbf{r}) = \frac{1}{\sqrt{N}}\sum_{j=1}^{N} e^{i\mathbf{k}\mathbf{R}_j}w_{nj}(\mathbf{r}). \tag{3.6}$$

Here j represents the lattice site $\mathbf{R}_j$ (N is the number of unit cells in the crystal) and **k** takes N discrete values in the Brillouin zone (BZ). The orthonormality relation for the Wannier functions

$$(w_{nj\sigma}, w_{n'j'\sigma'}) = \int w_{nj}^*(\mathbf{r})w_{n'j'}(\mathbf{r})\,d\mathbf{r}\,(\chi_\sigma, \chi_{\sigma'}) = \delta_{nn'}\delta_{jj'}\delta_{\sigma\sigma'}$$

can be easily confirmed by using that of the Bloch functions. The Wannier functions form a complete set, as do the Bloch functions.

Note that the Wannier functions are not eigenstates of the Hamiltonian, because each $w_{nj}(\mathbf{r})$ is a linear combination of $\varphi_{n\mathbf{k}}(\mathbf{r})$ corresponding to different eigenvalues $\varepsilon_{n\mathbf{k}}$. In particular, for the *free electron gas* (in the empty lattice), the Bloch functions are just the plane waves $\varphi_{n\mathbf{k}}(\mathbf{r}) = V^{-1/2}\mathrm{e}^{\mathrm{i}\mathbf{k}\mathbf{r}}$, and Wannier functions are the delta functions centred at the lattice sites $w_{nj}(\mathbf{r}) = \delta(\mathbf{r}-\mathbf{R}_j)$. When the crystal lattice potential (3.2) is present, as is the case of the DSFT, each Wannier function is still localized at a lattice site and we sometimes write $w_{nj}(\mathbf{r}) = w_n(\mathbf{r} - \mathbf{R}_j)$. Thus, the Wannier state $w_{nj\sigma}(\mathbf{r})$ describes a spin σ electron at the jth site in the nth energy band.

3.2 Many-Electron States

In discussing magnetism of a metal, we neglect the effect of the motion of its ionic lattice.[1] We thus consider only electronic subsystem in the electrostatic field produced by the periodic lattice. The Hamiltonian $\mathcal{H} = \mathcal{H}_0 + \mathcal{H}_\mathrm{I}$ of this system consists of the sum of kinetic and potential energy of the noninteracting electrons $\mathcal{H}_0$ and the Coulomb interaction $\mathcal{H}_\mathrm{I}$:

$$\mathcal{H}_0 = \sum_{i=1}^{N_\mathrm{e}}\left(-\frac{\hbar^2}{2m_\mathrm{e}}\nabla_i^2 + V(\mathbf{r}_i)\right), \qquad \mathcal{H}_\mathrm{I} = \frac{1}{2}\sum_{i\neq j}^{N_\mathrm{e}}\frac{e^2}{|\mathbf{r}_i - \mathbf{r}_j|}, \tag{3.7}$$

where $-e$ is the charge of an electron and N_e is the number of electrons.

According to quantum mechanics, the wave function $\Psi(\mathbf{r}_1, \ldots, \mathbf{r}_{N_\mathrm{e}})$ of a many-electron system must change sign when two electrons are interchanged:

$$\Psi(\mathbf{r}_1, \ldots, \mathbf{r}_i, \ldots, \mathbf{r}_j, \ldots, \mathbf{r}_{N_\mathrm{e}}) = -\Psi(\mathbf{r}_1, \ldots, \mathbf{r}_j, \ldots, \mathbf{r}_i, \ldots, \mathbf{r}_{N_\mathrm{e}}).$$

Therefore, a product $\varphi_{k_1}(\mathbf{r}_1)\ldots\varphi_{k_{N_\mathrm{e}}}(\mathbf{r}_{N_\mathrm{e}})$ of one-electron states $\varphi_k(\mathbf{r})$ $(k = 1,\ 2,\ \ldots)$ does not correctly represent a state of the many-electron system. Instead of a single product, we must take a *determinantal function*,[2]

$$\Phi_{k_1,\ldots,k_{N_\mathrm{e}}}(\mathbf{r}_1, \ldots, \mathbf{r}_{N_\mathrm{e}}) = \frac{1}{\sqrt{N_\mathrm{e}!}}\begin{vmatrix} \varphi_{k_1}(\mathbf{r}_1) & \cdots & \varphi_{k_1}(\mathbf{r}_{N_\mathrm{e}}) \\ \vdots & \ddots & \vdots \\ \varphi_{k_{N_\mathrm{e}}}(\mathbf{r}_1) & \cdots & \varphi_{k_{N_\mathrm{e}}}(\mathbf{r}_{N_\mathrm{e}}) \end{vmatrix}. \tag{3.8}$$

If we have a complete orthonormal system of one-electron functions,

$$\sum_k \varphi_k(\mathbf{r})\varphi_k(\mathbf{r}') = \delta(\mathbf{r} - \mathbf{r}'), \tag{3.9}$$

such as Bloch $\varphi_{n\mathbf{k}\sigma}(\mathbf{r})$ or Wannier $w_{nj\sigma}(\mathbf{r})$ functions, then the determinantal functions (3.8) form a complete and orthonormal system of N_e-electron states (for a proof, see, e.g. [4]). The states containing different number of electrons are defined to be orthogonal.

A determinantal function is specified by the one-electron functions which it contains. The only ambiguity is that of sign, and this can be removed by ordering the functions φ_k. However, instead of specifying the subscripts in the determinantal function, it can be more convenient to write (3.8) in the occupation number representation[3]

$$\Phi(n_1, n_2, \ldots),$$

where the occupation number n_k is equal to unity if this state appears in the determinant and zero otherwise.

---

[1]The lattice vibrations are taken into account in Chaps. 14 and 15.

[2]In the literature, this is also referred to as *Slater determinant* (see, e.g. [5, 7]).

[3]This is often written in the Dirac notation as $|n_1, n_2, \ldots\rangle$ but we will stick to the wave-mechanical notation.

3.3 Second Quantization

Dealing with the quantum-statistical average such as (2.23), one needs to calculate the matrix elements with respect to $N_e \times N_e$ determinantal functions. This calculation becomes practically impossible in metals because of the large number of electrons, $N_e \sim 10^{23}$. The method of second quantization allows to overcome this difficulty by introducing operators that are independent of the number of electrons in the system under study (the second quantization for phonons is briefly illustrated in Appendix G).

3.3.1 General Theory

We now introduce the creation operator $a_k^\dagger$ that adds the state φ_k to each determinantal function not containing this state. In other words, it converts a determinant of order N_e to a determinant of order $N_e + 1$ with the appropriate normalization factor and sign. In the occupation number representation this can be written as

$$a_k^\dagger \Phi(\dots, 0_k, \dots) = \theta_k \Phi(\dots, 1_k, \dots),$$
$$a_k^\dagger \Phi(\dots, 1_k, \dots) = 0.$$

Here

$$\theta_k = (-1)^{\sum\limits_{j<k} n_j},$$

where $\sum_{j<k} n_j$ is the number of occupied states that precede φ_k. Similarly, the annihilation operator a_k is defined by

$$a_k \Phi(\dots, 0_k, \dots) = 0,$$
$$a_k \Phi(\dots, 1_k, \dots) = \theta_k \Phi(\dots, 0_k, \dots).$$

The two definitions are written in a compact form as

$$a_k^\dagger \Phi(\dots, n_k, \dots) = \theta_k (1 - n_k) \Phi(\dots, 1_k, \dots), \tag{3.10}$$
$$a_k \Phi(\dots, n_k, \dots) = \theta_k n_k \Phi(\dots, 0_k, \dots). \tag{3.11}$$

If the annihilation operator acts on a one-electron state, i.e. the first-order determinant, the results will be a "zero-order" determinant with no electron state occupied, $\Phi(0, 0, \dots)$. This state is called "vacuum" and denoted by Φ_0. Any N_e-electron state can be constructed by acting on the vacuum state with appropriate creation operators:

$$\Phi_{k_1 \dots k_{N_e}} = a_{k_1}^\dagger \dots a_{k_{N_e}}^\dagger \Phi_0, \qquad k_1 < \dots < k_{N_e}.$$

The most important fact about the operators a_k and $a_k^\dagger$ is the anticommutation relations

$$\{a_k, a_{k'}\} = 0, \qquad \{a_k^\dagger, a_{k'}^\dagger\} = 0, \qquad \{a_k, a_{k'}^\dagger\} = \delta_{kk'}. \tag{3.12}$$

These relations are derived directly from the definitions (3.10) and (3.11) of a_k and $a_{k'}^\dagger$ (for details, see, e.g. [4]). It is also easy to check that the creation operator $a_k^\dagger$ is the Hermite conjugate to the annihilation operator a_k, which justifies the notation of the former.

We now proceed to representing physical quantities in a way that is independent of the number of electrons in the many-electron system. To write the expressions in the most compact form, we introduce the *field operator* $\psi(\mathbf{r})$ and its Hermite conjugate $\psi^\dagger(\mathbf{r})$:

$$\psi(\mathbf{r}) = \sum_k a_k \varphi_k(\mathbf{r}), \qquad \psi^\dagger(\mathbf{r}) = \sum_k a_k^\dagger \varphi_k^*(\mathbf{r}). \tag{3.13}$$

Using (3.12) and (3.9), one can verify the following anticommutation relations for $\psi(\mathbf{r})$ and $\psi^\dagger(\mathbf{r})$:

$$\{\psi(\mathbf{r}), \psi(\mathbf{r}')\} = 0,$$
$$\{\psi^\dagger(\mathbf{r}), \psi^\dagger(\mathbf{r}')\} = 0,$$
$$\{\psi(\mathbf{r}), \psi^\dagger(\mathbf{r}')\} = \delta(\mathbf{r} - \mathbf{r}').$$

Almost all physical quantities of the many-electron system are represented as either sums of one-electron quantities

$$\widetilde{\mathcal{A}}^{(1)} = \sum_{i=1}^{N_e} \mathcal{A}^{(1)}(\mathbf{r}_i)$$

(e.g. kinetic and potential energy; atomic spin and charge) or sums of two-electron quantities

$$\widetilde{\mathcal{A}}^{(2)} = \sum_{i\neq j}^{N_e} \mathcal{A}^{(2)}(\mathbf{r}_i, \mathbf{r}_j)$$

(e.g. Coulomb interaction). In the second-quantized form these quantities are written as follows (for a proof, see, e.g. [4]):

$$\widetilde{\mathcal{A}}^{(1)} = \int \psi^\dagger(\mathbf{r})\mathcal{A}^{(1)}(\mathbf{r})\psi(\mathbf{r})\,d\mathbf{r}, \tag{3.14}$$

$$\widetilde{\mathcal{A}}^{(2)} = \iint \psi^\dagger(\mathbf{r})\psi^\dagger(\mathbf{r}')\mathcal{A}^{(2)}(\mathbf{r}, \mathbf{r}')\psi(\mathbf{r}')\psi(\mathbf{r})\,d\mathbf{r}\,d\mathbf{r}'. \tag{3.15}$$

For a one-electron quantity that itself depends on $\mathbf{r}$:

$$\widetilde{\mathcal{A}}^{(1)}(\mathbf{r}) = \sum_{i=1}^{N_e} \mathcal{A}^{(1)}(\mathbf{r}_i, \mathbf{r})$$

(e.g. charge or spin density), we can rewrite (3.14) as

$$\widetilde{\mathcal{A}}^{(1)}(\mathbf{r}) = \int \psi^\dagger(\mathbf{r}')\mathcal{A}^{(1)}(\mathbf{r}', \mathbf{r})\psi(\mathbf{r}')\,d\mathbf{r}'. \tag{3.16}$$

3.3.2 Specific Operators

Charge Density

As the first example, we consider the charge density operator (in units of $-e$)

$$n(\mathbf{r}) = \sum_{i=1}^{N_e} \delta(\mathbf{r} - \mathbf{r}_i).$$

Substituting $\delta(\mathbf{r} - \mathbf{r}')$ for $\mathcal{A}^{(1)}(\mathbf{r}', \mathbf{r})$ in formula (3.16) and using $\delta(\mathbf{r}) = \delta(-\mathbf{r})$, we obtain

$$n(\mathbf{r}) = \int \psi^\dagger(\mathbf{r}')\delta(\mathbf{r}' - \mathbf{r})\psi(\mathbf{r}')\,d\mathbf{r}' = \psi^\dagger(\mathbf{r})\psi(\mathbf{r}).$$

Thus, the square of the amplitude of the field operator $|\psi(\mathbf{r})|^2 = \psi^\dagger(\mathbf{r})\psi(\mathbf{r})$ represents the spatial electron (charge) density.[4]

[4]In quantum mechanics, *quantization* means that one replaces physical quantities by operators that act on the wave function. The square of the wave function modulus gives the probability density. Here, in turn, the wave function is replaced by the field operator. Therefore, the term *second quantization* was created (for details, see [4, 5] and references therein).

The total number of electrons operator is given by

$$\mathcal{N}_e = \int \psi^\dagger(\mathbf{r})\psi(\mathbf{r})\,d\mathbf{r}. \tag{3.17}$$

In the Wannier representation, the field operators of type (3.13) are written as

$$\psi(\mathbf{r}) = \sum_{\nu j\sigma} a_{\nu j\sigma} w_{\nu j}(\mathbf{r})\chi_\sigma, \qquad \psi^\dagger(\mathbf{r}) = \sum_{\nu j\sigma} a^\dagger_{\nu j\sigma} w^*_{\nu j}(\mathbf{r})\chi^\dagger_\sigma, \tag{3.18}$$

where ν is the band index. Substituting (3.18) in (3.17), we obtain

$$\mathcal{N}_e = \sum_{\nu\nu' jj'\sigma\sigma'} a^\dagger_{\nu j\sigma} a_{\nu' j'\sigma'} \int w^*_{\nu j}(\mathbf{r}) w_{\nu' j'}(\mathbf{r})\,d\mathbf{r}\,(\chi_\sigma, \chi_{\sigma'}).$$

Using the orthonormality of the Wannier functions and spin states, we have

$$\mathcal{N}_e = \sum_{\nu j\sigma} a^\dagger_{\nu j\sigma} a_{\nu j\sigma} = \sum_{\nu j\sigma} n_{\nu j\sigma},$$

where $n_{\nu j\sigma} = a^\dagger_{\nu j\sigma} a_{\nu j\sigma}$ is the number of particles operator in the state (ν, j, σ).

Spin Density: Wannier Representation

Similarly, the spin density operator

$$\mathbf{s}(\mathbf{r}) = \mathbf{s}\sum_{i=1}^{N_e} \delta(\mathbf{r} - \mathbf{r}_i) \tag{3.19}$$

in the second-quantized form is written as

$$\mathbf{s}(\mathbf{r}) = \psi^\dagger(\mathbf{r})\,\mathbf{s}\,\psi(\mathbf{r}). \tag{3.20}$$

Integrating $\mathbf{s}(\mathbf{r})$ over the whole volume of the crystal and substituting (3.18), we obtain the total spin

$$S_\alpha = \int \psi^\dagger(\mathbf{r})\, s_\alpha\, \psi(\mathbf{r})\,d\mathbf{r} = \sum_{\nu j} s^\alpha_{\nu j}. \tag{3.21}$$

Here

$$s^\alpha_{\nu j} = \sum_{\sigma\sigma'} s^\alpha_{\sigma\sigma'} a^\dagger_{\nu j\sigma} a_{\nu j\sigma'},$$

where $s^\alpha_{\sigma\sigma'} = (\chi_\sigma, s_\alpha \chi_{\sigma'})$ is the matrix element of $s_\alpha = \frac{1}{2}\sigma_\alpha$ in the two-dimensional spin space. Using the explicit form (3.5) of the Pauli matrices σ_α, we obtain

$$s^x_{\nu j} = \frac{1}{2}\left(a^\dagger_{\nu j\uparrow} a_{\nu j\downarrow} + a^\dagger_{\nu j\downarrow} a_{\nu j\uparrow}\right), \tag{3.22}$$

$$s^y_{\nu j} = \frac{1}{2\mathrm{i}}\left(a^\dagger_{\nu j\uparrow} a_{\nu j\downarrow} - a^\dagger_{\nu j\downarrow} a_{\nu j\uparrow}\right), \tag{3.23}$$

$$s^z_{\nu j} = \frac{1}{2}\left(a^\dagger_{\nu j\uparrow} a_{\nu j\uparrow} - a^\dagger_{\nu j\downarrow} a_{\nu j\downarrow}\right) \tag{3.24}$$

or, equivalently,

$$s^x_{\nu j} = \frac{1}{2}(s^+_{\nu j} + s^-_{\nu j}),$$

$$s^y_{\nu j} = \frac{1}{2\mathrm{i}}(s^+_{\nu j} - s^-_{\nu j}),$$
$$s^z_{\nu j} = \frac{1}{2}(n_{\nu j\uparrow} - n_{\nu j\downarrow}),$$

where $s^+_{\nu j} = a^\dagger_{\nu j\uparrow} a_{\nu j\downarrow}$ and $s^-_{\nu j} = a^\dagger_{\nu j\downarrow} a_{\nu j\uparrow}$ are the spin-flip operators.

Spin Density: Bloch Representation

Throughout the book we will often operate with the spin density in the momentum representation,

$$s_\alpha(\mathbf{q}) = \int s^\alpha(\mathbf{r})\, \mathrm{e}^{-\mathrm{i}\mathbf{qr}}\, d\mathbf{r}. \tag{3.25}$$

To obtain an expression for $s_\alpha(\mathbf{q})$ in the second-quantized form, we use the Bloch representation of the field operators:

$$\psi(\mathbf{r}) = \sum_{\mathbf{k}\sigma} a_{\mathbf{k}\sigma} \varphi_\mathbf{k}(\mathbf{r}) \chi_\sigma, \qquad \psi^\dagger(\mathbf{r}) = \sum_{\mathbf{k}\sigma} a^\dagger_{\mathbf{k}\sigma} \varphi^*_\mathbf{k}(\mathbf{r}) \chi^\dagger_\sigma \tag{3.26}$$

(for brevity we omit the band index here). Substituting (3.26) in (3.20), we write (3.25) as

$$s_\alpha(\mathbf{q}) = \sum_{\mathbf{kk}'\sigma\sigma'} s^\alpha_{\sigma\sigma'} a^\dagger_{\mathbf{k}\sigma} a_{\mathbf{k}'\sigma'} \int \varphi^*_\mathbf{k}(\mathbf{r}) \varphi_{\mathbf{k}'}(\mathbf{r})\, \mathrm{e}^{-\mathrm{i}\mathbf{qr}}\, d\mathbf{r}.$$

According to Bloch's theorem (3.3), we have $\varphi_\mathbf{k}(\mathbf{r} + \mathbf{R}_j) = \mathrm{e}^{\mathrm{i}\mathbf{kR}_j} \varphi_\mathbf{k}(\mathbf{r})$. Therefore, changing the dummy variable $\mathbf{r}$ in the integral to $\mathbf{r}' + \mathbf{R}_j$, we obtain

$$\int \varphi^*_\mathbf{k}(\mathbf{r}' + \mathbf{R}_j) \varphi_{\mathbf{k}'}(\mathbf{r}' + \mathbf{R}_j)\, \mathrm{e}^{-\mathrm{i}\mathbf{q}(\mathbf{r}'+\mathbf{R}_j)}\, d\mathbf{r}' = \mathrm{e}^{-\mathrm{i}(\mathbf{k}-\mathbf{k}'+\mathbf{q})\mathbf{R}_j} \int \varphi^*_\mathbf{k}(\mathbf{r}') \varphi_{\mathbf{k}'}(\mathbf{r}')\, \mathrm{e}^{-\mathrm{i}\mathbf{qr}'} d\mathbf{r}'.$$

Averaging over all $\mathbf{R}_j$ in the crystal lattice and using the identity $\sum_j \mathrm{e}^{\mathrm{i}\mathbf{kR}_j} = N\delta_{\mathbf{k}0}$, we have

$$s_\alpha(\mathbf{q}) = \sum_{\mathbf{k}\sigma\sigma'} F(\mathbf{q}, \mathbf{k}) s^\alpha_{\sigma\sigma'} a^\dagger_{\mathbf{k}\sigma} a_{\mathbf{k}+\mathbf{q},\sigma'}, \tag{3.27}$$

where

$$F(\mathbf{q}, \mathbf{k}) = \int \varphi^*_\mathbf{k}(\mathbf{r}) \varphi_{\mathbf{k}+\mathbf{q}}(\mathbf{r})\, \mathrm{e}^{-\mathrm{i}\mathbf{qr}}\, d\mathbf{r} \tag{3.28}$$

is the *magnetic form-factor*. In the Wannier representation (3.6), letting $w_j(\mathbf{r}) = w(\mathbf{r} - \mathbf{R}_j)$, we come to

$$F(\mathbf{q}, \mathbf{k}) = \sum_j \mathrm{e}^{-\mathrm{i}\mathbf{kR}_j} \int w^*(\mathbf{r}) w(\mathbf{r} + \mathbf{R}_j) \mathrm{e}^{-\mathrm{i}\mathbf{qr}} d\mathbf{r}.$$

Neglecting the overlap of the Wannier functions at different sites, we obtain

$$F(\mathbf{q}, \mathbf{k}) \approx F(q) = \int |w(\mathbf{r})|^2 \mathrm{e}^{-\mathrm{i}\mathbf{qr}}\, d\mathbf{r}.$$

Introducing the operator of the *approximate* spin density $\mathbf{s}'(\mathbf{q})$ such that $\mathbf{s}(\mathbf{q}) = F(q)\,\mathbf{s}'(\mathbf{q})$, we write the spin correlator as

$$\langle \Delta s_\alpha(\mathbf{q}) \Delta s_\alpha(-\mathbf{q}) \rangle = |F(q)|^2 \,\langle \Delta s'_\alpha(\mathbf{q}) \Delta s'_\alpha(-\mathbf{q}) \rangle.$$

Further on we will always work with the approximate spin density (omitting the prime), if the opposite is not explicitly stated. In the absence of the form-factor, formula (3.27) reduces to

$$s_\alpha(\mathbf{q}) = \sum_{\mathbf{k}\sigma\sigma'} s^\alpha_{\sigma\sigma'} a^\dagger_{\mathbf{k}\sigma} a_{\mathbf{k}+\mathbf{q},\sigma'}. \tag{3.29}$$

Single-Site Spin

Dealing with metals, it is useful to define the single-site spin

$$s^\alpha_j = \int_{\Omega_j} s_\alpha(\mathbf{r})\, \mathrm{d}\mathbf{r}, \tag{3.30}$$

where the integration is carried out over the Wigner-Seitz cell centred at the jth lattice site. Clearly, the total spin is the sum of all single-site spins: $S_\alpha = \sum_j s^\alpha_j$. Comparing with (3.21), we see that

$$s^\alpha_j = \sum_\nu s^\alpha_{\nu j} = \sum_{\nu\sigma\sigma'} s^\alpha_{\sigma\sigma'} a^\dagger_{\nu j\sigma} a_{\nu j\sigma'}. \tag{3.31}$$

The Fourier transform (3.25) can be written as the series. Indeed, if we approximate the exponential $\mathrm{e}^{-\mathrm{i}\mathbf{q}\mathbf{r}}$ inside the Wigner-Seitz cell Ω_j by the value $\mathrm{e}^{-\mathrm{i}\mathbf{q}\mathbf{R}_j}$, we obtain

$$s_\alpha(\mathbf{q}) = \sum_j s^\alpha_j\, \mathrm{e}^{-\mathrm{i}\mathbf{q}\mathbf{R}_j}. \tag{3.32}$$

Taking (3.31) into account, we have

$$s_\alpha(\mathbf{q}) = \sum_{\nu j\sigma\sigma'} s^\alpha_{\sigma\sigma'} a^\dagger_{\nu j\sigma} a_{\nu j\sigma'}\, \mathrm{e}^{-\mathrm{i}\mathbf{q}\mathbf{R}_j}. \tag{3.33}$$

Next, we want to express $s_\alpha(\mathbf{q})$ in terms of the Bloch functions. For this we derive the relations between the creation-annihilation operators in the Wannier and Bloch bases. Using (3.6), we obtain

$$a^\dagger_{\nu j\sigma}\Phi_0 = w_{\nu j\sigma} = \frac{1}{\sqrt{N}} \sum_{\mathbf{k}}^{\mathrm{BZ}} \mathrm{e}^{-\mathrm{i}\mathbf{k}\mathbf{R}_j} \varphi_{n\mathbf{k}\sigma} = \frac{1}{\sqrt{N}} \sum_{\mathbf{k}}^{\mathrm{BZ}} \mathrm{e}^{-\mathrm{i}\mathbf{k}\mathbf{R}_j} a^\dagger_{n\mathbf{k}\sigma}\Phi_0,$$

where Φ_0 is the "vacuum" state. Therefore,

$$a^\dagger_{\nu j\sigma} = \frac{1}{\sqrt{N}} \sum_{\mathbf{k}}^{\mathrm{BZ}} \mathrm{e}^{-\mathrm{i}\mathbf{k}\mathbf{R}_j} a^\dagger_{n\mathbf{k}\sigma}, \qquad a_{\nu j\sigma} = \frac{1}{\sqrt{N}} \sum_{\mathbf{k}}^{\mathrm{BZ}} \mathrm{e}^{\mathrm{i}\mathbf{k}\mathbf{R}_j} a_{n\mathbf{k}\sigma}. \tag{3.34}$$

Substituting these relations in (3.33) and making use of $\sum_j \mathrm{e}^{\mathrm{i}\mathbf{q}\mathbf{R}_j} = N\delta_{\mathbf{q}0}$, we come to Eq. (3.29). That means the approximation $\mathrm{e}^{-\mathrm{i}\mathbf{q}\mathbf{r}} \approx \mathrm{e}^{-\mathrm{i}\mathbf{q}\mathbf{R}_j}$ inside Ω_j, which leads to the Fourier series (3.32), is equivalent to setting the form-factor $F(q)$ to unity. Therefore, the use of the Fourier series (3.32) in the DSFT is consistent with the fact that we ignore the form-factor.

Hamiltonian

Next, we write the Hamiltonian $\mathcal{H} = \mathcal{H}_0 + \mathcal{H}_\mathrm{I}$ in the second-quantized form. Applying formulae (3.14) and (3.15) to (3.7), we come to

$$\mathcal{H} = \sum_{\nu\nu' ii'\sigma} t_{ii'}^{\nu\nu'} a_{\nu i\sigma}^{\dagger} a_{\nu' i'\sigma} + \frac{1}{2} \sum_{\substack{\mu\nu i j\sigma \\ \mu'\nu' i' j'\sigma'}} U_{iji'j'}^{\mu\nu\mu'\nu'} a_{\mu i\sigma}^{\dagger} a_{\nu j\sigma'}^{\dagger} a_{\nu' j'\sigma'} a_{\mu' i'\sigma}, \tag{3.35}$$

where the *transfer energy* $t_{i\,i'}^{\nu\nu'}$ and *interaction coefficient* $U_{i\,j\,i'\,j'}^{\mu\nu\mu'\nu'}$ are given by the integrals

$$t_{ii'}^{\nu\nu'} = \int w_{\nu i}^{*}(\mathbf{r}) \left[-\frac{\hbar^2}{2m}\nabla^2 + V(\mathbf{r}) \right] w_{\nu' i'}(\mathbf{r})\, \mathrm{d}\mathbf{r}, \tag{3.36}$$

$$U_{iji'j'}^{\mu\nu\mu'\nu'} = \iint w_{\mu i}^{*}(\mathbf{r}) w_{\nu j}^{*}(\mathbf{r}') \frac{e^2}{|\mathbf{r}-\mathbf{r}'|} w_{\nu' j'}(\mathbf{r}') w_{\mu' i'}(\mathbf{r})\, \mathrm{d}\mathbf{r}\, \mathrm{d}\mathbf{r}'. \tag{3.37}$$

Since $\mathcal{H}$ is spin independent, expression (3.35) is symmetrical with respect to the spin indices σ and σ'.

Finally, we write the Hamiltonian $\mathcal{H}$ in the momentum representation. The creation-annihilation operators in the Bloch and Wannier bases are related by the formulae (3.34). Due to homogeneity, the transfer energy (3.36) is translationally invariant, $t_{ii'} = t_{i-i'}$ (for simplicity of notation, in the rest of the chapter we consider the single-band case). Hence we write the electrons energy $\mathcal{H}_0$ as

$$\mathcal{H}_0 = \sum_{\mathbf{k}\sigma} \varepsilon_{\mathbf{k}} a_{\mathbf{k}\sigma}^{\dagger} a_{\mathbf{k}\sigma} = \sum_{\mathbf{k}\sigma} \varepsilon_{\mathbf{k}} n_{\mathbf{k}\sigma}, \tag{3.38}$$

where

$$\varepsilon_{\mathbf{k}} = \sum_{j} t_{ij} \mathrm{e}^{-\mathrm{i}\mathbf{k}(\mathbf{R}_i - \mathbf{R}_j)}$$

is the nonmagnetic energy spectrum. The interaction term $\mathcal{H}_\mathrm{I}$ in the momentum representation is written as

$$\mathcal{H}_\mathrm{I} = \frac{1}{2} \sum_{\substack{\mathbf{klk'l'} \\ \sigma\sigma'}} U_{\mathbf{klk'l'}} a_{\mathbf{k}\sigma}^{\dagger} a_{\mathbf{l}\sigma'}^{\dagger} a_{\mathbf{l}'\sigma'} a_{\mathbf{k}'\sigma}, \tag{3.39}$$

where

$$U_{\mathbf{klk'l'}} = \iint \varphi_{\mathbf{k}}^{*}(\mathbf{r})\, \varphi_{\mathbf{l}}^{*}(\mathbf{r}') \frac{e^2}{|\mathbf{r}-\mathbf{r}'|} \varphi_{\mathbf{l}'}(\mathbf{r}')\, \varphi_{\mathbf{k}'}(\mathbf{r})\, \mathrm{d}\mathbf{r}\, \mathrm{d}\mathbf{r}'. \tag{3.40}$$

3.4 Noninteracting Electrons

As another illustration of the second-quantization technique, we calculate the magnetic susceptibility of noninteracting electrons in the field of the crystal lattice.

We calculate the (paramagnetic) susceptibility $\chi_{zz}^{0}(\mathbf{q}, \omega)$. Recall that we consider only the magnetic moment associated with the spin of an electron: $\mathbf{m} = -g\mu_\mathrm{B}\mathbf{s}$, where $g \approx 2$ is the electron g-factor and

$$\mu_\mathrm{B} = \frac{e\hbar}{2m_\mathrm{e}c} = 0.927 \times 10^{-20}\,\mathrm{erg/G} \tag{3.41}$$

is the *Bohr magneton* (c is the velocity of light). Then comparing formulae (2.22) and (3.19), we see that the magnetic moment at the point $\mathbf{r}$ is given by

$$\mathcal{M}(\mathbf{r}) = -g\mu_\mathrm{B}\mathbf{s}(\mathbf{r}),$$

and formula (2.38) can be written as

$$\chi_{zz}^{0}(\mathbf{q}, \omega) = g^2\mu_\mathrm{B}^2 \frac{\mathrm{i}}{\hbar} \int_0^\infty \langle [s_z(\mathbf{q}, t), s_z(-\mathbf{q})] \rangle \mathrm{e}^{\mathrm{i}\omega t}\, \mathrm{d}t. \tag{3.42}$$

Here $\langle\ldots\rangle$ is the grand canonical average with the Hamiltonian $\mathcal{H}_0' = \mathcal{H}_0 - \mu\mathcal{N}_e$ and time dependence means the Heisenberg representation,

$$s_\alpha(\mathbf{q}, t) = e^{i\mathcal{H}_0' t/\hbar} s_\alpha(\mathbf{q}) e^{-i\mathcal{H}_0' t/\hbar}.$$

For noninteracting electrons the time dependence of the creation-annihilation operators can be obtained explicitly. Indeed, differentiating the expression $a_{\mathbf{k}\sigma}(t) = e^{i\mathcal{H}_0' t/\hbar} a_{\mathbf{k}\sigma} e^{-i\mathcal{H}_0' t/\hbar}$, we obtain

$$i\hbar \frac{d}{dt} a_{\mathbf{k}\sigma}(t) = [a_{\mathbf{k}\sigma}, \mathcal{H}_0'](t). \tag{3.43}$$

To calculate the commutator, we write $\mathcal{H}_0$ in the momentum representation (3.38). Then Eq. (3.43) becomes

$$i\hbar \frac{d}{dt} a_{\mathbf{k}\sigma}(t) = (\varepsilon_{\mathbf{k}} - \mu) a_{\mathbf{k}\sigma}(t).$$

Integrating the latter, we obtain

$$a_{\mathbf{k}\sigma}(t) = a_{\mathbf{k}\sigma}\, e^{-i(\varepsilon_{\mathbf{k}} - \mu)t/\hbar}, \qquad a^\dagger_{\mathbf{k}\sigma}(t) = a^\dagger_{\mathbf{k}\sigma}\, e^{i(\varepsilon_{\mathbf{k}} - \mu)t/\hbar}. \tag{3.44}$$

Using formula (3.29), we write

$$s_z(\mathbf{q}) = \frac{1}{2} \sum_{\mathbf{k}} \left(a^\dagger_{\mathbf{k}\uparrow} a_{\mathbf{k}+\mathbf{q},\uparrow} - a^\dagger_{\mathbf{k}\downarrow} a_{\mathbf{k}+\mathbf{q},\downarrow} \right) \tag{3.45}$$

and

$$s_z(\mathbf{q}, t) = \frac{1}{2} \sum_{\mathbf{k}} \left(a^\dagger_{\mathbf{k}\uparrow} a_{\mathbf{k}+\mathbf{q},\uparrow} - a^\dagger_{\mathbf{k}\downarrow} a_{\mathbf{k}+\mathbf{q},\downarrow} \right) e^{i(\varepsilon_{\mathbf{k}} - \varepsilon_{\mathbf{k}+\mathbf{q}})t/\hbar}.$$

Hence the commutator in formula (3.42) becomes

$$[s_z(\mathbf{q}, t), s_z(-\mathbf{q})] = \frac{1}{4} \sum_{\mathbf{k}\mathbf{k}'\sigma\sigma'} \sigma\sigma' \left[a^\dagger_{\mathbf{k}\sigma} a_{\mathbf{k}+\mathbf{q},\sigma}, a^\dagger_{\mathbf{k}'\sigma'} a_{\mathbf{k}'-\mathbf{q},\sigma'} \right] e^{i(\varepsilon_{\mathbf{k}} - \varepsilon_{\mathbf{k}+\mathbf{q}})t/\hbar},$$

where σ is equal to $\uparrow$, $\downarrow$ or ± 1. Using the anticommutation relations (3.12), it is easy to verify the general commutation rule

$$\left[a^\dagger_\nu a_\mu, a^\dagger_{\nu'} a_{\mu'} \right] = a^\dagger_\nu a_{\mu'} \delta_{\mu\nu'} - a^\dagger_{\nu'} a_\mu \delta_{\mu'\nu}. \tag{3.46}$$

From the latter we have

$$[s_z(\mathbf{q}, t), s_z(-\mathbf{q})] = \frac{1}{4} \sum_{\mathbf{k}\sigma} \left(a^\dagger_{\mathbf{k}\sigma} a_{\mathbf{k}\sigma} - a^\dagger_{\mathbf{k}+\mathbf{q},\sigma} a_{\mathbf{k}+\mathbf{q},\sigma} \right) e^{i(\varepsilon_{\mathbf{k}} - \varepsilon_{\mathbf{k}+\mathbf{q}})t/\hbar}. \tag{3.47}$$

Recalling that $\omega = \omega + i0^+$, we write (3.42) as

$$\chi^0_{zz}(\mathbf{q}, \omega) = g^2 \mu_B^2 \lim_{\eta \to 0+} \frac{i}{\hbar} \int_0^\infty \left\langle [s_z(\mathbf{q}, t), s_z(-\mathbf{q})] \right\rangle e^{i\omega t} e^{-\eta t}\, dt.$$

Substituting (3.47) and integrating, we obtain

$$\chi^0_{zz}(\mathbf{q}, \omega) = \frac{1}{2} g^2 \mu_B^2 \sum_{\mathbf{k}} \frac{\langle n_{\mathbf{k}\sigma} \rangle - \langle n_{\mathbf{k}+\mathbf{q},\sigma} \rangle}{\varepsilon_{\mathbf{k}+\mathbf{q}} - \varepsilon_{\mathbf{k}} - \hbar\omega - i0^+}. \tag{3.48}$$

Here we used that $\langle n_{\mathbf{k}\sigma} \rangle = \langle a^\dagger_{\mathbf{k}\sigma} a_{\mathbf{k}\sigma} \rangle$ is spin-independent, and summation over σ gives the multiplier two.

For noninteracting electrons, the average number of electrons $\langle n_{\mathbf{k}\sigma} \rangle$ is calculated explicitly. To do this it is convenient to use the thermodynamic potential

$$\Omega = -T \ln \Xi, \qquad \Xi = \mathrm{Tr}\, e^{-\mathcal{H}_0'/T}. \tag{3.49}$$

The energy of noninteracting electrons is given by

$$\mathcal{H}_0' = \sum_q (\varepsilon_q - \mu)\hat{n}_q, \tag{3.50}$$

where $\hat{n}_q$ is the number of electrons operator in the state $q = (\mathbf{q}, \sigma)$. From (3.50) and (3.49) it is easy to see that

$$\frac{\partial \Omega}{\partial \varepsilon_k} = \frac{\mathrm{Tr}(\hat{n}_k\, \mathrm{e}^{-\mathcal{H}_0'/T})}{\mathrm{Tr} \mathrm{e}^{-\mathcal{H}_0'/T}} = \langle \hat{n}_k \rangle. \tag{3.51}$$

To calculate the grand partition function Ξ, we recall that the trace Tr is the sum of the matrix elements $(\Phi_{N_e}, \mathrm{e}^{-\mathcal{H}_0'/T}\Phi_{N_e})$ over all determinantal states Φ_{N_e}:

$$\Xi = \sum_{N_e=0}^{\infty} \sum_{\Phi_{N_e}} (\Phi_{N_e}, \mathrm{e}^{-\mathcal{H}_0'/T}\Phi_{N_e}).$$

The exponential on the right-hand side can be written as the product

$$\mathrm{e}^{-\mathcal{H}_0'/T} = \prod_q \mathrm{e}^{-(\varepsilon_q-\mu)\hat{n}_q/T},$$

because the operators $\hat{n}_q$ commute. Each Φ_{N_e} is an eigenstate of the number of electrons operator: $\hat{n}_q\Phi_{N_e} = n_q\Phi_{N_e}$. Therefore, taking the orthonormality of Φ_{N_e} into account, we have

$$\Xi = \sum_{n_q=0,1} \prod_q \mathrm{e}^{-(\varepsilon_q-\mu)n_q/T}.$$

The latter can be written as the product

$$\Xi = \prod_q \left[1 + \mathrm{e}^{-(\varepsilon_q-\mu)/T}\right].$$

Substituting this in the thermodynamic potential (3.49) and using (3.51), we obtain

$$\langle n_k \rangle = \frac{1}{\mathrm{e}^{(\varepsilon_k-\mu)/T} + 1} \equiv f(\varepsilon_k), \tag{3.52}$$

where $f(\varepsilon)$ is the *Fermi function* (Fig. 3.1).

At $T = 0$ the Fermi function is given by the step function (dashed line at Fig. 3.1), where $\mu = \varepsilon_F$ is the Fermi energy, and the negative of its derivative is the delta function $\delta(\varepsilon - \varepsilon_F)$. For $T > 0$, the Fermi function deviate from the step function in a neighbourhood of the order T around the chemical potential μ. However, the step function is a good approximation for the Fermi function in the temperature region $T \ll \varepsilon_F$ (for details, see [5]). In the ferromagnetic metals the Fermi energy ε_F is about 10 eV.

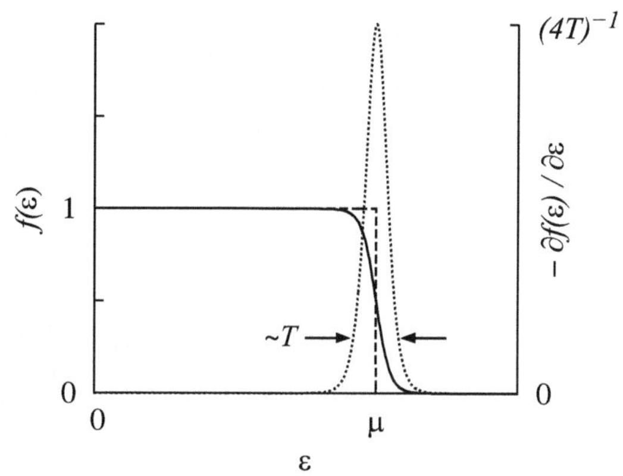

Fig. 3.1 Sketch of the Fermi function $f(\varepsilon)$ (solid line) and the negative of its derivative (dotted line) for $T > 0$. Dashed line shows the limit of the Fermi function at $T = 0$

Applying the above result (3.52), we write the magnetic susceptibility (3.48) as

$$\chi^0_{zz}(\mathbf{q}, \omega) = \frac{1}{2} g^2 \mu_B^2 F(\mathbf{q}, \omega), \tag{3.53}$$

where

$$F(\mathbf{q}, \omega) = \sum_{\mathbf{k}} \frac{f(\varepsilon_{\mathbf{k}}) - f(\varepsilon_{\mathbf{k+q}})}{\varepsilon_{\mathbf{k+q}} - \varepsilon_{\mathbf{k}} - \hbar\omega}$$

is called the *Lindhard function*. At $\mathbf{q} = 0$ and $\omega = 0$, we have

$$F(0, 0) = -\lim_{q \to 0} \sum_{\mathbf{k}} \frac{f(\varepsilon_{\mathbf{k}}) - f(\varepsilon_{\mathbf{k+q}})}{\varepsilon_{\mathbf{k}} - \varepsilon_{\mathbf{k+q}}} = -\sum_{\mathbf{k}} \frac{\partial f(\varepsilon_{\mathbf{k}})}{\partial \varepsilon}.$$

Then the uniform static susceptibility is

$$\chi^0_{zz}(0, 0) = -\frac{1}{2} g^2 \mu_B^2 \sum_{\mathbf{k}} \frac{\partial f(\varepsilon_{\mathbf{k}})}{\partial \varepsilon}. \tag{3.54}$$

To convert the sum over the Brillouin zone into the integral over energies, we make use of the sum rule

$$\sum_{\mathbf{k}} \ldots = N \int \ldots \nu(\varepsilon)\, \mathrm{d}\varepsilon, \tag{3.55}$$

where

$$\nu(\varepsilon) \equiv \frac{1}{N} \sum_{\mathbf{k}} \delta(\varepsilon - \varepsilon_{\mathbf{k}}) \tag{3.56}$$

is the nonmagnetic density of states per site and spin. Formula (3.54) becomes

$$\chi^0_{zz}(0, 0) = -\frac{1}{2} g^2 \mu_B^2 N \int \nu(\varepsilon) \frac{\partial f(\varepsilon)}{\partial \varepsilon}\, \mathrm{d}\varepsilon. \tag{3.57}$$

At low temperatures, replacing the derivative of the Fermi function $-\partial f(\varepsilon)/\partial \varepsilon$ by the delta function localized at the Fermi level ε_F (see Fig. 3.1), we finally obtain

$$\chi_P = \frac{1}{2} g^2 \mu_B^2 \nu(\varepsilon_F). \tag{3.58}$$

where $\chi_P \equiv N^{-1} \chi^0_{zz}(0, 0)$ is called the *Pauli susceptibility*.

As an application of the above result, we calculate the magnetization induced in an electron gas by a static magnetic field of the delta function type applied at the origin of the coordinates:

$$H_z(\mathbf{r}) = \frac{V}{N} H \delta(\mathbf{r}).$$

Then

$$\langle \mathcal{M}_z(\mathbf{r}) \rangle = \frac{1}{V} \sum_{\mathbf{q}} \langle \mathcal{M}_z(\mathbf{q}) \rangle\, \mathrm{e}^{\mathrm{i}\mathbf{qr}} = \frac{1}{V} \sum_{\mathbf{q}} \chi^0_{zz}(\mathbf{q}, 0) H_z(\mathbf{q}) \mathrm{e}^{\mathrm{i}\mathbf{qr}}.$$

Using (3.53) and

$$H_z(\mathbf{r}) = \sum_{\mathbf{q}} H_z(\mathbf{q})e^{i\mathbf{q}\mathbf{r}} = \sum_{\mathbf{q}} \frac{H}{N} e^{i\mathbf{q}\mathbf{r}},$$

we have

$$\langle \mathcal{M}_z(\mathbf{r}) \rangle = \frac{1}{2} g^2 \mu_B^2 \frac{H}{NV} \sum_{\mathbf{q}} F(\mathbf{q}, 0)e^{i\mathbf{q}\mathbf{r}} \equiv \frac{1}{2} g^2 \mu_B^2 \frac{H}{N} F(\mathbf{r}).$$

For a free electron gas, the static Lindhard function at $T = 0$ has the well-known form (see, e.g. [5])

$$F(\mathbf{q}, 0) = \nu(\varepsilon_F) \left(\frac{1}{2} + \frac{1 - x^2}{4x} \ln \left| \frac{1 + x}{1 - x} \right| \right)$$

with $x = q/(2k_F)$. The inverse Fourier transform is given by

$$F(\mathbf{r}) = 6\pi n_e \nu(\varepsilon_F) \frac{\sin(2k_F r) - k_F \cos(2k_F r)}{(2k_F r)^4},$$

where $n_e = N_e/N$ is the number of electrons per unit cell and $\hbar^2 k_F^2/(2m) = \varepsilon_F$ (for details of the calculation, see, e.g. [8, Appendix H]). Thus, a localized magnetic field produces oscillating magnetization within metallic electrons, which is called the *Ruderman–Kittel–Kasuya–Yosida (RKKI) oscillation*.[5]

References

1. S. Raimes, *The Wave Mechanics of Electrons in Metals* (North-Holland, Amsterdam, 1970)
2. C. Kittel, *Quantum Theory of Solids*, 2nd edn. (Wiley, New York, 1987)
3. C. Kittel, *Introduction to Solid State Physics*, 8th edn. (Wiley, New York, 2005)
4. S. Raimes, *Many-Electron Theory* (North-Holland, Amsterdam, 1972)
5. D.J. Kim, *New Perspectives in Magnetism of Metals* (Kluwer/Plenum, New York, 1999)
6. H. Bruus, K. Flensberg, *Many-Body Quantum Theory in Condensed Matter Physics*, 3rd edn. (Oxford University Press, Oxford, 2004)
7. R.M. White, *Quantum Theory of Magnetism*, 3rd edn. (Springer, Berlin, 2007)
8. Y. Kakehashi, *Modern Theory of Magnetism in Metals and Alloys* (Springer, Berlin, 2012)

[5] For a discussion of the RKKI interaction between *localized* spins see, e.g. [5, 7, 8].

4 Mean-Field Theory

Exact agreement with experiment... is, indeed, not attained; nor should it be, for it is known that the premises are an oversimplification. The character of the agreement or disagreement may, however, give a valuable guide... Without such guidance, in view of the complexity of the total problem, the most intensive theoretical efforts and computational labour may well be largely in vain. (E.C. Stoner, J. Phys. Radium ***12****, 372 (1951))*

This chapter is devoted to applications of the mean-field approximation in metals. We confine the presentation to the Hubbard model, which is relevant to the narrow-band metals. This way we avoid unnecessary technicalities and make a link with further chapters. We start with explaining the idea of the Hatree-Fock method in the single-band case, which is used to obtain the equations of the Stoner theory. Then we derive the uniform susceptibility and discuss implications of the Stoner theory for metals. For a more extensive introduction to the mean-field approach and its applications to magnetism, see, for instance, Refs. [1–6].

4.1 The Hubbard Model

The *Hubbard model* [7] is relevant if we consider narrow d bands of transition metals. The charge density of d electrons is localized around the nuclei of the solid, making it possible to speak about electrons on a particular lattice site. For the sake of simplicity, we consider the single-band case first. In this case the second-quantized Hamiltonian (3.35) reduces to

$$\mathcal{H} = \sum_{ii'\sigma} t_{ii'} a^{\dagger}_{i\sigma} a_{i'\sigma} + \frac{1}{2} \sum_{ii'\,jj'\,\sigma\sigma'} U_{iji'j'} a^{\dagger}_{i\sigma} a^{\dagger}_{j\sigma'} a_{j'\sigma'} a_{i'\sigma}.$$

Following [7], we assume that electrons interact only at the same site:

$$U_{iji'j'} = U \delta_{ij} \delta_{ii'} \delta_{jj'}. \tag{4.1}$$

Then the model is described by the Hamiltonian $\mathcal{H} = \mathcal{H}_0 + \mathcal{H}_{\mathrm{I}}$, where the electrons energy $\mathcal{H}_0$ is

$$\mathcal{H}_0 = \sum_{ii'\sigma} t_{ii'} a^{\dagger}_{i\sigma} a_{i'\sigma}$$

and $\mathcal{H}_{\mathrm{I}}$ is the single-site Coulomb interaction,

$$\mathcal{H}_{\mathrm{I}} = \frac{1}{2} U \sum_{i\sigma\sigma'} a^{\dagger}_{i\sigma} a^{\dagger}_{i\sigma'} a_{i\sigma'} a_{i\sigma}.$$

Using the anticommutation relations (3.12), we have

$$a^{\dagger}_{i\sigma} a^{\dagger}_{i\sigma'} a_{i\sigma'} a_{i\sigma} = a^{\dagger}_{i\sigma} a_{i\sigma} a^{\dagger}_{i\sigma'} a_{i\sigma'} - a^{\dagger}_{i\sigma} a_{i\sigma'} \delta_{\sigma\sigma'} = n_{i\sigma} n_{i\sigma'} - n_{i\sigma'} \delta_{\sigma\sigma'}.$$

N. B. Melnikov, B. I. Reser, *Dynamic Spin Fluctuation Theory of Metallic Magnetism*,
https://doi.org/10.1007/978-3-319-92974-3_4

Then

$$\mathcal{H}_{\rm I} = \frac{1}{2} U \sum_{i\sigma\sigma'} n_{i\sigma} n_{i\sigma'} - \frac{1}{2} U \sum_{i\sigma} n_{i\sigma} . \tag{4.2}$$

Since $n_{i\sigma}^2 = n_{i\sigma}$ and $n_{i\uparrow}$ and $n_{i\downarrow}$ commute, we write (4.2) as

$$\mathcal{H}_{\rm I} = \frac{1}{2} U \sum_{i\sigma} n_{i\sigma} n_{i\bar{\sigma}} = U \sum_{i} n_{i\uparrow} n_{i\downarrow},$$

where $\bar{\sigma}$ is the opposite spin to σ.

There are several ways of expressing the local interaction by the atomic charge $n_j = n_{j\uparrow} + n_{j\downarrow}$ and spin $s_j^z = \frac{1}{2}(n_{j\uparrow} - n_{j\downarrow})$. Using the simple identity

$$n_{j\uparrow} n_{j\downarrow} = \frac{1}{4}(n_{j\uparrow} + n_{j\downarrow})^2 - \frac{1}{4}(n_{j\uparrow} - n_{j\downarrow})^2,$$

we write

$$\mathcal{H}_{\rm I} = U \sum_{j} \left(\frac{1}{4} n_j^2 - (s_j^z)^2 \right) . \tag{4.3}$$

Taking into account $(s_j^z)^2 = \frac{1}{3}\mathbf{s}_j^2$, we obtain the spin-rotational form

$$\mathcal{H}_{\rm I} = U \sum_{j} \left(\frac{1}{4} n_j^2 - \frac{1}{3} \mathbf{s}_j^2 \right) . \tag{4.4}$$

Similarly, using the relation $(s_j^z)^2 = (\mathbf{s}_j \mathbf{e}_j)^2$, where $\mathbf{e}_j$ is an arbitrary unit vector, we come to

$$\mathcal{H}_{\rm I} = U \sum_{j} \left(\frac{1}{4} n_j^2 - (\mathbf{s}_j \mathbf{e}_j)^2 \right) . \tag{4.5}$$

In the momentum representation the interaction term $\mathcal{H}_{\rm I}$ is given by (3.39). Substituting the expression for the Bloch function (3.6) in formula (3.40) and taking the local interaction (4.1) into account, we have

$$U_{\mathbf{klk'l'}} = \frac{U}{N^2} \sum_{j} \mathrm{e}^{-\mathrm{i}(\mathbf{k}+\mathbf{l}-\mathbf{k'}-\mathbf{l'})\mathbf{R}_j} .$$

From the relation $\sum_j \mathrm{e}^{\mathrm{i}\mathbf{k}\mathbf{R}_j} = N\delta_{\mathbf{k}0}$, it follows that $U_{\mathbf{klk'l'}}$ is nonzero iff the momentum is conserved: $\mathbf{k} + \mathbf{l} = \mathbf{k'} + \mathbf{l'}$. Thus, we finally obtain

$$\mathcal{H}_{\rm I} = \frac{1}{2} \tilde{U} \sum_{\mathbf{klq}\sigma\sigma'} a_{\mathbf{k}\sigma}^{\dagger} a_{\mathbf{l}\sigma'}^{\dagger} a_{\mathbf{l}+\mathbf{q},\sigma'} a_{\mathbf{k}-\mathbf{q},\sigma} , \tag{4.6}$$

where $\tilde{U} = U/N$. Recalling the expression for $\mathcal{H}_0$ in the momentum representation (3.38), we write the Hubbard Hamiltonian in the momentum representation as

$$\mathcal{H} = \sum_{\mathbf{k}\sigma} \varepsilon_{\mathbf{k}} a_{\mathbf{k}\sigma}^{\dagger} a_{\mathbf{k}\sigma} + \frac{1}{2} \tilde{U} \sum_{\mathbf{klq}\sigma\sigma'} a_{\mathbf{k}\sigma}^{\dagger} a_{\mathbf{l}\sigma'}^{\dagger} a_{\mathbf{l}+\mathbf{q},\sigma'} a_{\mathbf{k}-\mathbf{q},\sigma} .$$

Even in the simple Hubbard model, the magnetic susceptibility cannot be calculated explicitly without further approximations. For the sake of further comparison with SFT results, we start with investigating the Hartree-Fock theory of metallic ferromagnets [8, 9], often called the Stoner theory of ferromagnetism.

4.2 Stoner Mean-Field Theory

4.2.1 Hartree-Fock Approximation

The general idea of the underlying mean-field approximation is to simplify the Hamiltonian $\mathcal{H} = \mathcal{AB}$ given by the product of two operators $\mathcal{A}$ and $\mathcal{B}$ using partial averaging:

$$\mathcal{H}_{\mathrm{MF}} = \mathcal{A}\langle\mathcal{B}\rangle + \langle\mathcal{A}\rangle\mathcal{B} - \langle\mathcal{A}\rangle\langle\mathcal{B}\rangle. \tag{4.7}$$

Here we write the last term to ensure the correct average: $\langle\mathcal{H}_{\mathrm{MF}}\rangle = \langle\mathcal{A}\rangle\langle\mathcal{B}\rangle$. The average is calculated self-consistently in such a way that it minimizes the free energy

$$F_{\mathrm{MF}} = -T \ln Z_{\mathrm{MF}}, \qquad Z_{\mathrm{MF}} = \mathrm{Tr} e^{-\mathcal{H}_{\mathrm{MF}}/T}$$

(for details, see, e.g. [4]). To find the minimum, we consider the free energy as a function of the parameters $\bar{\mathcal{A}} \equiv \langle\mathcal{A}\rangle$ and $\bar{\mathcal{B}} \equiv \langle\mathcal{B}\rangle$. Then the minimum satisfies the conditions

$$\frac{\partial F_{\mathrm{MF}}}{\partial \bar{\mathcal{A}}} = 0, \qquad \frac{\partial F_{\mathrm{MF}}}{\partial \bar{\mathcal{B}}} = 0.$$

Differentiating F_{MF}, we have

$$\frac{\partial F_{\mathrm{MF}}}{\partial \bar{\mathcal{A}}} = \frac{1}{Z_{\mathrm{MF}}} \mathrm{Tr}\left(\mathrm{e}^{-\mathcal{H}_{\mathrm{MF}}/T} \frac{\partial}{\partial \bar{\mathcal{A}}} \mathcal{H}_{\mathrm{MF}}\right) = \frac{1}{Z_{\mathrm{MF}}} \mathrm{Tr}\left(\mathrm{e}^{-\mathcal{H}_{\mathrm{MF}}/T} (\mathcal{B} - \bar{\mathcal{B}})\right),$$

Thus, the mean-field average of $\mathcal{B}$ is defined as the canonical average with the mean-field Hamiltonian:

$$\langle\mathcal{B}\rangle = \frac{1}{Z_{\mathrm{MF}}} \mathrm{Tr}\left(\mathcal{B} \mathrm{e}^{-\mathcal{H}_{\mathrm{MF}}/T}\right), \tag{4.8}$$

and the same holds for $\mathcal{A}$.[1]

We apply the above technique to the interaction term (4.6) by partially averaging different pairs of creation-annihilation operators $a^{\dagger}_{\mathbf{k}\sigma} a_{\mathbf{k}'\sigma'}$. The interaction part in the *Hartree-Fock approximation* consists of two parts:

$$\mathcal{H}_{\mathrm{I}}^{\mathrm{MF}} = \mathcal{H}_{\mathrm{I}}^{\mathrm{Hartree}} + \mathcal{H}_{\mathrm{I}}^{\mathrm{Fock}}.$$

The Hartree approximation gives the terms with the same spin (*Coulomb interaction*). Commuting the creation and annihilation operators in $\mathcal{H}_{\mathrm{I}}$, we have

$$\mathcal{H}_{\mathrm{I}} = \frac{1}{2}\tilde{U} \sum_{\mathbf{klq}\sigma\sigma'} a^{\dagger}_{\mathbf{k}\sigma} a_{\mathbf{k}-\mathbf{q},\sigma} a^{\dagger}_{\mathbf{l}\sigma'} a_{\mathbf{l}+\mathbf{q},\sigma'} - \frac{1}{2}\tilde{U} N \sum_{\mathbf{k}\sigma} a^{\dagger}_{\mathbf{k}\sigma} a_{\mathbf{k}\sigma}. \tag{4.9}$$

---

[1] Strictly speaking, use of the second-quantized representation implies that the grand canonical ensemble is employed. This requires changing the free energy F by the thermodynamic potential Ω in the above procedure and yields essentially the same result as in (4.8) but with $\mathcal{H}'_{\mathrm{MF}} = \mathcal{H}_{\mathrm{MF}} - \mu\mathcal{N}_{\mathrm{e}}$ instead of $\mathcal{H}_{\mathrm{MF}}$ and the grand canonical partition function $\Xi = \mathrm{Tr} e^{-\mathcal{H}'_{\mathrm{MF}}/T}$ instead of the canonical one $Z = \mathrm{Tr} e^{-\mathcal{H}_{\mathrm{MF}}/T}$. In this and the next chapters, we often omit the prime in the Hamiltonians where it does not lead to confusion.

The last term on the right-hand side leads to a shift of the energy spectrum and can be ignored. Applying (4.7) to the first term of (4.9), we obtain

$$\mathcal{H}_{\mathrm{I}}^{\mathrm{Hartree}} = \frac{1}{2}\tilde{U} \sum_{\mathbf{klq}\sigma\sigma'} \langle a_{\mathbf{k}\sigma}^{\dagger} a_{\mathbf{k}-\mathbf{q},\sigma} \rangle a_{\mathbf{l}\sigma'}^{\dagger} a_{\mathbf{l}+\mathbf{q},\sigma'} + \frac{1}{2}\tilde{U} \sum_{\mathbf{klq}\sigma\sigma'} \langle a_{\mathbf{l}\sigma'}^{\dagger} a_{\mathbf{l}+\mathbf{q},\sigma'} \rangle a_{\mathbf{k}\sigma}^{\dagger} a_{\mathbf{k}-\mathbf{q},\sigma} .$$

The terms on the right-hand side are equal, and hence

$$\mathcal{H}_{\mathrm{I}}^{\mathrm{Hartree}} = \tilde{U} \sum_{\mathbf{klq}\sigma\sigma'} \langle a_{\mathbf{l}\sigma'}^{\dagger} a_{\mathbf{l}+\mathbf{q},\sigma'} \rangle a_{\mathbf{k}\sigma}^{\dagger} a_{\mathbf{k}-\mathbf{q},\sigma} . \tag{4.10}$$

Here and hereafter we omit the scalar term in (4.7), because we already know the expression for the average (4.8).

The Fock approximation gives the terms with different spins (*exchange interaction*). Similar to the Hartree term, we obtain

$$\mathcal{H}_{\mathrm{I}}^{\mathrm{Fock}} = -\frac{1}{2}\tilde{U} \sum_{\mathbf{klq}\sigma\sigma'} \langle a_{\mathbf{k}\sigma}^{\dagger} a_{\mathbf{l}+\mathbf{q},\sigma'} \rangle a_{\mathbf{l}\sigma'}^{\dagger} a_{\mathbf{k}-\mathbf{q},\sigma} - \frac{1}{2}\tilde{U} \sum_{\mathbf{klq}\sigma\sigma'} \langle a_{\mathbf{l}\sigma'}^{\dagger} a_{\mathbf{k}-\mathbf{q},\sigma} \rangle a_{\mathbf{k}\sigma}^{\dagger} a_{\mathbf{l}+\mathbf{q},\sigma'} ,$$

where the negative sign results from commuting $a_{\mathbf{l}\sigma'}^{\dagger}$ and $a_{\mathbf{l}+\mathbf{q},\sigma'}$. Since the terms on the right-hand side are again equal, we write

$$\mathcal{H}_{\mathrm{I}}^{\mathrm{Fock}} = -\tilde{U} \sum_{\mathbf{klq}\sigma\sigma'} \langle a_{\mathbf{k}\sigma}^{\dagger} a_{\mathbf{l}+\mathbf{q},\sigma'} \rangle a_{\mathbf{l}\sigma'}^{\dagger} a_{\mathbf{k}-\mathbf{q},\sigma} . \tag{4.11}$$

As we show below, the exchange interaction implies that the energy of a pair of electrons with parallel spins is lower than the energy of a pair of electrons with antiparallel spins. Since the exchange energy is of the same order as the transfer energy, exchange interaction plays a dominant role in establishing the ferromagnetic ordering in metals, just as in the ferromagnetic insulators.

In the spatially uniform situation, we have

$$\langle a_{\mathbf{k}\sigma}^{\dagger} a_{\mathbf{k}'\sigma'} \rangle = \langle a_{\mathbf{k}\sigma}^{\dagger} a_{\mathbf{k}\sigma} \rangle \delta_{\mathbf{kk}'} \delta_{\sigma\sigma'} . \tag{4.12}$$

Then

$$\mathcal{H}_{\mathrm{I}}^{\mathrm{Hartree}} = \tilde{U} \sum_{\mathbf{k}\sigma} (N_{\uparrow} + N_{\downarrow}) a_{\mathbf{k}\sigma}^{\dagger} a_{\mathbf{k}\sigma} \tag{4.13}$$

and

$$\mathcal{H}_{\mathrm{I}}^{\mathrm{Fock}} = -\tilde{U} \sum_{\mathbf{k}\sigma} N_{\sigma} a_{\mathbf{k}\sigma}^{\dagger} a_{\mathbf{k}\sigma} , \tag{4.14}$$

where

$$N_{\sigma} = \sum_{\mathbf{k}} \langle a_{\mathbf{k}\sigma}^{\dagger} a_{\mathbf{k}\sigma} \rangle \tag{4.15}$$

is the total number of electrons with the spin σ. The interaction term finally becomes

$$\mathcal{H}_{\mathrm{I}}^{\mathrm{MF}} = \tilde{U} \sum_{\mathbf{k}\sigma} (N_{\uparrow} + N_{\downarrow} - N_{\sigma}) a_{\mathbf{k}\sigma}^{\dagger} a_{\mathbf{k}\sigma} = \tilde{U} \sum_{\mathbf{k}\sigma} \Big(\frac{1}{2} N_{\mathrm{e}} - \sigma \bar{S}_z \Big) a_{\mathbf{k}\sigma}^{\dagger} a_{\mathbf{k}\sigma} .$$

Here

$$N_e = N_\uparrow + N_\downarrow, \qquad \bar{S}_z = \frac{1}{2}(N_\uparrow - N_\downarrow) \tag{4.16}$$

are the total charge and spin.

Note that the Coulomb interaction leads to a uniform shift of the energy spectrum $\frac{1}{2}\tilde{U} N_e$. Therefore, the Coulomb interaction does not have an impact on the uniform static susceptibility and can be ignored. It is the exchange interaction contribution $-\sigma \tilde{U} \bar{S}_z$ that is responsible for the magnetism in metals, as we will see below. Thus, the mean-field Hamiltonian is

$$\mathcal{H}_{MF} = \mathcal{H}_0 + \mathcal{H}_I^{MF} = \sum_{\mathbf{k}\sigma} \varepsilon_{\mathbf{k}\sigma} a^\dagger_{\mathbf{k}\sigma} a_{\mathbf{k}\sigma}, \tag{4.17}$$

where

$$\varepsilon_{\mathbf{k}\sigma} = \varepsilon_{\mathbf{k}} - \sigma \tilde{U} \bar{S}_z \tag{4.18}$$

is the spin-polarized energy spectrum.

The Hamiltonian $\mathcal{H}_{MF}$ describes the system of *noninteracting* electrons, where each electron interacts with a static exchange field created by all the other electrons. This exchange field acts on top of the Coulomb field of the crystal lattice. Recall that our Bloch states are not just plane waves, as is the case in most of the textbooks (see, e.g. [3–5]).

4.2.2 Magnetization: The T^2 Law

If we apply a uniform static magnetic field H in the z-direction, the electrons acquire an additional energy

$$\mathcal{H}_M = -\mathcal{M}_z H_z,$$

where $\mathcal{M}_z = -g\mu_B S_z$ is the magnetic moment operator. By formula (3.20), we have

$$S_z = \int s_z(\mathbf{r})\, d\mathbf{r} = \int \psi^\dagger(\mathbf{r})\, s_z\, \psi(\mathbf{r})\, d\mathbf{r}.$$

Then, writing the field operators $\psi^\dagger(\mathbf{r})$ and $\psi(\mathbf{r})$ in the Bloch basis (3.26), we obtain

$$S_z = \frac{1}{2} \sum_{\mathbf{k}} \left(a^\dagger_{\mathbf{k}\uparrow} a_{\mathbf{k}\uparrow} - a^\dagger_{\mathbf{k}\downarrow} a_{\mathbf{k}\downarrow} \right). \tag{4.19}$$

Combining the magnetic energy

$$\mathcal{H}_M = -\frac{g\mu_B}{2} H_z \sum_{\mathbf{k}\sigma} \sigma a^\dagger_{\mathbf{k}\sigma} a_{\mathbf{k}\sigma}$$

with the energy of electrons (4.17), we have

$$\mathcal{H}_{MF} = \mathcal{H}_0 + \mathcal{H}_M + \mathcal{H}_I^{MF} = \sum_{\mathbf{k}\sigma} \varepsilon_{\mathbf{k}\sigma} a^\dagger_{\mathbf{k}\sigma} a_{\mathbf{k}\sigma}, \tag{4.20}$$

where the spin-polarized spectrum now is

$$\varepsilon_{\mathbf{k}\sigma} = \varepsilon_{\mathbf{k}} + \sigma \frac{g\mu_B}{2} H_z - \sigma \tilde{U} \bar{S}_z \tag{4.21}$$

(we omit the constant term $\frac{1}{2}\tilde{U} N_e$ just as before).

We now want to calculate the total mean spin $\bar{S}_z$ given by (4.16). As soon as we know the energy spectrum $\varepsilon_{\mathbf{k}\sigma}$ the average number of the spin-up and spin-down electrons (4.15) is calculated by formula (4.8):

$$N_\sigma = \sum_{\mathbf{k}} \langle n_{\mathbf{k}\sigma} \rangle = \sum_{\mathbf{k}} f(\varepsilon_{\mathbf{k}\sigma}). \tag{4.22}$$

Converting the sum over the Brillouin zone into the integral over energies according to (3.55), at $H_z = 0$ we obtain

$$\bar{n}_\sigma = \int \nu(\varepsilon) f(\varepsilon - \sigma U \bar{s}_z)\, \mathrm{d}\varepsilon = \int \nu_\sigma(\varepsilon, T) f(\varepsilon)\, \mathrm{d}\varepsilon, \tag{4.23}$$

where $\bar{n}_\sigma = N_\sigma/N$ is the number of electrons with the spin σ per site and $\bar{s}_z = \bar{S}_z/N$ is the single-site spin. Here $\nu(\varepsilon)$ is the nonmagnetic DOS of d electrons (3.56) that corresponds to the Hamiltonian $\mathcal{H}_0$, and

$$\nu_\sigma(\varepsilon, T) = \nu(\varepsilon + \sigma U \bar{s}_z) = \frac{1}{N} \sum_{\mathbf{k}} \delta(\varepsilon - \varepsilon_{\mathbf{k}\sigma}) \tag{4.24}$$

is the spin-polarized DOS of electrons that is determined by the mean-field Hamiltonian $\mathcal{H}_0 + \mathcal{H}_{\mathrm{I}}^{\mathrm{MF}}$. As temperature increases, the mean spin $\bar{s}_z(T)$ decreases, thus shifting the spin-polarized DOSs $\nu_\sigma(\varepsilon, T)$ towards each other. But the shape of the DOSs is not changed. In the paramagnetic region, the spin-up and spin-down DOSs coincide: $\nu_\uparrow(\varepsilon, T) = \nu_\downarrow(\varepsilon, T)$ for $T \geq T_{\mathrm{C}}$.

So far we have obtained the system of two equations (4.23) with respect to the unknowns $\bar{n}_\uparrow$ and $\bar{n}_\downarrow$ or, equivalently, with respect to

$$n_{\mathrm{e}} = \bar{n}_\uparrow + \bar{n}_\downarrow, \qquad \bar{s}_z = \frac{1}{2}(\bar{n}_\uparrow - \bar{n}_\downarrow), \tag{4.25}$$

where $n_{\mathrm{e}} = N_{\mathrm{e}}/N$ is the (mean) number of electrons per site. In calculations it is more convenient to use canonical average instead of the grand canonical average. That means, we consider the system with a fixed number of electrons N_{e} and make the chemical potential μ an unknown. Thus, we solve (4.25), where

$$\bar{n}_\sigma = \int \nu(\varepsilon + \sigma U \bar{s}_z) f(\varepsilon)\, \mathrm{d}\varepsilon,$$

with respect to the chemical potential μ and mean spin $\bar{s}_z$ at each temperature T. The input data are the nonpolarized DOS at $T = 0$, $\nu(\varepsilon)$, and interaction constant U. This is usually called the *Stoner model*.

In the electron gas model, one can obtain a low-temperature expansion of magnetization as follows. Integrating by parts the integral of an arbitrary function $g(\varepsilon)$ with the Fermi function, we write

$$\int_{-\infty}^{\infty} g(\varepsilon) f(\varepsilon)\, \mathrm{d}\varepsilon = \left[G(\varepsilon) f(\varepsilon)\right]_{-\infty}^{\infty} - \int_{-\infty}^{\infty} G(\varepsilon) \frac{\partial f(\varepsilon)}{\partial \varepsilon}\, \mathrm{d}\varepsilon, \tag{4.26}$$

where

$$G(\varepsilon) \equiv \int_{-\infty}^{\varepsilon} g(\varepsilon')\, \mathrm{d}\varepsilon'.$$

Since $G(-\infty) = 0$ and $f(\infty) = 0$, the integrated terms on the right-hand side of (4.26) vanish. At temperatures much lower than the Fermi level $T \ll \varepsilon_{\mathrm{F}}$, the integral on the right-hand side of (4.26) can be expanded as

$$-\int_{-\infty}^{\infty} G(\varepsilon) \frac{\partial f(\varepsilon)}{\partial \varepsilon}\, \mathrm{d}\varepsilon = G(\mu) + \frac{\pi^2}{6} G''(\mu) T^2 + \cdots, \tag{4.27}$$

where the prime stands for the derivative. For the chemical potential we have (see, e.g. [3])

$$\mu(T) = \varepsilon_{\mathrm{F}} + \frac{\pi^2}{6} \frac{\nu'(\varepsilon_{\mathrm{F}})}{\nu(\varepsilon_{\mathrm{F}})} T^2 + \cdots$$

Then (4.27) becomes

$$-\int_{-\infty}^{\infty} G(\varepsilon) \frac{\partial f(\varepsilon)}{\partial \varepsilon}\, \mathrm{d}\varepsilon = G(\varepsilon_{\mathrm{F}}) + \frac{\pi^2}{6} \left(G''(\varepsilon_{\mathrm{F}}) - G'(\varepsilon_{\mathrm{F}}) \frac{\nu'(\varepsilon_{\mathrm{F}})}{\nu(\varepsilon_{\mathrm{F}})} \right) T^2 + \cdots \tag{4.28}$$

If we take $g(\varepsilon) = \nu_\sigma(\varepsilon)$, where $\nu_\sigma(\varepsilon)$ is the DOS at $T = 0$, then $G(\varepsilon) = N_\sigma(\varepsilon)$ is the number of states with the spin σ and energy less or equal to ε. Using (4.26) and (4.28), we write (4.23) as

$$\bar{n}_\sigma = \int_{-\infty}^{\infty} \nu_\sigma(\varepsilon) f(\varepsilon)\, d\varepsilon = N_\sigma(\varepsilon_F) + \frac{\pi^2}{6}\left(\nu'_\sigma(\varepsilon_F) - \nu_\sigma(\varepsilon_F)\frac{\nu'(\varepsilon_F)}{\nu(\varepsilon_F)}\right) T^2 + \cdots$$

Hence from (4.25) we obtain

$$\bar{s}_z(T) = \bar{s}_z(0) - \alpha T^2 + \cdots$$

where $\alpha > 0$ is a constant. The expansion is valid at low temperatures, $T \ll \varepsilon_F$.

4.2.3 Uniform Static Susceptibility

Next, we calculate the uniform static susceptibility

$$\chi = \left.\frac{\partial\langle M_z\rangle}{\partial H}\right|_{H=0} = -g\mu_B \left.\frac{\partial\langle S_z\rangle}{\partial H}\right|_{H=0}$$

in the mean-field approximation. Using (4.21) and (4.22), we write the mean spin as

$$\bar{S}_z = \frac{1}{2}\sum_{\mathbf{k}}\left[f\left(\varepsilon_{\mathbf{k}} + \frac{g\mu_B}{2}H - \tilde{U}\bar{S}_z\right) - f\left(\varepsilon_{\mathbf{k}} - \frac{g\mu_B}{2}H + \tilde{U}\bar{S}_z\right)\right].$$

Differentiating this equation with respect to H, at $H = 0$ we obtain

$$\frac{\partial \bar{S}_z}{\partial H} = \frac{1}{2}\sum_{\mathbf{k}\sigma}\frac{\partial f(\varepsilon_{\mathbf{k}} - \sigma\tilde{U}\bar{S}_z)}{\partial\varepsilon}\left(\frac{g\mu_B}{2} - \tilde{U}\frac{\partial\bar{S}_z}{\partial H}\right).$$

Thus,

$$\chi = \frac{\chi^0}{1 - \dfrac{2\tilde{U}}{g^2\mu_B^2}\chi^0}, \tag{4.29}$$

where

$$\chi^0 = -\frac{g^2\mu_B^2}{2}\frac{1}{2}\sum_{\mathbf{k}\sigma}\frac{\partial f(\varepsilon_{\mathbf{k}\sigma})}{\partial\varepsilon}$$

is the susceptibility of noninteracting electrons (3.57). Using the sum rule (3.55), we obtain

$$\chi^0 = -\frac{g^2\mu_B^2}{2}\frac{1}{2}\sum_\sigma\int \nu(\varepsilon)\frac{\partial f(\varepsilon - \sigma U\bar{s}_z)}{\partial\varepsilon}\, d\varepsilon. \tag{4.30}$$

The latter can be written as

$$\chi^0 = -\frac{g^2\mu_B^2}{2}\frac{1}{2}\sum_\sigma\int \nu_\sigma(\varepsilon, T)\frac{\partial f(\varepsilon)}{\partial\varepsilon}\, d\varepsilon \equiv -\frac{g^2\mu_B^2}{2}\int \nu(\varepsilon, T)\frac{\partial f(\varepsilon)}{\partial\varepsilon}\, d\varepsilon,$$

where

$$\nu(\varepsilon, T) = \frac{1}{2}\sum_\sigma \nu(\varepsilon + \sigma U\bar{s}_z) = \frac{1}{2}\sum_\sigma \frac{1}{N}\sum_{\mathbf{k}} \delta(\varepsilon - \varepsilon_{\mathbf{k}\sigma}) \tag{4.31}$$

is the total DOS at temperature T (per site and spin).

In the paramagnetic state, formula (4.30) for χ^0 reduces to (3.57). Formula (4.29) for χ becomes

$$\chi_{\mathrm{S}} = \frac{\chi_{\mathrm{P}}}{1 - \dfrac{2U}{g^2\mu_{\mathrm{B}}^2}\chi_{\mathrm{P}}}, \tag{4.32}$$

where $\chi_{\mathrm{P}} = N^{-1}\chi^0$ is the Pauli susceptibility (3.54) and $\chi_{\mathrm{S}} \equiv N^{-1}\chi$ is the *Stoner susceptibility*. In formula (4.32) the Pauli susceptibility itself is much smaller than the Stoner susceptibility but is enhanced by the effect of the exchange interaction (for an estimate in the free electron gas, see, e.g. [3]). The multiplier $(1 - 2U/(g^2\mu_{\mathrm{B}}^2)\chi_{\mathrm{P}})^{-1}$ is called the *Stoner enhancement factor*.

At $T = 0$, using the expression for the Pauli susceptibility (3.58), we write the enhanced susceptibility (4.32) as

$$\chi_{\mathrm{S}} = \frac{\nu(\varepsilon_{\mathrm{F}})}{1 - U\nu(\varepsilon_{\mathrm{F}})}\frac{g^2\mu_{\mathrm{B}}^2}{2}. \tag{4.33}$$

If $U\nu(\varepsilon_{\mathrm{F}}) > 1$, then $\chi_{\mathrm{S}} < 0$, which implies that the paramagnetic state is unstable. Thus,

$$U\nu(\varepsilon_{\mathrm{F}}) > 1$$

gives the *Stoner condition* for ferromagnetism in metals. The Stoner condition is not a condition for ferromagnetism in a rigorous sense because it is derived within the mean-field approximation, i.e. does not take correlations into account.[2]

4.3 Band Calculations in Metals

Description of exchange and correlation effects was improved by the density-functional theory (DFT) [10–13]. A review of calculations using the Stoner model together with a realistic band structure of ferromagnetic metals is given in [14, 15].

The Stoner susceptibility (4.33) at $T = 0$ can be written as

$$\chi_{\mathrm{S}} = \frac{n(\varepsilon_{\mathrm{F}})}{1 - In(\varepsilon_{\mathrm{F}})}\frac{g^2\mu_{\mathrm{B}}^2}{2}, \tag{4.34}$$

where $n(\varepsilon)$ is the DOS per atom and spin and I is the Stoner parameter. Formula (4.34) can be obtained in the Hartree-Fock approximation of the *multiband* Hubbard Hamiltonian that we derive in Chap. 8 (see also [6]). The corresponding Stoner parameter is only weakly dependent on the wavevector thus giving a justification for the Hubbard model. However, the Stoner parameter calculated in the DFT differs from the Hubbard constant, because some of the correlation has already been taken into account in $n(\varepsilon)$. The Stoner condition $In(\varepsilon_{\mathrm{F}}) > 1$ serves as a useful qualitative measure of the magnetic tendency of a metal (see Table 4.1).

Table 4.1 Stoner parameter and Stoner condition in transition metals calculated in the density functional theory (from [16, 17])

Metal	V	Fe	Co	Ni	Pd	Pt
I (eV)	0.80	0.92	0.99	1.01	0.70	0.63
$In(\varepsilon_{\mathrm{F}})$	0.9	1.6	1.7	2.1	0.8	0.5

[2]Historically, by correlations one means the part of the electron-electron interaction that is not described by the Hartree-Fock approximation.

Table 4.2 Curie temperature of Fe, Co and Ni calculated in the density functional theory [16] and observed in experiment [18]

Metal	Fe	Co	Ni
T_C^{cal} (K)	4400–6200	3300–4800	2900
T_C^{exp} (K)	1044	1390	631

Attempts to describe *temperature* dependence of magnetic properties within the DFT do not lead to satisfactory results. The Curie temperature T_C^{cal}, obtained from the condition that the Stoner susceptibility (4.34) diverges:

$$0 = 1 - In(\varepsilon_F, T_C) = 1 + I \int n(\varepsilon) \frac{\partial f(\varepsilon, T_C)}{\partial \varepsilon} \, d\varepsilon,$$

is several times higher than the experimental one T_C^{exp} (Table 4.2), and the Curie-Weiss law (2.59) is violated [16, 17]. These facts are not surprising, because even with the realistic density of states the Fermi function is the only source of temperature dependence in the Stoner model.

Above the Curie temperature, the Stoner model cannot explain the existence of the local moments in ferromagnetic metals, which follows from the neutron scattering experiments (see, e.g. [19–22]).

At low temperatures the Stoner theory has a difficulty in describing magnetization. As it was shown in the previous section, in the electron gas, the Stoner model predicts

$$M(T)/M(0) - 1 \propto T^2.$$

Calculations for real metals also confirm these results (for details, see Chap. 10). Experimentally, however, the change in magnetization is proportional to $T^{3/2}$ (for a review, see [14, 15]), which implies the existence of spin waves in a metallic ferromagnet. The presence of spin waves in metals has been firmly demonstrated by inelastic neutron scattering (see, e.g. [22]). Spin waves do not appear within the Stoner theory, but they can be obtained in the random-phase approximation (RPA), which we describe next.

References

1. J.S. Smart, *Effective Field Theories of Magnetism* (Saunders, Philadelphia, 1966)
2. T. Moriya, *Spin Fluctuations in Itinerant Electron Magnetism* (Springer, Berlin, 1985)
3. D.J. Kim, *New Perspectives in Magnetism of Metals* (Kluwer/Plenum, New York, 1999)
4. H. Bruus, K. Flensberg, *Many-Body Quantum Theory in Condensed Matter Physics*, 3rd edn. (Oxford University Press, Oxford, 2004)
5. P. Mohn, *Magnetism in the Solid State*, 2nd edn. (Springer, Berlin, 2006)
6. R.M. White, *Quantum Theory of Magnetism*, 3rd edn. (Springer, Berlin, 2007)
7. J. Hubbard, Proc. R. Soc. Lond. A **276**, 238 (1963)
8. J.C. Slater, Phys. Rev. **49**, 537 (1936)
9. E.C. Stoner, Proc. R. Soc. Lond. A **165**, 372 (1938)
10. U. von Barth, L. Hedin, J. Phys. C Solid State Phys. **5**, 1629 (1972)
11. O. Gunnarsson, B.I. Lundqvist, Phys. Rev. B **13**, 4274 (1976)
12. W. Kohn, P. Vashishta, in *Theory of the Inhomogeneous Electron Gas*, ed. by S. Lundqvist, N.H. March (Plenum, New York, 1983), pp. 79–147
13. R.O. Jones, O. Gunnarsson, Rev. Mod. Phys. **61**, 689 (1989)
14. E.P. Wohlfarth, in *Ferromagnetic Materials*, vol. 1, ed. by E.P. Wohlfarth (North-Holland, Amsterdam, 1980), pp. 1–70
15. M. Shimizu, Rep. Prog. Phys. **44**, 329 (1981)
16. O. Gunnarsson, J. Phys. F: Met. Phys. **6**, 587 (1976)
17. O. Gunnarsson, Physica **91B**, 329 (1977)
18. J. Crangle, G.M. Goodman, Proc. R. Soc. Lond. A **321**, 477 (1971)
19. P.J. Brown, H. Capellmann, J. Déportes, D. Givord, K.R.A. Ziebeck, J. Magn. Magn. Mater. **30**, 243 (1982)
20. P.J. Brown, H. Capellmann, J. Déportes, D. Givord, K.R.A. Ziebeck, J. Magn. Magn. Mater. **31–34**, 295 (1983)
21. A.J. Holden, V. Heine, J.H. Samson, J. Phys. F: Met. Phys. **14**, 1005 (1984)
22. H.A. Mook, in *Spin Waves and Magnetic Excitations 1*, ed. by A.S. Borovik-Romanov, S.K. Sinha (Elsevier, Amsterdam, 1988), pp. 425–478

Random-Phase Approximation

5

The next approximation... goes under a wide variety of names (random phase approximation, independent-pair approximation, self-consistent field approximation, time-dependent Hartree–Fock approximation, etc.); in fact, there are almost as many names as there are ways of deriving the answer. We shall arbitrarily refer to it as the random phase approximation, or RPA. (D. Pines, Elementary Excitations in Solids, *Benjamin, New York, 1964)*

The random-phase approximation (RPA) has been widely used for studying linear responses to external perturbations and describing elementary excitations (see, e.g. [1–5]). In magnetic systems, the static susceptibility of the electron gas in the paramagnetic state was obtained by Wolff [6]. Dynamic longitudinal and transverse susceptibilities of the electron gas in the ferromagnetic states were obtained by Izuyama et al. [7]. Both [6] and [7] used the Hubbard model. In the electron gas model with the real Coulomb interaction, various linear responses to electric and magnetic fields in the ferromagnetic state were obtained by Kim et al. [8]. The band structure of real metals in the RPA was taken into account by a number of authors (see, e.g. [9, 10]).

In this chapter we calculate linear response to the space- and time-dependent magnetic field using the RPA. We obtain the longitudinal and transverse susceptibilities. These results lead to a discussion of the spin-flip excitations and spin waves and their effects on magnetization (see also [11–13]).

5.1 Magnetic Susceptibilities

Here we derive magnetic susceptibilities using the method suggested in [8] and simplified in [3]. This method can be thought of as a generalization of the mean-field approximation. It allows to obtain the response to a space- and time-dependent magnetic field. Instead of using formula (2.38) for the linear response susceptibility, we rather repeat some of the steps made in its derivation. First, we use an equation of motion for the magnetization in presence of an external magnetic field. The susceptibility is obtained by differentiating the magnetization and letting the magnetic field tend to zero.

5.1.1 Longitudinal Susceptibility

As usual we assume that the unperturbed state is ferromagnetic and the z-axis is aligned along the magnetization. To calculate the longitudinal susceptibility, we consider the linearly polarized magnetic field applied along the z-axis with the wavevector $\mathbf{q}$ and frequency ω,

$$\mathbf{H}(\mathbf{r}, t) = \hat{\mathbf{z}} H_z(\mathbf{q})\, \mathrm{e}^{\mathrm{i}(\mathbf{qr}-\omega t)},$$

where $\hat{\mathbf{z}}$ is the unit vector along the z-axis and $H_z(\mathbf{q}) > 0$ is the amplitude.[1] Then the external perturbation (2.36) is written as

$$\mathcal{H}_{\mathrm{ext}}(t) = -\int \mathbf{H}(\mathbf{r}, t) \boldsymbol{\mathcal{M}}(\mathbf{r})\, \mathrm{d}\mathbf{r} = -H_z(\mathbf{q}) \mathcal{M}_z(-\mathbf{q}) \mathrm{e}^{-\mathrm{i}\omega t}, \tag{5.1}$$

where

$$\mathcal{M}_z(\mathbf{q}) = \int \mathcal{M}_z(\mathbf{r})\, \mathrm{e}^{-\mathrm{i}\mathbf{qr}}\, \mathrm{d}\mathbf{r}$$

[1]Use of the complex field rather than real one $\mathbf{H}(\mathbf{r}, t) = \hat{\mathbf{z}} H_z(\mathbf{q}) \cos(\mathbf{qr}) \cos \omega t$ (see, e.g. [5]) largely simplifies the derivation.

N. B. Melnikov, B. I. Reser, *Dynamic Spin Fluctuation Theory of Metallic Magnetism*,
https://doi.org/10.1007/978-3-319-92974-3_5

is the spatial Fourier transform. According to (2.33) magnetization is given by

$$\langle \mathcal{M}_z(\mathbf{q})\rangle = \mathrm{Tr}(\rho_{\mathrm{tot}}(t)\mathcal{M}_z(\mathbf{q})),$$

where $\rho_{\mathrm{tot}}(t)$ is determined by $\mathcal{H}_{\mathrm{tot}}(t) = \mathcal{H} + \mathcal{H}_{\mathrm{ext}}(t)$. Using the equation of motion (2.30), we obtain

$$\mathrm{i}\hbar\frac{\mathrm{d}}{\mathrm{d}t}\langle \mathcal{M}_z(\mathbf{q})\rangle = \mathrm{Tr}\left(\mathrm{i}\hbar\frac{\mathrm{d}}{\mathrm{d}t}\rho_{\mathrm{tot}}(t)\mathcal{M}_z(\mathbf{q})\right) = \mathrm{Tr}([\mathcal{H}_{\mathrm{tot}}(t), \rho_{\mathrm{tot}}(t)]\mathcal{M}_z(\mathbf{q})).$$

Taking the cyclic property of trace into account, we write

$$\mathrm{i}\hbar\frac{\mathrm{d}}{\mathrm{d}t}\langle \mathcal{M}_z(\mathbf{q})\rangle = \mathrm{Tr}(\rho_{\mathrm{tot}}[\mathcal{M}_z(\mathbf{q}), \mathcal{H}_{\mathrm{tot}}]) = \langle[\mathcal{M}_z(\mathbf{q}), \mathcal{H}_{\mathrm{tot}}]\rangle.$$

The linear response occurs at the same frequency as the magnetic field, $\langle \mathcal{M}_z(\mathbf{q})\rangle = \langle \mathcal{M}_z(\mathbf{q}, \omega)\rangle\, \mathrm{e}^{-\mathrm{i}\omega t}$ (see Sect. 2.1.1). Therefore,

$$\hbar\omega\langle \mathcal{M}_z(\mathbf{q})\rangle = \langle[\mathcal{M}_z(\mathbf{q}), \mathcal{H}]\rangle + \langle[\mathcal{M}_z(\mathbf{q}), \mathcal{H}_{\mathrm{ext}}]\rangle.$$

Since the second quantization representation of $\mathcal{M}_z(\mathbf{q})$ is given by

$$\mathcal{M}_z(\mathbf{q}) = -g\mu_{\mathrm{B}}\frac{1}{2}\sum_{\mathbf{k}\sigma}\sigma a^\dagger_{\mathbf{k}\sigma}a_{\mathbf{k}+\mathbf{q},\sigma}, \tag{5.2}$$

we consider the equation

$$\hbar\omega\big\langle a^\dagger_{\mathbf{k}\sigma}a_{\mathbf{k}+\mathbf{q},\sigma}\big\rangle = \big\langle\big[a^\dagger_{\mathbf{k}\sigma}a_{\mathbf{k}+\mathbf{q},\sigma}, \mathcal{H}\big]\big\rangle + \big\langle\big[a^\dagger_{\mathbf{k}\sigma}a_{\mathbf{k}+\mathbf{q},\sigma}, \mathcal{H}_{\mathrm{ext}}\big]\big\rangle. \tag{5.3}$$

The latter can be solved if we replace the Hamiltonian $\mathcal{H}$ by its mean-field approximation (4.10) and (4.11), which reads

$$\mathcal{H}_{\mathrm{I}}^{\mathrm{MF}} = \tilde{U}\sum_{\mathbf{klq}'\sigma\sigma'}\big\langle a^\dagger_{\mathbf{l}\sigma'}a_{\mathbf{l}+\mathbf{q}',\sigma'}\big\rangle a^\dagger_{\mathbf{k}\sigma}a_{\mathbf{k}-\mathbf{q}',\sigma} - \tilde{U}\sum_{\mathbf{klq}'\sigma\sigma'}\big\langle a^\dagger_{\mathbf{k}\sigma}a_{\mathbf{l}+\mathbf{q}',\sigma'}\big\rangle a^\dagger_{\mathbf{l}\sigma'}a_{\mathbf{k}-\mathbf{q}',\sigma}. \tag{5.4}$$

In the presence of the external field the system is not spatially homogeneous and requirement (4.12) is no longer adequate. In the spin-polarized case, we assume that there are two kinds of nonzero averages:

$$\big\langle a^\dagger_{\mathbf{k}\sigma}a_{\mathbf{k}\sigma}\big\rangle \neq 0 \qquad \text{and} \qquad \big\langle a^\dagger_{\mathbf{k}\sigma}a_{\mathbf{k}+\mathbf{q},\sigma}\big\rangle \neq 0.$$

Then, keeping only the terms with $\mathbf{q}' = 0$ and $\mathbf{q}' = \mathbf{q}$ and the same spins in (5.4), we obtain

$$\mathcal{H}_{\mathrm{I}}^{\mathrm{MF}} = \tilde{U}\sum_{\mathbf{k}\sigma}(N_{\mathrm{e}} - N_\sigma)a^\dagger_{\mathbf{k}\sigma}a_{\mathbf{k}\sigma} + \tilde{U}\sum_{\mathbf{k}\sigma}(N_{\mathbf{q}} - N_{\mathbf{q}\sigma})a^\dagger_{\mathbf{k}\sigma}a_{\mathbf{k}-\mathbf{q},\sigma}, \tag{5.5}$$

where

$$N_{\mathbf{q}} = \sum_{\mathbf{k}\sigma}\big\langle a^\dagger_{\mathbf{k}\sigma}a_{\mathbf{k}+\mathbf{q},\sigma}\big\rangle, \qquad N_{\mathbf{q}\sigma} = \sum_{\mathbf{k}}\big\langle a^\dagger_{\mathbf{k}\sigma}a_{\mathbf{k}+\mathbf{q},\sigma}\big\rangle.$$

Now calculating the commutators, we have

$$\big[a^\dagger_{\mathbf{k}\sigma}a_{\mathbf{k}+\mathbf{q},\sigma}, \mathcal{H}_0\big] = \left[a^\dagger_{\mathbf{k}\sigma}a_{\mathbf{k}+\mathbf{q},\sigma}, \sum_{\mathbf{k}'\sigma'}\varepsilon_{\mathbf{k}'}a^\dagger_{\mathbf{k}'\sigma'}a_{\mathbf{k}'\sigma'}\right] = (\varepsilon_{\mathbf{k}+\mathbf{q}} - \varepsilon_{\mathbf{k}})a^\dagger_{\mathbf{k}\sigma}a_{\mathbf{k}+\mathbf{q},\sigma} \tag{5.6}$$

and

$$\begin{aligned}\big[a^\dagger_{\mathbf{k}\sigma}a_{\mathbf{k}+\mathbf{q},\sigma}, \mathcal{H}_{\mathrm{I}}^{\mathrm{MF}}\big] &= \left[a^\dagger_{\mathbf{k}\sigma}a_{\mathbf{k}+\mathbf{q},\sigma}, \tilde{U}\sum_{\mathbf{k}'\sigma'}(N_{\mathbf{q}} - N_{\mathbf{q}\sigma'})a^\dagger_{\mathbf{k}'\sigma'}a_{\mathbf{k}'-\mathbf{q},\sigma'}\right]\\ &= \tilde{U}(N_{\mathbf{q}} - N_{\mathbf{q}\sigma})\big(a^\dagger_{\mathbf{k}\sigma}a_{\mathbf{k}\sigma} - a^\dagger_{\mathbf{k}+\mathbf{q},\sigma}a_{\mathbf{k}+\mathbf{q},\sigma}\big).\end{aligned} \tag{5.7}$$

Here we used the fact that $a^{\dagger}_{\mathbf{k}\sigma}a_{\mathbf{k}+\mathbf{q},\sigma}$ commutes with the first term on the right-hand of (5.5). Similarly, taking (5.1) and (5.2) into account, we obtain

$$\begin{aligned}\left[a^{\dagger}_{\mathbf{k}\sigma}a_{\mathbf{k}+\mathbf{q},\sigma}, \mathcal{H}_{\text{ext}}\right] &= H_z(\mathbf{q})\mathrm{e}^{-\mathrm{i}\omega t}\frac{g\mu_{\mathrm{B}}}{2}\left[a^{\dagger}_{\mathbf{k}\sigma}a_{\mathbf{k}+\mathbf{q},\sigma}, \sum_{\mathbf{k}'\sigma'}\sigma' a^{\dagger}_{\mathbf{k}'\sigma'}a_{\mathbf{k}'-\mathbf{q},\sigma'}\right]\\ &= \sigma H_z(\mathbf{q})\mathrm{e}^{-\mathrm{i}\omega t}\frac{g\mu_{\mathrm{B}}}{2}\left(a^{\dagger}_{\mathbf{k}\sigma}a_{\mathbf{k}\sigma} - a^{\dagger}_{\mathbf{k}+\mathbf{q},\sigma}a_{\mathbf{k}+\mathbf{q},\sigma}\right).\end{aligned} \tag{5.8}$$

Substitution of the commutators (5.6)–(5.8) back into Eq. (5.3) gives

$$\begin{aligned}\hbar\omega\langle a^{\dagger}_{\mathbf{k}\sigma}a_{\mathbf{k}+\mathbf{q},\sigma}\rangle &= (\varepsilon_{\mathbf{k}+\mathbf{q}} - \varepsilon_{\mathbf{k}})\langle a^{\dagger}_{\mathbf{k}\sigma}a_{\mathbf{k}+\mathbf{q},\sigma}\rangle\\ &\quad + \left(\tilde{U}(N_{\mathbf{q}} - N_{\mathbf{q}\sigma}) + \sigma H_z(\mathbf{q})\mathrm{e}^{-\mathrm{i}\omega t}\frac{g\mu_{\mathrm{B}}}{2}\right)\left(\langle a^{\dagger}_{\mathbf{k}\sigma}a_{\mathbf{k}\sigma}\rangle - \langle a^{\dagger}_{\mathbf{k}+\mathbf{q},\sigma}a_{\mathbf{k}+\mathbf{q},\sigma}\rangle\right).\end{aligned}$$

Rearranging and summing over $\mathbf{k}$, we come to

$$N_{\mathbf{q}\sigma} = -F_{\sigma}(\mathbf{q}, \omega)\left(\tilde{U}N_{\mathbf{q}\bar{\sigma}} + \sigma H_z(\mathbf{q})\mathrm{e}^{-\mathrm{i}\omega t}\frac{g\mu_{\mathrm{B}}}{2}\right), \tag{5.9}$$

where

$$F_{\sigma}(\mathbf{q}, \omega) = \sum_{\mathbf{k}}\frac{\langle n_{\mathbf{k}\sigma}\rangle - \langle n_{\mathbf{k}+\mathbf{q},\sigma}\rangle}{\varepsilon_{\mathbf{k}+\mathbf{q}} - \varepsilon_{\mathbf{k}} - \hbar\omega}.$$

Solving the system of two linear equations (5.9) with respect to $N_{\mathbf{q}\uparrow}$ and $N_{\mathbf{q}\downarrow}$, we obtain

$$N_{\mathbf{q}\sigma} = -\sigma F_{\sigma}(\mathbf{q}, \omega)H_z(\mathbf{q})\mathrm{e}^{-\mathrm{i}\omega t}\frac{1 + \tilde{U}F_{\bar{\sigma}}(\mathbf{q}, \omega)}{1 - \tilde{U}^2F_{+}(\mathbf{q}, \omega)F_{-}(\mathbf{q}, \omega)}\frac{g\mu_{\mathrm{B}}}{2}.$$

Finally, using the relation

$$\langle M_z(\mathbf{q})\rangle = -g\mu_{\mathrm{B}}\frac{1}{2}(N_{\mathbf{q}\uparrow} - N_{\mathbf{q}\downarrow}) = \chi_{zz}(\mathbf{q}, \omega)H_z(\mathbf{q})\mathrm{e}^{-\mathrm{i}\omega t} \tag{5.10}$$

(compare with (2.5)), we come to

$$\chi_{zz}(\mathbf{q}, \omega) = \frac{F_{+}(\mathbf{q}, \omega) + F_{-}(\mathbf{q}, \omega) + 2\tilde{U}F_{+}(\mathbf{q}, \omega)F_{-}(\mathbf{q}, \omega)}{1 - \tilde{U}^2F_{+}(\mathbf{q}, \omega)F_{-}(\mathbf{q}, \omega)}\left(\frac{g\mu_{\mathrm{B}}}{2}\right)^2. \tag{5.11}$$

In the vanishing magnetic field we have $\langle a^{\dagger}_{\mathbf{k}\sigma}a_{\mathbf{k}\sigma}\rangle = f(\varepsilon_{\mathbf{k}\sigma})$, where $\varepsilon_{\mathbf{k}\sigma} = \varepsilon_{\mathbf{k}} - \sigma\tilde{U}\bar{S}_z$ (we omit the constant shift $\frac{1}{2}\tilde{U}N_{\mathrm{e}}$ just as before). Hence

$$F_{\sigma}(\mathbf{q}, \omega) = \sum_{\mathbf{k}}\frac{f(\varepsilon_{\mathbf{k}\sigma}) - f(\varepsilon_{\mathbf{k}+\mathbf{q},\sigma})}{\varepsilon_{\mathbf{k}+\mathbf{q}} - \varepsilon_{\mathbf{k}} - \hbar\omega}, \tag{5.12}$$

where $\sigma = \pm$ or $\uparrow,\downarrow$.

In the paramagnetic state, we have $F_{+}(\mathbf{q}, \omega) = F_{-}(\mathbf{q}, \omega) = F(\mathbf{q}, \omega)$, and (5.11) becomes

$$\chi_{zz}(\mathbf{q}, \omega) = \frac{F(\mathbf{q}, \omega)}{1 - \tilde{U}F(\mathbf{q}, \omega)}\frac{g^2\mu_{\mathrm{B}}^2}{2}. \tag{5.13}$$

The value $N^{-1}\chi_{zz}(0, 0)$ reduces to the Stoner susceptibility (4.32), and for noninteracting electrons ($\tilde{U} = 0$) it reduces to the Pauli susceptibility (3.58).

Expression (5.11) coincides with the one of [7] in units of $g^2\mu_B^2$ (it was obtained there by a different method, which we discuss in the next chapter). To recover the result of [3, 8], we introduce the notation

$$\tilde{F}_\sigma(\mathbf{q},\omega) = \frac{F_\sigma(\mathbf{q},\omega)}{1-\tilde{U}F_\sigma(\mathbf{q},\omega)}.$$

Then formula (5.11) can be written as

$$\chi_{zz}(\mathbf{q},\omega) = \frac{\tilde{F}_+(\mathbf{q},\omega)+\tilde{F}_-(\mathbf{q},\omega)+4\tilde{U}\tilde{F}_+(\mathbf{q},\omega)\tilde{F}_-(\mathbf{q},\omega)}{1+\tilde{U}[\tilde{F}_+(\mathbf{q},\omega)+\tilde{F}_-(\mathbf{q},\omega)]}\left(\frac{g\mu_B}{2}\right)^2.$$

In the paramagnetic state, we have $\tilde{F}_+(\mathbf{q},\omega) = \tilde{F}_-(\mathbf{q},\omega) = \tilde{F}(\mathbf{q},\omega)$, and (5.13) becomes

$$\chi_{zz}(\mathbf{q},\omega) = \tilde{F}(\mathbf{q},\omega)\frac{g^2\mu_B^2}{2}.$$

5.1.2 Transverse Susceptibility

In the previous section we dealt with a linearly polarized magnetic field. To create a spin wave precessing (anticlockwise) in the xy-plane, we consider the circularly polarized magnetic field with the wavevector $\mathbf{q}$ and frequency ω:

$$\mathbf{H}(\mathbf{r},t) = -H(\hat{\mathbf{x}}\cos(\mathbf{qr}-\omega t) - \hat{\mathbf{y}}\sin(\mathbf{qr}-\omega t)), \tag{5.14}$$

where $\hat{\mathbf{x}}$ and $\hat{\mathbf{y}}$ are the unit vectors along the x- and y-axes, respectively, and $H > 0$. Using the circular components (2.14), we write the external perturbation (2.39) as

$$\mathcal{H}_{\rm ext}(t) = -\int \mathbf{H}(\mathbf{r},t)\boldsymbol{\mathcal{M}}(\mathbf{r})\,\mathrm{d}\mathbf{r} = -\frac{1}{2}\int \big(H_-(\mathbf{r},t)\mathcal{M}_+(\mathbf{r}) + H_+(\mathbf{r},t)\mathcal{M}_-(\mathbf{r})\big)\,\mathrm{d}\mathbf{r}.$$

From (5.14) we obtain

$$H_\pm(\mathbf{r},t) = H_x(\mathbf{r},t) \pm \mathrm{i}H_y(\mathbf{r},t) = -H\mathrm{e}^{\mp\mathrm{i}(\mathbf{qr}-\omega t)}.$$

Then in the momentum-frequency representation, we have

$$\mathcal{H}_{\rm ext}(t) = -\frac{1}{2}(H_-(\mathbf{q})\mathcal{M}_+(-\mathbf{q})\mathrm{e}^{-\mathrm{i}\omega t} + H_+(-\mathbf{q})\mathcal{M}_-(\mathbf{q})\mathrm{e}^{\mathrm{i}\omega t}), \tag{5.15}$$

where $H_-(\mathbf{q}) = H_+(-\mathbf{q}) = -H$. Since the two terms on the right-hand side of (5.15) are Hermite conjugate to each other, it suffices to consider only one of them; we keep the first one,

$$\mathcal{H}_{\rm ext}(t) = -\frac{1}{2}H_-(\mathbf{q})\mathcal{M}_+(-\mathbf{q})\mathrm{e}^{-\mathrm{i}\omega t}. \tag{5.16}$$

Analogously to the longitudinal susceptibility, we write the equation of motion

$$\mathrm{i}\hbar\frac{\mathrm{d}}{\mathrm{d}t}\langle\mathcal{M}_-(\mathbf{q})\rangle = \langle[\mathcal{M}_-(\mathbf{q}),\mathcal{H}_{\rm tot}]\rangle.$$

Since $\langle\mathcal{M}_-(\mathbf{q})\rangle = \langle\mathcal{M}_-(\mathbf{q},\omega)\rangle\,\mathrm{e}^{-\mathrm{i}\omega t}$, we have

$$\hbar\omega\langle\mathcal{M}_-(\mathbf{q})\rangle = \langle[\mathcal{M}_-(\mathbf{q}),\mathcal{H}]\rangle + \langle[\mathcal{M}_-(\mathbf{q}),\mathcal{H}_{\rm ext}]\rangle,$$

where $\mathcal{M}_-(\mathbf{q}) = -g\mu_B s_-(\mathbf{q})$ and $\mathcal{M}_+(-\mathbf{q}) = -g\mu_B s_+(-\mathbf{q})$. Using (3.29), we write

$$s_+(-\mathbf{q}) = \sum_{\mathbf{k}} a^\dagger_{\mathbf{k}\uparrow} a_{\mathbf{k}-\mathbf{q},\downarrow}, \qquad s_-(\mathbf{q}) = \sum_{\mathbf{k}} a^\dagger_{\mathbf{k}\downarrow} a_{\mathbf{k}+\mathbf{q},\uparrow}.$$

Therefore, we consider the equation

$$\hbar\omega\langle a^\dagger_{\mathbf{k}\downarrow} a_{\mathbf{k}+\mathbf{q},\uparrow}\rangle = \langle[a^\dagger_{\mathbf{k}\downarrow} a_{\mathbf{k}+\mathbf{q},\uparrow}, \mathcal{H}_0]\rangle + \langle[a^\dagger_{\mathbf{k}\downarrow} a_{\mathbf{k}+\mathbf{q},\uparrow}, \mathcal{H}_I^{MF}]\rangle + \langle[a^\dagger_{\mathbf{k}\downarrow} a_{\mathbf{k}+\mathbf{q},\uparrow}, \mathcal{H}_{ext}]\rangle. \tag{5.17}$$

In the presence of the external perturbation (5.16), we assume that

$$\langle a^\dagger_{\mathbf{k}\sigma} a_{\mathbf{k}\sigma}\rangle \neq 0, \qquad \langle a^\dagger_{\mathbf{k}\downarrow} a_{\mathbf{k}+\mathbf{q},\uparrow}\rangle \neq 0.$$

Then, using (4.10) and (4.11), we write the interaction term in the mean-field approximation as

$$\mathcal{H}_I^{MF} = \tilde{U}\sum_{\mathbf{k}'\sigma'}(N_e - N_{\sigma'}) a^\dagger_{\mathbf{k}'\sigma'} a_{\mathbf{k}'\sigma'} - \tilde{U}\langle s_-(\mathbf{q})\rangle s_+(-\mathbf{q}). \tag{5.18}$$

By adding the first term on the right-hand side of (5.18) to $\mathcal{H}_0$, we replace the spectrum $\varepsilon_{\mathbf{k}}$ by the spin-polarized spectrum $\varepsilon_{\mathbf{k}\sigma}$. Therefore, substituting (5.16) and (5.18) in (5.17), we obtain

$$\begin{aligned}\hbar\omega\langle a^\dagger_{\mathbf{k}\downarrow} a_{\mathbf{k}+\mathbf{q},\uparrow}\rangle &= \Big[a^\dagger_{\mathbf{k}\downarrow} a_{\mathbf{k}+\mathbf{q},\uparrow}, \sum_{\mathbf{k}'\sigma'} \varepsilon_{\mathbf{k}'\sigma'} a^\dagger_{\mathbf{k}'\sigma'} a_{\mathbf{k}'\sigma'}\Big] \\ &\quad + \Big(\frac{g\mu_B}{2} H_-(\mathbf{q}) e^{-i\omega t} - \tilde{U}\langle s_-(\mathbf{q})\rangle\Big)\big[a^\dagger_{\mathbf{k}\downarrow} a_{\mathbf{k}+\mathbf{q},\uparrow}, s_+(-\mathbf{q})\big].\end{aligned} \tag{5.19}$$

Taking (3.46) into account, we calculate the first commutator

$$\left[a^\dagger_{\mathbf{k}\downarrow} a_{\mathbf{k}+\mathbf{q},\uparrow}, \sum_{\mathbf{k}'\sigma'} \varepsilon_{\mathbf{k}'\sigma'} a^\dagger_{\mathbf{k}'\sigma'} a_{\mathbf{k}'\sigma'}\right] = (\varepsilon_{\mathbf{k}+\mathbf{q},\uparrow} - \varepsilon_{\mathbf{k}\downarrow}) a^\dagger_{\mathbf{k}\downarrow} a_{\mathbf{k}+\mathbf{q},\uparrow}.$$

Similarly calculating the second commutator

$$\big[a^\dagger_{\mathbf{k}\downarrow} a_{\mathbf{k}+\mathbf{q},\uparrow}, s_+(-\mathbf{q})\big] = a^\dagger_{\mathbf{k}\downarrow} a_{\mathbf{k}\downarrow} - a^\dagger_{\mathbf{k}+\mathbf{q},\uparrow} a_{\mathbf{k}+\mathbf{q},\uparrow},$$

we write Eq. (5.19) as

$$\begin{aligned}&(\varepsilon_{\mathbf{k}+\mathbf{q},\uparrow} - \varepsilon_{\mathbf{k}\downarrow} - \hbar\omega)\langle a^\dagger_{\mathbf{k}\downarrow} a_{\mathbf{k}+\mathbf{q},\uparrow}\rangle \\ &\quad = \Big(-\frac{g\mu_B}{2} H_-(\mathbf{q}) e^{-i\omega t} + \tilde{U}\langle s_-(\mathbf{q})\rangle\Big)\big(\langle a^\dagger_{\mathbf{k}\downarrow} a_{\mathbf{k}\downarrow}\rangle - \langle a^\dagger_{\mathbf{k}+\mathbf{q},\uparrow} a_{\mathbf{k}+\mathbf{q},\uparrow}\rangle\big).\end{aligned}$$

Rearranging, we obtain

$$\langle s_-(\mathbf{q})\rangle = -\frac{F_{-+}(\mathbf{q},\omega)}{1 - \tilde{U}F_{-+}(\mathbf{q},\omega)} \frac{g\mu_B}{2} H_-(\mathbf{q}) e^{-i\omega t},$$

where

$$F_{-+}(\mathbf{q},\omega) = \sum_{\mathbf{k}} \frac{\langle n_{\mathbf{k}\downarrow}\rangle - \langle n_{\mathbf{k}+\mathbf{q},\uparrow}\rangle}{\varepsilon_{\mathbf{k}+\mathbf{q},\uparrow} - \varepsilon_{\mathbf{k}\downarrow} - \hbar\omega}.$$

Using the relation (2.40), we have

$$\langle \mathcal{M}_-(\mathbf{q})\rangle = -g\mu_B\langle s_-(\mathbf{q})\rangle = \frac{1}{2}\chi_{-+}(\mathbf{q},\omega) H_-(\mathbf{q}) e^{-i\omega t}.$$

From this we come to

$$\chi_{-+}(\mathbf{q},\omega)=\frac{F_{-+}(\mathbf{q},\omega)}{1-\tilde{U}F_{-+}(\mathbf{q},\omega)}g^2\mu_B^2, \tag{5.20}$$

where

$$F_{-+}(\mathbf{q},\omega)=\sum_{\mathbf{k}}\frac{f(\varepsilon_{\mathbf{k}\downarrow})-f(\varepsilon_{\mathbf{k}+\mathbf{q},\uparrow})}{\varepsilon_{\mathbf{k}+\mathbf{q},\uparrow}-\varepsilon_{\mathbf{k}\downarrow}-\hbar\omega} \tag{5.21}$$

is the *transverse Lindhard function*.

The other transverse susceptibility $\chi_{+-}(\mathbf{q},\omega)$ is obtained simply by swapping the spin subscripts in (5.20) and (5.21):

$$\chi_{+-}(\mathbf{q},\omega)=\frac{F_{+-}(\mathbf{q},\omega)}{1-\tilde{U}F_{+-}(\mathbf{q},\omega)}g^2\mu_B^2, \tag{5.22}$$

where

$$F_{+-}(\mathbf{q},\omega)=\sum_{\mathbf{k}}\frac{f(\varepsilon_{\mathbf{k}\uparrow})-f(\varepsilon_{\mathbf{k}+\mathbf{q},\downarrow})}{\varepsilon_{\mathbf{k}+\mathbf{q},\downarrow}-\varepsilon_{\mathbf{k}\uparrow}-\hbar\omega}. \tag{5.23}$$

In the paramagnetic state, we have $F_{+-}(\mathbf{q},\omega)=F_{-+}(\mathbf{q},\omega)=F(\mathbf{q},\omega)$ and

$$\frac{1}{2}\chi_{-+}(\mathbf{q},\omega)=\frac{1}{2}\chi_{+-}(\mathbf{q},\omega)=\chi_{zz}(\mathbf{q},\omega).$$

5.2 Magnetic Excitations

Singularities of the transverse susceptibility determine magnetic excitations of two types: spin-density waves and spin-flip excitations (see, e.g. [3]).

5.2.1 Spin-Density Waves

According to the classical spin-wave theory [14, 15], at low temperatures the magnetization of ferromagnets with localized spins follows the $T^{3/2}$ law:

$$M(T)=M(0)(1-a_{3/2}T^{3/2}).$$

In itinerant ferromagnets, the $T^{3/2}$ law was obtained for the electron gas model [16]. Here, following [3, 17], we give a detailed derivation of the spin-wave spectrum and $T^{3/2}$ law in itinerant ferromagnets without any assumptions about the electron spectrum.

The divergence of the transverse susceptibility (5.22) implies that spin-density wave with a frequency ω and wavevector $\mathbf{q}$ can persist without a magnetic field. This is an excitation mode of the spin system with an energy $\varepsilon=\hbar\omega$ and wavevector $\mathbf{q}$. We obtain the spin-wave spectrum in a metal from the poles of the transverse susceptibility $\chi_{+-}(\mathbf{q},\omega)$.[2]

By formula (5.22) the condition

$$\chi_{+-}(\mathbf{q},\omega)=\infty$$

is satisfied if

$$1-\tilde{U}F_{+-}(\mathbf{q},\omega)=0. \tag{5.24}$$

Using formulae (4.18) and (5.23), we obtain

$$F_{+-}(\mathbf{q},\omega)=\sum_{\mathbf{k}}\frac{f(\varepsilon_{\mathbf{k}\uparrow})-f(\varepsilon_{\mathbf{k}+\mathbf{q},\downarrow})}{\varepsilon_{\mathbf{k}+\mathbf{q}}-\varepsilon_{\mathbf{k}}+2\tilde{U}\bar{S}_z-\hbar\omega}. \tag{5.25}$$

[2]Poles of $\chi_{-+}(\mathbf{q},\omega)$ give an equation that corresponds to the spin waves precessing in the opposite direction.

Our goal is to find the relation between ω and $\mathbf{q}$ such that (5.24) is satisfied. At $\mathbf{q} = 0$ formula (5.24) becomes

$$1 + \frac{\tilde{U}}{2\tilde{U}\bar{S}_z - \hbar\omega} \sum_{\mathbf{k}} \left[f(\varepsilon_{\mathbf{k}\uparrow}) - f(\varepsilon_{\mathbf{k}\downarrow}) \right] = 0 \,. \tag{5.26}$$

Using (4.16) and (4.18), we have

$$\bar{S}_z = \frac{1}{2} \sum_{\mathbf{k}} \left[f(\varepsilon_{\mathbf{k}\uparrow}) - f(\varepsilon_{\mathbf{k}\downarrow}) \right].$$

Substituting the latter in (5.26), we see that $\omega = 0$ for $\mathbf{q} = 0$. The unenhanced susceptibility $F_{+-}(\mathbf{q}, \omega)$ is invariant with respect to the inversion $\mathbf{q} \to -\mathbf{q}$, just as the enhanced one (2.8). Therefore, for small $\mathbf{q}$ the spin-wave spectrum equation (5.24) has the form

$$\hbar\omega = (\mathbf{q}, A\mathbf{q}), \tag{5.27}$$

where $(\mathbf{q}, A\mathbf{q})$ is a quadratic form with the matrix A.

To derive an explicit expression for the right-hand of (5.27), we expand (5.24) keeping only linear term in ω and quadratic terms in $\mathbf{q}$. Other terms: $\mathbf{q}\omega$, ω^2, etc., have higher order in $\mathbf{q}$ when $\hbar\omega = (\mathbf{q}, A\mathbf{q})$, and can be ignored. We begin by writing the unenhanced susceptibility (5.25) as

$$F_{+-}(\mathbf{q}, \omega) = \frac{1}{2\tilde{U}\bar{S}_z} \sum_{\mathbf{k}} \left[\left(1 - \frac{\hbar\omega}{2\tilde{U}\bar{S}_z} + \frac{\varepsilon_{\mathbf{k}+\mathbf{q}} - \varepsilon_{\mathbf{k}}}{2\tilde{U}\bar{S}_z} \right)^{-1} \left(f(\varepsilon_{\mathbf{k}\uparrow}) - f(\varepsilon_{\mathbf{k}+\mathbf{q},\downarrow}) \right) \right]. \tag{5.28}$$

For small $\mathbf{q}$ such that

$$\hbar\omega = (\mathbf{q}, A\mathbf{q}) \ll 2\tilde{U}\bar{S}_z \,, \qquad |\varepsilon_{\mathbf{k}+\mathbf{q}} - \varepsilon_{\mathbf{k}}| \ll 2\tilde{U}\bar{S}_z \,,$$

we expand the fraction in (5.28) as a geometric series: $(1 - x)^{-1} = 1 + x + x^2 + \cdots$, $|x| < 1$. Retaining the second partial sum of the series, we obtain

$$\begin{aligned} F_{+-}(\mathbf{q}, \omega) = \frac{1}{2\tilde{U}\bar{S}_z} \Bigg\{ & \left(1 + \frac{\hbar\omega}{2\tilde{U}\bar{S}_z} \right) \sum_{\mathbf{k}} \left[f(\varepsilon_{\mathbf{k}\uparrow}) - f(\varepsilon_{\mathbf{k}+\mathbf{q},\downarrow}) \right] \\ & - \frac{1}{2\tilde{U}\bar{S}_z} \sum_{\mathbf{k}} (\varepsilon_{\mathbf{k}+\mathbf{q}} - \varepsilon_{\mathbf{k}}) \left[f(\varepsilon_{\mathbf{k}\uparrow}) - f(\varepsilon_{\mathbf{k}+\mathbf{q},\downarrow}) \right] \\ & + \frac{1}{(2\tilde{U}\bar{S}_z)^2} \sum_{\mathbf{k}} \left(\hbar\omega - (\varepsilon_{\mathbf{k}+\mathbf{q}} - \varepsilon_{\mathbf{k}}) \right)^2 \left[f(\varepsilon_{\mathbf{k}\uparrow}) - f(\varepsilon_{\mathbf{k}+\mathbf{q},\downarrow}) \right] \Bigg\}. \end{aligned} \tag{5.29}$$

In the first term of the expansion (5.29), we have

$$\sum_{\mathbf{k}} \left[f(\varepsilon_{\mathbf{k}\uparrow}) - f(\varepsilon_{\mathbf{k}+\mathbf{q},\downarrow}) \right] = \sum_{\mathbf{k}} \left[f(\varepsilon_{\mathbf{k}\uparrow}) - f(\varepsilon_{\mathbf{k}\downarrow}) \right] = 2\bar{S}_z. \tag{5.30}$$

In the second term of expression (5.29), we rearrange the sum as

$$\sum_{\mathbf{k}} (\varepsilon_{\mathbf{k}+\mathbf{q}} - \varepsilon_{\mathbf{k}}) \left[f(\varepsilon_{\mathbf{k}\uparrow}) - f(\varepsilon_{\mathbf{k}+\mathbf{q},\downarrow}) \right] = \sum_{\mathbf{k}} (\varepsilon_{\mathbf{k}+\mathbf{q}} - \varepsilon_{\mathbf{k}}) f(\varepsilon_{\mathbf{k}\uparrow}) + \sum_{\mathbf{k}} (\varepsilon_{\mathbf{k}-\mathbf{q}} - \varepsilon_{\mathbf{k}}) f(\varepsilon_{\mathbf{k}\downarrow}).$$

Using the Taylor expansion of the energy spectrum $\varepsilon_{\mathbf{k}+\mathbf{q}}$ up to quadratic terms in $\mathbf{q}$:

$$\varepsilon_{\mathbf{k}+\mathbf{q}} = \varepsilon_{\mathbf{k}} + \mathbf{q}\nabla\varepsilon_{\mathbf{k}} + \frac{1}{2}(\mathbf{q}\nabla)^2\varepsilon_{\mathbf{k}} + \cdots \,, \tag{5.31}$$

where $\nabla = (\partial/\partial k_1, \partial/\partial k_2, \partial/\partial k_3)$, we have

$$\sum_{\mathbf{k}}(\varepsilon_{\mathbf{k+q}}-\varepsilon_{\mathbf{k}})\big[f(\varepsilon_{\mathbf{k}\uparrow})-f(\varepsilon_{\mathbf{k+q},\downarrow})\big]$$
$$=\sum_{\mathbf{k}}(\mathbf{q}\nabla\varepsilon_{\mathbf{k}})\big[f(\varepsilon_{\mathbf{k}\uparrow})-f(\varepsilon_{\mathbf{k}\downarrow})\big]+\sum_{\mathbf{k}}\frac{1}{2}((\mathbf{q}\nabla)^2\varepsilon_{\mathbf{k}})\big[f(\varepsilon_{\mathbf{k}\uparrow})+f(\varepsilon_{\mathbf{k}\downarrow})\big]. \quad (5.32)$$

From the inversion symmetry of the energy spectrum $\varepsilon_{\mathbf{k}}=\varepsilon_{-\mathbf{k}}$ it follows that $\nabla\varepsilon_{-\mathbf{k}}=-\nabla\varepsilon_{\mathbf{k}}$. Then the first sum on the right-hand side of (5.32) vanishes, and we obtain

$$\sum_{\mathbf{k}}(\varepsilon_{\mathbf{k+q}}-\varepsilon_{\mathbf{k}})\big[f(\varepsilon_{\mathbf{k}\uparrow})-f(\varepsilon_{\mathbf{k+q},\downarrow})\big]=\sum_{\mathbf{k}}\frac{1}{2}((\mathbf{q}\nabla)^2\varepsilon_{\mathbf{k}})\big[f(\varepsilon_{\mathbf{k}\uparrow})+f(\varepsilon_{\mathbf{k}\downarrow})\big]. \quad (5.33)$$

Finally, we transform the sum in the third term of expression (5.29). Using the expansion (5.31) up to quadratic terms in $\mathbf{q}$ and taking $\hbar\omega=(\mathbf{q},A\mathbf{q})$ into account, we have

$$\sum_{\mathbf{k}}\big(\hbar\omega-(\varepsilon_{\mathbf{k-q}}-\varepsilon_{\mathbf{k}})\big)^2\big[f(\varepsilon_{\mathbf{k}\uparrow})-f(\varepsilon_{\mathbf{k-q},\downarrow})\big]=\sum_{\mathbf{k}}(\mathbf{q}\nabla\varepsilon_{\mathbf{k}})^2\big[f(\varepsilon_{\mathbf{k}\uparrow})-f(\varepsilon_{\mathbf{k}\downarrow})\big]. \quad (5.34)$$

Substitution of (5.30), (5.33) and (5.34) in (5.29) yields

$$F_{+-}(\mathbf{q},\omega)=\frac{1}{\tilde{U}}\Big(1+\frac{\hbar\omega}{2\tilde{U}\bar{S}_z}\Big)-\frac{1}{(2\tilde{U}\bar{S}_z)^2}\sum_{\mathbf{k}}\frac{1}{2}((\mathbf{q}\nabla)^2\varepsilon_{\mathbf{k}})\big[f(\varepsilon_{\mathbf{k}\uparrow})+f(\varepsilon_{\mathbf{k}\downarrow})\big]$$
$$+\frac{1}{(2\tilde{U}\bar{S}_z)^3}\sum_{\mathbf{k}}(\mathbf{q}\nabla\varepsilon_{\mathbf{k}})^2\big[f(\varepsilon_{\mathbf{k}\uparrow})-f(\varepsilon_{\mathbf{k}\downarrow})\big]. \quad (5.35)$$

Substituting the latter in (5.24), we write the spin-wave spectrum equation (5.27) as

$$\hbar\omega=\frac{1}{2\bar{S}_z}\sum_{\mathbf{k}}\frac{1}{2}((\mathbf{q}\nabla)^2\varepsilon_{\mathbf{k}})\big[f(\varepsilon_{\mathbf{k}\uparrow})+f(\varepsilon_{\mathbf{k}\downarrow})\big]$$
$$-\frac{1}{\tilde{U}(2\bar{S}_z)^2}\sum_{\mathbf{k}}(\mathbf{q}\nabla\varepsilon_{\mathbf{k}})^2\big[f(\varepsilon_{\mathbf{k}\uparrow})-f(\varepsilon_{\mathbf{k}\downarrow})\big]. \quad (5.36)$$

Usually the quadratic form in the spin-wave spectrum equation (5.36) is replaced by a spherically symmetric one[3]:

$$\hbar\omega=Dq^2, \quad (5.37)$$

where the coefficient D is called the *spin-wave stiffness constant*. Then, the spin-wave limit of the unenhanced susceptibility (5.35) is

$$F_{+-}(\mathbf{q},\omega)=\frac{1}{\tilde{U}}\Big(1+\frac{\hbar\omega-Dq^2}{2\tilde{U}\bar{S}_z}\Big), \quad (5.38)$$

and the enhanced susceptibility (5.22) takes the form

$$\chi_{+-}(\mathbf{q},\omega)=\frac{1}{\tilde{U}}\Big(\frac{1}{1-\tilde{U}F_{+-}(\mathbf{q},\omega)}-1\Big)g^2\mu_{\mathrm{B}}^2$$
$$=\frac{1}{\tilde{U}}\Big(\frac{2\tilde{U}\bar{S}_z}{Dq^2-\hbar\omega}-1\Big)g^2\mu_{\mathrm{B}}^2, \quad (5.39)$$

[3]This simple form of the spin-wave spectrum is in good agreement with neutron scattering experiments (see, e.g. [18]).

where ω stands for $\omega + \mathrm{i}0^+$. Using the Sokhotsky formula (A.44) and delta function property $\delta(ax) = \delta(x)/|a|$ (see Appendix A.2.1), we have

$$\mathrm{Im}\chi_{+-}(\mathbf{q}, \omega) = 2\bar{S}_z \pi \delta(\hbar\omega - Dq^2) g^2 \mu_\mathrm{B}^2. \tag{5.40}$$

Energy dissipation is proportional to $\mathrm{Im}\chi_{+-}(\mathbf{q}, \omega)$ (see, e.g. [19]). Therefore, dissipation takes nonzero values for $\mathbf{q}$ and ω such that $\hbar\omega = Dq^2$. Vice versa, the excitation energy $\hbar\omega = Dq^2$ creates a spin wave with a wavevector modulus q.

5.2.2 Magnetization: The $T^{3/2}$ Law

The magnetization is related to the local moment and local fluctuation. Indeed,

$$\langle \mathcal{M}(\mathbf{r})\mathcal{M}(\mathbf{r}')\rangle = \langle \mathcal{M}(\mathbf{r})\rangle\langle \mathcal{M}(\mathbf{r}')\rangle + \langle \Delta\mathcal{M}(\mathbf{r})\Delta\mathcal{M}(\mathbf{r}')\rangle,$$

where $\Delta\mathcal{M}(\mathbf{r}) = \mathcal{M}(\mathbf{r}) - \langle \mathcal{M}(\mathbf{r})\rangle$. Integrating $\mathbf{r}$ and $\mathbf{r}'$ over a Wigner-Seitz cell, we obtain

$$m_\mathrm{L}^2 = m_z^2 + \Delta m^2.$$

At low temperatures, we use the standard assumption that the magnetic field changes the *orientation* of spins but does not change their magnitude. Then, discarding the longitudinal fluctuation, we write the local magnetic moment as

$$m_\mathrm{L}^2 = m_z^2 + \Delta m_\perp^2. \tag{5.41}$$

Here

$$\Delta m_\perp^2 \equiv \iint_\mathrm{WS} \left(\langle \Delta\mathcal{M}_x(\mathbf{r})\Delta\mathcal{M}_x(\mathbf{r}')\rangle + \langle \Delta\mathcal{M}_y(\mathbf{r})\Delta\mathcal{M}_y(\mathbf{r}')\rangle \right) \mathrm{d}\mathbf{r}\,\mathrm{d}\mathbf{r}' \tag{5.42}$$

is the local transverse fluctuation, where both integrals are taken over the same Wigner-Seitz cell.

In the momentum representation, the square of the transverse spin fluctuation (5.42) is given by

$$\Delta m_\perp^2 = \frac{1}{N^2} \sum_\mathbf{q} \left(\langle \Delta\mathcal{M}_x(\mathbf{q})\Delta\mathcal{M}_x(-\mathbf{q})\rangle + \langle \Delta\mathcal{M}_y(\mathbf{q})\Delta\mathcal{M}_y(-\mathbf{q})\rangle \right).$$

Using the fluctuation-dissipation theorem (2.52), we come to

$$\Delta m_\perp^2 = -\frac{1}{N^2} \sum_\mathbf{q} \frac{1}{\pi} \int \frac{\hbar}{\mathrm{e}^{-\hbar\omega/T} - 1} \,\mathrm{Im}\,(\chi_{xx}(\mathbf{q}, \omega) + \chi_{yy}(\mathbf{q}, \omega))\,\mathrm{d}\omega. \tag{5.43}$$

Recalling relations (2.38), (2.41) and (2.42), we have

$$\chi_{xx}(\mathbf{q}, \omega) + \chi_{yy}(\mathbf{q}, \omega) = \frac{1}{2}[\chi_{-+}(\mathbf{q}, \omega) + \chi_{+-}(\mathbf{q}, \omega)]. \tag{5.44}$$

Therefore,

$$\Delta m_\perp^2 = -\frac{1}{2N^2} \sum_\mathbf{q} \frac{1}{\pi} \int \frac{\hbar}{\mathrm{e}^{-\hbar\omega/T} - 1} \,\mathrm{Im}\,(\chi_{-+}(\mathbf{q}, \omega) + \chi_{+-}(\mathbf{q}, \omega))\,\mathrm{d}\omega. \tag{5.45}$$

Next, we use the long-wave approximation, i.e. assume that the main contribution to the sum over the Brillouin zone comes from the terms with small $\mathbf{q}$. Hence we can replace the transverse susceptibilities by the spin-wave expression (5.40). Using formula (5.40), we write

$$\mathrm{Im}\chi_{+-}(\mathbf{q}, \omega) = 2\bar{S}_z \frac{\pi}{\hbar} \delta\left(\omega - \frac{Dq^2}{\hbar}\right) g^2 \mu_\mathrm{B}^2. \tag{5.46}$$

By the property (2.44), we obtain

$$\mathrm{Im}\chi_{-+}(\mathbf{q}, \omega) = -2\bar{S}_z \frac{\pi}{\hbar} \delta\left(\omega + \frac{Dq^2}{\hbar}\right) g^2 \mu_\mathrm{B}^2. \tag{5.47}$$

Substituting (5.46) and (5.47) in (5.45), and using the property of the delta-function (A.36), we have

$$\Delta m_{\perp}^2 = \bar{s}_z g^2 \mu_{\mathrm{B}}^2 \frac{1}{N} \sum_{\mathbf{q}} [B(Dq^2) - B(-Dq^2)], \tag{5.48}$$

where $B(\varepsilon) = (e^{\varepsilon/T} - 1)^{-1}$ is the *Bose function*. Using the identity $B(\varepsilon) - B(-\varepsilon) = 2B(\varepsilon) + 1$, we obtain

$$\Delta m_{\perp}^2 = m_z(0) g\mu_{\mathrm{B}} \left(1 + \frac{2}{N} \sum_{\mathbf{q}} B(Dq^2)\right),$$

where $m_z(0) = g\mu_{\mathrm{B}}\bar{s}_z$.

Finally, we replace the sum over the wavevector by the integral over the Brillouin zone, which is approximated for simplicity by the sphere of the same volume. Hence the fluctuation takes the form

$$\Delta m_{\perp}^2 = m_z(0) g\mu_{\mathrm{B}} \left(1 + 2\frac{4\pi}{\Omega_{\mathrm{BZ}}} \int_0^{q_{\mathrm{B}}} \frac{1}{\exp(Dq^2/T) - 1} q^2 \,\mathrm{d}q\right),$$

where Ω_{BZ} is the volume of the Brillouin zone and q_{B} is radius of the equal-volume sphere. Changing the variable to $x = (D/T)^{\frac{1}{2}} q$ and setting the upper limit of the integral equal to infinity (at low temperatures), we come to

$$\Delta m_{\perp}^2 = m_z(0) g\mu_{\mathrm{B}} \left(1 + \frac{2}{\Omega_{\mathrm{BZ}}} \zeta(3/2) \left(\frac{\pi}{D}\right)^{\frac{3}{2}} T^{3/2}\right), \tag{5.49}$$

where $\zeta(3/2) \approx 2.612$ is the value of the Riemann zeta function. Using $\Omega_{\mathrm{BZ}} = (2\pi)^3/\Omega_{\mathrm{WS}}$, where Ω_{WS} is the volume of the Wigner-Seitz cell, we obtain

$$\Delta m_{\perp}^2 = m_z(0) g\mu_{\mathrm{B}} + 2m_z^2(0) a_{3/2} T^{3/2}\,. \tag{5.50}$$

Here the coefficient $a_{3/2}$ is given by

$$a_{3/2} = \frac{g\mu_{\mathrm{B}}\Omega_{\mathrm{WS}}}{m_z(0)} \zeta(3/2)(4\pi D)^{-\frac{3}{2}}. \tag{5.51}$$

Since the square of the local magnetic moment (5.41) remains constant with temperature:

$$m_z^2(0) + \Delta m_{\perp}^2(0) = m_z^2(T) + \Delta m_{\perp}^2(T),$$

from (5.50) we obtain

$$m_z^2(T) = m_z^2(0)(1 - 2a_{3/2}T^{3/2}).$$

At low temperatures, we come to the $T^{3/2}$ law

$$\frac{m_z(T)}{m_z(0)} = 1 - a_{3/2}T^{3/2}. \tag{5.52}$$

Note that at $T = 0$ the square of the transverse spin fluctuation (5.50) is given by

$$\Delta s_{\perp}^2 = \bar{s}_z.$$

Taking formula (5.41) into account, we have $s_{\mathrm{L}}^2 = \bar{s}_z^2 + \Delta s_{\perp}^2$. Hence, the square of the local spin in itinerant-electron magnets at $T = 0$ is written as

$$s_{\mathrm{L}}^2 = \bar{s}_z(\bar{s}_z + 1), \tag{5.53}$$

just as for ferromagnets with localized spins. But here the value of the mean spin $\bar{s}_z$ can be *noninteger*.

In localized spin magnets, the quant of spin waves is a spin one quasi-particle, called *magnon* (see, e.g. [3, 5, 20]). Then, at low temperature, magnetization is written as

$$M_z(T) = M_z(0) - g\mu_B \sum_{\mathbf{q}} \langle \tilde{n}_{\mathbf{q}} \rangle,$$

where $\langle \tilde{n}_{\mathbf{q}} \rangle$ is thermal expectation of the number of excited magnons. The same result holds for itinerant electron magnets (see, e.g. [3, 21]). In the case of noninteracting magnons, the average is described by the Bose statistics, and one obtains the $T^{3/2}$ law. When interaction is present the situation becomes different. In particular, when electron–phonon interaction plays an important role, the $T^{3/2}$ law can fail [3].

5.2.3 Stoner Spin-Flip Excitations

The spin waves are not the only magnetic excitations that are described by the transverse susceptibility. The *Stoner excitations* are determined by singularities[4] of the numerator of the enhanced susceptibility (5.20):

$$F_{+-}(\mathbf{q}, \omega) = \infty.$$

Indeed, applying the Sokhotsky formula (A.44) to (5.23), we have

$$\mathrm{Im} F_{+-}(\mathbf{q}, \omega) = \pi \sum_{\mathbf{k}} (f(\varepsilon_{\mathbf{k}\uparrow}) - f(\varepsilon_{\mathbf{k}+\mathbf{q},\downarrow}))\, \delta(\hbar\omega - \varepsilon_{\mathbf{k}+\mathbf{q},\downarrow} + \varepsilon_{\mathbf{k}\uparrow}).$$

The right-hand side is nonzero at energies such that

$$\hbar\omega = \varepsilon_{\mathbf{k}+\mathbf{q},\downarrow} - \varepsilon_{\mathbf{k}\uparrow}. \tag{5.54}$$

By its definition, an excitation is the *addition* of a discrete amount of energy to a system that results in its transition from a state of lower energy to one of higher energy (excited state). Since the Fermi function is monotone decreasing (Fig. 3.1), the energy (5.54) is positive if and only if

$$f(\varepsilon_{\mathbf{k}\uparrow}) - f(\varepsilon_{\mathbf{k}+\mathbf{q},\downarrow}) > 0.$$

At low temperatures the Fermi function is close to the step function, and the latter inequality reduces to

$$\varepsilon_{\mathbf{k}\uparrow} < \mu, \qquad \varepsilon_{\mathbf{k}+\mathbf{q},\downarrow} = \varepsilon_{\mathbf{k}\uparrow} + \hbar\omega > \mu,$$

where μ is the chemical potential (see Fig. 3.1). In other words, the excitation removes a spin-up electron from the inside of the Fermi surface and puts it outside of the Fermi surface with the spin reversal. This single-electron spin-flip excitation has the same effect on the magnetization as the spin wave, i.e. magnetization is decreased by $g\mu_B$.

The Stoner excitations have energies in a large range of values for each q, unlike the spin waves that have only one energy corresponding to each q. We briefly illustrate the picture in the electron gas, where $\varepsilon_{\mathbf{k}} = (\hbar^2 k^2)/2m$. Recalling that $\varepsilon_{\mathbf{k}\sigma} = \varepsilon_{\mathbf{k}} - \sigma \tilde{U} \bar{S}_z$, we write relation (5.54) as

$$\hbar\omega = \frac{\hbar^2}{2m}(2\mathbf{kq} + q^2) + 2\tilde{U}\bar{S}_z. \tag{5.55}$$

If, for simplicity, we replace the Brillouin zone by an equal-volume sphere of radius q_B, then k lies in the range 0 to q_B, and the scalar product $\mathbf{kq}$ satisfies the inequality

$$-q_B q \le \mathbf{kq} \le q_B q.$$

[4] These singularities are called the *branch cuts* (for details, see Appendix A.2.4).

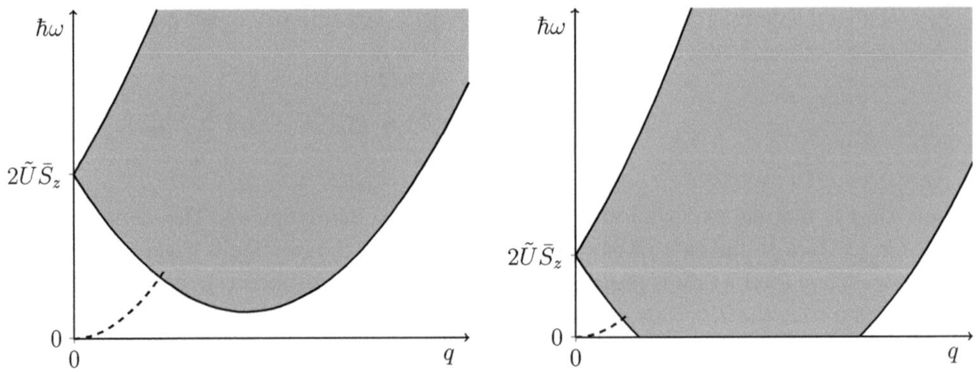

Fig. 5.1 Sketch of the spin-wave spectrum (dashed line) and Stoner continuum (grey area) at temperatures close to $T = 0$ (left) and higher temperatures (right) for the parabolic energy spectrum

The domain bounded by two parabolas (5.55) with $\mathbf{kq} = \pm q_B q$ is called the *Stoner continuum* (Fig. 5.1). There is an energy gap for Stoner excitations at $q = 0$, which is equal to $2\tilde{U}\bar{S}_z$. For small temperatures and small wavevectors, spin waves require less energy (Fig. 5.1, left), and hence we observe the $T^{3/2}$ law for magnetization. As temperature increases, the mean spin $\bar{S}_z$ decreases. Therefore, the Stoner continuum absorbs larger part of the spin-waves dispersion curve $\hbar\omega = Dq^2$, and the T^2 contribution becomes dominant (Fig. 5.1, right). As we show in Chap. 12, the switching temperature in metals is rather difficult to locate.

At the q-point where the spin-wave spectrum curve merges with the Stoner continuum the spin waves attenuate. This process, referred to as *Landau damping*, remains an open problem both from the experimental and the theoretical point of view (see, e.g. [3, 13, 18, 22, 23]).

Above the Curie temperature, the peak of the imaginary part of the transverse susceptibility in iron [24] and nickel [25] was attributed to the damped spin waves (paramagnons). But this interpretation has been criticized, because the height of the peak and its width are about equal [26, 27]. The RPA cannot be applied here because it was derived at low temperatures. In SFT [22] it was shown that well-defined "sloppy" spin waves can arise in itinerant ferromagnets above the Curie point in a limited range of wavevectors. We will come back to this question later in Chap. 15.

The Curie temperature calculated from the poles of the RPA longitudinal susceptibility is lower than the one in the Stoner theory, because the spin-wave excitations require less energy than the spin flips in the Stoner theory (for details, see Chap. 12). But the Curie temperature in the RPA is still too high compared with the experimental one. All this indicates that one should consider spin fluctuations beyond the Stoner theory and RPA.

References

1. D. Pines, *Elementary Excitations in Solids* (Benjamin, New York, 1964)
2. P. Nozieres, *Theory of Interacting Fermi Systems* (Benjamin, New York, 1964)
3. D.J. Kim, *New Perspectives in Magnetism of Metals* (Kluwer/Plenum, New York, 1999)
4. H. Bruus, K. Flensberg, *Many-Body Quantum Theory in Condensed Matter Physics*, 3rd edn. (Oxford University Press, Oxford, 2004)
5. R.M. White, *Quantum Theory of Magnetism*, 3rd edn. (Springer, Berlin, 2007)
6. P.A. Wolff, Phys. Rev. **120**, 814 (1960)
7. T. Izuyama, D.J. Kim, R. Kubo, J. Phys. Soc. Jpn. **18**(7), 1025 (1963)
8. D.J. Kim, B.B. Schwartz, H.C. Praddaude, Phys. Rev. B **7**, 205 (1973)
9. J.F. Cooke, J.W. Lynn, H.L. Davis, Phys. Rev. B **21**, 4118 (1980)
10. J. Callaway, A.K. Chatterjee, S.P. Singhal, A. Ziegler, Phys. Rev. B **28**, 3818 (1983)
11. D. Mattis, *The Theory of Magnetism* (Springer, Berlin, 1981)
12. K. Yosida, *Theory of Magnetism*, 2nd edn. (Springer, Berlin, 1998)
13. P. Mohn, *Magnetism in the Solid State*, 2nd edn. (Springer, Berlin, 2006)
14. F. Bloch, Z. Physik **61**, 206 (1930)
15. F. Bloch, Z. Physik **74**, 295 (1932)
16. T. Moriya, A. Kawabata, J. Phys. Soc. Jpn. **35**(3), 669 (1973)
17. N.B. Melnikov, B.I. Reser, Theor. Math. Phys. **181**(2), 1435 (2014)
18. H.A. Mook, in *Spin Waves and Magnetic Excitations 1*, ed. by A.S. Borovik-Romanov, S.K. Sinha (Elsevier, Amsterdam, 1988), pp. 425–478
19. L.D. Landau, E.M. Lifshitz, *Statistical Physics, Course on Theoretical Physics*, vol. 5, 3rd edn. (Pergamon, Oxford, 1985)
20. C. Kittel, *Introduction to Solid State Physics*, 8th edn. (Wiley, New York, 2005)
21. Y. Kakehashi, *Modern Theory of Magnetism in Metals and Alloys* (Springer, Berlin, 2012)
22. J.A. Hertz, M.A. Klenin, Physica B&C **91B**, 49 (1977)
23. P. Buczek, A. Ernst, L.M. Sandratskii, Phys. Rev. B **84**, 174418 (2011)

24. J.W. Lynn, Phys. Rev. B **11**, 2624 (1975)
25. H.A. Mook, J.W. Lynn, R.M. Nicklow, Phys. Rev. Lett. **30**, 556 (1973)
26. J. Hubbard, in *Physics of Transition Metals 1980 (Inst. Phys. Conf. Ser., No. 55)*, ed. by P. Rhodes (IOP, Bristol, 1981), pp. 669–687
27. J. Wicksted, G. Shirane, P. Böni, Phys. Rev. B **30**, 3655 (1984)

6 Green Functions at Finite Temperatures

What I cannot create, I do not understand. (R. Feynman)

The Green functions is an indispensable tool for studying interacting particles. The usefulness of the Green functions will be shown in the succeeding chapters. Here we give a brief self-contained introduction to the Green functions, their basic properties and applications. A comprehensive treatment is given in a number of well-known textbooks such as [1–7]. A readable introduction to the subject from a somewhat different perspective than ours can be found in [8, Chap. 10 and 11], [9, Chap. 8] and [10, Chap. 8 and 11]. We consider two types of Green functions: the fermion-type Green functions that describe propagation of electrons and boson-type Green functions that describe correlations of the spin density.

6.1 Fermion-Type Green Functions

We use fermion-type Green functions to obtain the density of states, mean charge and mean spin of interacting electrons at *finite* temperatures. There are two different kinds of the Green functions: the real-time and temperature Green functions. The former are related to observed characteristics and the latter can be calculated either by the diagram technique or by the functional integral method. We present both real-time and temperature Green functions and explain how they are linked to one another.

6.1.1 Real-Time Green Function

General Properties

First we introduce the *retarded* real-time Green function

$$G^{\mathrm{r}}_{jj'\sigma\sigma'}(t,t') = -\frac{\mathrm{i}}{\hbar}\langle\{a_{j\sigma}(t), a^{\dagger}_{j'\sigma'}(t')\}\rangle\theta(t-t'), \tag{6.1}$$

where $\theta(t)$ is zero for $t < 0$ and is unity for $t > 0$. Here

$$\langle\ldots\rangle = \Xi^{-1}\mathrm{Tr}(\ldots\mathrm{e}^{-\mathcal{H}'/T}), \qquad \mathcal{H}' = \mathcal{H} - \mu\mathcal{N}_{\mathrm{e}},$$

is the grand canonical ensemble average and

$$a^{\dagger}_{j\sigma}(t) = \mathrm{e}^{\mathrm{i}\mathcal{H}'t/\hbar}a^{\dagger}_{j\sigma}\mathrm{e}^{-\mathrm{i}\mathcal{H}'t/\hbar}, \qquad a_{j\sigma}(t) = \mathrm{e}^{\mathrm{i}\mathcal{H}'t/\hbar}a_{j\sigma}\mathrm{e}^{-\mathrm{i}\mathcal{H}'t/\hbar}$$

are the Wannier creation-annihilation operators in the Heisenberg representation (for simplicity of notation we omit the energy band index). The Green function (6.1) is called retarded, because it represents the response at the time t of a system to a perturbation applied at the time $t' < t$.

N. B. Melnikov, B. I. Reser, *Dynamic Spin Fluctuation Theory of Metallic Magnetism*,
https://doi.org/10.1007/978-3-319-92974-3_6

For a time-stationary system, it is easy to show using the cyclic property of trace that $G^{\mathrm{r}}_{jj'\sigma\sigma'}(t,t')$ is a function of $t-t'$:

$$G^{\mathrm{r}}_{jj'\sigma\sigma'}(t-t') = -\frac{\mathrm{i}}{\hbar}\langle\{a_{j\sigma}(t-t'), a^{\dagger}_{j'\sigma'}\}\rangle\theta(t-t').$$

Therefore, it is convenient to consider the Fourier transform with respect to time,

$$G^{\mathrm{r}}_{jj'\sigma\sigma'}(\omega) = \int_{-\infty}^{\infty} G^{\mathrm{r}}_{jj'\sigma\sigma'}(t)\mathrm{e}^{\mathrm{i}\omega t}\,\mathrm{d}t = -\frac{\mathrm{i}}{\hbar}\int_{0}^{\infty}\langle\{a_{j\sigma}(t), a^{\dagger}_{j'\sigma'}\}\rangle\mathrm{e}^{\mathrm{i}\omega t}\,\mathrm{d}t.$$

The inverse transformation is given by

$$G^{\mathrm{r}}_{jj'\sigma\sigma'}(t) = \frac{1}{2\pi}\int G_{jj'\sigma\sigma'}(\omega)\mathrm{e}^{-\mathrm{i}\omega t}\,\mathrm{d}\omega.$$

To ensure that $G^{\mathrm{r}}_{jj'\sigma\sigma'}(t) = 0$ for $t < 0$, the function $G^{\mathrm{r}}_{jj'\sigma\sigma'}(\omega)$ must be analytic in the upper half-plane of the complex ω-plane (see Appendix A.2.4).

If the system is translationally invariant, the Green function depends only on the distance between the electrons $j - j'$. Then, using the spatial Fourier transformation, we have

$$G^{\mathrm{r}}_{\mathbf{k}\sigma\sigma'}(t) = -\frac{\mathrm{i}}{\hbar}\langle\{a_{\mathbf{k}\sigma}(t), a^{\dagger}_{\mathbf{k}\sigma'}\}\rangle\theta(t).$$

If the Hamiltonian is spin-independent, then the Green function is spin diagonal and its diagonal element is

$$G^{\mathrm{r}}_{\mathbf{k}\sigma}(t) = -\frac{\mathrm{i}}{\hbar}\langle\{a_{\mathbf{k}\sigma}(t), a^{\dagger}_{\mathbf{k}\sigma}\}\rangle\theta(t). \tag{6.2}$$

Equation of Motion

A direct method of calculating the Green function is to solve the equation of motion. For a single annihilation operator, we have

$$\mathrm{i}\hbar\frac{\mathrm{d}}{\mathrm{d}t}a_{\mathbf{k}\sigma}(t) = [a_{\mathbf{k}\sigma}, \mathcal{H}'](t). \tag{6.3}$$

Differentiating both sides of (6.2) with respect to t and using Eq. (6.3) and property $\mathrm{d}\theta(t)/\mathrm{d}t = \delta(t)$, we have

$$\mathrm{i}\hbar\frac{\mathrm{d}}{\mathrm{d}t}G^{\mathrm{r}}_{\mathbf{k}\sigma}(t) = -\frac{\mathrm{i}}{\hbar}\langle\{[a_{\mathbf{k}\sigma}, \mathcal{H}'](t), a^{\dagger}_{\mathbf{k}\sigma}\}\rangle\theta(t) + \langle\{a_{\mathbf{k}\sigma}(t), a^{\dagger}_{\mathbf{k}\sigma}\}\rangle\delta(t). \tag{6.4}$$

Taking into account the property $f(t)\delta(t) = f(0)\delta(t)$ and anticommutation relation $\{a_{\mathbf{k}\sigma}, a^{\dagger}_{\mathbf{k}\sigma}\} = 1$, we obtain

$$\mathrm{i}\hbar\frac{\mathrm{d}}{\mathrm{d}t}G^{\mathrm{r}}_{\mathbf{k}\sigma}(t) = -\frac{\mathrm{i}}{\hbar}\langle\{[a_{\mathbf{k}\sigma}, \mathcal{H}'](t), a^{\dagger}_{\mathbf{k}\sigma}\}\rangle\theta(t) + \delta(t). \tag{6.5}$$

In the general case, the first term on the right-hand side will give us a more complicated Green function, which is an average of four creation-annihilation operators. The time derivative of this new Green function will give averages of more creation-annihilation operators, etc. An approximate solution can be obtained by truncating this chain of equations of motion (for the boson-type Green functions, this will be shown in Sect. 6.2 when calculating the RPA susceptibility).

In the case of *noninteracting* electrons with the Hamiltonian $\mathcal{H}_0$, the equation of motion (6.5) can be solved explicitly. Indeed, using the anticommutation relations for the creation-annihilation operators just as in Sect. 3.4, we write the right-hand side of (6.3) as

$$[a_{\mathbf{k}\sigma}, \mathcal{H}'_0](t) = (\varepsilon_{\mathbf{k}} - \mu)a_{\mathbf{k}\sigma}(t). \tag{6.6}$$

Substituting the latter in (6.5), we have

$$\mathrm{i}\hbar\frac{\mathrm{d}}{\mathrm{d}t}G^{\mathrm{r}0}_{\mathbf{k}}(t) = (\varepsilon_{\mathbf{k}} - \mu)G^{\mathrm{r}0}_{\mathbf{k}}(t) + \delta(t). \tag{6.7}$$

Applying the Fourier transformation, we obtain

$$G_{\mathbf{k}}^{\mathrm{r}0}(\omega) = \frac{1}{\hbar\omega + \mu - \varepsilon_{\mathbf{k}} + \mathrm{i}0^+} \tag{6.8}$$

(the appearance of the term $\mathrm{i}0^+$ is explained in Appendix A.2.5). Thus, poles of the Fourier transform of the Green function determine the energy spectrum. Taking the inverse Fourier transformation, we get the Green function itself:

$$G_{\mathbf{k}}^{\mathrm{r}0}(t) = -\mathrm{i}\mathrm{e}^{-\mathrm{i}(\varepsilon_{\mathbf{k}}-\mu)t}\theta(t). \tag{6.9}$$

For interpretation of the Green function $G_{\mathbf{k}}^{\mathrm{r}0}(t)$ as the probability amplitude of the electron propagation, see, e.g. [3, 6].

Spectral Function

Following the same argument as in Sect. 2.1.3, we obtain the relation between the real and imaginary parts of the Green function:

$$\mathrm{Re}G_{\mathbf{k}\sigma}^{\mathrm{r}}(\omega) = \frac{1}{\pi}\mathcal{P}\int \frac{\mathrm{Im}G_{\mathbf{k}\sigma}^{\mathrm{r}}(\omega')}{\omega' - \omega}\,\mathrm{d}\omega', \tag{6.10}$$

$$\mathrm{Im}G_{\mathbf{k}\sigma}^{\mathrm{r}}(\omega) = -\frac{1}{\pi}\mathcal{P}\int \frac{\mathrm{Re}G_{\mathbf{k}\sigma}^{\mathrm{r}}(\omega')}{\omega' - \omega}\,\mathrm{d}\omega', \tag{6.11}$$

where $\mathcal{P}$ means the Cauchy principle value. Using (6.10), we can write $G_{\mathbf{k}\sigma}^{\mathrm{r}}(\omega) = \mathrm{Re}G_{\mathbf{k}\sigma}^{\mathrm{r}}(\omega) + \mathrm{i}\mathrm{Im}G_{\mathbf{k}\sigma}^{\mathrm{r}}(\omega)$ as

$$G_{\mathbf{k}\sigma}^{\mathrm{r}}(\omega) = \frac{1}{\pi}\mathcal{P}\int \frac{\mathrm{Im}G_{\mathbf{k}\sigma}^{\mathrm{r}}(\omega')}{\omega' - \omega}\,\mathrm{d}\omega' + \mathrm{i}\mathrm{Im}G_{\mathbf{k}\sigma}^{\mathrm{r}}(\omega).$$

Taking into account the Sokhotsky formula (A.44):

$$\int \frac{f(x')}{x' - x \pm \mathrm{i}0^+}\,\mathrm{d}x' = \mathcal{P}\int \frac{f(x')}{x' - x}\,\mathrm{d}x' \mp \mathrm{i}\pi f(x),$$

we have

$$G_{\mathbf{k}\sigma}^{\mathrm{r}}(\omega) = \frac{1}{\pi}\int \frac{\mathrm{Im}G_{\mathbf{k}\sigma}^{\mathrm{r}}(\omega')}{\omega' - \omega - \mathrm{i}0^+}\,\mathrm{d}\omega' \equiv \int \frac{A_{\mathbf{k}\sigma}^{\mathrm{r}}(\omega')}{\omega - \omega' + \mathrm{i}0^+}\,\mathrm{d}\omega', \tag{6.12}$$

where

$$A_{\mathbf{k}\sigma}^{\mathrm{r}}(\omega) = -\frac{1}{\pi}\mathrm{Im}G_{\mathbf{k}\sigma}^{\mathrm{r}}(\omega)$$

is called the (retarded) *spectral function*. In particular, for noninteracting electrons, writing the Green function (6.8) as

$$G_{\mathbf{k}}^{\mathrm{r}0}(\omega) = \int \frac{\delta(\varepsilon' + \mu - \varepsilon_{\mathbf{k}})}{\hbar\omega - \varepsilon' + \mathrm{i}0^+}\,\mathrm{d}\varepsilon'$$

and using (6.12), we obtain

$$A_{\mathbf{k}}^{\mathrm{r}0}(\omega) = \delta(\hbar\omega + \mu - \varepsilon_{\mathbf{k}}). \tag{6.13}$$

Similar to the retarded Green function we introduce the *advanced* Green function

$$G_{jj'\sigma\sigma'}^{\mathrm{a}}(t) = \frac{\mathrm{i}}{\hbar}\langle\{a_{j\sigma}(t), a_{j'\sigma'}^{\dagger}\}\rangle\theta(-t). \tag{6.14}$$

Doing as above, we prove that its Fourier transform

$$G_{\mathbf{k}\sigma}^{\mathrm{a}}(\omega) = \int_{-\infty}^{0} G_{\mathbf{k}\sigma}^{\mathrm{a}}(t)\mathrm{e}^{\mathrm{i}\omega t}\,\mathrm{d}t$$

satisfies the following relations:

$$\mathrm{Re}G^{\mathrm{a}}_{\mathbf{k}\sigma}(\omega) = -\frac{1}{\pi}\mathcal{P}\int \frac{\mathrm{Im}G^{\mathrm{a}}_{\mathbf{k}\sigma}(\omega')}{\omega' - \omega}\,\mathrm{d}\omega', \tag{6.15}$$

$$\mathrm{Im}G^{\mathrm{a}}_{\mathbf{k}\sigma}(\omega) = \frac{1}{\pi}\mathcal{P}\int \frac{\mathrm{Re}G^{\mathrm{a}}_{\mathbf{k}\sigma}(\omega')}{\omega' - \omega}\,\mathrm{d}\omega'. \tag{6.16}$$

Hence

$$G^{\mathrm{a}}_{\mathbf{k}\sigma}(\omega) = \int \frac{A^{\mathrm{a}}_{\mathbf{k}\sigma}(\omega')}{\omega - \omega' - \mathrm{i}0^{+}}\,\mathrm{d}\omega', \tag{6.17}$$

where

$$A^{\mathrm{a}}_{\mathbf{k}\sigma}(\omega) = \frac{1}{\pi}\mathrm{Im}G^{\mathrm{a}}_{\mathbf{k}\sigma}(\omega). \tag{6.18}$$

Comparing the results for the retarded (6.12) and advanced (6.17) Green functions, we see that

$$G^{\mathrm{a}}_{\mathbf{k}\sigma}(\omega) = G_{\mathbf{k}\sigma}(\omega - \mathrm{i}0^{+}), \qquad G^{\mathrm{r}}_{\mathbf{k}\sigma}(\omega) = G_{\mathbf{k}\sigma}(\omega + \mathrm{i}0^{+}),$$

where

$$G_{\mathbf{k}\sigma}(z) = \int \frac{A_{\mathbf{k}\sigma}(\omega')}{z - \omega'}\,\mathrm{d}\omega' \tag{6.19}$$

is an analytic function except on the real axis. Thus, the Fourier transforms of the advanced and retarded Green functions are just complex conjugate of each other:

$$G^{\mathrm{a}}_{\mathbf{k}\sigma}(\omega) = (G^{\mathrm{r}}_{\mathbf{k}\sigma}(\omega))^{*}. \tag{6.20}$$

The Green function of the interacting electrons in a crystal can be written as

$$G_{\mathbf{k}\sigma}(z) = \frac{1}{\hbar z + \mu - \varepsilon_{\mathbf{k}} - \Sigma_{\mathbf{k}\sigma}(z)}, \tag{6.21}$$

where $\Sigma_{\mathbf{k}\sigma}(z)$ is called the *self-energy*. The meaning of the self-energy becomes clear if we look at formulae (6.17) and (6.18). Indeed, for noninteracting electrons, similar to (6.8) we have

$$G^{\mathrm{a}0}_{\mathbf{k}}(\omega) = \frac{1}{\hbar\omega + \mu - \varepsilon_{\mathbf{k}} - \mathrm{i}0^{+}}, \tag{6.22}$$

and the spectral function is given by a single delta function peaked at $\varepsilon_{\mathbf{k}} - \mu$, just as in formula (6.13). Comparing with (6.21), we readily see that

$$\mathrm{Re}\Sigma^{0}_{\mathbf{k}\sigma}(\omega) = 0, \qquad \mathrm{Im}\Sigma^{0}_{\mathbf{k}\sigma}(\omega) = 0^{+},$$

where $\Sigma^{0}(\omega) = \Sigma^{0}(\omega - \mathrm{i}0^{+})$. For interacting electrons, $\Sigma_{\mathbf{k}\sigma}(\omega)$ is a nonzero complex quantity, and the spectral function is given by

$$A^{\mathrm{a}}_{\mathbf{k}\sigma}(\omega) = \frac{1}{\pi}\mathrm{Im}G^{\mathrm{a}}_{\mathbf{k}\sigma}(\omega) = \frac{1}{\pi}\frac{\mathrm{Im}\Sigma_{\mathbf{k}\sigma}(\omega)}{(\hbar\omega + \mu - \varepsilon_{\mathbf{k}} - \mathrm{Re}\Sigma_{\mathbf{k}\sigma}(\omega))^2 + (\mathrm{Im}\Sigma_{\mathbf{k}\sigma}(\omega))^2}.$$

Thus, the delta function peak of a one-electron excitation becomes spread out over an energy interval. The location of the peak is shifted by $\mathrm{Re}\Sigma$ and $(\mathrm{Im}\Sigma)^{-1}$ determines the finite life-time of the electron's state (see, e.g. [9, 10]).

6.1.2 Temperature Green Function

Relation with Charge and Spin Density

The *temperature* Green function[1] for fermions is defined by

$$G_{jj'\sigma\sigma'}(\tau,\tau') = -\langle T_\tau\, a_{j\sigma}(\tau) a^\dagger_{j'\sigma'}(\tau')\rangle \equiv \begin{cases} -\langle a_{j\sigma}(\tau) a^\dagger_{j'\sigma'}(\tau')\rangle, & \tau > \tau', \\ \langle a^\dagger_{j'\sigma'}(\tau') a_{j\sigma}(\tau)\rangle, & \tau < \tau'. \end{cases} \tag{6.23}$$

Here $\langle \dots \rangle$ is the average in the grand canonical ensemble, $a_{j\sigma}$ and $a^\dagger_{j'\sigma'}$ are the creation-annihilation operators of the Wannier states and the τ-dependence means the "Heisenberg" representation $a_{j\sigma}(\tau) = \mathrm{e}^{\mathcal{H}'\tau/\hbar} a_{j\sigma}\, \mathrm{e}^{-\mathcal{H}'\tau/\hbar}$ with respect to the "time" $\tau \in [0, \hbar/T]$. If we take $\tau = it$, we come to the usual Heisenberg representation. Therefore, τ is often called the imaginary time. This link between τ and t is not quite incidental. As we show later in this section, the temperature and real-time Green functions are related to each other in a very similar way.

The Green function (6.23) allows to obtain local characteristics such as the mean values of charge and spin at a site. To show this, it is convenient to introduce the *local density matrix* ρ_j with the elements

$$\rho_{j\sigma\sigma'} = a^\dagger_{j\sigma'} a_{j\sigma}. \tag{6.24}$$

As a Hermitian 2×2 matrix, ρ_j can be expressed in the form (for details, see Appendix A.1.4)

$$\rho_j = \rho_j^0 \sigma^0 + \boldsymbol{\rho}_j \boldsymbol{\sigma}, \qquad \rho_j^\mu = \frac{1}{2}\mathrm{Sp}\,(\sigma^\mu \rho_j), \qquad \mu = 0, x, y, z, \tag{6.25}$$

where σ^0 is the unity 2×2 matrix, σ^α ($\alpha = x, y, z$) are the Pauli matrices, and Sp denotes the sum of diagonal elements over the spin indices.[2] From formulae (6.25) it immediately follows that the scalar component in the expansion ρ_j^0 is equal to one half of the local charge operator:

$$\frac{1}{2} n_j = \frac{1}{2}(n_{j\uparrow} + n_{j\downarrow}) = \frac{1}{2}\left(a^\dagger_{j\uparrow} a_{j\uparrow} + a^\dagger_{j\downarrow} a_{j\downarrow}\right) = \rho_j^0, \tag{6.26}$$

and the vector component $\boldsymbol{\rho}_j$ is equal to the local spin operator: $\boldsymbol{\rho}_j = \mathbf{s}_j$. Indeed, if we use the second-quantized representation (3.20) of the $\alpha = x, y, z$ component of the spin operator $\mathbf{s}_j = \frac{1}{2}\boldsymbol{\sigma}$, then taking (6.25) into account, we write

$$s_j^\alpha = \sum_{\sigma\sigma'} s^\alpha_{j\sigma\sigma'} a^\dagger_{j\sigma} a_{j\sigma'} = \sum_{\sigma\sigma'} s^\alpha_{j\sigma\sigma'} \rho_{j\sigma'\sigma} = \mathrm{Sp}\left(s_j^\alpha \rho_j\right) = \rho_j^\alpha. \tag{6.27}$$

The Green function $G_{jj\sigma\sigma'}(\tau, \tau + 0^+)$ is equal to the average local density matrix $\langle \rho_{j\sigma\sigma'}\rangle$. Indeed, by the definition (6.23), for $\tau < \tau'$ we have

$$G_{jj'\sigma\sigma'}(\tau,\tau') = \langle a^\dagger_{j'\sigma'}(\tau')\, a_{j\sigma}(\tau)\rangle, \tag{6.28}$$

so that

$$G_{jj\sigma\sigma'}(\tau, \tau + 0^+) = \langle a^\dagger_{j\sigma'}\, a_{j\sigma}\rangle = \langle \rho_{j\sigma\sigma'}\rangle. \tag{6.29}$$

Introducing the Hermitian 2×2-matrix G_{jj} with the elements $G_{jj\sigma\sigma'}(\tau, \tau + 0^+)$, we can write

$$G_{jj} = \sum_\mu G^\mu_{jj} \sigma^\mu, \qquad G^\mu_{jj} = \frac{1}{2}\mathrm{Sp}(G_{jj}\sigma^\mu), \qquad \mu = 0, x, y, z. \tag{6.30}$$

---

[1] Also called Matsubara Green function since they appeared in the seminal paper [11].

[2] Here by $\rho_j^\mu \sigma^\mu$ we mean the operator tensor product $\rho_j^\mu \otimes \sigma^\mu$ (for details, see Appendix A.1.5). For brevity, we omit the tensor product notation throughout the book.

From relations (6.26) and (6.27) we have

$$G_{jj}^{\mu=0} = \frac{1}{2}\langle n_j\rangle, \qquad G_{jj}^{\mu=\alpha} = \langle s_j^\alpha\rangle, \qquad \alpha = x, y, z. \tag{6.31}$$

Summing (6.30) over spins and taking into account $\mathrm{Sp}\sigma^\mu = 2\delta_{\mu 0}$, we obtain

$$\mathrm{Sp}G_{jj} = \mathrm{Sp}\Big(\sum_\mu G_{jj}^\mu \sigma^\mu\Big) = \sum_\mu G_{jj}^\mu \,\mathrm{Sp}\sigma^\mu = 2G_{jj}^{\mu=0}. \tag{6.32}$$

Then, taking relations (6.31) and (6.32) into account, we can express the average of the total number of electrons operator $\mathcal{N}_\mathrm{e} = \sum_{j\sigma} n_{j\sigma}$ and average of the total spin operator $\mathbf{S} = \sum_{j\sigma} \mathbf{s}_{j\sigma}$ as follows:

$$N_\mathrm{e} = \frac{T}{\hbar}\mathrm{Tr}G, \qquad \bar{S}^\alpha = \frac{T}{\hbar}\mathrm{Tr}G^\alpha, \tag{6.33}$$

where

$$\mathrm{Tr}G^\mu \equiv \sum_j \int_0^{\hbar/T} G_{jj}^\mu(\tau, \tau + 0^+)\,\mathrm{d}\tau, \tag{6.34}$$

$$\mathrm{Tr}G \equiv \sum_j \int_0^{\hbar/T} \mathrm{Sp}G_{jj}(\tau, \tau + 0^+)\,\mathrm{d}\tau.$$

From Eq. (6.29) we see that $G_{jj}^\mu(\tau, \tau + 0^+)$ is independent of j and τ.

Relation with the Real-Time Green Function

Due to translational invariance of the system in space and "time":

$$G_{jj'}(\tau, \tau') = G_{j-j'}(\tau - \tau'),$$

the spatial Fourier transform of the Green function is **k**-diagonal:

$$G_{\mathbf{k}\mathbf{k}'}(\tau) = G_\mathbf{k}(\tau)\delta_{\mathbf{k}\mathbf{k}'},$$

where

$$G_j(\tau) = \frac{1}{N}\sum_\mathbf{k} G_\mathbf{k}(\tau)\mathrm{e}^{\mathrm{i}\mathbf{k}\mathbf{R}_j}, \qquad G_\mathbf{k}(\tau) = \sum_j G_j(\tau)\mathrm{e}^{-\mathrm{i}\mathbf{k}\mathbf{R}_j}.$$

The Green function $G_\mathbf{k}(\tau)$ satisfies the following relation

$$G_\mathbf{k}(\tau - \hbar/T) = -G_\mathbf{k}(\tau), \qquad \tau > 0 \tag{6.35}$$

(for the proof, see Appendix C.2, where $\bar{G}_\mathbf{k}(\tau)$ is the same Green function of interacting electrons; the use of the bar in this context is explained in Chap. 8). Hence the function $G_\mathbf{k}(\tau)$ can be expanded in the *odd* thermodynamic "frequencies"[3] $\omega_n = (2n+1)\pi T/\hbar, \;\; n = 0, \pm 1, \pm 2, \ldots$:

$$G_\mathbf{k}(\tau) = \frac{T}{\hbar}\sum_n G_\mathbf{k}(\mathrm{i}\omega_n)\mathrm{e}^{-i\omega_n\tau},$$

where the Fourier coefficient is calculated as

$$G_\mathbf{k}(\mathrm{i}\omega_n) = \int_0^{\hbar/T} G_\mathbf{k}(\tau)\mathrm{e}^{\mathrm{i}\omega_n\tau}\,\mathrm{d}\tau$$

[3] The thermodynamic "frequencies" are also called Matsubara frequencies.

(for details, see Appendix C.2). Therefore, in the momentum-"frequency" representation we write (6.34) as

$$\mathrm{Tr}G^{\mu} = \lim_{\tau \to 0^+} \sum_{\mathbf{k}n} G^{\mu}_{\mathbf{k}}(\mathrm{i}\omega_n)\, \mathrm{e}^{\mathrm{i}\omega_n \tau}. \tag{6.36}$$

For brevity, we will often write

$$\mathrm{Tr}G^{\mu} = \sum_{\mathbf{k}n} G^{\mu}_{\mathbf{k}}(\mathrm{i}\omega_n).$$

The temperature Green function $G_{\mathbf{k}}(\mathrm{i}\omega_n)$ can be analytically continued from the points $z = \mathrm{i}\omega_n$ to the whole complex z-plane except for the real axis. At $z = \omega \pm \mathrm{i}0^+$ the result coincides with the real-time Green function $G_{\mathbf{k}}(\omega)$ (see, e.g. [9]). Shortly,

$$G_{\mathbf{k}}(\mathrm{i}\omega_n)\Big|_{\mathrm{i}\omega_n \to \omega + \mathrm{i}0^+} = G^{\mathrm{r}}_{\mathbf{k}}(\omega), \qquad G_{\mathbf{k}}(\mathrm{i}\omega_n)\Big|_{\mathrm{i}\omega_n \to \omega - \mathrm{i}0^+} = G^{\mathrm{a}}_{\mathbf{k}}(\omega).$$

In particular, using (6.22), for noninteracting electrons we have

$$G^0_{\mathbf{k}}(\mathrm{i}\omega_n) = \frac{1}{\mathrm{i}\hbar\omega_n + \mu - \varepsilon_{\mathbf{k}}}. \tag{6.37}$$

Introducing the operator $G^0(z)$ with the matrix elements $G^0_{\mathbf{k}}(z)$, we can write

$$G^0(z) = (\hbar z + \mu - \mathcal{H}_0)^{-1}. \tag{6.38}$$

If we apply a time independent external field, the Hamiltonian becomes $\mathcal{H}_0 + \hat{V}$, where in the second-quantized form

$$\hat{V} = \sum_{\mathbf{k}\mathbf{k}'\sigma} V_{\mathbf{k}\mathbf{k}'\sigma} a^{\dagger}_{\mathbf{k}\sigma} a_{\mathbf{k}'\sigma}.$$

Then the (operator) Green function is written as

$$G(z) = (\hbar z + \mu - \mathcal{H}_0 - \hat{V})^{-1}.$$

Note that a scalar relation like (6.37) is not valid anymore, because the matrix $V_{\mathbf{k}\mathbf{k}'\sigma}$ is generally nondiagonal in $\mathbf{k}$, and so is $G_{\mathbf{k}\mathbf{k}'\sigma}$. However, if the external field is spatially uniform, we have

$$G_{\mathbf{k}\sigma}(z) = \frac{1}{\hbar z + \mu - \varepsilon_{\mathbf{k}} - V_{\sigma}}. \tag{6.39}$$

6.2 Boson-Type Green Functions

In the linear response theory the dynamic susceptibility $\chi_{\alpha\beta}(\mathbf{q}, \omega)$ is given by the Fourier transform (2.46) of the function

$$\chi_{\alpha\beta}(\mathbf{q}, t) = \frac{\mathrm{i}}{\hbar}\big\langle[\Delta\mathcal{M}_{\alpha}(\mathbf{q}, t), \Delta\mathcal{M}_{\beta}(-\mathbf{q})]\big\rangle\theta(t). \tag{6.40}$$

The latter is called the *boson-type* retarded Green function, because each spin operator consists of two creation-annihilation operators. The boson-type Green function is defined by the commutator instead of the anticommutator in the fermion-type Green function (6.1).[4]

As we have already shown in the previous chapters, dynamic susceptibility (6.40) has a multitude of applications. In particular, it describes the spin-density correlations in metals (see Chap. 2). The spectrum of the longitudinal susceptibility determines the magnetic phase transition and spectrum of the transverse susceptibility determines magnetic excitations such as spin flip and spin waves (see Sect. 5.2).

In Sect. 3.4 we showed how to calculate the dynamic susceptibility for noninteracting electrons directly from expression (2.38) in the linear response theory. Here we obtain the dynamic susceptibility from the same expression (2.38) but for *interacting* electrons using the equation of motion method, just as in [12]. By applying the Hartree-Fock approximation to the right-hand side of the equation of motion, we obtain the same RPA solution that was calculated in Sect. 5.1.

6.2.1 Dynamic Susceptibility

Longitudinal Susceptibility

We need to obtain the Green function

$$\chi_{zz}(\mathbf{q},t)=\frac{\mathrm{i}}{\hbar}\big\langle[\Delta\mathcal{M}_z(\mathbf{q},t),\Delta\mathcal{M}_z(-\mathbf{q})]\big\rangle\theta(t).$$

The latter reduces to (for details, see Appendix A.1.2)

$$\chi_{zz}(\mathbf{q},t)=\frac{\mathrm{i}}{\hbar}\big\langle[\mathcal{M}_z(\mathbf{q},t),\mathcal{M}_z(-\mathbf{q})]\big\rangle\theta(t). \tag{6.41}$$

Here the magnetic moment operator $\mathcal{M}_z(\mathbf{q})$ can be written in the second-quantized form as

$$\mathcal{M}_z(\mathbf{q})=-g\mu_\mathrm{B}\frac{1}{2}\sum_{\mathbf{k}}\big(a^\dagger_{\mathbf{k}\uparrow}a_{\mathbf{k}+\mathbf{q},\uparrow}-a^\dagger_{\mathbf{k}\downarrow}a_{\mathbf{k}+\mathbf{q},\downarrow}\big).$$

We calculate (6.41) by writing the equation of motion for the Green function

$$\chi_{zz}(\mathbf{k}\sigma,\mathbf{q},t)=-g\mu_\mathrm{B}\frac{\mathrm{i}}{\hbar}\big\langle\big[(a^\dagger_{\mathbf{k}\sigma}a_{\mathbf{k}+\mathbf{q},\sigma})(t),\mathcal{M}_z(-\mathbf{q})\big]\big\rangle\theta(t). \tag{6.42}$$

Differentiating (6.42) and taking $\mathrm{d}\theta(t)/\mathrm{d}t=\delta(t)$ into account (see Appendix A.2.1), we obtain

$$\begin{aligned}\mathrm{i}\hbar\frac{\mathrm{d}}{\mathrm{d}t}\chi_{zz}(\mathbf{k}\sigma,\mathbf{q},t)&=g\mu_\mathrm{B}\big\langle\big[(a^\dagger_{\mathbf{k}\sigma}a_{\mathbf{k}+\mathbf{q},\sigma})(t),\mathcal{M}_z(-\mathbf{q})\big]\big\rangle\delta(t)\\&\quad-g\mu_\mathrm{B}\frac{\mathrm{i}}{\hbar}\Big\langle\Big[\mathrm{i}\hbar\frac{\mathrm{d}}{\mathrm{d}t}(a^\dagger_{\mathbf{k}\sigma}a_{\mathbf{k}+\mathbf{q},\sigma})(t),\mathcal{M}_z(-\mathbf{q})\Big]\Big\rangle\theta(t).\end{aligned}$$

Recall that the Heisenberg representation $\mathcal{A}(t)=\mathrm{e}^{\mathrm{i}\mathcal{H}t/\hbar}\mathcal{A}\,\mathrm{e}^{-\mathrm{i}\mathcal{H}t/\hbar}$ of an operator $\mathcal{A}$ satisfies

$$\mathrm{i}\hbar\frac{\mathrm{d}}{\mathrm{d}t}\mathcal{A}(t)=[\mathcal{A},\mathcal{H}](t). \tag{6.43}$$

Taking the property $f(t)\delta(t)=f(0)\delta(t)$ into account, we obtain

$$\begin{aligned}\mathrm{i}\hbar\frac{\mathrm{d}}{\mathrm{d}t}\chi_{zz}(\mathbf{k}\sigma,\mathbf{q},t)&=g\mu_\mathrm{B}\big\langle\big[a^\dagger_{\mathbf{k}\sigma}a_{\mathbf{k}+\mathbf{q},\sigma},\mathcal{M}_z(-\mathbf{q})\big]\big\rangle\delta(t)\\&\quad-g\mu_\mathrm{B}\frac{\mathrm{i}}{\hbar}\Big\langle\Big[[a^\dagger_{\mathbf{k}\sigma}a_{\mathbf{k}+\mathbf{q},\sigma},\mathcal{H}](t),\mathcal{M}_z(-\mathbf{q})\Big]\Big\rangle\theta(t).\end{aligned} \tag{6.44}$$

[4]We use the opposite signs for the fermion- and boson-type Green functions, just as in [9].

Using the general commutation rule (3.46), we calculate the first commutator

$$\begin{aligned}\left[a^\dagger_{\mathbf{k}\sigma}a_{\mathbf{k}+\mathbf{q},\sigma}, \mathcal{M}_z(-\mathbf{q})\right] &= -g\mu_\mathrm{B}\left[a^\dagger_{\mathbf{k}\sigma}a_{\mathbf{k}+\mathbf{q},\sigma}, \frac{1}{2}\sum_{\mathbf{k}'\sigma'}\sigma' a^\dagger_{\mathbf{k}'\sigma'}a_{\mathbf{k}'-\mathbf{q},\sigma'}\right] \\ &= -g\mu_\mathrm{B}\sigma\frac{1}{2}\big(a^\dagger_{\mathbf{k}\sigma}a_{\mathbf{k}\sigma} - a^\dagger_{\mathbf{k}+\mathbf{q},\sigma}a_{\mathbf{k}+\mathbf{q},\sigma}\big). \end{aligned} \tag{6.45}$$

In the second term on the right-hand side of (6.44), we have $\mathcal{H} = \mathcal{H}_0 + \mathcal{H}_\mathrm{I}$. Similar to (6.45) we obtain

$$\left[a^\dagger_{\mathbf{k}\sigma}a_{\mathbf{k}+\mathbf{q},\sigma}, \mathcal{H}_0\right] = \left[a^\dagger_{\mathbf{k}\sigma}a_{\mathbf{k}+\mathbf{q},\sigma}, \sum_{\mathbf{k}'\sigma'}\varepsilon_{\mathbf{k}'}a^\dagger_{\mathbf{k}'\sigma'}a_{\mathbf{k}'\sigma'}\right] = (\varepsilon_{\mathbf{k}+\mathbf{q}} - \varepsilon_\mathbf{k})a^\dagger_{\mathbf{k}\sigma}a_{\mathbf{k}+\mathbf{q},\sigma}. \tag{6.46}$$

If we discard the interaction term $\mathcal{H}_\mathrm{I}$ altogether, then equation of motion (6.44) reduces to

$$\mathrm{i}\hbar\frac{\mathrm{d}}{\mathrm{d}t}\chi^0_{zz}(\mathbf{k}\sigma, \mathbf{q}, t) = -g^2\mu_\mathrm{B}^2\sigma\frac{1}{2}(f(\varepsilon_\mathbf{k}) - f(\varepsilon_{\mathbf{k}+\mathbf{q}}))\delta(t) + (\varepsilon_{\mathbf{k}+\mathbf{q}} - \varepsilon_\mathbf{k})\chi^0_{zz}(\mathbf{k}\sigma, \mathbf{q}, t). \tag{6.47}$$

Carrying out the Fourier transform with respect to t, we write

$$(\hbar\omega - \varepsilon_{\mathbf{k}+\mathbf{q}} + \varepsilon_\mathbf{k})\chi^0_{zz}(\mathbf{k}\sigma, \mathbf{q}, \omega) = -g^2\mu_\mathrm{B}^2\frac{1}{2}\sigma(f(\varepsilon_\mathbf{k}) - f(\varepsilon_{\mathbf{k}+\mathbf{q}})).$$

Summing over $\mathbf{k}$ and taking into account

$$\chi^0_{zz}(\mathbf{q}, \omega) = \frac{1}{2}\sum_\mathbf{k}(\chi^0_{zz}(\mathbf{k}\uparrow, \mathbf{q}, \omega) - \chi^0_{zz}(\mathbf{k}\downarrow, \mathbf{q}, \omega)),$$

we come to expression (3.53):

$$\chi^0_{zz}(\mathbf{q}, \omega) = \frac{1}{2}\sum_\mathbf{k}\frac{f(\varepsilon_\mathbf{k}) - f(\varepsilon_{\mathbf{k}+\mathbf{q}})}{\varepsilon_{\mathbf{k}+\mathbf{q}} - \varepsilon_\mathbf{k} - \hbar\omega}g^2\mu_\mathrm{B}^2.$$

For interacting electrons, we obtain (for details, see Appendix D.1)

$$\begin{aligned}\left[a^\dagger_{\mathbf{k}\sigma}a_{\mathbf{k}+\mathbf{q},\sigma}, \mathcal{H}_\mathrm{I}\right] = \frac{1}{2}\tilde{U}\Big(&\sum_{\mathbf{k}''\mathbf{p}\sigma''} a^\dagger_{\mathbf{k}\sigma}a^\dagger_{\mathbf{k}''\sigma''}a_{\mathbf{k}''+\mathbf{p},\sigma''}a_{\mathbf{k}+\mathbf{q}-\mathbf{p},\sigma} \\ &- \sum_{\mathbf{k}'\mathbf{p}\sigma'} a^\dagger_{\mathbf{k}\sigma}a^\dagger_{\mathbf{k}'\sigma'}a_{\mathbf{k}+\mathbf{q}+\mathbf{p},\sigma}a_{\mathbf{k}'-\mathbf{p},\sigma'} + \sum_{\mathbf{k}'\mathbf{p}\sigma'} a^\dagger_{\mathbf{k}'\sigma'}a^\dagger_{\mathbf{k}-\mathbf{p},\sigma}a_{\mathbf{k}'-\mathbf{p},\sigma'}a_{\mathbf{k}+\mathbf{q},\sigma} \\ &- \sum_{\mathbf{k}''\mathbf{p}\sigma''} a^\dagger_{\mathbf{k}+\mathbf{p},\sigma}a^\dagger_{\mathbf{k}''\sigma''}a_{\mathbf{k}''+\mathbf{p},\sigma''}a_{\mathbf{k}+\mathbf{q},\sigma}\Big). \end{aligned} \tag{6.48}$$

Clearly, substitution of this expression in the equation of motion would lead to averages of six creation-annihilation operators. Their time derivatives would give averages of more creation-annihilation operators, etc.

To split up the chain of differential equations we apply the mean-field approximation to the right-hand side of (6.48). Calculations lead to

$$\left[a^\dagger_{\mathbf{k}\sigma}a_{\mathbf{k}+\mathbf{q},\sigma}, \mathcal{H}_\mathrm{I}\right] = \tilde{U}(\bar{n}_{\mathbf{k}\sigma} - \bar{n}_{\mathbf{k}+\mathbf{q},\sigma})\sum_{\mathbf{k}'}a^\dagger_{\mathbf{k}'\bar{\sigma}}a_{\mathbf{k}'+\mathbf{q},\bar{\sigma}}. \tag{6.49}$$

Substituting (6.45), (6.46) and (6.49) in (6.44), we come to

$$\mathrm{i}\hbar\frac{\mathrm{d}}{\mathrm{d}t}\chi_{zz}(\mathbf{k}\sigma, \mathbf{q}, t) = -g^2\mu_\mathrm{B}^2\sigma\frac{1}{2}(\bar{n}_{\mathbf{k}\sigma} - \bar{n}_{\mathbf{k}+\mathbf{q},\sigma})\delta(t)$$

$$- g\mu_B(\varepsilon_{\mathbf{k+q}} - \varepsilon_{\mathbf{k}})\frac{i}{\hbar}\left\langle\left[\left(a^{\dagger}_{\mathbf{k}\sigma}a_{\mathbf{k+q},\sigma}\right)(t), \mathcal{M}_z(-\mathbf{q})\right]\right\rangle\theta(t)$$

$$- g\mu_B\tilde{U}(\bar{n}_{\mathbf{k}\sigma} - \bar{n}_{\mathbf{k+q},\sigma})\frac{i}{\hbar}\left\langle\left[\sum_{\mathbf{k}'}\left(a^{\dagger}_{\mathbf{k}'\bar{\sigma}}a_{\mathbf{k}'+\mathbf{q},\bar{\sigma}}\right)(t), \mathcal{M}_z(-\mathbf{q})\right]\right\rangle\theta(t),$$

or briefly,

$$i\hbar\frac{d}{dt}\chi_{zz}(\mathbf{k}\sigma, \mathbf{q}, t) = -g^2\mu_B^2\sigma\frac{1}{2}(\bar{n}_{\mathbf{k}\sigma} - \bar{n}_{\mathbf{k+q},\sigma})\delta(t)$$
$$+ (\varepsilon_{\mathbf{k+q}} - \varepsilon_{\mathbf{k}})\chi_{zz}(\mathbf{k}\sigma, \mathbf{q}, t) + \tilde{U}(\bar{n}_{\mathbf{k}\sigma} - \bar{n}_{\mathbf{k+q},\sigma})\sum_{\mathbf{k}'}\chi_{zz}(\mathbf{k}'\bar{\sigma}, \mathbf{q}, t). \tag{6.50}$$

Making the Fourier transform with respect to t, we write (6.50) as

$$(\hbar\omega - \varepsilon_{\mathbf{k+q}} + \varepsilon_{\mathbf{k}})\chi_{zz}(\mathbf{k}\sigma, \mathbf{q}, \omega)$$
$$= (\bar{n}_{\mathbf{k}\sigma} - \bar{n}_{\mathbf{k+q},\sigma})\left(-\sigma\frac{1}{2}g^2\mu_B^2 + \tilde{U}\sum_{\mathbf{k}'}\chi_{zz}(\mathbf{k}'\bar{\sigma}, \mathbf{q}, \omega)\right).$$

Summing over $\mathbf{k}$, we obtain

$$\chi_{zz}(\sigma, \mathbf{q}, \omega) = F_\sigma(\mathbf{q}, \omega)\left(\sigma\frac{1}{2}g^2\mu_B^2 - \tilde{U}\chi_{zz}(\bar{\sigma}, \mathbf{q}, \omega)\right), \tag{6.51}$$

where

$$F_\sigma(\mathbf{q}, \omega) = \sum_{\mathbf{k}}\frac{f(\varepsilon_{\mathbf{k}\sigma}) - f(\varepsilon_{\mathbf{k+q},\sigma})}{\varepsilon_{\mathbf{k+q}} - \varepsilon_{\mathbf{k}} - \hbar\omega}. \tag{6.52}$$

Here we used $\bar{n}_{\mathbf{k}\sigma} = f(\varepsilon_{\mathbf{k}\sigma})$, where $\varepsilon_{\mathbf{k}\sigma} = \varepsilon_{\mathbf{k}} - \sigma\tilde{U}\bar{S}_z$. Solving the system of two linear equations (6.51), we have

$$\chi_{zz}(\sigma, \mathbf{q}, \omega) = \frac{1}{2}\sigma\frac{F_\sigma(\mathbf{q}, \omega)(1 + \tilde{U}F_{\bar{\sigma}}(\mathbf{q}, \omega))}{1 - \tilde{U}^2F_\sigma(\mathbf{q}, \omega)F_{\bar{\sigma}}(\mathbf{q}, \omega)}g^2\mu_B^2.$$

Using

$$\chi_{zz}(\mathbf{q}, \omega) = \frac{1}{2}(\chi_{zz}(\uparrow, \mathbf{q}, \omega) - \chi_{zz}(\downarrow, \mathbf{q}, \omega)),$$

we obtain the RPA expression (5.11):

$$\chi_{zz}(\mathbf{q}, \omega) = \frac{1}{4}\frac{F_+(\mathbf{q}, \omega) + F_-(\mathbf{q}, \omega) + 2\tilde{U}F_+(\mathbf{q}, \omega)F_-(\mathbf{q}, \omega)}{1 - \tilde{U}^2F_+(\mathbf{q}, \omega)F_-(\mathbf{q}, \omega)}g^2\mu_B^2.$$

In the paramagnetic state $F_+(\mathbf{q}, \omega) = F_-(\mathbf{q}, \omega) = F(\mathbf{q}, \omega)$, the latter reduces to

$$\chi_{zz}(\mathbf{q}, \omega) = \frac{1}{2}\frac{F(\mathbf{q}, \omega)}{1 - \tilde{U}F(\mathbf{q}, \omega)}g^2\mu_B^2.$$

Transverse Susceptibility

We now apply the above method to calculate the transverse susceptibility (2.41). Consider the corresponding Green function

$$\chi_{-+}(\mathbf{q}, t) = \frac{i}{\hbar}\langle[\mathcal{M}_-(\mathbf{q}, t), \mathcal{M}_+(-\mathbf{q})]\rangle\theta(t),$$

where by formulae (3.29) we have

$$\mathcal{M}_+(\mathbf{q}) = -g\mu_B\sum_{\mathbf{k}}a^{\dagger}_{\mathbf{k}\uparrow}a_{\mathbf{k+q},\downarrow}, \qquad \mathcal{M}_-(\mathbf{q}) = -g\mu_B\sum_{\mathbf{k}}a^{\dagger}_{\mathbf{k}\downarrow}a_{\mathbf{k+q},\uparrow}.$$

Differentiating the function

$$\chi_{-+}(\mathbf{k},\mathbf{q},t) = -g\mu_{\mathrm{B}}\frac{\mathrm{i}}{\hbar}\langle\!\langle\big[\big(a^{\dagger}_{\mathbf{k}\downarrow}a_{\mathbf{k}+\mathbf{q},\uparrow}\big)(t),\mathcal{M}_{+}(-\mathbf{q})\big]\rangle\!\rangle\theta(t),$$

we obtain the equation of motion

$$\begin{aligned}\mathrm{i}\hbar\frac{\mathrm{d}}{\mathrm{d}t}\chi_{-+}(\mathbf{k},\mathbf{q},t) &= g\mu_{\mathrm{B}}\langle\!\langle\big[\big(a^{\dagger}_{\mathbf{k}\downarrow}a_{\mathbf{k}+\mathbf{q},\uparrow}\big)(t),\mathcal{M}_{+}(-\mathbf{q})\big]\rangle\!\rangle\delta(t)\\ &- g\mu_{\mathrm{B}}\frac{\mathrm{i}}{\hbar}\left\langle\!\!\left\langle\left[\mathrm{i}\hbar\frac{\mathrm{d}}{\mathrm{d}t}\big(a^{\dagger}_{\mathbf{k}\downarrow}a_{\mathbf{k}+\mathbf{q},\uparrow}\big)(t),\mathcal{M}_{+}(-\mathbf{q})\right]\right\rangle\!\!\right\rangle\theta(t).\end{aligned}$$

Taking the property $f(t)\delta(t) = f(0)\delta(t)$ into account and using Eq. (6.43), we have

$$\begin{aligned}\mathrm{i}\hbar\frac{\mathrm{d}}{\mathrm{d}t}\chi_{-+}(\mathbf{k},\mathbf{q},t) &= g\mu_{\mathrm{B}}\langle\!\langle\big[a^{\dagger}_{\mathbf{k}\downarrow}a_{\mathbf{k}+\mathbf{q},\uparrow},\mathcal{M}_{+}(-\mathbf{q})\big]\rangle\!\rangle\delta(t)\\ &- g\mu_{\mathrm{B}}\frac{\mathrm{i}}{\hbar}\left\langle\!\!\left\langle\Big[\big[a^{\dagger}_{\mathbf{k}\downarrow}a_{\mathbf{k}+\mathbf{q},\uparrow},\mathcal{H}\big](t),\mathcal{M}_{+}(-\mathbf{q})\Big]\right\rangle\!\!\right\rangle\theta(t).\end{aligned}\tag{6.53}$$

The first commutator is

$$\big[a^{\dagger}_{\mathbf{k}\downarrow}a_{\mathbf{k}+\mathbf{q},\uparrow},s_{+}(-\mathbf{q})\big] = \left[a^{\dagger}_{\mathbf{k}\downarrow}a_{\mathbf{k}+\mathbf{q},\uparrow},\sum_{\mathbf{k}'}a^{\dagger}_{\mathbf{k}'\uparrow}a_{\mathbf{k}'-\mathbf{q},\downarrow}\right] = a^{\dagger}_{\mathbf{k}\downarrow}a_{\mathbf{k}\downarrow} - a^{\dagger}_{\mathbf{k}+\mathbf{q},\uparrow}a_{\mathbf{k}+\mathbf{q},\uparrow}.\tag{6.54}$$

In the second term, the commutator with $\mathcal{H}_0$ is given by

$$\big[a^{\dagger}_{\mathbf{k}\downarrow}a_{\mathbf{k}+\mathbf{q},\uparrow},\mathcal{H}_0\big] = \left[a^{\dagger}_{\mathbf{k}\downarrow}a_{\mathbf{k}+\mathbf{q},\uparrow},\sum_{\mathbf{k}'\sigma'}\varepsilon_{\mathbf{k}'}a^{\dagger}_{\mathbf{k}'\sigma'}a_{\mathbf{k}'\sigma'}\right] = (\varepsilon_{\mathbf{k}+\mathbf{q}}-\varepsilon_{\mathbf{k}})a^{\dagger}_{\mathbf{k}\downarrow}a_{\mathbf{k}+\mathbf{q},\uparrow}.\tag{6.55}$$

The commutator with $\mathcal{H}_{\mathrm{I}}$ is written as (for details, see Appendix D.2)

$$\begin{aligned}\big[a^{\dagger}_{\mathbf{k}\downarrow}a_{\mathbf{k}+\mathbf{q},\uparrow},\mathcal{H}_{\mathrm{I}}\big] = \frac{1}{2}\tilde{U}\Bigg(&\sum_{\mathbf{k}''\mathbf{p}\sigma''}a^{\dagger}_{\mathbf{k}\downarrow}a^{\dagger}_{\mathbf{k}''\sigma''}a_{\mathbf{k}''+\mathbf{p},\sigma''}a_{\mathbf{k}+\mathbf{q}-\mathbf{p},\uparrow}\\ &-\sum_{\mathbf{k}'\mathbf{p}\sigma'}a^{\dagger}_{\mathbf{k}\downarrow}a^{\dagger}_{\mathbf{k}'\sigma'}a_{\mathbf{k}+\mathbf{q}+\mathbf{p},\uparrow}a_{\mathbf{k}'-\mathbf{p},\sigma'}+\sum_{\mathbf{k}'\mathbf{p}\sigma'}a^{\dagger}_{\mathbf{k}'\sigma'}a^{\dagger}_{\mathbf{k}-\mathbf{p},\downarrow}a_{\mathbf{k}'-\mathbf{p},\sigma'}a_{\mathbf{k}+\mathbf{q},\uparrow}\\ &-\sum_{\mathbf{k}''\mathbf{p}\sigma''}a^{\dagger}_{\mathbf{k}+\mathbf{p},\downarrow}a^{\dagger}_{\mathbf{k}''\sigma''}a_{\mathbf{k}''+\mathbf{p},\sigma''}a_{\mathbf{k}+\mathbf{q},\uparrow}\Bigg).\end{aligned}\tag{6.56}$$

To split up this chain of differential equations we apply the mean-field approximation to the right-hand side of (6.56). Calculations lead to

$$\big[a^{\dagger}_{\mathbf{k}\downarrow}a_{\mathbf{k}+\mathbf{q},\uparrow},\mathcal{H}_{\mathrm{I}}\big] = -\tilde{U}(\bar{n}_{\mathbf{k}\downarrow}-\bar{n}_{\mathbf{k}+\mathbf{q},\uparrow})\sum_{\mathbf{k}'}a^{\dagger}_{\mathbf{k}'\downarrow}a_{\mathbf{k}'+\mathbf{q},\uparrow} + \tilde{U}(N_{\downarrow}-N_{\uparrow})a^{\dagger}_{\mathbf{k}\downarrow}a_{\mathbf{k}+\mathbf{q},\uparrow}.\tag{6.57}$$

Substituting (6.54), (6.55) and (6.57) in (6.53), we finally obtain

$$\begin{aligned}\mathrm{i}\hbar\frac{\mathrm{d}}{\mathrm{d}t}\chi_{-+}(\mathbf{k},\mathbf{q},t) &= -g^2\mu_{\mathrm{B}}^2(\bar{n}_{\mathbf{k}\downarrow}-\bar{n}_{\mathbf{k}+\mathbf{q},\uparrow})\delta(t)\\ &+(\varepsilon_{\mathbf{k}+\mathbf{q},\uparrow}-\varepsilon_{\mathbf{k}\downarrow})\chi_{-+}(\mathbf{k},\mathbf{q},t) - \tilde{U}(\bar{n}_{\mathbf{k}\downarrow}-\bar{n}_{\mathbf{k}+\mathbf{q},\uparrow})\sum_{\mathbf{k}'}\chi_{-+}(\mathbf{k}',\mathbf{q},t),\end{aligned}\tag{6.58}$$

where $\varepsilon_{\mathbf{k}\downarrow} = \varepsilon_{\mathbf{k}} + \tilde{U}N_{\uparrow}$ and $\varepsilon_{\mathbf{k}+\mathbf{q},\uparrow} = \varepsilon_{\mathbf{k}+\mathbf{q}} + \tilde{U}N_{\downarrow}$. Using (4.16), we write

$$\varepsilon_{\mathbf{k}\sigma} = \varepsilon_{\mathbf{k}} + \frac{1}{2}\tilde{U}N_{\mathrm{e}} - \sigma\tilde{U}\bar{S}_z.$$

Omitting the spin-independent term $\frac{1}{2}\tilde{U} N_e$, we have $\varepsilon_{\mathbf{k}\sigma} = \varepsilon_{\mathbf{k}} - \sigma \tilde{U} \bar{S}_z$. The Fourier transform of Eq. (6.58) is

$$(\hbar\omega - \varepsilon_{\mathbf{k}+\mathbf{q},\uparrow} + \varepsilon_{\mathbf{k}\downarrow})\chi_{-+}(\mathbf{k},\mathbf{q},\omega) = -(\bar{n}_{\mathbf{k}\downarrow} - \bar{n}_{\mathbf{k}+\mathbf{q},\uparrow}) \left(g^2\mu_B^2 + \tilde{U} \sum_{\mathbf{k}'} \chi_{-+}(\mathbf{k}',\mathbf{q},\omega) \right).$$

Summing over $\mathbf{k}$ and using $\bar{n}_{\mathbf{k}\sigma} = f(\varepsilon_{\mathbf{k}\sigma})$, we obtain the equation

$$\chi_{-+}(\mathbf{q},\omega) = F_{-+}(\mathbf{q},\omega) \left(g^2\mu_B^2 + \tilde{U} \chi_{-+}(\mathbf{q},\omega) \right),$$

where

$$F_{-+}(\mathbf{q},\omega) = \sum_{\mathbf{k}} \frac{f(\varepsilon_{\mathbf{k}\downarrow}) - f(\varepsilon_{\mathbf{k}+\mathbf{q},\uparrow})}{\varepsilon_{\mathbf{k}+\mathbf{q},\uparrow} - \varepsilon_{\mathbf{k}\downarrow} - \hbar\omega}.$$

Hence

$$\chi_{-+}(\mathbf{q},\omega) = \frac{F_{-+}(\mathbf{q},\omega)}{1 - \tilde{U} F_{-+}(\mathbf{q},\omega)} g^2\mu_B^2.$$

For noninteracting electrons ($\tilde{U} = 0$), we obtain

$$\chi^0_{-+}(\mathbf{q},\omega) = F(\mathbf{q},\omega)\, g^2\mu_B^2. \tag{6.59}$$

6.2.2 Thermodynamic Susceptibility

In the previous subsection we used the equation of motion method to calculate the RPA susceptibilities in dealing with the interaction effects. To go beyond the RPA in the following chapters, we use the notion of the boson-type *temperature* Green function. This leads to an auxiliary function of the imaginary time called the *thermodynamic susceptibility*. In this subsection, first we relate the thermodynamic susceptibility to the thermodynamic spin correlator. Then we show how the thermodynamic susceptibility is related to the real-time susceptibility. The relation between the two susceptibilities is illustrated by explicit calculation of the thermodynamic susceptibility for noninteracting electrons.

Relation with the Spin Correlator

As we will see in the next chapters, in calculations it is more convenient to use the boson-type *temperature* Green function

$$\begin{aligned}\chi_{\alpha\beta}(\mathbf{q},\tau,\tau') &= \langle T_\tau\, \Delta\mathcal{M}_\alpha(\mathbf{q},\tau)\, \Delta\mathcal{M}_\beta(-\mathbf{q},\tau') \rangle \\ &\equiv \begin{cases} \langle \Delta\mathcal{M}_\alpha(\mathbf{q},\tau)\, \Delta\mathcal{M}_\beta(-\mathbf{q},\tau') \rangle, & \tau > \tau', \\ \langle \Delta\mathcal{M}_\beta(-\mathbf{q},\tau')\, \Delta\mathcal{M}_\alpha(\mathbf{q},\tau) \rangle, & \tau' > \tau, \end{cases}\end{aligned} \tag{6.60}$$

where $\langle \ldots \rangle$ is the canonical average, $\Delta\mathcal{M}_\alpha(\mathbf{q},\tau) = \mathrm{e}^{\mathcal{H}\tau/\hbar}\, \Delta\mathcal{M}_\alpha(\mathbf{q})\, \mathrm{e}^{-\mathcal{H}\tau/\hbar}$ is the "Heisenberg" representation with respect to the "time" $\tau \in [0, \hbar/T]$ and T_τ is the "time"-ordering operator.[5] Due to the cyclic property of trace in formula (6.60), the temperature Green function depends only on the "time" difference,

$$\chi_{\alpha\beta}(\mathbf{q},\tau,\tau') = \chi_{\alpha\beta}(\mathbf{q},\tau - \tau'). \tag{6.61}$$

Moreover, the following relation holds:

$$\chi_{\alpha\beta}(\mathbf{q},\tau - \hbar/T) = \chi_{\alpha\beta}(\mathbf{q},\tau)$$

for $\tau \in [0, \hbar/T]$. Therefore, the Fourier coefficients are nonzero only for even thermodynamic "frequencies" $\omega_m = 2\pi m T/\hbar$ and are given by

$$\chi_{\alpha\beta}(\mathbf{q},\mathrm{i}\omega_m) = \int_0^{\hbar/T} \chi_{\alpha\beta}(\mathbf{q},\tau)\, \mathrm{e}^{\mathrm{i}\omega_m\tau}\, \mathrm{d}\tau \tag{6.62}$$

[5] In the fermion-type *temperature* Green function (6.23) the "time"-ordering operator had the minus sign. Here the "time"-ordering operator gets the plus sign because each $\Delta\mathcal{M}_\alpha(\mathbf{q},\tau)$ consist of two creation-annihilation operators.

(the proof is similar to the one for the fermion-type *temperature* Green function, see Appendix C.2). Recalling $\mathcal{M}_\alpha(\mathbf{q}) = -g\mu_B s_\alpha(\mathbf{q})$, we write (6.60) as

$$\chi_{\alpha\beta}(\mathbf{q}, i\omega_m) = g^2\mu_B^2 \int_0^{\hbar/T} \langle \Delta s_\alpha(\mathbf{q}, \tau)\, \Delta s_\beta(-\mathbf{q})\rangle\, e^{i\omega_m\tau}\, d\tau. \tag{6.63}$$

Calculating the spin correlator in the momentum-"frequency" representation, we obtain

$$\langle \Delta s_\alpha(\mathbf{q}, i\omega_m)\Delta s_\beta(-\mathbf{q}, -i\omega_m)\rangle = \frac{T}{\hbar}\int_0^{\hbar/T} \langle \Delta s_\alpha(\mathbf{q}, \tau)\Delta s_\beta(-\mathbf{q})\rangle\, e^{i\omega_m\tau}\, d\tau, \tag{6.64}$$

where the Fourier transform is defined as

$$\Delta s_\alpha(\mathbf{q}, i\omega_m) = \frac{T}{\hbar}\int_0^{\hbar/T} \Delta s_\alpha(\mathbf{q}, \tau)\, e^{i\omega_m\tau}\, d\tau.$$

Using (6.64), we write the thermodynamic susceptibility (6.63) in the form [13]

$$\chi_{\alpha\beta}(\mathbf{q}, i\omega_m) = \frac{\hbar}{T}\langle \Delta s_\alpha(\mathbf{q}, i\omega_m)\Delta s_\beta(-\mathbf{q}, -i\omega_m)\rangle\, g^2\mu_B^2. \tag{6.65}$$

Relation with the Dynamic Susceptibility

Now we establish the relation between the functions $\chi_{\alpha\beta}(\mathbf{q}, \omega)$ and $\chi_{\alpha\beta}(\mathbf{q}, i\omega_m)$ using the analytic continuation method (see, e.g. [1, 10]). First, we derive an expression for $\chi_{\alpha\beta}(\mathbf{q}, \omega)$ in the energy representation. Using (6.40), we have

$$\chi_{\alpha\beta}(\mathbf{q}, t) = g^2\mu_B^2 \frac{i}{\hbar Z}\Big(\sum_{kk'} e^{iE_k t/\hbar}\big(\Delta s_\alpha(\mathbf{q})\big)_{kk'} e^{-iE_{k'}t/\hbar}\big(\Delta s_\beta(-\mathbf{q})\big)_{k'k} e^{-E_k/T}$$
$$- \sum_{kk'}\big(\Delta s_\beta(-\mathbf{q})\big)_{kk'} e^{iE_{k'}t/\hbar}\big(\Delta s_\alpha(\mathbf{q})\big)_{k'k} e^{-iE_k t/\hbar} e^{-E_k/T}\Big)\,\theta(t),$$

where E_k is an eigenvalue of $\mathcal{H}$ and $(\ldots)_{kk'}$ denotes the matrix element. Interchanging the indices k and k' in the second sum, we come to

$$\chi_{\alpha\beta}(\mathbf{q}, t) = g^2\mu_B^2 \frac{i}{\hbar Z}\Big(\sum_{kk'}\big(\Delta s_\alpha(\mathbf{q})\big)_{kk'}\big(\Delta s_\beta(-\mathbf{q})\big)_{k'k}\big(e^{-E_k/T} - e^{-E_{k'}/T}\big)e^{i(E_k - E_{k'})t/\hbar}\Big)\,\theta(t).$$

Calculating the Fourier transform (2.46), we finally obtain

$$\chi_{\alpha\beta}(\mathbf{q}, \omega) = g^2\mu_B^2 \frac{i}{\hbar Z}\sum_{kk'}\big(\Delta s_\alpha(\mathbf{q})\big)_{kk'}\big(\Delta s_\beta(-\mathbf{q})\big)_{k'k}\big(e^{-E_k/T} - e^{-E_{k'}/T}\big)$$
$$\times \lim_{\eta\to 0^+}\int_0^\infty e^{i((E_k - E_{k'})/\hbar + \omega + i\eta)t}\, dt$$
$$= -g^2\mu_B^2 \frac{\hbar}{Z}\sum_{kk'}\big(\Delta s_\alpha(\mathbf{q})\big)_{kk'}\big(\Delta s_\beta(-\mathbf{q})\big)_{k'k}\frac{e^{-E_k/T} - e^{-E_{k'}/T}}{E_k - E_{k'} + \hbar\omega + i0^+}. \tag{6.66}$$

Similarly, we write the temperature Green function (6.61) in the eigenbasis of the Hamiltonian $\mathcal{H}$:

$$\chi_{\alpha\beta}(\mathbf{q},\tau) = g^2\mu_B^2 \frac{1}{Z}\sum_{kk'} \left(\Delta s_\alpha(\mathbf{q})\right)_{kk'} \left(\Delta s_\beta(-\mathbf{q})\right)_{k'k} \mathrm{e}^{-E_k/T}\, \mathrm{e}^{(E_k-E_{k'})\tau/\hbar}.$$

Calculating the Fourier transform (6.62), we have

$$\chi_{\alpha\beta}(\mathbf{q},\mathrm{i}\omega_m) = -g^2\mu_B^2 \frac{\hbar}{Z}\sum_{kk'} \left(\Delta s_\alpha(\mathbf{q})\right)_{kk'} \left(\Delta s_\beta(-\mathbf{q})\right)_{k'k} \frac{\mathrm{e}^{-E_k/T} - \mathrm{e}^{-E_{k'}/T}}{E_k - E_{k'} + \mathrm{i}\hbar\omega_m}. \tag{6.67}$$

Expressions (6.66) and (6.67) are special cases of the function $\chi_{\alpha\beta}(\mathbf{q}, z)$, which is analytic in the upper half-plane $\mathrm{Im} z > 0$. Thus, (6.66) is obtained from (6.67) by the analytic continuation

$$\chi_{\alpha\beta}(\mathbf{q},\mathrm{i}\omega_m)\Big|_{\mathrm{i}\omega_m\to\omega+\mathrm{i}0^+} = \chi_{\alpha\beta}(\mathbf{q},\omega). \tag{6.68}$$

Noninteracting Electrons

We illustrate the analytic continuation formula (6.68) by considering noninteracting electrons. The real-time susceptibility of noninteracting electrons $\chi^0_{zz}(\mathbf{q},\omega)$ was calculated in Chap. 3. Let us calculate the thermodynamic susceptibility $\chi^0_{zz}(\mathbf{q},\mathrm{i}\omega_m)$. Using (6.65), we write the inverse Fourier transform of $\chi^0_{zz}(\mathbf{q},\mathrm{i}\omega_m)$ as

$$\begin{aligned}\chi^0_{zz}(\mathbf{q},\tau) &= g^2\mu_B^2 \langle T_\tau\, \Delta s_z(\mathbf{q},\tau)\, \Delta s_z(-\mathbf{q})\rangle \\ &= g^2\mu_B^2 \left(\langle T_\tau\, s_z(\mathbf{q},\tau)\, s_z(-\mathbf{q})\rangle - \langle s_z(\mathbf{q},\tau)\rangle\langle s_z(-\mathbf{q})\rangle\right),\end{aligned}$$

where $\langle\ldots\rangle$ is the canonical average with the Hamiltonian $\mathcal{H}_0$. Averaging formula (3.32), we obtain $\langle s_z(\mathbf{q},\tau)\rangle = N\bar{s}_z\delta_{\mathbf{q}0}$. Now, using the expression for the spin operator (3.45), we have

$$\chi^0_{zz}(\mathbf{q},\tau) = g^2\mu_B^2 \left(\frac{1}{4}\sum_{\mathbf{k}\mathbf{k}'\sigma\sigma'} \sigma\sigma' \langle T_\tau\, a^\dagger_{\mathbf{k}\sigma}(\tau) a_{\mathbf{k}+\mathbf{q},\sigma}(\tau)\, a^\dagger_{\mathbf{k}'\sigma'} a_{\mathbf{k}'-\mathbf{q},\sigma'}\rangle - N^2\bar{s}_z^2\delta_{\mathbf{q}0}\right). \tag{6.69}$$

The problem now is to evaluate the thermal average of four creation and annihilation operators $\langle T_\tau\, a^\dagger_{\mathbf{k}\sigma}(\tau) a_{\mathbf{l},\sigma}(\tau)\, a^\dagger_{\mathbf{k}'\sigma'} a_{\mathbf{l}'\sigma'}\rangle$ appearing in the integrand. Wick's theorem tells us how to reduce this average to a sum of products of pair averages $\langle T_\tau\, a^\dagger_{\mathbf{k}\sigma}(\tau) a_{\mathbf{l},\sigma}\rangle$ (see, e.g. [9]). Using Wick's theorem, we obtain

$$\begin{aligned}\langle T_\tau\, a^\dagger_{\mathbf{k}\sigma}(\tau) a_{\mathbf{k}+\mathbf{q},\sigma}(\tau)\, a^\dagger_{\mathbf{k}'\sigma'} a_{\mathbf{k}'-\mathbf{q},\sigma'}\rangle &= \langle a^\dagger_{\mathbf{k}\sigma}(\tau) a_{\mathbf{k}+\mathbf{q},\sigma}(\tau)\rangle\langle a^\dagger_{\mathbf{k}'\sigma'} a_{\mathbf{k}'-\mathbf{q},\sigma'}\rangle \\ &\quad - \langle T_\tau\, a^\dagger_{\mathbf{k}\sigma}(\tau) a_{\mathbf{k}'-\mathbf{q},\sigma'}\rangle\langle T_\tau\, a^\dagger_{\mathbf{k}'\sigma'} a_{\mathbf{k}+\mathbf{q},\sigma}(\tau)\rangle.\end{aligned} \tag{6.70}$$

In the second term we had to swap the creation-annihilation operators three times, which gives us the minus sign; in the first term no swaps were needed. Substituting (6.70) in (6.69) and taking the translational invariance (4.12) into account, we have

$$\begin{aligned}\chi^0_{zz}(\mathbf{q},\tau) = g^2\mu_B^2 \Bigg(&\frac{1}{4}\delta_{\mathbf{q}0}\sum_{\mathbf{k}\sigma}\sigma\langle a^\dagger_{\mathbf{k}\sigma} a_{\mathbf{k}\sigma}\rangle \sum_{\mathbf{k}'\sigma'}\sigma'\langle a^\dagger_{\mathbf{k}'\sigma'} a_{\mathbf{k}'\sigma'}\rangle \\ &- \frac{1}{4}\sum_{\mathbf{k}\sigma}\langle T_\tau\, a^\dagger_{\mathbf{k}\sigma}(\tau) a_{\mathbf{k}\sigma}\rangle\langle T_\tau\, a^\dagger_{\mathbf{k}+\mathbf{q},\sigma}(-\tau) a_{\mathbf{k}+\mathbf{q},\sigma}\rangle - N^2\bar{s}_z^2\delta_{\mathbf{q}0}\Bigg).\end{aligned}$$

The first and last terms on the right-hand side cancel. In the second term, the operators can be treated as commuting, because they are acted upon by the operator T_τ. Hence, using the cyclic property of trace, we write

$$\chi_{zz}^0(\mathbf{q},\tau) = -g^2\mu_B^2\frac{1}{4}\sum_{\mathbf{k}\sigma}\langle T_\tau\, a_{\mathbf{k}\sigma}(-\tau)a_{\mathbf{k}\sigma}^\dagger\rangle\langle T_\tau\, a_{\mathbf{k}+\mathbf{q},\sigma}(\tau)a_{\mathbf{k}+\mathbf{q},\sigma}^\dagger\rangle.$$

Since for noninteracting electrons the averages are spin-independent, by the definition of the Green function (6.23), we have

$$\chi_{zz}^0(\mathbf{q},\tau) = -g^2\mu_B^2\frac{1}{2}\sum_{\mathbf{k}} G_{\mathbf{k}}^0(-\tau)G_{\mathbf{k}+\mathbf{q}}^0(\tau).$$

Performing the inverse Fourier transform of the Green functions by the formula (C.31), we obtain

$$\chi_{zz}^0(\mathbf{q},\mathrm{i}\omega_m) = -g^2\mu_B^2\frac{1}{2}\frac{T}{\hbar}\sum_{\mathbf{k}n} G_{\mathbf{k}}^0(\mathrm{i}\omega_n)G_{\mathbf{k}+\mathbf{q}}^0(\mathrm{i}\omega_n+\mathrm{i}\omega_m). \tag{6.71}$$

Summation over the odd "frequencies" ω_n in expression (6.71) leads to (for details, see Appendix A.2.5)

$$\chi_{zz}^0(\mathbf{q},\mathrm{i}\omega_m) = -g^2\mu_B^2\frac{1}{2}\sum_{\mathbf{k}}\left(\frac{f(\varepsilon_{\mathbf{k}})}{\varepsilon_{\mathbf{k}}-\varepsilon_{\mathbf{k}+\mathbf{q}}+\mathrm{i}\hbar\omega_m}+\frac{f(\varepsilon_{\mathbf{k}+\mathbf{q}}-\mathrm{i}\hbar\omega_m)}{\varepsilon_{\mathbf{k}+\mathbf{q}}-\mathrm{i}\hbar\omega_m-\varepsilon_{\mathbf{k}}}\right).$$

Taking into account the Fermi function property $f(\varepsilon-\mathrm{i}\hbar\omega_m) = f(\varepsilon)$, we obtain

$$\chi_{zz}^0(\mathbf{q},\mathrm{i}\omega_m) = \frac{1}{2}g^2\mu_B^2\sum_{\mathbf{k}}\frac{f(\varepsilon_{\mathbf{k}})-f(\varepsilon_{\mathbf{k}+\mathbf{q}})}{\varepsilon_{\mathbf{k}+\mathbf{q}}-\varepsilon_{\mathbf{k}}-\mathrm{i}\hbar\omega_m} = \frac{1}{2}g^2\mu_B^2F(\mathbf{q},\mathrm{i}\omega_m). \tag{6.72}$$

Replacing $\mathrm{i}\omega_m$ by $\omega+\mathrm{i}0^+$ in the expression (6.72), we come to the real-time unenhanced susceptibility (3.53). Here the analytic continuation consisted in just replacing $\mathrm{i}\omega_m$ by $\omega+\mathrm{i}0^+$, because the right-hand side is a rational function (a fraction of analytic functions). If the function is not rational, one has to approximate it by a rational function (e.g. by the Padé approximation).

Summation Rule

In the DSFT we deal with local characteristics, such as magnetization or local moment, and it suffices to use a particular form of the analytic continuation known as the summation rules. Here we derive the summation rule for the boson-type Green function (magnetic susceptibility). Summing (6.64) over the "frequencies" and taking into account $(T/\hbar)\sum_m \mathrm{e}^{\mathrm{i}\omega_m\tau} = \delta(\tau)$, we have

$$\begin{aligned}\sum_m\langle\Delta s_\alpha(\mathbf{q},\mathrm{i}\omega_m)\Delta s_\beta(-\mathbf{q},-\mathrm{i}\omega_m)\rangle &= \int_0^{\hbar/T}\langle\Delta s_\alpha((\mathbf{q}),\tau)\Delta s_\beta(-\mathbf{q})\rangle\delta(\tau)\,\mathrm{d}\tau\\ &= \langle\Delta s_\alpha(\mathbf{q})\Delta s_\beta(-\mathbf{q})\rangle.\end{aligned}$$

Using (6.65), we come to

$$g^2\mu_B^2\langle\Delta s_\alpha(\mathbf{q})\Delta s_\beta(-\mathbf{q})\rangle = \frac{T}{\hbar}\sum_m\chi_{\alpha\beta}(\mathbf{q},\mathrm{i}\omega_m). \tag{6.73}$$

In particular, the local spin fluctuation is given by

$$\langle(\Delta s_j^\alpha)^2\rangle = \frac{1}{N^2}\sum_{\mathbf{q}}\langle\Delta s_\alpha(\mathbf{q})\Delta s_\alpha(-\mathbf{q})\rangle = \frac{1}{g^2\mu_B^2}\frac{1}{N^2}\frac{T}{\hbar}\sum_{\mathbf{q}m}\chi_{\alpha\alpha}(\mathbf{q},\mathrm{i}\omega_m).$$

On the other hand, taking $t=0$ in the fluctuation-dissipation theorem (2.51), we have

$$g^2\mu_B^2\langle\Delta s_\alpha(\mathbf{q})\Delta s_\beta(\mathbf{q})\rangle = \frac{1}{2\pi}\int\frac{\mathrm{i}\hbar}{\mathrm{e}^{-\hbar\omega/T}-1}\left[\chi_{\alpha\beta}(\mathbf{q},\omega)-(\chi_{\beta\alpha}(\mathbf{q},\omega))^*\right]\mathrm{d}\omega. \tag{6.74}$$

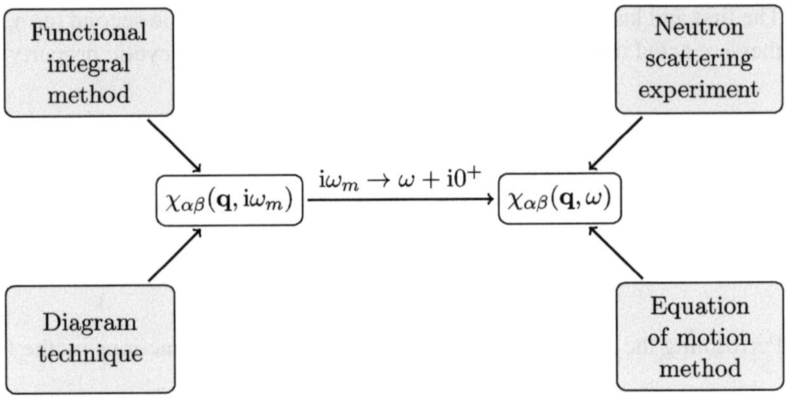

Fig. 6.1 Magnetic susceptibility $\chi_{\alpha\beta}(\mathbf{q}, \omega)$ measurement and calculation methods

Comparing (6.73) and (6.74), we write

$$\frac{T}{\hbar}\sum_m \chi_{\alpha\beta}(\mathbf{q}, \mathrm{i}\omega_m) = \frac{1}{2\pi}\int \frac{\mathrm{i}\hbar}{\mathrm{e}^{-\hbar\omega/T}-1}\left[\chi_{\alpha\beta}(\mathbf{q}, \omega) - \left(\chi_{\beta\alpha}(\mathbf{q}, \omega)\right)^*\right] \mathrm{d}\omega.$$

In particular, for $\alpha = \beta$, we have

$$\frac{T}{\hbar}\sum_m \chi_{\alpha\alpha}(\mathbf{q}, \mathrm{i}\omega_m) = -\frac{1}{\pi}\int \frac{\hbar}{\mathrm{e}^{-\hbar\omega/T}-1}\mathrm{Im}\chi_{\alpha\alpha}(\mathbf{q}, \omega)\, \mathrm{d}\omega. \tag{6.75}$$

As we have shown, imaginary part of the susceptibility is an odd function of frequency (2.10). Therefore, changing variables $\omega \to -\omega$ in the integral (6.75) and using (2.10), we obtain the sum rule over the even "frequencies":

$$\frac{T}{\hbar}\sum_m \chi_{\alpha\alpha}(\mathbf{q}, \mathrm{i}\omega_m) = \frac{1}{\pi}\int \frac{\hbar}{\mathrm{e}^{\hbar\omega/T}-1}\mathrm{Im}\chi_{\alpha\alpha}(\mathbf{q}, \omega)\, \mathrm{d}\omega.$$

Measurement and Calculation Methods

Closing this chapter, we briefly discuss methods for measuring and calculating the magnetic susceptibility (Fig. 6.1). Experimentally the magnetic susceptibility $\chi_{\alpha\beta}(\mathbf{q}, \omega)$ can be obtained from the neutron scattering measurements, which we discuss in Chaps. 14 and 15. One method for calculating $\chi_{\alpha\beta}(\mathbf{q}, \omega)$ is the equation of motion method. In practice the equation of motion method is rarely used beyond the RPA (for ferromagnets with localized spins, see, e.g. [14, 15]). Instead of the dynamic susceptibility $\chi_{\alpha\beta}(\mathbf{q}, \omega)$, we can calculate the thermodynamic susceptibility $\chi_{\alpha\beta}(\mathbf{q}, \mathrm{i}\omega_m)$ and use the analytic continuation (6.68). There are two general methods of calculating the thermodynamic susceptibility $\chi_{\alpha\beta}(\mathbf{q}, \mathrm{i}\omega_m)$. The first method is the *diagram technique* and the second one is the *functional integral method.*

The *diagram technique* is a perturbation method based on Wick's theorem (see, e.g. [3, 8–10]). It has been used for weak ferromagnets, which have small Curie temperature compared with the Fermi energy ($T_C \ll \varepsilon_F$). In weak ferromagnets one can go beyond the RPA by applying the Fermi liquid theory [16] or self-consistent renormalization (SCR) theory [17] and their generalizations (for a review, see, e.g. [18–20]); both approaches are phenomenological. In magnetism of metals at finite temperature these theories are insufficient.

To obtain a quantum-statistical description of metallic magnetism at finite temperatures the *functional integral method* is applied. The method is based on the Stratonovich-Hubbard transformation (see Chap. 8), and diagrams are used solely for pictorial purposes.

References

1. A.A. Abrikosov, L.P. Gorkov, I.E. Dzyaloshinski, *Methods of Quantum Field Theory in Statistical Physics* (Prentice-Hall, Englewood Cliffs, 1963)
2. A.L. Fetter, J.D. Walecka, *Quantum Theory of Many-Particle Systems* (McGraw-Hill, New York, 1971)
3. R.D. Mattuck, *A Guide to Feynman Diagrams in the Many-Body Problem*, 2nd edn. (McGraw-Hill, New York, 1976)
4. E.M. Lifshitz, L.P. Pitaevskii, *Statistical Physics, Part 2. In Landau and Lifshitz Course on Theoretical Physics*, vol. 9 (Pergamon, Oxford, 1980)
5. G. Rickayzen, *Green's Functions and Condensed Matter* (Academic, London, 1980)
6. S. Doniach, E.H. Sondheimer, *Green's Functions for Solid State Physicists*, 2nd edn. (Imperial College Press, London, 1998)
7. G.D. Mahan, *Many-Particle Physics*, 3rd edn. (Kluwer/Plenum, New York, 2000)
8. S. Raimes, *Many-Electron Theory* (North-Holland, Amsterdam, 1972)
9. D.J. Kim, *New Perspectives in Magnetism of Metals* (Kluwer/Plenum, New York, 1999)
10. H. Bruus, K. Flensberg, *Many-Body Quantum Theory in Condensed Matter Physics*, 3rd edn. (Oxford University Press, Oxford, 2004)
11. T. Matsubara, Prog. Theor. Phys. **14**(4), 351 (1955)
12. T. Izuyama, D.J. Kim, R. Kubo, J. Phys. Soc. Jpn. **18**(7), 1025 (1963)
13. N.B. Melnikov, B.I. Reser, Theor. Math. Phys. **181**(2), 1435 (2014)
14. S.V. Tyablikov, *Methods in the Quantum Theory of Magnetism* (Springer, New York, 1967)
15. N.M. Plakida, Theor. Math. Phys. **168**, 1303 (2011)
16. I.E. Dzyaloshinskii, P.S. Kondratenko, Sov. Phys. JETP **43**, 1036 (1976)
17. T. Moriya, A. Kawabata, J. Phys. Soc. Jpn. **35**(3), 669 (1973)
18. T. Moriya, *Spin Fluctuations in Itinerant Electron Magnetism* (Springer, Berlin, 1985)
19. R.M. White, *Quantum Theory of Magnetism*, 3rd edn. (Springer, Berlin, 2007)
20. Y. Takahashi, *Spin Fluctuation Theory of Itinerant Electron Magnetism* (Springer, Berlin, 2013)

Spin Fluctuation Theory in the Ising Model

7

Theories come and go, but examples stay forever. (I.M. Gelfand)

Here we illustrate spin fluctuation theory in the example of the Ising model (see, e.g. [1,2]), where spins are treated classically and the free energy has a simple analytic form. In spin fluctuation theory, we replace the pair interaction by the interaction of spins with a fluctuating exchange field. The calculation of magnetic characteristics, such as magnetization or local moment, is carried out in two steps. Firstly, we calculate the magnetic characteristic in the system of *noninteracting* spins with a fixed exchange field configuration. Secondly, we average over all possible field configurations with the probability density given by the free energy. Calculating the averages over field configurations requires an approximation of the free energy. The simplest one leads to the mean-field theory, which neglects the feedback of the spin fluctuations on the mean field. The effect of spin fluctuations on the mean field leads to a reduction of the Curie temperature. When the fluctuations become strong, a discontinuous first-order phase transition can appear in the Gaussian approximation. By taking into account higher-order terms of the free energy, we obtain the second-order phase transition, which is experimentally observed in metals (for details, see [3,4]).

7.1 Spins in the Fluctuating Field

The Hamiltonian of the Ising model is

$$\mathcal{H} = -\frac{1}{2}\sum_{jj'} J_{jj'} S_j S_{j'}, \tag{7.1}$$

where $S_j = \pm 1/2$ is the spin at the site $\mathbf{R}_j$ of a three-dimensional crystal lattice and $J_{jj'} = J_{j-j'}$ is the interaction coefficient (Fig. 7.1). Using the cyclic boundary conditions, we write Hamiltonian (7.1) in the Fourier representation:

$$\mathcal{H} = -\frac{1}{2N}\sum_{\mathbf{q}} J_{\mathbf{q}} |S_{\mathbf{q}}|^2, \tag{7.2}$$

where $\mathbf{q}$ is the wavevector taking values in the Brillouin zone, N is the number of lattice sites, and the discrete Fourier transform of an arbitrary function f_j is defined by the formulae

$$f_{\mathbf{q}} = \sum_j f_j \mathrm{e}^{-\mathrm{i}\mathbf{q}\mathbf{R}_j}, \qquad f_j = \frac{1}{N}\sum_{\mathbf{q}} f_{\mathbf{q}} \mathrm{e}^{\mathrm{i}\mathbf{q}\mathbf{R}_j}. \tag{7.3}$$

The partition function is given by

$$Z = \sum_{\mathbf{S}} \mathrm{e}^{-\mathcal{H}/T} \equiv \mathrm{Tr}\mathrm{e}^{-\mathcal{H}/T},$$

N. B. Melnikov, B. I. Reser, *Dynamic Spin Fluctuation Theory of Metallic Magnetism*,
https://doi.org/10.1007/978-3-319-92974-3_7

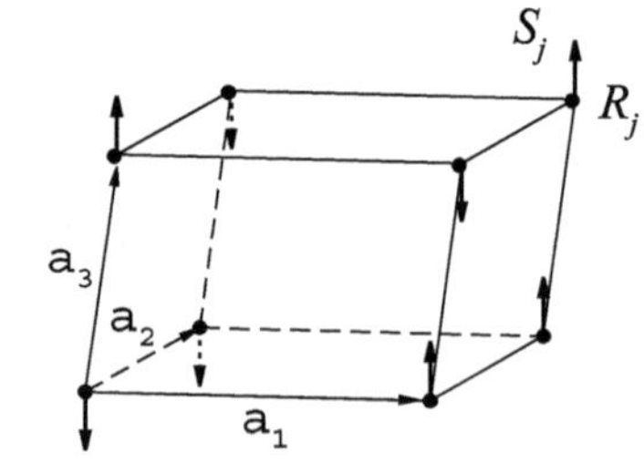

Fig. 7.1 Sketch of a unit cell in a three-dimensional crystal lattice for the Ising model

where $\mathbf{S} = (S_1, \ldots, S_N)$ is the spin configuration and Tr denotes the sum over 2^N possible spin configurations of the system. Using representation (7.2) of the Hamiltonian $\mathcal{H}$ as a sum of squares, we write the partition function as[1]

$$Z = \operatorname{Tr} \prod_{\mathbf{q}} \exp\Big(\frac{J_{\mathbf{q}}}{2NT}|S_{\mathbf{q}}|^2\Big). \tag{7.4}$$

To calculate the partition function (7.4) we use the Stratonovich-Hubbard transformation [6,7], which consists in replacing the pair interaction of spins with the interaction of spins with a fluctuating field $\mathbf{V} = (V_{\mathbf{q}_1}, \ldots, V_{\mathbf{q}_N})$. The key element of the Stratonovich-Hubbard transformation is the identity

$$\exp\Big(\frac{A^2}{a}\Big) = \sqrt{\frac{a}{\pi}} \int \exp\big(-ax^2 + 2Ax\big)\, \mathrm{d}x, \tag{7.5}$$

which is valid for any real A and $a > 0$. Applying identity (7.5) with $a = 1/2$ and $A = \frac{1}{2} S_{\mathbf{q}} \sqrt{J_{\mathbf{q}}/(NT)}$, we obtain[2]

$$Z = \Big[\int \exp\Big(-\sum_{\mathbf{q}} \frac{|z_{\mathbf{q}}|^2}{2}\Big) \mathrm{d}\mathbf{z}\Big]^{-1} \operatorname{Tr}\Big[\int \exp\Big(-\sum_{\mathbf{q}} \frac{|z_{\mathbf{q}}|^2}{2} + \sqrt{\frac{J_{\mathbf{q}}}{NT}} \sum_{\mathbf{q}} S_{\mathbf{q}} z_{-\mathbf{q}}\Big) \mathrm{d}\mathbf{z}\Big] \tag{7.6}$$

(for details, see Appendix B.3.1). Introducing the new variable $V_{\mathbf{q}} = z_{\mathbf{q}}\sqrt{NTJ_{\mathbf{q}}}$, which has dimensions of energy, we finally come to

$$Z = \Big(\int \mathrm{e}^{-F_0(\mathbf{V})/T}\, \mathrm{d}\mathbf{V}\Big)^{-1} \int \mathrm{e}^{-(F_0(\mathbf{V})+F_1(\mathbf{V}))/T}\, \mathrm{d}\mathbf{V}. \tag{7.7}$$

Here,

$$F_0(\mathbf{V}) = \frac{1}{2N} \sum_{\mathbf{q}} \frac{|V_{\mathbf{q}}|^2}{J_{\mathbf{q}}} \tag{7.8}$$

is the energy of the field and

$$F_1(\mathbf{V}) = -T \ln \operatorname{Tr} \mathrm{e}^{-\mathcal{H}(\mathbf{V})/T} \tag{7.9}$$

is the free energy of the system of the *noninteracting* spins in the field $\mathbf{V}$, where

$$\mathcal{H}(\mathbf{V}) = -\frac{1}{N} \sum_{\mathbf{q}} S_{\mathbf{q}} V_{-\mathbf{q}} \tag{7.10}$$

is the Hamiltonian of this system. Since the spins are now independent, the free energy $F_1(\mathbf{V})$ can be calculated explicitly. Rewriting Hamiltonian (7.10) in the site representation

$$\mathcal{H}(\mathbf{V}) = -\sum_j S_j V_j, \tag{7.11}$$

[1] Similar results in the Heisenberg model require additional approximations in the functional integral method [5].

[2] Here, the Fourier coefficients $J_{\mathbf{q}}$ are real. They can all be made positive by shifting energy of the system (7.2) by a constant amount, which is proportional to $\sum_{\mathbf{q}} |S_{\mathbf{q}}|^2$.

we calculate the trace on the right-hand side of (7.9):

$$\mathrm{Tr}\mathrm{e}^{-\mathcal{H}(\mathbf{V})/T} = \sum_{\mathbf{S}} \exp\Bigl(\sum_j \frac{S_j V_j}{T}\Bigr) = \prod_j \sum_{S_j=\pm 1/2} \exp\Bigl(\frac{S_j V_j}{T}\Bigr).$$

Thus, free energy (7.9) is given by

$$F_1(\mathbf{V}) = -T \sum_j \ln\Bigl(2\cosh\Bigl(\frac{V_j}{2T}\Bigr)\Bigr). \tag{7.12}$$

In spin fluctuation theory, an observable $\langle A \rangle \equiv \mathrm{Tr}(A\,\mathrm{e}^{-\mathcal{H}/T})/\mathrm{Tr}\mathrm{e}^{-\mathcal{H}/T}$ is calculated by

$$\langle A \rangle = \int A(\mathbf{V})\, p(\mathbf{V})\, \mathrm{d}\mathbf{V}, \tag{7.13}$$

where the quantum-mechanical average $A(\mathbf{V})$ in the system of independent spins in the presence of the field $\mathbf{V}$ is defined as

$$A(\mathbf{V}) = \mathrm{Tr}(A\,\mathrm{e}^{-\mathcal{H}(\mathbf{V})/T}) \big/ \mathrm{Tr}\mathrm{e}^{-\mathcal{H}(\mathbf{V})/T},$$

and the probability density $p(\mathbf{V})$ is given by

$$p(\mathbf{V}) \propto \exp(-F(\mathbf{V})/T), \qquad F(\mathbf{V}) = F_0(\mathbf{V}) + F_1(\mathbf{V}). \tag{7.14}$$

One of the main advantages of spin fluctuation theory is the possibility to express the mean spin and spin correlators in terms of the mean field and field correlators. The mean spin

$$\bar{S} = Z^{-1}\mathrm{Tr}(S_j\,\mathrm{e}^{-\mathcal{H}/T})$$

and mean field

$$\bar{V} = \Bigl(\int \mathrm{e}^{-F(\mathbf{V})/T}\,\mathrm{d}\mathbf{V}\Bigr)^{-1} \int V_j\,\mathrm{e}^{-F(\mathbf{V})/T}\,\mathrm{d}\mathbf{V}$$

are related by the formula $\bar{V} = J_0 \bar{S}$, where J_0 is the zeroth Fourier coefficient, and the spin and field correlators are related by the formula

$$\langle |\Delta S_{\mathbf{q}}|^2 \rangle = \frac{1}{J_{\mathbf{q}}^2} \langle |\Delta V_{\mathbf{q}}|^2 \rangle - \frac{NT}{J_{\mathbf{q}}},$$

where $\Delta S_{\mathbf{q}} = S_{\mathbf{q}} - \langle S_{\mathbf{q}} \rangle$ and $\Delta V_{\mathbf{q}} = V_{\mathbf{q}} - \langle V_{\mathbf{q}} \rangle$ (for details, see [8, Appendix A]). Calculating partition function (7.7) and mean values (7.13) in practice requires the Gaussian approximation.

7.2 Approximations of the Free Energy

7.2.1 Quadratic Part of the Free Energy

The Gaussian approximation of the fluctuating field $\mathbf{V}$ with probability density (7.14) is defined as follows. The function $F(\mathbf{V})$ is replaced with the translation-invariant quadratic form

$$F^{(2)}(\mathbf{V}) = \sum_{jj'} \Delta V_j A_{j-j'} \Delta V_{j'} = \sum_{\mathbf{q}} \Delta V_{\mathbf{q}} A_{\mathbf{q}} \Delta V_{-\mathbf{q}},$$

which determines the Gaussian fluctuating field with the probability density

$$p^{(2)}(\mathbf{V}) = \frac{1}{Z^{(2)}}\mathrm{e}^{-F^{(2)}(\mathbf{V})/T}, \qquad Z^{(2)} = \int \mathrm{e}^{-F^{(2)}(\mathbf{V})/T}\,\mathrm{d}\mathbf{V}.$$

The simplest Gaussian approximation is given by the saddle-point method (see, e.g. [9]). The mean field $\bar{\mathbf{V}}$ is obtained from the local minimum condition, and the value of $A_{\mathbf{q}}$ is determined by the second derivative of the function $F(\mathbf{V})$ at the mean field:

$$\frac{\partial F(\bar{\mathbf{V}})}{\partial V_j} = 0, \qquad A_{\mathbf{q}} = \frac{1}{2}\frac{\partial^2 F(\bar{\mathbf{V}})}{\partial V_{\mathbf{q}} \partial V_{-\mathbf{q}}}. \tag{7.15}$$

Using formula (7.12) in the first relation of (7.15), we obtain the mean-field equation

$$\bar{V} = \frac{J_0}{2} \tanh\left(\frac{\bar{V}}{2T}\right). \tag{7.16}$$

At small T this equation has a stable nonzero solution $\bar{V} > 0$, which corresponds to the ferromagnetic state. Paramagnetic solution $\bar{V} = 0$ always exists, and it is stable at high T. At the phase transition from the ferro- to paramagnetic state, the two solutions merge. That means that the functions on the left- and right-hand sides of Eq. (7.16) are tangent to each other at $\bar{V} = 0$. From the tangency condition, the phase transition temperature in the mean-field theory is $T_C^{MF} = J_0/4$. For fluctuating field integral (7.13), the saddle-point method gives $\langle A \rangle = A(\bar{\mathbf{V}})$, where $\bar{\mathbf{V}} = (\bar{V}, \ldots, \bar{V})$ is the mean-field configuration. The mean-field value is the leading term in the asymptotic expansion of the integral (7.13) as $T \to 0$.

The *optimal Gaussian approximation* [10–12] of the fluctuating field is applicable in a wider range of temperatures. The parameters $\bar{V}$ and $A_{\mathbf{q}}$ of the optimal Gaussian approximation are obtained from the system of nonlinear equations (see Appendix B.4.2)

$$\left\langle \frac{\partial F(\mathbf{V})}{\partial V_{\mathbf{q}}} \right\rangle_{(2)} = 0, \qquad A_{\mathbf{q}} = \frac{1}{2}\left\langle \frac{\partial^2 F(\mathbf{V})}{\partial V_{\mathbf{q}} \partial V_{-\mathbf{q}}} \right\rangle_{(2)}, \tag{7.17}$$

where the mean values are calculated by the formula

$$\langle \ldots \rangle_{(2)} = \int (\ldots) p^{(2)}(\mathbf{V})\, d\mathbf{V}.$$

In the Ising model, system of nonlinear equations (7.17) can be written as

$$\bar{V} = -J_0 \left\langle \frac{\partial F_1(\mathbf{V})}{\partial V_j} \right\rangle_{(2)}, \qquad A_{\mathbf{q}} = \frac{1}{2N}\left(\frac{1}{J_{\mathbf{q}}} + \left\langle \frac{\partial^2 F_1(\mathbf{V})}{\partial V_j^2} \right\rangle_{(2)} \right), \tag{7.18}$$

where the mean values are independent of the index j.

7.2.2 Higher-Order Terms of the Free Energy

At a high temperature, it is necessary to consider the fourth-order terms in the expansion of the function $F(\mathbf{V})$, as is done in the Landau phase transition theory (see, e.g. [13, 14]):

$$\begin{aligned} F(\mathbf{V}) \approx F(\tilde{\mathbf{V}}) &+ \sum_{\mathbf{q}} \frac{\partial F(\tilde{\mathbf{V}})}{\partial V_{\mathbf{q}}} \Delta V_{\mathbf{q}} + \frac{1}{2!} \sum_{\mathbf{q}\mathbf{q}'} \frac{\partial^2 F(\tilde{\mathbf{V}})}{\partial V_{\mathbf{q}} \partial V_{\mathbf{q}'}} \Delta V_{\mathbf{q}} \Delta V_{\mathbf{q}'} \\ &+ \frac{1}{4!} \sum_{\mathbf{p}\mathbf{p}'\mathbf{q}\mathbf{q}'} \frac{\partial^4 F(\tilde{\mathbf{V}})}{\partial V_{\mathbf{p}} \partial V_{\mathbf{p}'} \partial V_{\mathbf{q}} \partial V_{\mathbf{q}'}} \Delta V_{\mathbf{p}} \Delta V_{\mathbf{p}'} \Delta V_{\mathbf{q}} \Delta V_{\mathbf{q}'}, \end{aligned} \tag{7.19}$$

where $\Delta V_{\mathbf{q}} = V_{\mathbf{q}} - \tilde{V}_{\mathbf{q}}$. Since the Gaussian integral are the only ones that can be calculated in practice, we need to incorporate the fourth-order terms into the quadratic approximation.

We use the Gaussian decoupling, which is based on Wick's theorem (for details, see [3]). As a result, expansion (7.19) is written as

$$F(\mathbf{V}) \approx \sum_{\mathbf{q}} \frac{\partial F(\tilde{\mathbf{V}})}{\partial V_{\mathbf{q}}} \Delta V_{\mathbf{q}}$$
$$+ \frac{1}{2} \sum_{\mathbf{q}\mathbf{q}'} \left(\frac{\partial^2 F(\tilde{\mathbf{V}})}{\partial V_{\mathbf{q}} \partial V_{\mathbf{q}'}} + \frac{1}{4} \sum_{\mathbf{p}} \frac{\partial^4 F_1(\tilde{\mathbf{V}})}{\partial V_{\mathbf{p}} \partial V_{-\mathbf{p}} \partial V_{\mathbf{q}} \partial V_{\mathbf{q}'}} \langle \Delta V_{\mathbf{p}} \Delta V_{-\mathbf{p}} \rangle_{(2)} \right) \Delta V_{\mathbf{q}} \Delta V_{\mathbf{q}'}. \tag{7.20}$$

To obtain the parameters of the renormalized Gaussian approximation, $\bar{V}$ and $A_{\mathbf{q}}$, we use Eq. (7.17). The first equation in system (7.17) gives the same first equation as in the optimal Gaussian approximation (7.18). In second equation in (7.17), we obtain an additional term that depends on the 4th derivative

$$\frac{\partial^4 F_1(\mathbf{V})}{\partial V_{\mathbf{p}} \partial V_{-\mathbf{p}} \partial V_{\mathbf{q}} \partial V_{-\mathbf{q}}} = \frac{1}{N^4} \sum_j \frac{\partial^4 F_1(\mathbf{V})}{\partial V_j^4}.$$

Substituting the latter in (7.20), from Eq. (7.17) we finally obtain

$$\bar{V} = -J_0 \left\langle \frac{\partial F_1(\mathbf{V})}{\partial V_j} \right\rangle_{(2)}, \qquad A_{\mathbf{q}} = \frac{1}{2N} \left(\frac{1}{J_{\mathbf{q}}} + \left\langle \frac{\partial^2 F_1(\mathbf{V})}{\partial V_j^2} \right\rangle_{(2)} + \frac{D}{4} \left\langle \frac{\partial^4 F_1(\mathbf{V})}{\partial V_j^4} \right\rangle_{(2)} \right), \tag{7.21}$$

where the mean values are independent of the index j, and the local (single-site) fluctuation $D = \langle \Delta V_j^2 \rangle_{(2)}$ is given by the formula

$$D = \frac{1}{N^2} \sum_{\mathbf{q}} \langle |\Delta V_{\mathbf{q}}|^2 \rangle_{(2)}. \tag{7.22}$$

7.3 Local Fluctuating Field

The Gaussian field $\mathbf{V}$ is completely determined by the mean field $\bar{V}$ and coefficients $A_{\mathbf{q}}$, or alternatively by $\bar{V}$ and mean-square fluctuations $\langle |\Delta V_{\mathbf{q}}|^2 \rangle_{(2)} = T/(2A_{\mathbf{q}})$. But calculating those $N+1$ quantities for each T is an unnecessary complicated procedure for calculating the average magnetic characteristics such as the magnetization and local magnetic moment. Because these characteristics depend only on the mean field and the *local* fluctuation, the average $\langle \ldots \rangle_{(2)}$ in the equations of the optimal Gaussian approximation (7.18) is replaced with the average with the probability density function

$$p(V_j) = \frac{1}{\sqrt{2\pi D}} \exp\left(-\frac{(V_j - \bar{V})^2}{2D} \right). \tag{7.23}$$

As a result, the fluctuating fields at different sites become independent and identically distributed, but their parameters $\bar{V}$ and D are functions of all mean-field fluctuations $\langle |\Delta V_{\mathbf{q}}|^2 \rangle_{(2)}$. Taking (7.18) and (7.22) into account, we have

$$D = \frac{T}{N} \sum_{\mathbf{q}} J_{\mathbf{q}} \left(1 + J_{\mathbf{q}} \left\langle \frac{\partial^2 F_1(\mathbf{V})}{\partial V_j^2} \right\rangle \right)^{-1}. \tag{7.24}$$

Because we are interested in the qualitative character of the temperature dependence, a reasonable simplification is achieved if we replace the interaction coefficients $J_{\mathbf{q}}$ in (7.24) with a mean value J (we thus avoid the summation over the Brillouin zone). The values of $\bar{V}$ and D are then determined by the system of equations

$$\bar{V} = -J_0 \left\langle \frac{\partial F_1(\mathbf{V})}{\partial V_j} \right\rangle, \qquad D = JT \left(1 + J \left\langle \frac{\partial^2 F_1(\mathbf{V})}{\partial V_j^2} \right\rangle \right)^{-1}, \tag{7.25}$$

where the average $\langle \ldots \rangle$ is calculated with probability density (7.23). Similarly, we find the second equation in system (7.21) for the renormalized Gaussian approximation:

$$D = JT\left[1 + J\left(\left\langle\frac{\partial^2 F_1(\mathbf{V})}{\partial V_j^2}\right\rangle + \frac{D}{4}\left\langle\frac{\partial^4 F_1(\mathbf{V})}{\partial V_j^4}\right\rangle\right)\right]^{-1}. \tag{7.26}$$

7.4 Magnetic Phase Diagrams

We write the equations of the optimal and renormalized Gaussian approximations in an explicit form, using free energy formula (7.12). System of equations (7.25) is written as

$$\bar{V} = \frac{J_0}{2}\left\langle\tanh\left(\frac{V}{2T}\right)\right\rangle, \qquad D = JT\left(1 - \frac{J}{4T}\left\langle\cosh^{-2}\left(\frac{V}{2T}\right)\right\rangle\right)^{-1}, \tag{7.27}$$

and for Eq. (7.26) we have

$$D = JT\left\{1 - \frac{J}{4T}\left[\left\langle\cosh^{-2}\left(\frac{V}{2T}\right)\right\rangle - \frac{D}{8T^2}\left(1 - 4\left\langle\tanh^2\left(\frac{V}{2T}\right)\right\rangle + 3\left\langle\tanh^4\left(\frac{V}{2T}\right)\right\rangle\right)\right]\right\}^{-1}.$$

(For brevity, we omit the index j here and hereafter.)

The ratio $0 \le J/J_0 \le 1$ determines the character of the fluctuations. When the interaction is independent of the distance between spins ($J_{j-j'} = J_0/N$), using formula (7.3), we obtain $J_\mathbf{q} = J_0\delta_{\mathbf{q}0}$ and $J = J_0/N$ (weak fluctuations). When $N \to \infty$, we have $J/J_0 \to 0$, and the limit case $J = 0$ corresponds to the mean-field theory. In contrast, when only nearest-neighbouring spins interact, we have $J_{j-j'} \approx J_0\delta_{j,j'}$. Hence, using formula (7.3), we find that $J_\mathbf{q} \approx J$ and therefore $J/J_0 \approx 1$ (strong fluctuations).

As can be seen in Fig. 7.2, by taking fluctuations into account in the optimal Gaussian approximation, we decrease the phase transition temperature compared with the mean-field theory. But in the case of weak fluctuations $J/J_0 = 0.4$, the qualitative behaviour of the temperature dependence $\bar{V}(T)$ still has the second-order phase transition.

In the case of strong fluctuations $J/J_0 = 0.8$, the solution of the system of the optimal Gaussian approximation becomes nonunique (Fig. 7.3a): in addition to the ferromagnetic and paramagnetic solutions, an intermediate solution appears at high temperatures. Hence, as the temperature increases from zero, a discontinuous first-order phase transition from the ferromagnetic to the paramagnetic state occurs. With the reverse change of temperature, a jump from the paramagnetic up to the ferromagnetic state occurs at a smaller temperature value, and hence we have a temperature hysteresis. In this temperature interval the mean field $\bar{V}$ is used as a parameter, and the system of nonlinear equations (7.27) is solved with respect to the

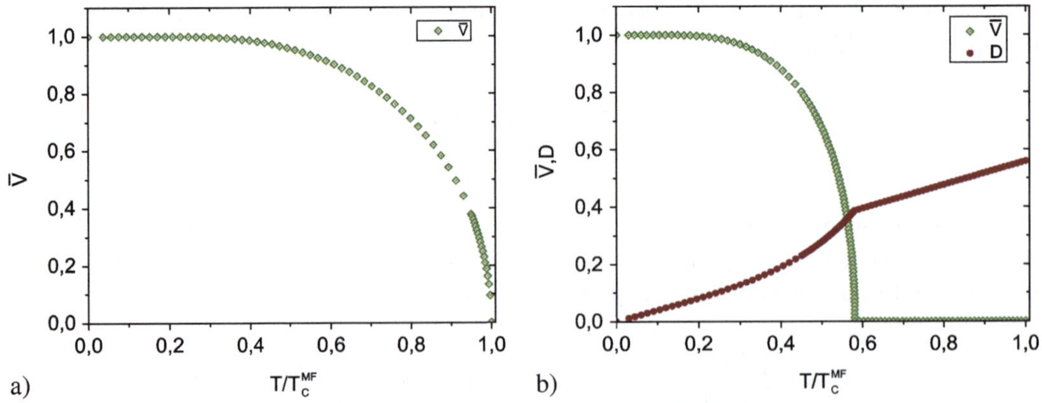

Fig. 7.2 The mean field $\bar{V}$ and local fluctuation D as functions of the reduced temperature T/T_C^{MF} in (**a**) the mean-field theory and (**b**) the optimal Gaussian approximation in the case of weak fluctuations, $J/J_0 = 0.4$

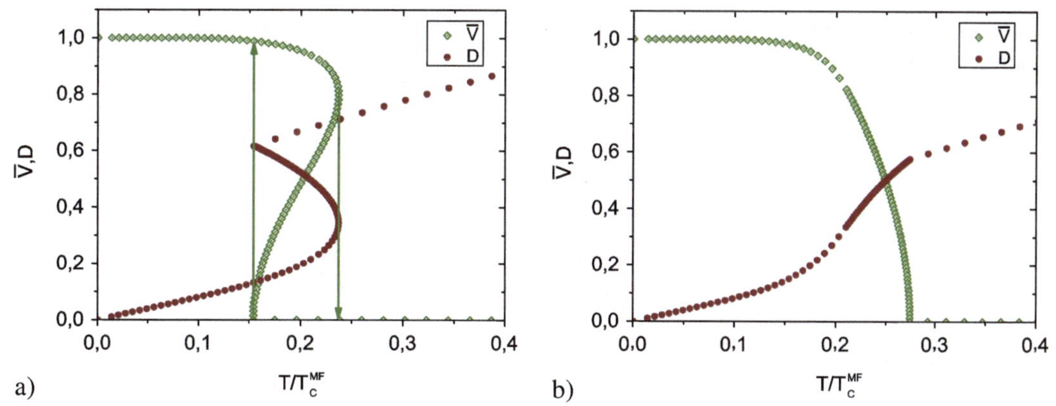

Fig. 7.3 The mean field $\bar{V}$ and local fluctuation D as functions of the reduced temperature T/T_C^{MF} (**a**) in the optimal and (**b**) in the renormalized Gaussian approximations in the case of strong fluctuations, $J/J_0 = 0.8$

local fluctuation D and temperature T. There is only one solution (D, T) of the system at each $\bar{V}$ (for details of the numerical methods see [8]).

Finally, the fourth-order terms in the renormalized Gaussian approximation eliminate the hysteresis in the case of strong fluctuations $J/J_0 = 0.8$ and give the second-order phase transition (Fig. 7.3b).

Analyzing Eqs. (7.25) and (7.26), we can explain the appearance of the first-order phase transition in the optimal Gaussian approximation and its disappearance in the renormalized Gaussian approximation.[3] Indeed, in the presence of a homogeneous magnetic field h (in energy units) the mean-field $\bar{V} = J_0\bar{S}$ is oriented in the opposite direction to h. Hence, $\bar{S}$ depends on $h - \bar{V}(h)$, and the mean-field equation becomes

$$\bar{V}(h) = J_0\bar{S}(h - \bar{V}(h)).$$

Differentiating both sides with respect to h, we obtain the enhanced magnetic susceptibility:

$$-\left.\frac{d\bar{S}}{dh}\right|_{h=0} = \frac{\chi_0}{1 - J_0\chi_0}, \tag{7.28}$$

where

$$\chi_0 = -\left.\frac{\partial\bar{S}}{\partial h}\right|_{h=0} = -\left\langle\frac{\partial^2 F_1(\mathbf{V})}{\partial V_j^2}\right\rangle \tag{7.29}$$

is the unenhanced magnetic susceptibility (with constant $\bar{V}$) of noninteracting spins. At the Curie temperature T_C, the enhanced susceptibility (7.28) tends to infinity, and the condition for the phase transition from the ferromagnetic to the paramagnetic state becomes

$$1 - J_0\chi_0 = 0. \tag{7.30}$$

To find the order of the phase transition, we investigate the derivative $d\bar{V}/dD$ as T approaches T_C from the left. Differentiating mean-field equation (7.25) with respect to D and taking (7.23) and (7.29) into account, we obtain

$$\frac{d\bar{V}}{dD} = -\frac{J_0}{2(1 - J_0\chi_0)}\left\langle\frac{\partial^3 F_1(\mathbf{V})}{\partial V_j^3}\right\rangle. \tag{7.31}$$

To find the average of the third derivative of the free energy, we write fluctuation (7.26) in the form

$$D = \frac{JT}{1 - J(1+\eta)\chi_0},$$

[3]For magnets with itinerant electrons, see [11].

where η is the higher-order correction coefficient. Differentiating this expression with respect to $\bar{V}$ and taking (7.23) and (7.29) into account, we obtain

$$\frac{\mathrm{d}D}{\mathrm{d}\bar{V}} = -\frac{J^2(1+\eta)T}{(1-J(1+\eta)\chi_0)^2}\left\langle \frac{\partial^3 F_1(\mathbf{V})}{\partial V_j^3} \right\rangle. \tag{7.32}$$

Equating (7.31) and reciprocal of (7.32), we finally obtain

$$\frac{\mathrm{d}\bar{V}}{\mathrm{d}D} = -\sqrt{\frac{J_0}{2(1-J_0\chi_0)}}\,\frac{(1-J(1+\eta)\chi_0)}{J\sqrt{(1+\eta)T}}. \tag{7.33}$$

In the case of weak fluctuations $J \ll J_0$, at phase transition point (7.30), we have

$$\frac{\mathrm{d}\bar{V}}{\mathrm{d}D} = -\frac{c}{\sqrt{1-J_0\chi_0}} = -\infty, \qquad c > 0. \tag{7.34}$$

Therefore, $\bar{V}$ continuously decreases to zero as T approaches T_C from the left, and a second-order phase transition occurs.

In the case of strong fluctuations, $J \approx J_0$, the situations in the optimal Gaussian approximation and renormalized Gaussian approximation differ.

Consider the Gaussian approximation ($\eta = 0$) first. At the phase transition point, expression (7.33) becomes

$$\frac{\mathrm{d}\bar{V}}{\mathrm{d}D} = -\sqrt{\frac{1-J_0\chi_0}{2J_0T_\mathrm{C}}} = 0.$$

As a result, the mean field remains at a nonzero level as T approaches T_C from the left. At the point $T = T_\mathrm{C}$ itself the mean-field jumps to $\bar{V} = 0$, and we have the first-order phase transition.

In the renormalized Gaussian approximation ($\eta < 0$), the fluctuations are weakened by the fourth-order terms, i.e. $1 - J(1+\eta)\chi_0$ is nonzero in formula (7.33), and we come to expression (7.34). This leads to the second-order phase transition, which is experimentally observed in metals.

References

1. K. Huang, *Statistical Mechanics*, 2nd edn. (Wiley, New York, 1987)
2. D.J. Amit, *Field Theory, the Renormalization Group, and Critical Phenomena* (McGraw-Hill, London, 1978)
3. N.B. Melnikov, G.V. Paradezhenko, Theor. Math. Phys. **183**(3), 486 (2015)
4. N.B. Melnikov, G.V. Paradezhenko, Phys. Procedia **75**, 731 (2015)
5. V.I. Grebennikov, Solid State Phenom. **152–153**, 563 (2009)
6. R.L. Stratonovich, Sov. Phys. Dokl. **2**, 416 (1958)
7. J. Hubbard, Phys. Rev. Lett. **3**(2), 77 (1959)
8. N.B. Melnikov, G.V. Paradezhenko, Vychisl. Metody Programm. **15**, 461 (2014) [in Russian]
9. J. Zinn-Justin, *Path Integrals in Quantum Mechanics* (Oxford University Press, Oxford, 2004)
10. B.I. Reser, V.I. Grebennikov, Phys. Met. Metallogr. **85**, 20 (1998)
11. N.B. Melnikov, B.I. Reser, V.I. Grebennikov, J. Phys. A Math. Theor. **43**, 195004 (2010)
12. N.B. Melnikov, B.I. Reser, Procs. Steklov Inst. Math. **271**, 149 (2010)
13. A.Z. Patashinskii, V.L. Pokrovskii, *Fluctuation Theory of Phase Transitions* (Pergamon, Oxford, 1979)
14. R.M. White, T.H. Geballe, *Long Range Order in Solids* (Academic, New York, 1979)

Functional Integral Method

8

The functional integral method has won us some ground in the long struggle to understand the magnetism of iron and nickel... Present theories could be improved by a proper treatment of the multi-orbital aspect of the d band. (J. Hubbard, in Electron Correlation and Magnetism in Narrow-Band Systems, Springer, Berlin, 1981)

To go beyond the RPA in spin fluctuation theory we use the functional integral method. The method is based on the Stratonovich-Hubbard transformation, which requires a special form of the model Hamiltonian. In Sect. 8.1, we introduce a multiband Hubbard Hamiltonian and express it in terms of the atomic charge and spin [1]. In Sect. 8.2 we explain the functional integral method itself. In Sect. 8.3 we obtain expressions for the free energy, mean spin and spin-density correlator.

8.1 Multiband Hubbard Hamiltonian

The main assumption about the model Hamiltonian is that d electrons interact only at the same site. The first model of this kind was the tight-binding model with intraatomic exchange interaction, known as the Stoner model [2, 3]. A single-band version of this model, called the Hubbard model [4], was described in Chap. 4. In the static single-site spin fluctuation theory [5–9], the tight-binding approximation of the one-electron part was replaced by a model density of states but still the single-band Hubbard Hamiltonian was used (for an overview, see, e.g. [10]). In the DSFT we use the real density of states to describe the one-electron part and multiband Hubbard Hamiltonian to describe the interaction part.

8.1.1 Intraatomic Interaction and Hund's Rule

In the interaction part of the Hamiltonian $\mathcal{H}_{\mathrm{I}}$, we keep only the intraatomic Coulomb U and exchange J interactions (the same for all degenerate d bands).[1] Our goal is to express $\mathcal{H}_{\mathrm{I}}$ in terms of the atomic charge and spin operators:

$$n_j = \sum_{\nu\sigma} a^{\dagger}_{\nu j\sigma} a_{\nu j\sigma} = \sum_{\nu\sigma} n_{\nu j\sigma}, \qquad \mathbf{s}_j = \sum_{\nu\sigma\sigma'} \left(\frac{1}{2}\boldsymbol{\sigma}\right)_{\sigma\sigma'} a^{\dagger}_{\nu j\sigma} a_{\nu j\sigma'}. \tag{8.1}$$

We consider the second-quantized form of the interacting electrons Hamiltonian (3.35). As the first step, we neglect all the interaction coefficients (3.37) but the Coulomb interaction

$$U = U^{\nu\nu'\nu\nu'}_{iiii} = \iint |w_{\nu' i}(\mathbf{r})|^2 |w^*_{\nu i}(\mathbf{r}')|^2 \frac{e^2}{|\mathbf{r}-\mathbf{r}'|}\,\mathrm{d}\mathbf{r}\,\mathrm{d}\mathbf{r}' \tag{8.2}$$

[1] Although the interatomic terms are not necessarily negligible, this simplification should make subsequent theoretical treatment exceedingly simple without losing the most essential parts of the physics (see the discussion in [10]).

N. B. Melnikov, B. I. Reser, *Dynamic Spin Fluctuation Theory of Metallic Magnetism*,
https://doi.org/10.1007/978-3-319-92974-3_8

and exchange interaction

$$J = U_{iiii}^{\nu\nu'\nu'\nu} = \iint w_{\nu' i}^*(\mathbf{r}) w_{\nu i}^*(\mathbf{r}') \frac{e^2}{|\mathbf{r}-\mathbf{r}'|} w_{\nu' i}(\mathbf{r}') w_{\nu i}(\mathbf{r}) \, d\mathbf{r} \, d\mathbf{r}'. \tag{8.3}$$

Then the second term on the right-hand side of (3.35) becomes

$$\begin{aligned} \mathcal{H}_\mathrm{I} &= \frac{1}{2}U \sum_{\nu i\sigma\sigma'} a_{\nu i\sigma}^\dagger a_{\nu i\sigma'}^\dagger a_{\nu i\sigma'} a_{\nu i\sigma} + \frac{1}{2}U \sum_{\substack{\nu\neq\nu' \\ i\sigma\sigma'}} a_{\nu i\sigma}^\dagger a_{\nu' i\sigma'}^\dagger a_{\nu' i\sigma'} a_{\nu i\sigma} \\ &+ \frac{1}{2}J \sum_{\substack{\nu\neq\nu' \\ i\sigma\sigma'}} a_{\nu i\sigma}^\dagger a_{\nu' i\sigma'}^\dagger a_{\nu i\sigma'} a_{\nu' i\sigma}. \end{aligned} \tag{8.4}$$

Using anticommutation relations (3.12) as in the single-band case (4.2), we write the first term on the right-hand side of (8.4) as

$$\frac{1}{2}U \sum_{\nu i\sigma\sigma'} a_{\nu i\sigma}^\dagger a_{\nu i\sigma'}^\dagger a_{\nu i\sigma'} a_{\nu i\sigma} = \frac{1}{2}U \sum_{\nu i\sigma} n_{\nu i\sigma} n_{\nu i\bar{\sigma}}, \tag{8.5}$$

where $\bar{\sigma}$ means the opposite spin to σ. Similarly, the second term on the right-hand side of (8.4) becomes

$$\frac{1}{2}U \sum_{\substack{\nu\neq\nu' \\ i\sigma\sigma'}} a_{\nu i\sigma}^\dagger a_{\nu' i\sigma'}^\dagger a_{\nu' i\sigma'} a_{\nu i\sigma} = \frac{1}{2}U \sum_{\substack{\nu\neq\nu' \\ i\sigma\sigma'}} n_{\nu i\sigma} n_{\nu' i\sigma'}. \tag{8.6}$$

The third term of (8.4) is transformed as follows:

$$\begin{aligned} \frac{1}{2}J \sum_{\substack{\nu\neq\nu' \\ i\sigma\sigma'}} a_{\nu i\sigma}^\dagger a_{\nu' i\sigma'}^\dagger a_{\nu i\sigma'} a_{\nu' i\sigma} &= -\frac{1}{2}J \sum_{\substack{\nu\neq\nu' \\ i\sigma\sigma'}} a_{\nu i\sigma}^\dagger a_{\nu i\sigma'} a_{\nu' i\sigma'}^\dagger a_{\nu' i\sigma} \\ &= -\frac{1}{2}J \sum_{\nu\neq\nu', i\sigma} \left(a_{\nu i\sigma}^\dagger a_{\nu i\sigma} a_{\nu' i\sigma}^\dagger a_{\nu' i\sigma} + a_{\nu i\sigma}^\dagger a_{\nu i\bar{\sigma}} a_{\nu' i\bar{\sigma}}^\dagger a_{\nu' i\sigma} \right). \end{aligned} \tag{8.7}$$

Using the spin-flip operators

$$s_{\nu i}^+ = a_{\nu i\uparrow}^\dagger a_{\nu i\downarrow}, \qquad s_{\nu i}^- = a_{\nu i\downarrow}^\dagger a_{\nu i\uparrow},$$

we have

$$\sum_\sigma a_{\nu i\sigma}^\dagger a_{\nu i\bar{\sigma}} a_{\nu' i\bar{\sigma}}^\dagger a_{\nu' i\sigma} = s_{\nu i}^+ s_{\nu' i}^- + s_{\nu i}^- s_{\nu' i}^+.$$

Since $s_{\nu i}^+$ and $s_{\nu' i}^-$ for $\nu \neq \nu'$ commute, we write (8.7) as

$$\frac{1}{2}J \sum_{\substack{\nu\neq\nu' \\ i\sigma\sigma'}} a_{\nu i\sigma}^\dagger a_{\nu' i\sigma'}^\dagger a_{\nu i\sigma'} a_{\nu' i\sigma} = -\frac{1}{2}J \sum_{\nu\neq\nu', i\sigma} n_{\nu i\sigma} n_{\nu' i\sigma} - J \sum_{\nu\neq\nu', i} s_{\nu i}^+ s_{\nu' i}^-. \tag{8.8}$$

Substituting formulae (8.5), (8.6) and (8.8) in the right-hand side of (8.4), we obtain

$$\mathcal{H}_\mathrm{I} = \frac{1}{2}U \sum_{\nu\nu' i\sigma} n_{\nu i\sigma} n_{\nu' i\bar{\sigma}} + \frac{1}{2}(U-J) \sum_{\nu\neq\nu', i\sigma} n_{\nu i\sigma} n_{\nu' i\sigma} - J \sum_{\nu\neq\nu', i} s_{\nu i}^+ s_{\nu' i}^-. \tag{8.9}$$

The operator $s_{\nu j}^+ s_{\nu' j}^-$, $\nu \neq \nu'$, flips spins of two electrons in different bands, keeping the total spin of the atom $\mathbf{s}_j$ unchanged. Since the charge n_j and spin $\mathbf{s}_j$ of an *atom* are the only variables we are going to consider, the simultaneous spin flip is neglected. Then

$$\mathcal{H}_{\mathrm{I}} = \frac{1}{2}U\sum_{\nu\nu' j\sigma} n_{\nu j\sigma} n_{\nu' j\bar{\sigma}} + \frac{1}{2}(U-J)\sum_{\nu\neq\nu', j\sigma} n_{\nu j\sigma} n_{\nu' j\sigma}. \tag{8.10}$$

Formula (8.10) can be interpreted as follows. If on the average there are two electrons at the site j in the bands ν and ν' with the *opposite* spins σ and $\bar{\sigma}$, then the average of their interaction energy

$$\frac{1}{2}U n_{\nu j\sigma} n_{\nu' j\bar{\sigma}} + \frac{1}{2}U n_{\nu' j\bar{\sigma}} n_{\nu j\sigma}$$

is equal to U. Similarly, if on the average two electrons at the site j in the bands ν and ν' have the *same* spin σ, they contribute to the average of the interaction energy (8.10) only if the bands are different: $\nu \neq \nu'$ (in full agreement with the Pauli exclusion principle). In this case, the average of their interaction energy

$$\frac{1}{2}(U-J) n_{\nu j\sigma} n_{\nu' j\sigma} + \frac{1}{2}(U-J) n_{\nu' j\sigma} n_{\nu j\sigma}$$

is equal to $U - J$. This leads to a *Hund's rule* of coupling the spins at the same site: *parallel spin configurations have lower energy than antiparallel.* Thus, the intraatomic exchange correlation favours large local spin in metals.

In ferromagnetic metals and alloys, the 3d shell holds 10 electrons as a maximum. According to Hund's rule, the first 5 electrons with the same spin tend to populate all 5 different d bands. The rest of the 3d electrons must have the opposite spin. Therefore, the magnetization (per site) is equal to 10 minus the average number of 3d electrons. For iron this is about $10 - 7.3 = 2.7$. A similar calculation gives 1.3 for cobalt and 0.6 for nickel. These numbers are in reasonable agreement with the experiment (see Table 10.2). The above rule regarding magnetization holds remarkably well for the whole series of ferromagnetic binary alloys of iron, cobalt and nickel, and some neighbouring elements [11, 12].

8.1.2 Atomic Charge and Spin Density

The first term on the right-hand side of (8.10) appears already in single-band SFT models [5–9]. Indeed, using the notation $n_{j\sigma} = \sum_\nu n_{\nu j\sigma}$, we come to

$$\frac{1}{2}U\sum_{\nu\nu' j\sigma} n_{\nu j\sigma} n_{\nu' j\bar{\sigma}} = \frac{1}{2}U\sum_{j\sigma} n_{j\sigma} n_{j\bar{\sigma}} = U\sum_{j} n_{j\uparrow} n_{j\downarrow}.$$

The second term in expression (8.10) can be written as

$$\frac{1}{2}(U-J)\sum_{\nu\neq\nu', j\sigma} n_{\nu j\sigma} n_{\nu' j\sigma} = \frac{1}{2}(U-J)\sum_{j\sigma}\Big(n_{j\sigma} n_{j\sigma} - \sum_{\nu} n_{\nu j\sigma} n_{\nu j\sigma}\Big),$$

Approximating

$$\sum_{\nu} n_{\nu j\sigma} n_{\nu j\sigma} \approx \frac{1}{N_{\mathrm{d}}}\sum_{\nu\nu'} n_{\nu j\sigma} n_{\nu' j\sigma} = \frac{1}{N_{\mathrm{d}}} n_{j\sigma} n_{j\sigma},$$

where $N_{\mathrm{d}} = 5$ is the number of degenerate d band in the ferromagnetic metals, we obtain

$$\frac{1}{2}(U-J)\sum_{\nu\neq\nu', j\sigma} n_{\nu j\sigma} n_{\nu' j\sigma} = \frac{1}{2}(U-J)\frac{N_{\mathrm{d}}-1}{N_{\mathrm{d}}}\sum_{j\sigma} n_{j\sigma} n_{j\sigma}.$$

Hence the interaction term (8.10) becomes

$$\mathcal{H}_\mathrm{I} = U\sum_j n_{j\uparrow}n_{j\downarrow} + \frac{1}{2}(U-J)\frac{N_\mathrm{d}-1}{N_\mathrm{d}}\sum_{j\sigma} n_{j\sigma}^2. \tag{8.11}$$

There are several alternative forms of expressing $\mathcal{H}_\mathrm{I}$ as a sum of squares of the charge and spin operators. The first form is obtained using the equalities

$$n_{j\uparrow}n_{j\downarrow} = \frac{1}{4}n_j^2 - (s_j^z)^2, \qquad \sum_\sigma n_{j\sigma}^2 = \frac{1}{2}n_j^2 + 2(s_j^z)^2, \tag{8.12}$$

which follow from the expressions

$$\begin{aligned} \frac{1}{4}n_j^2 &= \frac{1}{4}(n_{j\uparrow}^2 + n_{j\downarrow}^2) + \frac{1}{2}n_{j\uparrow}n_{j\downarrow}, \\ (s_j^z)^2 &= \frac{1}{4}(n_{j\uparrow}^2 + n_{j\downarrow}^2) - \frac{1}{2}n_{j\uparrow}n_{j\downarrow}. \end{aligned} \tag{8.13}$$

Substituting equalities (8.12) into formula (8.11), we come to

$$\mathcal{H}_\mathrm{I} = \sum_j\left(\frac{u_0}{4}n_j^2 - u_z(s_j^z)^2\right), \tag{8.14}$$

where

$$u_0 = \frac{(2N_\mathrm{d}-1)U - (N_\mathrm{d}-1)J}{N_\mathrm{d}}, \qquad u_z = \frac{U + (N_\mathrm{d}-1)J}{N_\mathrm{d}}. \tag{8.15}$$

In the ferromagnetic metals we have $N_\mathrm{d} = 5$, and the value of the interaction constant

$$u_z = \frac{U+4J}{5} \tag{8.16}$$

turns out to be about 1 eV (see Table 8.1).

To obtain a spin-rotationally invariant representation of $\mathcal{H}_\mathrm{I}$, we need an expression for the square of the spin density operator $\mathbf{s}_j^2$. Calculating $(s_j^x)^2$ and $(s_j^y)^2$ similar to (8.13), we obtain

$$\mathbf{s}_j^2 = 3(s_j^z)^2 + \sum_{\nu\neq\nu'} s_{\nu j}^+ s_{\nu' j}^-.$$

Discarding the last term just as in formula (8.9), we come to the approximate relation $(s_j^z)^2 \approx \frac{1}{3}\mathbf{s}_j^2$. From (8.14) we immediately obtain

Table 8.1 The Coulomb interaction U and exchange interaction J constants used by Kakehashi and Patoary [13] and interaction constant u_z calculated by formula (8.16) in the ferromagnetic metals

Metal	U (eV)	J (eV)	u_z (eV)
Fe	2.30	0.90	1.18
Co	3.33	0.94	1.42
Ni	3.01	0.90	1.32

$$\mathcal{H}_{\rm I} = \sum_j \left(\frac{u_0}{4} n_j^2 - \frac{u_z}{3} \mathbf{s}_j^2 \right). \tag{8.17}$$

Similarly, using the approximate relation $(s_j^z)^2 \approx (\mathbf{s}_j \mathbf{e}_j)^2$, where $\mathbf{e}_j$ is an arbitrary unit vector, we come to

$$\mathcal{H}_{\rm I} = \sum_j \left(\frac{u_0}{4} n_j^2 - u_z (\mathbf{s}_j \mathbf{e}_j)^2 \right). \tag{8.18}$$

In the single-band model, approximate formulae (8.14), (8.17) and (8.18) reduce to (4.3), (4.4) and (4.5), respectively.

An exact evaluation of the partition function must always give the same result, no matter which particular form of the interaction part we choose. But the exact solution is neither possible nor desired, because we start with an approximate Hamiltonian and seek for a generalization of the Stoner theory (see, e.g. [14, 15]).

Due to approximations in the SFT, formulae (8.14), (8.17) and (8.18) lead to different results. Expressions (8.14) and (8.18) are not spin-rotationally invariant. In the latter, the rotational invariance can be restored by integrating the partition function over all directions of each $\mathbf{e}_j$ (see, e.g. [5, 6, 14]). However, even after that, the form (8.18) does not allow one to perform Gaussian integrals in the DSFT.[2] We use the rotationally spin-invariant form of the *multiband Hubbard Hamiltonian* (8.17). At $T = 0$ this form leads to the Stoner equation with $u = u_z/3$ instead of u_z, just as in the single-band case [5, 14]. Therefore, in SFT the value of u is often obtained from the Stoner equation with the known magnetic moment $m_z(0)$ at $T = 0$.

8.2 Functional Integral over Fluctuating Fields

8.2.1 Thermodynamic "Time" Dependence

Recall that the thermodynamics of the system is determined by the grand partition function (2.25):

$$\Xi = {\rm Tr} e^{-\mathcal{H}'/T}, \qquad \mathcal{H}' = \mathcal{H}_0' + \mathcal{H}_{\rm I}. \tag{8.19}$$

In classical statistical mechanics, $\mathcal{H}_0'$ and $\mathcal{H}_{\rm I}$ are ordinary functions and hence $e^{-\mathcal{H}'/T} = e^{-\mathcal{H}_0'/T} e^{-\mathcal{H}_{\rm I}/T}$. Then we could apply the Stratonovich-Hubbard transformation to $e^{-\mathcal{H}_{\rm I}/T}$ just as we did in the Ising model.

In quantum statistical mechanics, the operators $\mathcal{H}_0' = \mathcal{H}_0 - \mu \mathcal{N}_{\rm e}$ and $\mathcal{H}_{\rm I}$ do not commute, therefore the exponential $e^{-\mathcal{H}'/T}$ in expression (8.19) cannot be transformed to the product of the exponentials $e^{-\mathcal{H}_0'/T}$ and $e^{-\mathcal{H}_{\rm I}/T}$. One can overcome this difficulty by making use of the "time"-ordering trick (for details, see Appendix B.1):

$$\Xi = {\rm Tr} \left[e^{-\mathcal{H}_0'/T}\, T_\tau \exp\left(- \int_0^{1/T} \mathcal{H}_{\rm I}(\tau)\, d\tau \right) \right], \tag{8.20}$$

where

$$\mathcal{H}_{\rm I}(\tau) = e^{\mathcal{H}_0'\tau}\, \mathcal{H}_{\rm I}\, e^{-\mathcal{H}_0'\tau} \tag{8.21}$$

is the "interaction" representation with respect to the "time" $\tau \in [0, 1/T]$ (from now on we set $\hbar = 1$ unless the contrary is explicitly stated). The "time"-ordering operator T_τ rearranges operators in the product in such a way that the "times" of the operators decrease from left to right. Then the *"time"-ordered exponential* can be understood as the limit [16]

$$T_\tau \exp\left(- \int_0^{1/T} \mathcal{H}_{\rm I}(\tau)\, d\tau \right) = \lim_{N \to \infty} \exp(-\mathcal{H}_{\rm I}(\tau_N) \Delta\tau) \ldots \exp(-\mathcal{H}_{\rm I}(\tau_1) \Delta\tau),$$

[2] Instead of $\prod_j {\rm D}\mathbf{V}_j$ in the functional integral (8.25), which we obtain using (8.17), the form (8.18) leads to $\prod_j (1/V_j^2) {\rm D}\mathbf{V}_j$. This form is tractable only in the single-site static approximation that does not require the Gaussian approximation.

where $\tau_n = n/(NT)$ and $\Delta\tau = 1/(NT)$. Thus, it is the quantum nature of the system that brings in the dynamics. The classical approach in the functional integral method leads to the *static approximation*.

8.2.2 Electrons in the Fluctuating Field

In the Stoner mean-field theory the interaction between electrons is understood as the interaction with a spin-dependent exchange field. This leads to a rigid shift of spin-up and spin-down electron DOSs. The nature of the exchange field postulated by the Stoner model is illustrated schematically in Fig. 8.1, left. The exchange field is the same at all sites,[3] has its maximum value at $T = 0$, decreases as T increases and finally becomes zero at and above the Curie temperature T_C.

The SFT recognizes that the exchange field at a site depends upon the spin at that site, which is a vector quantity. Therefore, the exchange field can vary both in direction and absolute value from site to site [5, 6, 14]. This opens up the possibility that above T_C the exchange fields may not vanish but produce a zero magnetization (Fig. 8.1, right).

The picture presented at Fig. 8.1 corresponds to a classical treatment of spins as vectors in the three-dimensional space. Due to the quantum nature of spin, the exchange field depends on an additional "dynamic" variable. As a result, in the DSFT the exchange field fluctuates both in space and in "time". This leads to important differences with the static (classical) SFT, as we show in the following chapters.

We introduce the fluctuating exchange field by using the Stratonovich-Hubbard transformation [17, 18]. The latter is based on the following identity for an arbitrary operator $\mathcal{A}$:

$$\exp\left(\frac{\mathcal{A}^2}{a}\right) = \sqrt{\frac{a}{\pi}} \int \exp(-ax^2 + 2\mathcal{A}x)\,\mathrm{d}x, \qquad a > 0. \tag{8.22}$$

We expressed the interaction term $\mathcal{H}_\mathrm{I}$ as a sum of squares of the local charge and spin. If the exponential of a sum of operators could be represented as a product of exponentials, we would apply the above identity to each exponential in the product. As we showed in the previous section this is generally not the case, because the operators in the sum usually do not commute with each other. Therefore, we apply the Stratonovich-Hubbard transformation to the "time"-ordered exponential.

Similar to the single-band case (6.24), we introduce the local density matrix ρ_j with the elements

$$\rho_{j\sigma\sigma'} = \sum_\nu a^\dagger_{\nu j\sigma'} a_{\nu j\sigma}.$$

From formulae (8.1) it immediately follows that the scalar component ρ_j^0 in the expansion (6.25) is equal to one half of the local charge operator, and the vector component $\boldsymbol{\rho}_j$ is equal to the local spin operator:

$$\rho_j^0 = \frac{1}{2} n_j, \qquad \boldsymbol{\rho}_j = \mathbf{s}_j, \tag{8.23}$$

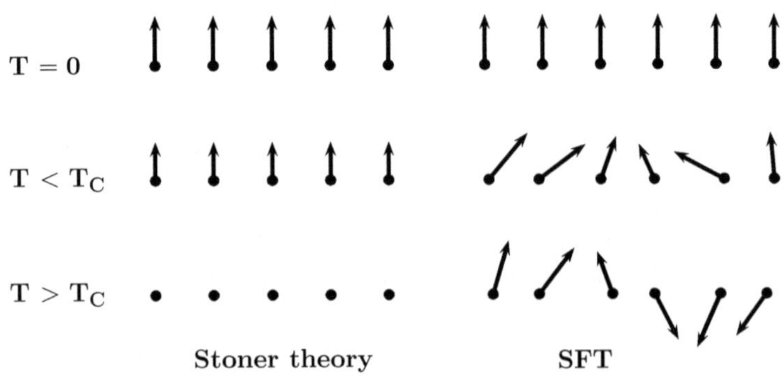

Fig. 8.1 Sketch of the exchange field configurations in the Stoner model and in SFT at different temperatures

[3]By saying the "exchange field at a site" in the itinerant electron model, we mean the integral value of the exchange field over the Wigner-Seitz cell centered at this lattice site.

just as in formulae (6.26) and (6.27).

The interaction term $\mathcal{H}_{\mathrm{I}}$ is expressed as a sum of squares of the operators ρ_j^μ. Using the relations for the local density matrix (8.23), we write expression (8.17) as

$$\mathcal{H}_{\mathrm{I}} = \sum_j \left(u_0 (\rho_j^0)^2 - u \boldsymbol{\rho}_j^2 \right),$$

where $u = u_z/3$. Substituting the latter in the grand partition function (8.20), we obtain[4]

$$\begin{aligned}
\Xi &= \operatorname{Tr}\left[\mathrm{e}^{-\mathcal{H}_0'/T}\, T_\tau \exp\left(-\int_0^{1/T} \sum_j \left(u_0 (\rho_j^0(\tau))^2 - u \boldsymbol{\rho}_j^2(\tau) \right) \mathrm{d}\tau \right) \right] \\
&= \operatorname{Tr}\left[\mathrm{e}^{-\mathcal{H}_0'/T}\, T_\tau \exp\left(\sum_j \int_0^{1/T} \left(u_0 (\mathrm{i}\rho_j^0(\tau))^2 + u \boldsymbol{\rho}_j^2(\tau) \right) \mathrm{d}\tau \right) \right] \\
&= \operatorname{Tr}\left[\mathrm{e}^{-\mathcal{H}_0'/T}\, T_\tau \prod_j \exp\left(u_0 \int_0^{1/T} (\mathrm{i}\rho_j^0(\tau))^2\, \mathrm{d}\tau \right) \exp\left(u \int_0^{1/T} \boldsymbol{\rho}_j^2(\tau)\, \mathrm{d}\tau \right) \right],
\end{aligned} \tag{8.24}$$

where the "time"-dependence means the "interaction" representation (8.21). Applying the Stratonovich-Hubbard transformation to the squares of the operators $i\rho_j^0(\tau)$ and $\boldsymbol{\rho}_j(\tau)$ in the Hamiltonian (8.24), we obtain [5,6,19]

$$\begin{aligned}
\Xi = &\left(\int T_\tau \exp\left(-\sum_j \int_0^{1/T} \left(\frac{1}{u} \mathbf{V}_j^2(\tau) + \frac{1}{u_0} \Phi_j^2(\tau) \right) \mathrm{d}\tau \right) \prod_j \left[\mathrm{D}\mathbf{V}_j(\tau) \mathrm{D}\Phi_j(\tau) \right] \right)^{-1} \\
&\times \int \exp\left(-\sum_j \int_0^{1/T} \left(\frac{1}{u} \mathbf{V}_j^2(\tau) + \frac{1}{u_0} \Phi_j^2(\tau) \right) \mathrm{d}\tau \right) \operatorname{Tr}\left[\exp(-\mathcal{H}_0'/T) \right. \\
&\left. \times T_\tau \exp\left(-\int_0^{1/T} \sum_j 2(\mathbf{V}_j(\tau) \boldsymbol{\rho}_j(\tau) + \mathrm{i}\Phi_j(\tau) \rho_j^0(\tau))\, \mathrm{d}\tau \right) \right] \prod_j \left[\mathrm{D}\mathbf{V}_j(\tau) \mathrm{D}\Phi_j(\tau) \right]
\end{aligned} \tag{8.25}$$

(for details, see Appendix B.3.2). The integration variables $\mathbf{V}_j(\tau)$ and $\Phi_j(\tau)$ are called the *exchange* and *charge* fields, respectively.

Next, introducing the notation $V_j^0(\tau) = \mathrm{i}\Phi_j(\tau)$, we join the charge and exchange components into one 2×2 matrix $V_j(\tau) = V_j^0(\tau)\sigma^0 + \mathbf{V}_j(\tau)\boldsymbol{\sigma}$. Then the grand partition function (8.25) can be rewritten as the *functional integral*

$$\Xi = \left(\int \mathrm{e}^{-F_0(V)/T}\, \mathrm{D}V \right)^{-1} \int \mathrm{e}^{-(F_0(V) + \Omega_1(V))/T}\, \mathrm{D}V. \tag{8.26}$$

Here $\mathrm{D}V \equiv \prod_j [\mathrm{D}\mathbf{V}_j(\tau) \mathrm{D}V_j^0(\tau)]$ indicates the functional integration over $\mathbf{V}_j(\tau)$ and $V_j^0(\tau)$ on the "time" interval $[0, 1/T]$,

$$F_0(V) = T \int_0^{1/T} \sum_j \left(\frac{1}{u} \mathbf{V}_j^2(\tau) - \frac{1}{u_0} (V_j^0(\tau))^2 \right) \mathrm{d}\tau \tag{8.27}$$

is the energy of the fluctuating field $V \equiv (V_1(\tau), V_2(\tau), \dots)$ and

[4] The operators $\rho_j^\mu(\tau)$ can be treated as commuting as long as they are under the T_τ sign.

$$\Omega_1(V) = -T \ln \mathrm{Tr}\Big[T_\tau \exp\Big(-\int_0^{1/T} \mathcal{H}'(V)\,\mathrm{d}\tau \Big)\Big] \tag{8.28}$$

is the thermodynamic potential of electrons in the field, where

$$\mathcal{H}'(V) = \mathcal{H}_0' + 2\sum_j \Big(\mathbf{V}_j(\tau)\boldsymbol{\rho}_j(\tau) + V_j^0(\tau)\rho_j^0(\tau) \Big). \tag{8.29}$$

Rewriting the latter as

$$\mathcal{H}'(V) = \mathcal{H}_0' + \sum_j \mathrm{Sp}\big(V_j(\tau)\rho_j(\tau) \big)$$

and substituting $a_{j\sigma}^{+} a_{j\sigma'}$ for $\rho_{j\sigma'\sigma}$, we see that $\mathcal{H}'(V) = \mathcal{H}_0' + \hat{V}$ is the Hamiltonian of *noninteracting* electrons in the "time"-dependent external field

$$\hat{V} = \sum_{\nu j \sigma \sigma'} V_{j\sigma\sigma'}(\tau) a_{\nu j \sigma}^{\dagger}(\tau) a_{\nu j \sigma'}(\tau). \tag{8.30}$$

8.2.3 Charge Fluctuations

Charge fluctuations possess high energy and hence have a small probability. Therefore, the integrals over the charge field in formula (8.26) can be evaluated by the saddle-point method (see, e.g. [20,21]). Namely, taking the functional derivative with respect to $V_j^0(\tau)$ of the integrand in (8.26), we obtain (for details, see Appendix B.2)

$$\frac{\delta}{\delta V_j^0(\tau)} \mathrm{e}^{-(F_0(V)+\Omega_1(V))/T} = \left(\frac{2V_j^0(\tau)}{u_0} - 2\rho_j^0(V) \right) \mathrm{e}^{-(F_0(V)+\Omega_1(V))/T}, \tag{8.31}$$

where

$$\rho_j^0(V) \equiv \mathrm{Tr}\Big[T_\tau\, \rho_j^0(\tau) \exp \int_0^{1/T} \Big(\Omega_1(V) - \mathcal{H}'(V) \Big)\,\mathrm{d}\tau' \Big].$$

Equating the functional derivative to zero, we come to the equation

$$V_j^0(\tau) = u_0 \rho_j^0(V). \tag{8.32}$$

(Note that both sides depend on $V_j^0(\tau)$ here.) To calculate the grand partition function (8.26), one needs to solve equation (8.32) with respect to the charge field $V_j^0(\tau)$ and then substitute the result, as a function of the exchange field $\mathbf{V}_j(\tau)$, in integral (8.26). Thus, our formulation of the theory takes into account both charge and spin fluctuations. However, the characteristic time of the charge fluctuations is substantially smaller than that of the spin fluctuations and, upon averaging, their contribution to magnetic characteristics is not significant. Moreover, the interatomic interaction, which is neglected in the Hubbard model, leads to a considerable effective screening of the charge fluctuations. Therefore, here we take the charge field $V_j^0(\tau)$ to be equal to its average value

$$\bar{V}^0 = u_0 \langle \rho_j^0(V) \rangle.$$

Recalling that $\rho_j^0(V) = n_j(V)/2$ and $\langle n_j(V) \rangle = \langle n_j \rangle$, we obtain

$$\bar{V}^0 = \frac{u_0}{2N} \Big\langle \sum_j n_j(V) \Big\rangle = \frac{u_0}{2} n_{\mathrm{e}},$$

where n_e is the number of d electrons per atom.[5] Substituting the mean value $\bar{V}^0$ for $V_j^0(\tau)$ in (8.25), we have

$$\begin{aligned}\Xi = &\left(\int T_\tau \exp\Big(-\int_0^{1/T} \sum_j \Big(\frac{1}{u}\mathbf{V}_j^2(\tau) - \frac{1}{u_0}(\bar{V}^0)^2\Big)\,\mathrm{d}\tau\Big)\,\mathrm{D}V\right)^{-1}\\ &\int T_\tau \exp\Big(-\int_0^{1/T} \sum_j \Big(\frac{1}{u}\mathbf{V}_j^2(\tau) - \frac{1}{u_0}(\bar{V}^0)^2\Big)\,\mathrm{d}\tau\Big)\\ &\times \mathrm{Tr}\Big[T_\tau \exp\Big(-\int_0^{1/T}\Big(\mathcal{H}_0' + 2\sum_j[\mathbf{V}_j(\tau)\boldsymbol{\rho}_j(\tau) + \bar{V}^0\rho_j^0(\tau)]\Big)\,\mathrm{d}\tau\Big)\Big]\,\mathrm{D}V.\end{aligned} \tag{8.33}$$

It is easy to see that

$$2\sum_j \bar{V}^0\rho_j^0(\tau) = \bar{V}^0\sum_j n_j(\tau) = \bar{V}^0\mathcal{N}_e(\tau).$$

Moreover, the operators

$$\mathcal{N}_e = \sum_{\nu\mathbf{k}\sigma} n_{\nu\mathbf{k}\sigma}, \qquad \mathcal{H}_0' = \sum_{\nu\mathbf{k}\sigma}(\varepsilon_{\mathbf{k}} - \mu)n_{\nu\mathbf{k}\sigma}$$

commute, so that $\mathcal{N}_e(\tau) = \mathrm{e}^{\mathcal{H}_0'\tau}\mathcal{N}_e\,\mathrm{e}^{-\mathcal{H}_0'\tau} = \mathcal{N}_e$. Therefore,

$$2\sum_j \bar{V}^0\rho_j^0(\tau) = \bar{V}^0\mathcal{N}_e,$$

and we can write the partition function (8.33) as

$$\begin{aligned}\Xi = &\left(\int \exp\Big(-\frac{1}{u}\int_0^{1/T}\sum_j \mathbf{V}_j^2(\tau)\,\mathrm{d}\tau\Big)\,\mathrm{D}V\right)^{-1}\int \exp\Big(-\frac{1}{u}\int_0^{1/T}\sum_j \mathbf{V}_j^2(\tau)\,\mathrm{d}\tau\Big)\\ &\times \mathrm{Tr}\Big[T_\tau \exp\Big(-\int_0^{1/T}\Big(\mathcal{H}_0' + \bar{V}^0\mathcal{N}_e + 2\sum_j \mathbf{V}_j(\tau)\boldsymbol{\rho}_j(\tau)\Big)\,\mathrm{d}\tau\Big)\Big]\,\mathrm{D}V.\end{aligned}$$

The term $\bar{V}^0\mathcal{N}_e$ shifts the chemical potential in the original Hamiltonian $\mathcal{H}_0'$:

$$\mathcal{H}_0' + \bar{V}^0\mathcal{N}_e = \sum_{\nu\mathbf{k}\sigma}(\varepsilon_{\mathbf{k}} - \mu + \bar{V}_0)a^\dagger_{\nu\mathbf{k}\sigma}a_{\nu\mathbf{k}\sigma}.$$

So, redefining the Hamiltonian $\mathcal{H}_0'$, we can assume $V_j^0(\tau) \equiv 0$. In the absence of the charge component, the energy of the fluctuating field (8.27) becomes

$$\begin{aligned}F_0(V) &= \frac{T}{u}\int_0^{1/T}\sum_j \mathbf{V}_j^2(\tau)\,\mathrm{d}\tau\\ &= \frac{1}{2uN_d}T\int_0^{1/T}\sum_{\nu j}\mathrm{Sp}(V_j^2(\tau))\,\mathrm{d}\tau \equiv \frac{1}{2uN_d}\mathrm{Tr}(V^2),\end{aligned} \tag{8.34}$$

and Hamiltonian (8.29) of the noninteracting electrons in the field V is given by

[5] In [5, 6] the charge and spin fluctuations were separated for the first time, and the dominant role of the spin fluctuations was emphasized. The charge field V_j^0 was determined from the charge neutrality condition $\rho_j^0(V) = \langle\rho_j^0(V)\rangle$, as a function of the exchange field $\mathbf{V}_j$.

$$\mathcal{H}'(V) = \mathcal{H}_0' + 2\sum_j \mathbf{V}_j(\tau)\boldsymbol{\rho}_j(\tau). \tag{8.35}$$

It is now easy to verify that the grand canonical average

$$\langle \mathcal{A} \rangle = \frac{1}{\Xi}\mathrm{Tr}\left[\mathcal{A}\,\mathrm{e}^{-\mathcal{H}'/T}\right]$$

of a physical quantity described by the operator $\mathcal{A}$ can be calculated in two steps. First, we calculate the quantum-statistical average $\mathcal{A}(V)$ in the system of noninteracting electrons in the field:

$$\begin{aligned}\mathcal{A}(V) &= \frac{\mathrm{Tr}\left[\mathcal{A}\,T_\tau \exp\left(-\int_0^{1/T} \mathcal{H}'(V)\,\mathrm{d}\tau\right)\right]}{\mathrm{Tr}T_\tau \exp\left(-\int_0^{1/T} \mathcal{H}'(V)\,\mathrm{d}\tau\right)} \\ &= \mathrm{Tr}\left[\mathcal{A}T_\tau \exp\int_0^{1/T} \left(\Omega_1(V) - \mathcal{H}'(V)\right)\mathrm{d}\tau\right]. \end{aligned} \tag{8.36}$$

Then we average over the fluctuating field configurations

$$\langle \mathcal{A} \rangle = \int \mathcal{A}(V)p(V)\mathrm{D}V \equiv \langle \mathcal{A}(V) \rangle, \tag{8.37}$$

with the probability density

$$p(V) = \left(\int \mathrm{e}^{-(F_0(V)+\Omega_1(V))/T}\,\mathrm{D}V\right)^{-1} \mathrm{e}^{-(F_0(V)+\Omega_1(V))/T}.$$

8.3 Exact Relations

8.3.1 Field-Dependent Thermodynamic Potential

In order to determine the thermodynamic potential $\Omega_1(V)$, we need to consider the fermion-type *temperature* Green function

$$\bar{G}_{\nu jj'\sigma\sigma'}(\tau,\tau') = -\langle T_\tau\, a_{\nu j\sigma}(\tau) a^\dagger_{\nu j'\sigma'}(\tau')\rangle \equiv \begin{cases} -\langle a_{\nu j\sigma}(\tau) a^\dagger_{\nu j'\sigma'}(\tau')\rangle, & \tau > \tau', \\ \langle a^\dagger_{\nu j'\sigma'}(\tau') a_{\nu j\sigma}(\tau)\rangle, & \tau < \tau'. \end{cases}$$

Here $\langle \ldots \rangle$ is the grand canonical average of *interacting* electrons, $a^\dagger_{\nu j\sigma}(\tau)$ and $a_{\nu j\sigma}(\tau)$ are the Wannier creation-annihilation operators in the "Heisenberg" representation and $\tau \in [0, 1/T]$. Note that this is the same Green function of interacting electrons as (6.23) but with the additional band index. The reason why we use bar in the DSFT is the following.

According to formula (8.37) the Green function $\bar{G}$ can be expressed as the average of the Green function of noninteracting electrons in the field $G(V)$ over the fluctuating field configurations [22, 23]:

$$\bar{G}_{\nu jj'\sigma\sigma'}(\tau,\tau') = \int G_{\nu jj'\sigma\sigma'}(\tau,\tau',V)\,p(V)\,\mathrm{D}V \equiv \langle\langle G_{\nu jj'\sigma\sigma'}(\tau,\tau')\rangle_V\rangle,$$

or briefly $\bar{G} = \langle G(V)\rangle$, where

$$G_{\nu jj'\sigma\sigma'}(\tau,\tau',V) = -\frac{\mathrm{Tr}\Big[T_\tau a_{\nu j\sigma}(\tau)a^\dagger_{\nu j'\sigma'}(\tau')T_\tau \exp\Big(-\int_0^{1/T}\mathcal{H}'(V)\,\mathrm{d}\tau''\Big)\Big]}{\mathrm{Tr}T_\tau \exp\Big(-\int_0^{1/T}\mathcal{H}'(V)\,\mathrm{d}\tau''\Big)}$$
$$\equiv -\big\langle T_\tau a_{\nu j\sigma}(\tau)a^\dagger_{\nu j'\sigma'}(\tau')\big\rangle_V. \qquad (8.38)$$

In particular, the Green function in the absence of the fluctuating field (zeroth Green function) is

$$G^0_{\nu jj'\sigma}(\tau,\tau') = -\frac{\mathrm{Tr}\Big[T_\tau a_{\nu j\sigma}(\tau)a^\dagger_{\nu j'\sigma}(\tau')\,\mathrm{e}^{-\mathcal{H}'_0/T}\Big]}{\mathrm{Tr}\mathrm{e}^{-\mathcal{H}'_0/T}}$$
$$\equiv -\big\langle T_\tau a_{\nu j\sigma}(\tau)a^\dagger_{\nu j'\sigma}(\tau')\big\rangle_0. \qquad (8.39)$$

The Green function $G(V)$ satisfies the integral equation [22]

$$G_{\nu jj'\sigma\sigma'}(\tau,\tau',V) = G^0_{\nu jj'\sigma}(\tau,\tau')\delta_{\sigma\sigma'}$$
$$+\int\sum_{j''\sigma''} G^0_{\nu jj''\sigma}(\tau,\tau'')V_{j''\sigma\sigma''}(\tau'')G_{\nu j''j'\sigma''\sigma'}(\tau'',\tau',V)\,\mathrm{d}\tau'',$$

or in a compact operator form

$$G = G^0 + G^0VG, \qquad (8.40)$$

where $G \equiv G(V)$ and V is the diagonal matrix with the elements

$$V_{jj'\sigma\sigma'}(\tau,\tau') = V_{j\sigma\sigma'}(\tau)\delta_{jj'}\delta(\tau-\tau').$$

Equation (8.40) is sometimes called the *Dyson equation.* (A simple derivation of the Dyson equation for a static field V is given in Appendix A.1.1.)

A method to relate $\Omega_1(V)$ to the Green function $G(V)$ is to vary the strength of the field V (see [24] and references therein). We consider the Hamiltonian

$$\mathcal{H}'(\lambda) = \mathcal{H}'_0 + \lambda\hat{V}, \qquad (8.41)$$

where the interaction $\hat{V}$ is given by (8.30) and parameter λ increases adiabatically from 0 to 1. Then $\mathcal{H}'(0) = \mathcal{H}'_0$ and $\mathcal{H}'(1) = \mathcal{H}'(V)$. The thermodynamic potential corresponding to $\mathcal{H}'(\lambda)$ is given by

$$\Omega_1(\lambda) = -T\ln\mathrm{Tr}\Big[T_\tau \exp\Big(-\int_0^{1/T}\mathcal{H}'(\lambda)\,\mathrm{d}\tau\Big)\Big].$$

The derivative of the potential with respect to λ is

$$\frac{\partial\Omega_1(\lambda)}{\partial\lambda} = \Big\langle T\int_0^{1/T} V\,\mathrm{d}\tau\Big\rangle_\lambda, \qquad (8.42)$$

where the average $\langle\ldots\rangle_\lambda$ of an arbitrary operator $\mathcal{O}$ is defined as

$$\langle\mathcal{O}\rangle_\lambda \equiv \frac{\mathrm{Tr}\Big(\mathcal{O}\,T_\tau \exp(-\int_0^{1/T}\mathcal{H}'(\lambda)\,\mathrm{d}\tau)\Big)}{\mathrm{Tr}\Big(T_\tau \exp(-\int_0^{1/T}\mathcal{H}'(\lambda)\,\mathrm{d}\tau)\Big)}. \qquad (8.43)$$

Substituting (8.30) in (8.42), we have

$$\frac{\partial \Omega_1(\lambda)}{\partial \lambda} = \left\langle T \int_0^{1/T} \Big(\sum_{\nu j \sigma \sigma'} V_{j\sigma\sigma'}(\tau) a_{\nu j\sigma}^\dagger(\tau) a_{\nu j\sigma'}(\tau) \Big) \, \mathrm{d}\tau \right\rangle_\lambda$$

$$= T \int_0^{1/T} \sum_{\nu j \sigma \sigma'} \Big(V_{j\sigma\sigma'}(\tau) \langle a_{\nu j\sigma}^\dagger(\tau) a_{\nu j\sigma'}(\tau) \rangle_\lambda \Big) \, \mathrm{d}\tau. \tag{8.44}$$

Introducing the Green function

$$G^\lambda_{\nu j j \sigma\sigma'}(\tau, \tau) = -\langle T_\tau a_{\nu j\sigma}(\tau) \, a^+_{\nu j\sigma'}(\tau + 0^+) \rangle_\lambda,$$

we write (8.44) as

$$\frac{\partial \Omega_1(\lambda)}{\partial \lambda} = T \mathrm{Tr}(V G^\lambda).$$

Integrating over λ between 0 and 1, we obtain

$$\Omega_1(1) - \Omega_1(0) = \int_0^1 T \mathrm{Tr}(V G^\lambda) \, \mathrm{d}\lambda, \tag{8.45}$$

where $\Omega_1(1) = \Omega_1(V)$ and $\Omega_1(0) = -T \ln \mathrm{Tr} \exp(-\mathcal{H}'_0/T)$.

The Green function G^λ satisfies the equation similar to (8.40):

$$G^\lambda = G^0 + \lambda G^0 V G^\lambda.$$

Solving this equation for G^λ yields

$$G^\lambda = (1 - \lambda G^0 V)^{-1} G^0. \tag{8.46}$$

Substituting in (8.45), we obtain

$$\Omega_1(1) - \Omega_1(0) = T \mathrm{Tr} \int_0^1 (1 - \lambda G^0 V)^{-1} G^0 V \, \mathrm{d}\lambda$$

$$= -T \mathrm{Tr} \int_0^1 \frac{\mathrm{d}}{\mathrm{d}\lambda} \ln(1 - \lambda G^0 V) \, \mathrm{d}\lambda = -T \mathrm{Tr} \ln(1 - G^0 V). \tag{8.47}$$

From (8.46) at $\lambda = 1$ we have $1 - G^0 V = G^0 (G(V))^{-1}$. Substituting and using formula (A.20):

$$\mathrm{Tr} \ln(AB) = \mathrm{Tr} \ln A + \mathrm{Tr} \ln B,$$

which is valid for any Hermitian matrices A and B, we finally obtain [24]

$$\Omega_1(V) = -T \ln \mathrm{Tr} \, \mathrm{e}^{-\mathcal{H}'_0/T} - T \mathrm{Tr} \ln G^0 + T \mathrm{Tr} \ln G(V). \tag{8.48}$$

Now, we proceed to the *canonical* ensemble, replacing the grand partition function (8.26) by the canonical partition function

$$Z = \left(\int \mathrm{e}^{-F_0(V)/T} \, \mathrm{D}V \right)^{-1} \int \mathrm{e}^{-F(V)/T} \, \mathrm{D}V, \tag{8.49}$$

where $F(V) = F_0(V) + F_1(V)$. Using formula (2.26) that relates the thermodynamic potential with the free energy, in place of (8.48) we have

$$F_1(V) = -T \ln \mathrm{Tr} \, \mathrm{e}^{-\mathcal{H}_0/T} - T \mathrm{Tr} \ln G^0 + T \mathrm{Tr} \ln G(V). \tag{8.50}$$

The canonical ensemble average can be calculated by the same formula (8.37):

$$\langle \mathcal{A} \rangle = \int A(V)p(V)\mathrm{D}V \equiv \langle A(V) \rangle. \tag{8.51}$$

Here

$$A(V) = \frac{\mathrm{Tr}\Big[\mathcal{A}\, T_\tau \exp\Big(-\int_0^{1/T} \mathcal{H}(V)\,\mathrm{d}\tau\Big)\Big]}{\mathrm{Tr} T_\tau \exp\Big(-\int_0^{1/T} \mathcal{H}(V)\,\mathrm{d}\tau\Big)} = \mathrm{Tr}\Big[\mathcal{A}\, T_\tau \exp \int_0^{1/T} \big(F_1(V) - \mathcal{H}(V)\big)\,\mathrm{d}\tau\Big] \tag{8.52}$$

and the probability density of the fluctuating field V is given by

$$p(V) = Q^{-1}\,\mathrm{e}^{-F(V)/T}, \tag{8.53}$$

where

$$Q = \int \mathrm{e}^{-F(V)/T}\,\mathrm{D}V$$

is the normalizing factor.

8.3.2 Mean Spin and Spin-Density Correlator

The functional integral formalism allows to relate the mean and correlators of the spin density with the mean and correlators of the fluctuating field, respectively [1]. We start with the mean-field equation

$$\bar{s}_z = -\frac{1}{u}\bar{V}_z, \tag{8.54}$$

which relates the mean spin $\bar{s}_z = \langle s_j^z \rangle$ with the mean field

$$\bar{V}_z = Q^{-1} \int V_j^z(\tau)\mathrm{e}^{-F(V)/T}\,\mathrm{D}V. \tag{8.55}$$

Due to translational invariance both $\bar{s}_z$ and $\bar{V}_z$ are independent of the site j, and the right-hand side of (8.55) is also independent of the "time" τ.

We carry out the derivation of (8.54) in the momentum-"frequency" representation. Consider the average

$$\left\langle \frac{\partial F_1(V)}{\partial V_{\mathbf{q}m}^\alpha} \right\rangle = Q^{-1} \int \frac{\partial F_1(V)}{\partial V_{\mathbf{q}m}^\alpha}\,\mathrm{e}^{-(F_0(V)+F_1(V))/T}\,\mathrm{D}V.$$

Rewriting the latter as

$$\left\langle \frac{\partial F_1(V)}{\partial V_{\mathbf{q}m}^\alpha} \right\rangle = -T Q^{-1} \int \mathrm{e}^{-F_0(V)/T} \frac{\partial}{\partial V_{\mathbf{q}m}^\alpha}\Big(\mathrm{e}^{-F_1(V)/T}\Big)\mathrm{D}V$$

and integrating by parts, we obtain

$$\left\langle \frac{\partial F_1(V)}{\partial V_{\mathbf{q}m}^\alpha} \right\rangle = -Q^{-1} \int \frac{\partial F_0(V)}{\partial V_{\mathbf{q}m}^\alpha}\,\mathrm{e}^{-(F_0(V)+F_1(V))/T}\,\mathrm{D}V = -\left\langle \frac{\partial F_0(V)}{\partial V_{\mathbf{q}m}^\alpha} \right\rangle.$$

That means

$$\left\langle \frac{\partial F(V)}{\partial V_{\mathbf{q}m}^\alpha} \right\rangle = \left\langle \frac{\partial F_0(V)}{\partial V_{\mathbf{q}m}^\alpha} \right\rangle + \left\langle \frac{\partial F_1(V)}{\partial V_{\mathbf{q}m}^\alpha} \right\rangle = 0. \tag{8.56}$$

Note that relation (8.56) is independent of a particular form of the functions $F_0(V)$ and $F_1(V)$.

To obtain the explicit form (8.54), we start with the first term in (8.56). Using the Fourier transformations (C.19) and (C.29), we write the energy of the field as

$$F_0(V) = \frac{T}{u}\int_0^{1/T}\sum_j \mathbf{V}_j^2(\tau)\,\mathrm{d}\tau = \frac{N}{u}\sum_{\mathbf{q}m\alpha}|V_{\mathbf{q}m}^{\alpha}|^2. \tag{8.57}$$

Differentiating and averaging, in the ferromagnetic state we obtain

$$\left\langle\frac{\partial F_0(V)}{\partial V_{\mathbf{q}m}^{\alpha}}\right\rangle = \frac{2N}{u}\bar{V}_z\delta_{\mathbf{q}0}\delta_{m0}\delta_{\alpha z}. \tag{8.58}$$

Next, we transform the second term in equation (8.56). According to (8.28) and (8.29) the free energy in the field V is given by the expression

$$F_1(V) = -T\ln\mathrm{Tr}\,T_\tau\,\exp\left(-\int_0^{1/T}\mathcal{H}(V)\,\mathrm{d}\tau\right), \tag{8.59}$$

where the Hamiltonian is

$$\mathcal{H}(V) = \mathcal{H}_0 + 2\sum_{j\alpha}V_j^{\alpha}(\tau)s_j^{\alpha}(\tau).$$

Using the Fourier transformations (C.16) and (C.18), we come to

$$F_1(V) = -T\ln\mathrm{Tr}\,\exp\left(-\frac{1}{T}\left(\mathcal{H}_0 + 2\sum_{\mathbf{q}m\alpha}V_{\mathbf{q}m}^{\alpha}s_{-\mathbf{q}-m}^{\alpha}\right)\right). \tag{8.60}$$

Differentiating and averaging, we obtain

$$\left\langle\frac{\partial F_1(V)}{\partial V_{\mathbf{q}m}^{\alpha}}\right\rangle = 2\bar{s}_zN\delta_{\mathbf{q}0}\delta_{m0}\delta_{\alpha z}. \tag{8.61}$$

Substitution of (8.58) and (8.61) in (8.56) gives the mean-field equation (8.54).

Similarly, for the spin-density correlator we prove the relation

$$\left\langle\Delta s_j^{\alpha}(\tau)\Delta s_{j'}^{\beta}(\tau')\right\rangle = \frac{1}{u^2}\left\langle\Delta V_j^{\alpha}(\tau)\Delta V_{j'}^{\beta}(\tau')\right\rangle - \frac{1}{2u}\delta_{jj'}\delta(\tau-\tau')\delta_{\alpha\beta}, \tag{8.62}$$

where $\Delta s_j^{\alpha}(\tau) = s_j^{\alpha}(\tau) - \bar{s}_\alpha$ and $\Delta V_j^{\alpha}(\tau) = V_j^{\alpha}(\tau) - \bar{V}_\alpha$ (for details, see Appendix C.4). Integration of relation (8.62) with respect to "time" yields the equation for the *single-site* spin fluctuation $\langle(\Delta s_j^{\alpha})^2\rangle \equiv \langle\Delta s_\alpha^2\rangle$,

$$\left\langle\Delta s_\alpha^2\right\rangle = \frac{1}{u^2}\left\langle\Delta V_\alpha^2\right\rangle - \frac{T}{2u}, \tag{8.63}$$

where

$$\left\langle\Delta V_\alpha^2\right\rangle = Q^{-1}\int(\Delta V_j^{\alpha}(\tau))^2\mathrm{e}^{-F(V)/T}\,\mathrm{D}V$$

is the mean-square fluctuation of the field.

References

1. N.B. Melnikov, B.I. Reser, V.I. Grebennikov, J. Phys. Condens. Matter **23**, 276003 (2011)
2. J.C. Slater, Phys. Rev. **49**, 537 (1936)
3. E.C. Stoner, Proc. R. Soc. Lond. A **165**, 372 (1938)
4. J. Hubbard, Proc. R. Soc. Lond. A **276**, 238 (1963)
5. J. Hubbard, Phys. Rev. B **19**, 2626 (1979)
6. J. Hubbard, Phys. Rev. B **20**, 4584 (1979)
7. H. Hasegawa, J. Phys. Soc. Jpn. **46**, 1504 (1979)

8. H. Hasegawa, J. Phys. Soc. Jpn. **49**, 178 (1980)
9. H. Hasegawa, J. Phys. Soc. Jpn. **49**, 963 (1980)
10. T. Moriya, *Spin Fluctuations in Itinerant Electron Magnetism* (Springer, Berlin, 1985)
11. J.C. Slater, J. Appl. Phys. **8**, 385 (1937)
12. J.C. Slater, *Quantum Theory of Molecules and Solids, Vol. 4: The Self-consistent Field for Molecules and Solids* (McGraw-Hill, New York, 1974)
13. Y. Kakehashi, M.A.R. Patoary, J. Phys. Soc. Jpn. **80**, 034706 (2011)
14. J. Hubbard, in *Electron Correlation and Magnetism in Narrow-Band Systems*, ed. by T. Moriya (Springer, Berlin, 1981), p. 29
15. R.E. Prange, in *Electron Correlation and Magnetism in Narrow-Band Systems*, ed. by T. Moriya (Springer, Berlin, 1981), p. 55
16. B.V. Medvedev, *Principles of Theoretical Physics* (Nauka, Moscow, 1977) [in Russian]
17. R.L. Stratonovich, Sov. Phys. Dokl. **2**, 416 (1958)
18. J. Hubbard, Phys. Rev. Lett. **3**(2), 77 (1959)
19. V.I. Grebennikov, Y.I. Prokopjev, O.B. Sokolov, E.A. Turov, Phys. Met. Metallogr. **52**, 1 (1981)
20. L.H. Ryder, *Quantum Field Theory*, 2nd edn. (Cambridge University Press, Cambridge, 1996)
21. J. Zinn-Justin, *Path Integrals in Quantum Mechanics* (Oxford University Press, Oxford, 2004)
22. J.A. Hertz, M.A. Klenin, Phys. Rev. B **10**(3), 1084 (1974)
23. J.A. Hertz, M.A. Klenin, Physica B&C **91B**, 49 (1977)
24. N.B. Melnikov, B.I. Reser, V.I. Grebennikov, J. Phys. A Math. Theor. **43**, 195004 (2010)

9 Gaussian Approximation

The only form of functional integrals we can evaluate is the Gaussian quadrature... (N.N. Bogoliubov and D.V. Shirkov, Introduction to the Theory of Quantized Fields, *3rd edn., Wiley, New York, 1980)*

In this chapter we describe the Gaussian approximation of the fluctuating field in the functional integral method. First, we present the simplest Gaussian approximation based on the saddle-point method. This approximation leads to the Stoner mean-field equations and RPA dynamic susceptibility. The *optimal* Gaussian approximation in the DSFT utilizes a quadratic approximation of the free energy based on a variational principle, which we describe in a rather general form here. The optimal Gaussian approximation allows to take both quantum nature (dynamics) and spatial correlation (nonlocality) of thermal fluctuations of the electron spin density.

9.1 Motivation

The task of SFT is to evaluate the partition function (8.49) and averages of the form (8.51). The integrals in (8.49) and (8.51) depend on the sum $F(V) = F_0(V) + F_1(V)$ of the energy of the field $F_0(V)$ and free energy of noninteracting electrons in the field $F_1(V)$. Using formulae (8.34) and (8.50) and omitting the V-independent term, which is unimportant, we can write

$$F(V) = (2uN_{\rm d})^{-1}{\rm Tr}V^2 + T{\rm Tr}\ln G(V). \tag{9.1}$$

Formally, the Green function $G(V)$ can be expressed in terms of G^0 and V by means of Eq. (8.40):

$$G(V) = G^0(1 - VG^0)^{-1}. \tag{9.2}$$

However, the matrices G^0 and V cannot be diagonalized simultaneously in either the site-"time" or momentum-"frequency" representations. Indeed, in the site-"time" representation, the exchange field is diagonal and the zeroth Green function is translationally invariant,[1]

$$V_{jj'}(\tau,\tau') = V_j(\tau)\delta_{jj'}\delta(\tau-\tau'), \qquad G^0_{jj'}(\tau,\tau') = G^0_{j-j'}(\tau-\tau'). \tag{9.3}$$

In the momentum-"frequency" representation it is the opposite,

$$V_{\mathbf{k}\mathbf{k}'nn'} = V_{\mathbf{k}-\mathbf{k}',n-n'}, \qquad G^0_{\mathbf{k}\mathbf{k}'nn'} = G^0_{\mathbf{k}n}\delta_{\mathbf{k}\mathbf{k}'}\delta_{nn'}, \tag{9.4}$$

i.e. the exchange field is translationally invariant and the zeroth Green function is diagonal (for details, see Appendix C.2). Thus, practical use of the functional integral method requires a suitable approximation of the exact expression (9.1).

[1] In the DSFT we omit the band index ν in the Green functions, because we consider $N_{\rm d}$ *degenerate* d bands.

N. B. Melnikov, B. I. Reser, *Dynamic Spin Fluctuation Theory of Metallic Magnetism*,
https://doi.org/10.1007/978-3-319-92974-3_9

If the fluctuations $\Delta V = V - \bar{V}$ around the mean field $\bar{V}$ are not too large, we can utilize a quadratic approximation $F^{(2)}(V)$ of the function $F(V)$. This implies that the probability density function $p(V) \propto \mathrm{e}^{-F(V)/T}$ of the fluctuating field V is reduced to the Gaussian probability density

$$p^{(2)}(V) \propto \mathrm{e}^{-F^{(2)}(V)/T}. \tag{9.5}$$

Due to the space and "time" translational invariance, the quadratic form $F^{(2)}(V)$ can be written in the momentum-"frequency" representation as

$$F^{(2)}(V) = \sum_{\mathbf{q}m\alpha\beta} \Delta V^{\alpha}_{\mathbf{q}m} A^{\alpha\beta}_{\mathbf{q}m} \Delta V^{\beta}_{-\mathbf{q}-m}, \tag{9.6}$$

where A is the Hermitian matrix of the quadratic form. By the translational invariance, the matrix A is diagonal in momenta $\mathbf{q}$ and "frequencies" m but is not diagonal with respect to $\alpha, \beta = x, y, z$ (for details, see Appendix A.3.3).

9.2 Saddle-Point Approximation

By formula (8.51), an observable $\langle \mathcal{A} \rangle$ is determined by the functional integral

$$\int A(V)p(V)\,\mathrm{D}V \propto \int A(V)\mathrm{e}^{-F(V)/T}\,\mathrm{D}V, \tag{9.7}$$

where the function $A(V)$ is given by formula (8.52). As we already mentioned in Chap. 7, the simplest approximation of the integral is obtained by the saddle-point method. The underlying idea of the method is as follows. At low temperatures, the main contribution to the integral (9.7) comes from the neighbourhood of $\bar{V}$ that minimizes $F(V)$ (see, e.g. [1]). The minimum is obtained from the condition that the linear part vanishes:

$$\frac{\partial F(\bar{V})}{\partial V}\Delta V = 0. \tag{9.8}$$

Near $\bar{V}$ the function $F(V)$ is replaced by the quadratic part of its Taylor series,

$$F^{(2)}(\bar{V}) = \frac{1}{2}\frac{\partial^2 F(\bar{V})}{\partial V^2}\Delta V^2 \tag{9.9}$$

(we omit the unimportant field-independent term). If T is small, the probability density (9.5) is close to the delta function (A.36), and the functional integral is approximated as

$$\int A(V)p^{(2)}(V)\,\mathrm{D}V \approx A(\bar{V}).$$

To write Eq. (9.8) explicitly, we linearize expression (9.1). Using the expansion (A.14), we write (9.8) as

$$T\,\mathrm{Tr}\left(\frac{\bar{V}\Delta V}{uN_{\mathrm{d}}T} + G(\bar{V})\Delta V\right) = 0. \tag{9.10}$$

The Green function $G(\bar{V}) = (z + \mu - \mathcal{H}_0 - \bar{V})^{-1}$ is diagonal in the momentum-"frequency" representation. Choosing the z-axis along the direction of the mean field: $\bar{V} = \bar{V}^z\sigma^z$, we have

$$G_{\mathbf{k}n\sigma}(\bar{V}) = \frac{1}{\mathrm{i}\omega_n + \mu - \varepsilon_{\mathbf{k}} - \sigma\bar{V}_z}, \tag{9.11}$$

where $\sigma = \pm 1$. Using $G^x(\bar{V}) = 0$ and $G^y(\bar{V}) = 0$, we rewrite (9.10) as

$$(uN_\mathrm{d})^{-1}\mathrm{Tr}(\bar{V}^z\Delta V^z) + T\mathrm{Tr}(G^z(\bar{V})\Delta V^z) = 0. \tag{9.12}$$

Since the mean field has only one nonzero Fourier coefficient $\bar{V}^z_{\mathbf{q}m} = \bar{V}_z\delta_{\mathbf{q}0}\delta_{m0}$, the first term on the left-hand side of (9.12) reduces to

$$(uN_\mathrm{d})^{-1}\mathrm{Tr}\big(\bar{V}^z\Delta V^z\big) = \tilde{u}^{-1}\bar{V}_z\Delta V^z_{00}. \tag{9.13}$$

Using relations (9.4), we can write the second term of (9.12) as

$$\mathrm{Tr}\big(G^z(\bar{V})\Delta V^z\big) = N_\mathrm{d}\sum_{\mathbf{k}n} G^z_{\mathbf{k}n}(\bar{V})\Delta V^z_{\mathbf{kk},nn} = \big(\mathrm{Tr}G^z(\bar{V})\big)\Delta V^z_{00}. \tag{9.14}$$

Substituting (9.13) and (9.14) in (9.12), we come to

$$\bar{V}_z = -u\bar{s}_z, \tag{9.15}$$

where the mean spin $\bar{s}_z$ is given by

$$\bar{s}_z = N^{-1}N_\mathrm{d}T\sum_{\mathbf{q}n} G^z_{\mathbf{q}n}(\bar{V}) \tag{9.16}$$

(cf. with the single-band case (6.33) and (6.36)). The condition that the total number of electrons is conserved leads to the chemical potential equation

$$N_\mathrm{e} = T\mathrm{Tr}\, G(\bar{V}). \tag{9.17}$$

Calculating sums over the thermodynamic "frequencies", we obtain the Stoner mean-field equations

$$\bar{s}_z = \frac{N_\mathrm{d}}{N}\frac{1}{2}\sum_{\mathbf{k}}(f(\varepsilon_{\mathbf{k}\uparrow}) - f(\varepsilon_{\mathbf{k}\downarrow})), \tag{9.18}$$

$$N_\mathrm{e} = N_\mathrm{d}\sum_{\mathbf{k}}(f(\varepsilon_{\mathbf{k}\uparrow}) - f(\varepsilon_{\mathbf{k}\downarrow})), \tag{9.19}$$

where $\varepsilon_{\mathbf{k}\sigma} = \varepsilon_{\mathbf{k}} - \sigma u\bar{s}_z$ (for details, see Appendix A.2.5). In the single-band case, Eqs. (9.18) and (9.19) reduce to (4.16).

Now, we write explicitly the quadratic form (9.9) that approximates the function $F(V) = F_0(V) + F_1(V)$. Using (8.57), we write the first term as

$$F_0(V) = \frac{1}{\tilde{u}}\sum_{\mathbf{q}m\alpha}|V^\alpha_{\mathbf{q}m}|^2 = \frac{1}{\tilde{u}}\sum_{\mathbf{q}m\alpha}V^\alpha_{\mathbf{q}m}V^\alpha_{-\mathbf{q}-m},$$

where $\tilde{u} = u/N$ is the Fourier transform of the effective interaction constant u. Taking the second derivative of $F(V)$, we obtain

$$F^{(2)}(V) = \sum_{\mathbf{q}m\alpha\beta}\Delta V^\alpha_{\mathbf{q}m}\left(\frac{\delta_{\alpha\beta}}{\tilde{u}} - \chi^{0\alpha\beta}_{\mathbf{q}m}\right)\Delta V^\beta_{-\mathbf{q}-m}, \tag{9.20}$$

where

$$\chi^{0\alpha\beta}_{\mathbf{q}m} = -\frac{1}{2}\frac{\partial^2 F_1(\bar{V})}{\partial V^\alpha_{\mathbf{q}m}\partial V^\beta_{-\mathbf{q}-m}} \tag{9.21}$$

is the *unenhanced* (zeroth) susceptibility in units of $g^2\mu_\mathrm{B}^2/2$. Expanding the free energy $F_1(V) = T\mathrm{Tr}\ln G(V)$ up to the second order (see Appendix A.1.2):

$$F_1(V) = T\mathrm{Tr}(G(\bar{V})\Delta V) + \frac{1}{2}T\mathrm{Tr}(G(\bar{V})\Delta V G(\bar{V})\Delta V) + \cdots,$$

we write formula (9.21) in the form

$$\chi^{0\alpha\beta}_{\mathbf{q}m} = -\frac{T}{4}\frac{\partial^2\mathrm{Tr}(G(\bar{V})\Delta V G(\bar{V})\Delta V)}{\partial V^\alpha_{\mathbf{q}m}\partial V^\beta_{-\mathbf{q}-m}}. \tag{9.22}$$

Since the Green function $G(\bar{V})$ is diagonal in the momentum-"frequency" representation (9.11), we have (see Appendix C.2)

$$\mathrm{Tr}\big(G(\bar{V})\varDelta V G(\bar{V})\varDelta V\big) = N_{\mathrm{d}} \sum_{\mathbf{qk}mn} \mathrm{Sp}\big(G_{\mathbf{k}n}(\bar{V})\varDelta V_{\mathbf{q}m} G_{\mathbf{k-q},\,n-m}(\bar{V})\varDelta V_{-\mathbf{q}-m}\big).$$

Using the expansion of the 2×2 matrices $G_{\mathbf{k}n}(\bar{V})$ and $V_{\mathbf{q}m}$ in terms of the Pauli matrices and taking into account the axial symmetry ($G^x(\bar{V}) = 0$ and $G^y(\bar{V}) = 0$), we calculate (9.22):

$$\chi^{0\alpha\beta}_{\mathbf{q}m} = -\frac{N_{\mathrm{d}}}{2} T \sum_{\mathbf{k}n\gamma_1\gamma_2} G^{\gamma_1}_{\mathbf{k}n}(\bar{V}) G^{\gamma_2}_{\mathbf{k-q},n-m}(\bar{V}) \mathrm{Sp}\big(\sigma^{\gamma_1}\sigma^{\alpha}\sigma^{\gamma_2}\sigma^{\beta}\big), \tag{9.23}$$

where $\gamma = 0, z$. Using formulae (A.22) for trace of the Pauli matrices, we obtain

$$\chi^{0}_{\mathbf{q}m} = \begin{pmatrix} \chi^{0xx}_{\mathbf{q}m} & \chi^{0xy}_{\mathbf{q}m} & 0 \\ -\chi^{0xy}_{\mathbf{q}m} & \chi^{0xx}_{\mathbf{q}m} & 0 \\ 0 & 0 & \chi^{0zz}_{\mathbf{q}m} \end{pmatrix}, \tag{9.24}$$

in full agreement with the general form of the susceptibility tensor (2.13) in a system with axial symmetry. In Sect. 10.4, we show that the *enhanced* susceptibility in the saddle-point approximation has an RPA form.

9.3 Free Energy Minimum Principle

The approximation of the functional integral in the DSFT is based on a variational principle, which we describe in a rather general form here. Let $F_{\mathrm{mod}}(V)$ be the "modelling" free energy, which approximates the free energy $F(V)$. Then the identity

$$\int \mathrm{e}^{-F(V)/T}\,\mathrm{D}V = \int \mathrm{e}^{-(F(V)-F_{\mathrm{mod}}(V))/T}\,\mathrm{e}^{-F_{\mathrm{mod}}(V)/T}\,\mathrm{D}V$$

can be rewritten as

$$\left(\int \mathrm{e}^{-F_{\mathrm{mod}}(V)/T}\,\mathrm{D}V\right)^{-1} \int \mathrm{e}^{-F(V)/T}\,\mathrm{D}V = \left\langle \mathrm{e}^{-(F(V)-F_{\mathrm{mod}}(V))/T}\right\rangle_{\mathrm{mod}},$$

where the average is defined by

$$\langle \ldots \rangle_{\mathrm{mod}} \equiv \left(\int \mathrm{e}^{-F_{\mathrm{mod}}(V)/T}\,\mathrm{D}V\right)^{-1} \int \ldots\, \mathrm{e}^{-F_{\mathrm{mod}}(V)/T}\,\mathrm{D}V.$$

Applying the inequality $\langle \exp f\rangle \geq \exp\langle f\rangle$ with f being a real function and taking the logarithm, we come to the upper bound for the total free energy:

$$\mathcal{F} \leq \mathcal{F}_{\mathrm{mod}} + \big\langle F(V) - F_{\mathrm{mod}}(V)\big\rangle_{\mathrm{mod}}, \tag{9.25}$$

where[2]

$$\mathcal{F} = -T \ln \int \mathrm{e}^{-F(V)/T}\,\mathrm{D}V, \qquad \mathcal{F}_{\mathrm{mod}} = -T \ln \int \mathrm{e}^{-F_{\mathrm{mod}}(V)/T}\,\mathrm{D}V. \tag{9.26}$$

[2]By formula (8.49) the free energy $\mathcal{F} = -T \ln Z$ should be written as

$$\mathcal{F} = T \ln \int \mathrm{e}^{-F_0(V)/T}\,\mathrm{D}V - T \ln \int \mathrm{e}^{-F(V)/T}\,\mathrm{D}V,$$

and the same first term should appear in $\mathcal{F}_{\mathrm{mod}}$. Since the extra terms cancel in (9.25), we omit them for brevity.

Formula (9.25) was obtained by Feynman [2] and is sometimes called the *Feynman inequality* (see also [3]). An operator formulation of the free energy minimum principle and a proof, based on the Peierls variational principle [4], are given by Tyablikov [5].

By the *optimal* approximation we call the one that minimizes the right-hand side of inequality (9.25) in a certain class of functions $F_{\text{mod}}(V)$. Using the "modelling" function of the form $F_{\text{mod}}(V) = \sum_j f_j(V_j)$, where $f_j(V_j)$ is an arbitrary function of a single variable V_j, and applying the free energy minimum principle, one obtains the local approximation of SFT [6,7].

Dynamic nonlocal approximation of SFT is based on the *optimal* Gaussian approximation of the fluctuating field, i.e. the "modelling" function $F_{\text{mod}}(V)$ is chosen from the class of quadratic functions. In the *paramagnetic* region the optimal Gaussian approximation was obtained in [8,9] and applied to iron with a model density of states in [10]. In the *ferromagnetic* region, the optimal Gaussian approximation was applied in [11,12]. For an *arbitrary magnetic ordering*, the optimal Gaussian approximation was obtained in [13].

9.4 Optimal Gaussian Approximation

9.4.1 General Formulation

In the optimal Gaussian approximation the parameters $\bar{V}$ and $A^{\alpha\beta}_{\mathbf{q}m}$ of the quadratic form (9.6) are evaluated from the system of nonlinear equations

$$\left\langle \frac{\partial F(V)}{\partial V^{\alpha}_{\mathbf{q}m}} \right\rangle = 0, \qquad A^{\alpha\beta}_{\mathbf{q}m} = \frac{1}{2} \left\langle \frac{\partial^2 F(V)}{\partial V^{\alpha}_{\mathbf{q}m}\, \partial V^{\beta}_{-\mathbf{q}-m}} \right\rangle, \tag{9.27}$$

where the average $\langle \ldots \rangle$ is calculated self-consistently with the Gaussian probability density (9.5) (Eq. (9.27) are derived in Appendix B.4.2). The optimal Gaussian approximation is quite general and can be used to describe not only ferromagnets but also antiferromagnets or ferrimagnets (see, e.g. [14, 15]).

In the ferromagnetic state, the mean field $\bar{V}$ is independent of site and hence its Fourier transform has only one nonzero coefficient: $\bar{V}^{z}_{\mathbf{q}m} = \bar{V}_z \delta_{\mathbf{q}0} \delta_{m0}$. Therefore, we need to consider the first equation in (9.27) only at $\mathbf{q} = 0$ and $m = 0$. In the paramagnetic state the first equation in (9.27) is redundant, because the mean field vanishes, but the coefficients $A^{\alpha\beta}_{\mathbf{q}m}$ remain and the second equation in (9.27) must be written for all $\mathbf{q}$ and m, just as in the ferromagnetic state.

Note that the first equation has the same form as (8.56). The only difference is the form of average. The second equation in (9.27) is specific to the Gaussian approximation. The optimal Gaussian approximation (9.27) should be contrasted to the saddle-point approximation:

$$\frac{\partial F(\bar{V})}{\partial V^{\alpha}_{\mathbf{q}m}} = 0, \qquad A^{\alpha\beta}_{\mathbf{q}m} = \frac{1}{2} \frac{\partial^2 F(\bar{V})}{\partial V^{\alpha}_{\mathbf{q}m}\, \partial V^{\beta}_{-\mathbf{q}-m}}, \tag{9.28}$$

where both derivatives are taken at the mean field and there is no feedback from the field fluctuations.

9.4.2 Ferromagnetic State

We now develop the equations of the optimal Gaussian approximation (9.27). Similar to the saddle-point approximation, we write the first equation in (9.27) as

$$\bar{V}_z = -u \bar{s}_z. \tag{9.29}$$

Here the mean spin is given by

$$\bar{s}_z = N^{-1} N_{\text{d}} T \sum_{\mathbf{q}n} \bar{G}^{z}_{\mathbf{q}n}, \tag{9.30}$$

where the mean Green function $\bar{G}$ is calculated using the optimal Gaussian average. The quadratic form (9.6) of the optimal Gaussian approximation (9.27) is given by

$$F^{(2)}(V) = \sum_{\mathbf{q}m\alpha\beta} \Delta V^{\alpha}_{\mathbf{q}m} \left(\frac{\delta_{\alpha\beta}}{\tilde{u}} - \chi^{0\alpha\beta}_{\mathbf{q}m} \right) \Delta V^{\beta}_{-\mathbf{q}-m}, \tag{9.31}$$

where the *unenhanced* (zeroth) susceptibility is

$$\chi^{0\alpha\beta}_{\mathbf{q}m} = -\frac{1}{2} \left\langle \frac{\partial^2 F_1(V)}{\partial V^{\alpha}_{\mathbf{q}m} \partial V^{\beta}_{-\mathbf{q}-m}} \right\rangle. \tag{9.32}$$

Expanding the free energy $F_1(V) = T\mathrm{Tr}\ln G(V)$ up to the second order,

$$F_1(V) = T\mathrm{Tr}(G(\tilde{V})\Delta V) + \frac{1}{2} T\mathrm{Tr}(G(\tilde{V})\Delta V G(\tilde{V})\Delta V) + \cdots,$$

and replacing the mean of a product of the Green functions $G(V)$ by the product of the mean Green functions $\langle G(\tilde{V})\rangle \equiv \bar{G}$, we write formula (9.32) in the form

$$\chi^{0\alpha\beta}_{\mathbf{q}m} = -\frac{T}{4} \frac{\partial^2 \mathrm{Tr}(\bar{G}\Delta V \bar{G}\Delta V)}{\partial V^{\alpha}_{\mathbf{q}m} \partial V^{\beta}_{-\mathbf{q}-m}}. \tag{9.33}$$

Similar to the saddle-point approximation, we obtain

$$\chi^{0\alpha\beta}_{\mathbf{q}m} = -\frac{N_{\mathrm{d}}}{2} T \sum_{\mathbf{k}n\gamma_1\gamma_2} \bar{G}^{\gamma_1}_{\mathbf{k}n} \bar{G}^{\gamma_2}_{\mathbf{k}-\mathbf{q},n-m} \mathrm{Sp}(\sigma^{\gamma_1}\sigma^{\alpha}\sigma^{\gamma_2}\sigma^{\beta}), \tag{9.34}$$

where $\gamma = 0, z$. Using formulae (A.22) for trace of products of the Pauli matrices, it is easy to check that the unenhanced susceptibility tensor $\chi^0_{\mathbf{q}m}$ has the form (9.24), as it should be in a system with axial symmetry.

In the diagonal approximation [11, 12, 16] of the unenhanced susceptibility: $\chi^{0\alpha\beta}_{\mathbf{q}m} = \chi^{0\alpha}_{\mathbf{q}m}\delta_{\alpha\beta}$, the quadratic form (9.31) becomes

$$F^{(2)}(V) = \sum_{\mathbf{q}m\alpha} \Delta V^{\alpha}_{\mathbf{q}m} \left(\tilde{u}^{-1} - \chi^{0\alpha}_{\mathbf{q}m} \right) \Delta V^{\alpha}_{-\mathbf{q}-m}.$$

Then the mean-square fluctuation $\langle \Delta V^{\alpha}_{\mathbf{q}m} \Delta V^{\alpha}_{-\mathbf{q}-m} \rangle = \langle |\Delta V^{\alpha}_{\mathbf{q}m}|^2 \rangle$ of the Gaussian distribution (9.5) can be written down explicitly (see Appendix A.3.3):

$$\langle |\Delta V^{\alpha}_{\mathbf{q}m}|^2 \rangle = \frac{T}{2(\tilde{u}^{-1} - \chi^{0\alpha}_{\mathbf{q}m})}. \tag{9.35}$$

The condition that the total number of electrons is conserved (6.33) leads to the equation

$$N_{\mathrm{e}} = T\mathrm{Tr}\,\bar{G}. \tag{9.36}$$

9.4.3 Self-Energy Equation

As we showed in Chap. 6, it is convenient to consider the Green function of *interacting* electrons $\bar{G} = \langle G(V) \rangle$ as a function of a certain "effective medium" so that the following relation holds (see, e.g. [17, 18])

$$\bar{G}(z) = (z + \mu - \mathcal{H}_0 - \Sigma(z))^{-1}, \tag{9.37}$$

where $\Sigma(z)$ is the self-energy. Due to translational invariance, the matrices of $\mathcal{H}_0$, $\bar{G}$ and Σ depend on the difference of the site indices $j - j'$. Hence in the momentum representation all of them become $\mathbf{q}$-diagonal, and the diagonal element of the Green function (9.37) is

$$\bar{G}_{\mathbf{q}}(z) = (z + \mu - \varepsilon_{\mathbf{q}} - \Sigma_{\mathbf{q}}(z))^{-1}.$$

To calculate the self-energy Σ, we apply the *quasistatic approximation*. This means we deal with $G(V)$ as if the fluctuating field V was static but the average $\langle G(V) \rangle$ is calculated with the probability density that takes into account contributions from different momenta (nonlocality) and "frequencies" (dynamics). In this approximation it is easy to see the physical meaning of the self-energy Σ.

In the presence of the static exchange field V, the Green function of noninteracting electrons (8.38) reduces to

$$G(V) = (z + \mu - \mathcal{H}_0 - V)^{-1}, \tag{9.38}$$

where V is independent of z. Using $(\mathcal{AB})^{-1} = \mathcal{B}^{-1}\mathcal{A}^{-1}$, we can write (9.38) in the form

$$G(V) = [1 - (z + \mu - \mathcal{H}_0 - \Sigma)^{-1}(V - \Sigma)]^{-1}(z + \mu - \mathcal{H}_0 - \Sigma)^{-1},$$

or shortly,

$$G = (1 - \bar{G}(V - \Sigma))^{-1}\bar{G}. \tag{9.39}$$

Multiplying both sides by $(1 - \bar{G}(V - \Sigma))$ from the left and rearranging, we come to

$$G = \bar{G} + \bar{G}(V - \Sigma)G. \tag{9.40}$$

Substituting (9.39) in the right-hand side of (9.40), we obtain the equation

$$G = \bar{G} + \bar{G}T\bar{G}, \tag{9.41}$$

where the scattering T-matrix is defined by

$$T = (V - \Sigma)(1 - \bar{G}(V - \Sigma))^{-1}.$$

Averaging both sides over V and assuming that the average of the T-matrix is zero (see, e.g. [19]), we come to the equation

$$\bar{T} = \big\langle (V - \Sigma)(1 - \bar{G}(V - \Sigma))^{-1} \big\rangle = 0. \tag{9.42}$$

Similarly, averaging both sides of (9.39), we see that

$$\big\langle (1 - \bar{G}(V - \Sigma))^{-1} \big\rangle = 1. \tag{9.43}$$

Next, we develop the second-order perturbation theory. Introducing $\Delta V = V - \bar{V}$ and $\Delta\Sigma = \Sigma - \bar{V}$, we write Eq. (9.42) as

$$\big\langle (\Delta V - \Delta\Sigma)(1 - \bar{G}(\Delta V - \Delta\Sigma))^{-1} \big\rangle = 0.$$

Taking (9.43) into account, we obtain

$$\Delta\Sigma = \big\langle \Delta V(1 - \bar{G}(\Delta V - \Delta\Sigma))^{-1} \big\rangle. \tag{9.44}$$

Expanding $(1 - \bar{G}(\Delta V - \Delta\Sigma))^{-1}$ in the geometric series (see Appendix A.1.1), we come to

$$\Delta\Sigma = \big\langle \Delta V(1 + \bar{G}(\Delta V - \Delta\Sigma) + \cdots) \big\rangle. \tag{9.45}$$

The term linear in ΔV vanishes because $\langle \Delta V \rangle = 0$. Thus, in the second-order perturbation theory with respect to ΔV, we have

$$\Delta\Sigma = \langle \Delta V \bar{G} \Delta V \rangle. \tag{9.46}$$

In the momentum-"frequency" representation, taking into account

$$\begin{aligned} \bar{G}_{\mathbf{kk}'nn'\sigma\sigma'} &= \bar{G}_{\mathbf{k}n\sigma}\delta_{\mathbf{kk}'}\delta_{nn'}\delta_{\sigma\sigma'}, \\ V_{\mathbf{kk}'nn'\sigma\sigma'} &= V_{\mathbf{k}-\mathbf{k}',n-n',\sigma}\delta_{\sigma\sigma'}, \end{aligned}$$

we obtain

$$\Delta\Sigma_{\mathbf{k}n} = \sum_{\mathbf{k}'n'} \langle \Delta V_{\mathbf{k}-\mathbf{k}',n-n'} \bar{G}_{\mathbf{k}'n'} \Delta V_{\mathbf{k}'-\mathbf{k},n'-n} \rangle .$$

The latter is finally written as

$$\Delta\Sigma_{\mathbf{k}n} = \sum_{\mathbf{q}m} \langle \Delta V_{\mathbf{q}m} \bar{G}_{\mathbf{k}-\mathbf{q},n-m} \Delta V_{-\mathbf{q}-m} \rangle , \tag{9.47}$$

where

$$\bar{G}_{\mathbf{k}n} = (\mathrm{i}\omega_n + \mu - \varepsilon_{\mathbf{k}} - \sigma \bar{V}_z - \Delta\Sigma_{\mathbf{k}n})^{-1} . \tag{9.48}$$

Thus, the optimal Gaussian approximation yields the system of nonlinear equations (9.29), (9.35), (9.36), (9.47) and (9.48) with respect to μ, $\bar{V}$ and $\langle |\Delta V^{\alpha}_{\mathbf{q}m}|^2 \rangle$. However, calculation of all the fluctuations $\langle |\Delta V^{\alpha}_{\mathbf{q}m}|^2 \rangle$ is an excessively complicated procedure since magnetic characteristics, like the magnetization and local magnetic moment, use only their sums over $\mathbf{q}$ and m.

References

1. J. Zinn-Justin, *Path Integrals in Quantum Mechanics* (Oxford University Press, Oxford, 2004)
2. R.P. Feynman, Phys. Rev. **97**, 660 (1955)
3. R.P. Feynman, *Statistical Mechanics*, 2nd edn. (Benjamin, Reading, 1998)
4. R. Peierls, Phys. Rev. **54**, 918 (1938)
5. S.V. Tyablikov, *Methods in the Quantum Theory of Magnetism* (Springer, New York, 1967)
6. J. Hubbard, Phys. Rev. B **19**, 2626 (1979)
7. J. Hubbard, Phys. Rev. B **20**, 4584 (1979)
8. J.A. Hertz, M.A. Klenin, Phys. Rev. B **10**(3), 1084 (1974)
9. J.A. Hertz, M.A. Klenin, Physica B&C **91B**, 49 (1977)
10. V.I. Grebennikov, J. Magn. Magn. Mater. **84**, 59 (1990)
11. V.I. Grebennikov, Phys. Solid State **40**, 79 (1998)
12. B.I. Reser, V.I. Grebennikov, Phys. Met. Metallogr. **85**, 20 (1998)
13. N.B. Melnikov, B.I. Reser, Procs. Steklov Inst. Math. **271**, 149 (2010)
14. V.I. Grebennikov, O.B. Sokolov, J. Phys. Condens. Matter **4**, 3283 (1992)
15. V.I. Grebennikov, D.I. Radzivonchik, Solid State Phenom. **233–234**, 25 (2015)
16. N.B. Melnikov, B.I. Reser, V.I. Grebennikov, J. Phys. Condens. Matter **23**, 276003 (2011)
17. H. Ehrenreich, L.M. Schwartz, in *Solid State Physics: Advances in Research and Applications*, vol. 31, ed. by H. Ehrenreich, F.S.D. Turnbull (Academic, New York, 1976), p. 149
18. D.J. Kim, *New Perspectives in Magnetism of Metals* (Kluwer/Plenum, New York, 1999)
19. Y.A. Izyumov, Y.N. Skryabin, *Statistical Mechanics of Magnetically Ordered Systems* (Springer, Berlin, 1988)

Single-Site Gaussian Approximation

10

Anything that begins well, ends badly. Anything that begins badly, ends worse. (Murphy's Law)

The starting point of SFT is the Hubbard Hamiltonian, which relies on the assumption that the electron–electron interaction is local. Therefore, it seemed quite logical at the beginning to assume that the local spin fluctuations dominate [1–5]. However, the static single-site SFT has a number of problems, which we mentioned in the Introduction. We can do better than treating the fluctuations as purely local. The idea is to consider the local fluctuating field whose mean value and mean-square fluctuations are calculated self-consistently and depend on *all* momenta and "frequencies" [6, 7]. Application of this idea to real metals requires elaborate approximations. In Sects. 10.1–10.3 we develop the DSFT in detail. In Sect. 10.4 we show how to calculate various magnetic characteristics in the functional integral method. In Sect. 10.5, we calculate temperature dependency of magnetic characteristics in the ferromagnetic metals and compare results of different approximations in the DSFT.

10.1 Coherent Potential Equation

Electron–electron interaction generates fluctuations of the charge and spin density both at $T = 0$ and at finite temperatures. The DSFT separates the zero-point fluctuations (ground-state fluctuations) and thermal fluctuations. We assume that correlation effects caused by zero-point fluctuations are already taken into account in some way in the calculation of the electron DOS at $T = 0$ and in the renormalization of the interaction constant u. It remains for us to consider only the temperature-dependent part of the correlation effects, in other words the thermal fluctuations (for details, see [8–11]). The thermal fluctuations possess low energies $\hbar\omega_m$ [6, 12]. Therefore, we can write $\bar{G}_{\mathbf{k}-\mathbf{q},n-m} \approx \bar{G}_{\mathbf{k}-\mathbf{q},n}$ in Eq. (9.47). Then

$$\Delta\Sigma_{\mathbf{k}}(z) = \sum_{\mathbf{q}m} \langle \Delta V_{\mathbf{q}m} \bar{G}_{\mathbf{k}-\mathbf{q}}(z) \Delta V_{-\mathbf{q}-m} \rangle. \tag{10.1}$$

Two extreme estimates are usually used to calculate the sum over the Brillouin zone in Eq. (10.1). The first type of estimate: $\bar{G}_{\mathbf{k}-\mathbf{q}}(z) \approx \bar{G}_{\mathbf{k}}(z)$ implies that the main contribution comes from the quasihomogeneous fluctuations with $q \approx 0$. This is the so-called long-wavelength limit (see, e.g. [6, 7]). The second type of estimate assumes that the Green function $\bar{G}_{\mathbf{k}-\mathbf{q}}(z)$ is replaced by the single-site Green function

$$g(z) = \frac{1}{N} \sum_{\mathbf{k}} \bar{G}_{\mathbf{k}}(z). \tag{10.2}$$

Then (10.1) becomes

$$\Delta\Sigma(z) = \sum_{\mathbf{q}m} \langle \Delta V_{\mathbf{q}m}\, g(z) \Delta V_{-\mathbf{q}-m} \rangle, \tag{10.3}$$

where

$$\Sigma(z) = \frac{1}{N} \sum_{\mathbf{k}} \Sigma_{\mathbf{k}}(z). \tag{10.4}$$

N. B. Melnikov, B. I. Reser, *Dynamic Spin Fluctuation Theory of Metallic Magnetism*,
https://doi.org/10.1007/978-3-319-92974-3_10

The approximation (10.2) assumes that excitations with any $\mathbf{q}$ are roughly the same and is called the *disordered local moment* approach [1, 2]. The self-energy $\Sigma_{\mathbf{k}}(z)$ reduces to the *coherent potential* $\Sigma(z)$, which is uniform in space. The single-site Green function $g(\varepsilon)$ and coherent potential $\Sigma(\varepsilon)$ in Eq. (10.3) are spin-diagonal matrices,

$$g(\varepsilon) = \begin{pmatrix} g_{\uparrow}(\varepsilon) & 0 \\ 0 & g_{\downarrow}(\varepsilon) \end{pmatrix}, \qquad \Sigma(\varepsilon) = \begin{pmatrix} \Sigma_{\uparrow}(\varepsilon) & 0 \\ 0 & \Sigma_{\downarrow}(\varepsilon) \end{pmatrix}. \tag{10.5}$$

The single-site Green function can be expressed in terms of the electron DOS. Indeed, substituting (6.17) in (10.2), we obtain[1]

$$g_{\sigma}(\varepsilon) = \int \frac{\nu_{\sigma}(\varepsilon')}{\varepsilon - \varepsilon'} \mathrm{d}\varepsilon', \tag{10.6}$$

where $\varepsilon = \varepsilon - \mathrm{i}0^{+}$ and

$$\nu_{\sigma}(\varepsilon) = \frac{1}{N} \sum_{\mathbf{k}} A_{\mathbf{k}\sigma}(\varepsilon)$$

is the spin-polarized DOS (per band, site and spin). Taking (6.18) into account, we write

$$\nu_{\sigma}(\varepsilon) = \frac{1}{\pi} \mathrm{Im} g_{\sigma}(\varepsilon). \tag{10.7}$$

In particular, for *noninteracting* electrons in the presence of the exchange field $\sigma \bar{V}_z$, the Green function $\bar{G}_{\mathbf{k}}(\varepsilon)$ is given by (6.39). Therefore, (10.2) yields

$$g_{\sigma}^{0}(\varepsilon) = \frac{1}{N} \sum_{\mathbf{k}} \frac{1}{\varepsilon - \xi_{\mathbf{k}} - \sigma \bar{V}_z}, \tag{10.8}$$

where $\xi_{\mathbf{k}} = \varepsilon_{\mathbf{k}} - \mu$, and (10.7) is written as

$$\nu_{\sigma}^{0}(\varepsilon) = \frac{1}{\pi} \mathrm{Im} \left(\frac{1}{N} \sum_{\mathbf{k}} \frac{1}{\varepsilon - \xi_{\mathbf{k}} - \sigma \bar{V}_z} \right).$$

Using the Sokhotsky formula (A.43), we have

$$\nu_{\sigma}^{0}(\varepsilon) = \frac{1}{N} \sum_{\mathbf{k}} \frac{1}{\pi} \mathrm{Im} \frac{1}{\varepsilon - \xi_{\mathbf{k}\sigma}} = \frac{1}{N} \sum_{\mathbf{k}} \delta(\varepsilon - \xi_{\mathbf{k}\sigma}), \tag{10.9}$$

where $\xi_{\mathbf{k}\sigma} = \xi_{\mathbf{k}} + \sigma \bar{V}_z$. This is the Hartree-Fock DOS (4.24) but written in the *grand* canonical ensemble.

For *interacting* electrons, replacing the self-energy $\Sigma_{\mathbf{k}\sigma}(z)$ by the coherent potential (10.4) in formula (6.21), we obtain

$$\nu_{\sigma}(\varepsilon) = \frac{1}{\pi} \mathrm{Im} \left(\frac{1}{N} \sum_{\mathbf{k}} \frac{1}{\varepsilon - \xi_{\mathbf{k}} - \Sigma_{\sigma}(\varepsilon)} \right).$$

As we explained in Chap. 6, the self-energy smears the delta-peaks of the DOS (10.9). Using (6.21), (10.2) and the definition of the delta function, we have (see, e.g. [13])

$$g_{\sigma}(\varepsilon) = \frac{1}{N} \sum_{\mathbf{k}} \frac{1}{\varepsilon - \xi_{\mathbf{k}} - \Sigma_{\sigma}(\varepsilon)} = \frac{1}{N} \sum_{\mathbf{k}} \int \frac{\delta(\varepsilon' - \xi_{\mathbf{k}})}{\varepsilon - \varepsilon' - \Sigma_{\sigma}(\varepsilon)} \mathrm{d}\varepsilon'.$$

Finally, we write

[1] Henceforth we deal with the *advanced* Green function and omit the superscript.

$$g_\sigma(\varepsilon) = \int \frac{\nu(\varepsilon')}{\varepsilon - \varepsilon' - \Sigma_\sigma(\varepsilon)} \mathrm{d}\varepsilon', \tag{10.10}$$

where

$$\nu(\varepsilon) = \frac{1}{\pi}\mathrm{Im}\left(\frac{1}{N}\sum_{\mathbf{k}} \frac{1}{\varepsilon - \xi_{\mathbf{k}}}\right) = \frac{1}{N}\sum_{\mathbf{k}} \delta(\varepsilon - \xi_{\mathbf{k}}) \tag{10.11}$$

is the nonmagnetic DOS at $T = 0$ (per unit cell, band and spin) in the grand canonical ensemble.

10.2 Single-Site Gaussian Fluctuating Field

As we see from formula (10.10), the primary task is the calculation of the coherent potential $\Sigma_\sigma(\varepsilon)$. Taking the matrix expressions (10.5) into account, we write the coherent potential equation (10.3) as

$$\Delta\Sigma_\sigma(\varepsilon) = g_\sigma(\varepsilon)\langle\Delta V_z^2\rangle + g_{\bar{\sigma}}(\varepsilon)\langle\Delta V_\perp^2\rangle, \tag{10.12}$$

where $\Delta V_\perp^2 = \Delta V_x^2 + \Delta V_y^2$. Here

$$\langle\Delta V_\alpha^2\rangle = \sum_{\mathbf{q}m}\langle|\Delta V_{\mathbf{q}m}^\alpha|^2\rangle, \qquad \alpha = x, y, z, \tag{10.13}$$

is the mean-square fluctuation of the fluctuating field

$$V_\alpha = \sum_{\mathbf{q}m} V_{\mathbf{q}m}^\alpha. \tag{10.14}$$

Using the Fourier transformation (see Appendix C.3), we can verify that

$$V_\alpha = V_{j=0}^\alpha(\tau = 0)$$

is the single-site Gaussian fluctuating field independent of site and "time".

Instead of the fluctuation (10.13), we consider

$$\langle\Delta V_\alpha^2\rangle' = \sum_{\mathbf{q}m}\langle|\Delta V_{\mathbf{q}m}^\alpha|^2\rangle', \tag{10.15}$$

where $\langle|\Delta V_{\mathbf{q}m}^\alpha|^2\rangle' \equiv \langle|\Delta V_{\mathbf{q}m}^\alpha|^2\rangle - \langle|\Delta V_{\mathbf{q}m}^\alpha|^2\rangle_0$ is the mean-square fluctuation (9.35) minus the intrinsic mean-square fluctuation of the field

$$\langle|\Delta V_{\mathbf{q}m}^\alpha|^2\rangle_0 \equiv \frac{\int |\Delta V_{\mathbf{q}m}^\alpha|^2 \mathrm{e}^{-F_0(V)/T}\,\mathrm{D}V}{\int \mathrm{e}^{-F_0(V)/T}\,\mathrm{D}V} = \frac{\tilde{u}T}{2}.$$

The latter is independent of $\mathbf{q}$ and m, and gives no contribution to observable characteristics (see [11,12]). Then, using (9.35), we write (10.15) as

$$\langle\Delta V_\alpha^2\rangle' = \sum_{\mathbf{q}m} \frac{\tilde{u}T}{2} \frac{\tilde{u}\chi_{\mathbf{q}m}^{0\alpha}}{1 - \tilde{u}\chi_{\mathbf{q}m}^{0\alpha}}. \tag{10.16}$$

Obtaining computational formulae for the single-site fluctuation (10.16) is not an easy task. First, the calculation was carried out in the static ($\omega_m = 0$) long-wave ($\mathbf{q} = 0$) approximation for paramagnets [6, 7] and ferromagnets [14, 15]. The dynamic nonlocal approximation (DNA) was developed and applied for paramagnets in [12] and for ferromagnets in [8–10].

The sum over the even "frequencies" $\omega_m = 2\pi mT$ is replaced by the integral over the energy variable using analytic continuation (6.75) (for details, see Appendix A.2.5):

$$\sum_m \langle |\Delta V^\alpha_{\mathbf{q}m}|^2\rangle' = \frac{\tilde{u}}{\pi}\int_0^\infty \left(B(\varepsilon)+\frac{1}{2}\right) \mathrm{Im}\,\frac{1}{1-\tilde{u}\chi^{0\alpha}_{\mathbf{q}}(\varepsilon)}\,\mathrm{d}\varepsilon, \tag{10.17}$$

where $B(\varepsilon) = (\exp(\varepsilon/T)-1)^{-1}$ is the Bose function. The second term in the integral describes the contribution of the zero-point fluctuations that should be discarded. Using the Tailor expansion $\chi^{0\alpha}_{\mathbf{q}}(\varepsilon) \approx \chi^{0\alpha}_{\mathbf{q}}(0) + \mathrm{i}\varphi^\alpha_{\mathbf{q}}\varepsilon$ and approximation

$$\frac{1}{\mathrm{e}^{\varepsilon/T}-1} \approx \begin{cases} T/\varepsilon, & \varepsilon < \varepsilon_0 = (\pi^2/6)T, \\ 0, & \varepsilon > \varepsilon_0, \end{cases} \tag{10.18}$$

for the Bose function, we come to

$$\sum_m \langle |\Delta V^\alpha_{\mathbf{q}m}|^2\rangle' = \frac{\tilde{u}T}{2\lambda^\alpha_{\mathbf{q}}}\frac{2}{\pi}\arctan\frac{\tilde{u}\varphi^\alpha_{\mathbf{q}}\pi^2 T}{6\lambda^\alpha_{\mathbf{q}}}, \tag{10.19}$$

where

$$\lambda^\alpha_{\mathbf{q}} = 1-\tilde{u}\chi^{0\alpha}_{\mathbf{q}}(0), \qquad \varphi^\alpha_{\mathbf{q}} = \left.\frac{d\mathrm{Im}\chi^{0\alpha}_{\mathbf{q}}(\varepsilon)}{d\varepsilon}\right|_{\varepsilon=0}. \tag{10.20}$$

The approximation (10.18) not only reproduces the behaviour of the Bose function $B(\varepsilon)$ with respect to thermal energies, but also has the same first moment $\int_0^\infty \varepsilon B(\varepsilon)\,\mathrm{d}\varepsilon = (\pi T)^2/6$, which essentially defines the upper bound ε_0. Thus, the approximation (10.18) is well justified. Its another advantage is the possibility of the straight-forward proceeding to the static limit at high temperatures, when the argument of the arctangent in (10.19) is much larger than unity.

The function $\lambda^\alpha_{\mathbf{q}}$ is calculated by the formula [9]

$$\lambda^\alpha_{\mathbf{q}} = \lambda^\alpha_0 + (\lambda^\alpha_{\mathrm{L}} - \lambda^\alpha_0)q^2/\overline{q^2}. \tag{10.21}$$

Here

$$\lambda^\alpha_{\mathrm{L}} = 1 - u\chi^{0\alpha}_{\mathrm{L}}(0), \tag{10.22}$$

where the local unenhanced susceptibility $\chi^{0\alpha}_{\mathrm{L}}(0)$ is defined as

$$\chi^{0\alpha}_{\mathrm{L}}(0) = \frac{1}{N}\sum_j \chi^{0\alpha}_j(0) = \frac{1}{N^2}\sum_{\mathbf{q}} \chi^{0\alpha}_{\mathbf{q}}(0), \tag{10.23}$$

and $\overline{q^2} = N^{-1}\sum_{\mathbf{q}} q^2$ is the average of q^2 over the Brillouin zone. For simplicity the function $\varphi^\alpha_{\mathbf{q}}$ is approximated by its mean value $N\varphi^\alpha_{\mathrm{L}}$, where $\varphi^\alpha_{\mathrm{L}} = N^{-2}\sum_{\mathbf{q}}\varphi^\alpha_{\mathbf{q}}$.

The summation over $\mathbf{q}$ is carried out by the integration over the Brillouin zone, approximated for simplicity by the equal-volume sphere of the radius q_{B}. Using (10.19), (10.21) and $\overline{q^2} = 0.6q^2_{\mathrm{B}}$, for the single-site fluctuation (10.16) in the DNA we finally obtain [9]

$$\langle \Delta V^2_\alpha\rangle' = \frac{uT}{2\lambda^\alpha_{\mathrm{L}}}\int_0^1 \frac{1}{a^2_\alpha + b^2_\alpha k^2}\frac{2}{\pi}\arctan\frac{c_\alpha}{a^2_\alpha + b^2_\alpha k^2}3k^2\,\mathrm{d}k, \tag{10.24}$$

where $k = q/q_{\mathrm{B}},\ q_{\mathrm{B}} = (3/(4\pi)\Omega_{\mathrm{BZ}})^{1/3}$, and

$$a^2_\alpha = \lambda^\alpha_0/\lambda^\alpha_{\mathrm{L}}, \qquad b^2_\alpha = (1-a^2_\alpha)/0.6, \qquad c_\alpha = u\varphi^\alpha_{\mathrm{L}}\pi^2 T/(6\lambda^\alpha_{\mathrm{L}}). \tag{10.25}$$

An approximate formula for evaluating the integral (10.24) is given in Appendix H.4.

The quantities λ^α_0 in formula (10.21) are calculated as follows. In the absence of magnetic anisotropy, any small external magnetic field causes rotation of the large spontaneous magnetization of the ferromagnet, i.e. the enhanced susceptibility $\chi^\alpha_0(0) = \chi^{0\alpha}_0(0)/\lambda^\alpha_0$ diverges. Therefore, we assume that $\lambda^\alpha_0(0)$ are infinitesimal at $T < T_{\mathrm{C}}$. In the paramagnetic region $(T > T_{\mathrm{C}})$ the quantities λ^x_0 and λ^z_0 are equal to each other, and the static uniform susceptibility $\chi^{0z}_0(0)$ is obtained by the numerical differentiation of the spin $\bar{s}_z$ with respect to the magnetic field h (in energy units):

$$\chi_0^{0z}(0) = -\frac{\partial \bar{s}_z}{\partial h} \simeq -\frac{\bar{s}_z(\Delta h/2) - \bar{s}_z(-\Delta h/2)}{\Delta h}. \tag{10.26}$$

10.3 Mean Single-Site Green Function

To calculate the local susceptibility $\chi_L^{0\alpha}(\varepsilon) = \chi_L^{0\alpha}(0) + i\varphi_L^\alpha \varepsilon$, we replace the mean Green function $\bar{G}_{\mathbf{k}}(z)$ in (9.34) by the mean single-site Green function $g(z)$, just as we did in the coherent potential equation (10.3). Using (10.2) and (10.23), we rewrite (9.34) as

$$\chi_L^{0\alpha}(i\omega_m) = -\frac{N_d}{2} T \sum_n \sum_{\gamma_1\gamma_2} g^{\gamma_1}(i\omega_n) g^{\gamma_2}(i\omega_n - i\omega_m) \mathrm{Sp}\big(\sigma^{\gamma_1}\sigma^\alpha\sigma^{\gamma_2}\sigma^\alpha\big). \tag{10.27}$$

Replacing the sum over the "frequencies" $\omega_n = (2n+1)\pi T$ by the integral over the energy variable (see Appendix A.2.5), we obtain

$$\chi_L^{0\alpha}(z) = -\frac{N_d}{2\pi} \sum_{\gamma_1\gamma_2} \int \mathrm{Im}\Big(g^{\gamma_1}(\varepsilon)\big(g^{\gamma_2}(\varepsilon - z) + g^{\gamma_2}(\varepsilon + z)\big) \\ \times \mathrm{Sp}\big(\sigma^{\gamma_1}\sigma^\alpha\sigma^{\gamma_2}\sigma^\alpha\big)\Big) f(\varepsilon)\,\mathrm{d}\varepsilon, \tag{10.28}$$

where $f(\varepsilon) = [\exp((\varepsilon - \mu)/T) + 1]^{-1}$ is the Fermi function, $g(\varepsilon)$ is the mean single-site Green function in the *canonical* ensemble and $\varepsilon = \varepsilon - i0^+$. By formulae (10.5) the 2×2 matrix $g(\varepsilon)$ is spin-diagonal. Hence $g^x(\varepsilon) = 0$ and $g^y(\varepsilon) = 0$. The two nonzero components are given by

$$g^0(\varepsilon) = \frac{1}{2}(g_\uparrow(\varepsilon) + g_\downarrow(\varepsilon)), \qquad g^z(\varepsilon) = \frac{1}{2}(g_\uparrow(\varepsilon) - g_\downarrow(\varepsilon)).$$

Calculating the trace of products of the Pauli matrices by formula (A.22), we obtain

$$\chi_L^{0x}(0) = -\frac{N_d}{\pi} \int \mathrm{Im}(g_\uparrow g_\downarrow)\, f\,\mathrm{d}\varepsilon, \tag{10.29}$$

$$\varphi_L^x = \frac{N_d}{\pi} \int \mathrm{Im}g_\uparrow\, \mathrm{Im}g_\downarrow \left(-\frac{\partial f}{\partial \varepsilon}\right) \mathrm{d}\varepsilon, \tag{10.30}$$

$$\chi_L^{0z}(0) = -\frac{N_d}{2\pi} \int \big(\mathrm{Im}g_\uparrow^2 + \mathrm{Im}g_\downarrow^2\big)\, f\,\mathrm{d}\varepsilon, \tag{10.31}$$

$$\varphi_L^z = \frac{N_d}{2\pi} \int \Big[\big(\mathrm{Im}g_\uparrow\big)^2 + \big(\mathrm{Im}g_\downarrow\big)^2\Big] \left(-\frac{\partial f}{\partial \varepsilon}\right) \mathrm{d}\varepsilon. \tag{10.32}$$

A general numerical method for calculating integrals with the Fermi function was developed in [16] and a simple method for calculating integrals with the derivative of the Fermi function was given in [10] (see Appendices H.2 and H.3).

Similarly, replacing the summation over the "frequencies" $\omega_n = (2n+1)\pi T$ by integration over the energy variable, we rewrite the mean field (9.29) as

$$\bar{V}_z = -u\frac{N_d}{2\pi} \int \mathrm{Im}\big(g_\uparrow - g_\downarrow\big) f\,\mathrm{d}\varepsilon \tag{10.33}$$

and the electrons number conservation equation (9.36) as

$$n_e = \frac{N_d}{\pi} \int \mathrm{Im}\big(g_\uparrow + g_\downarrow\big) f\,\mathrm{d}\varepsilon. \tag{10.34}$$

In the canonical ensemble the single-site Green function $g_\sigma(\varepsilon)$ is given by the same formula (10.10):

$$g_\sigma(\varepsilon) = \int \frac{\nu(\varepsilon')}{\varepsilon - \varepsilon' - \Sigma_\sigma(\varepsilon)}\,\mathrm{d}\varepsilon',$$

where the electron DOS $\nu(\varepsilon)$ and coherent potential $\Sigma_\sigma(\varepsilon)$ are in the *canonical* ensemble (for details, see Appendix E.1). Representing the self-energy as $\Sigma_\sigma(\varepsilon) = \sigma\bar{V}_z + \Delta\Sigma_\sigma(\varepsilon)$, we finally have

$$g_\sigma(\varepsilon) = \int \frac{\nu(\varepsilon')}{\varepsilon - \sigma\bar{V}_z - \Delta\Sigma_\sigma(\varepsilon) - \varepsilon'}\,\mathrm{d}\varepsilon' \tag{10.35}$$

(for numerical calculation of the integral (10.35), see Appendix H.1). The fluctuational contribution to the self-energy $\Delta\Sigma_\sigma(\varepsilon)$ is obtained from Eq. (9.44) using the single-site approximation:

$$\Delta\Sigma = \big\langle [1 - (\Delta V - \Delta\Sigma)g]^{-1}\Delta V \big\rangle. \tag{10.36}$$

In the second order with respect to ΔV and *infinite order* with respect to $\bar{V}_z$, we have (for derivation, see Appendix E.2)

$$\Delta\Sigma_\sigma(\varepsilon) = \frac{g_\sigma(\varepsilon)\langle\Delta V_z^2\rangle'}{1 + 2\sigma\bar{V}_z g_\sigma(\varepsilon)} + 2g_{\bar{\sigma}}(\varepsilon)\langle\Delta V_x^2\rangle', \tag{10.37}$$

where $\bar{\sigma} \equiv -\sigma$. This result is slightly more general than the second-order approximation (10.12) of Eq. (10.3).[2]

To calculate $\chi_0^{0z}(0)$ by formula (10.26), we use the same expression for $\bar{s}_z$ as the one on the right-hand side of Eq. (10.33):

$$\bar{s}_z(\Delta h/2) = \frac{N_\mathrm{d}}{2\pi}\int \mathrm{Im}\big(g_\uparrow - g_\downarrow\big) f\,\mathrm{d}\varepsilon, \tag{10.38}$$

but the exchange field $\sigma\bar{V}_z$ in formula (10.35) is replaced by a small magnetic field $\Delta h/2$ (in energy units).

Thus, we obtain a closed system of equations in four unknowns: the mean field $\bar{V}_z$, mean-square transverse $\langle\Delta V_x^2\rangle' = \langle\Delta V_y^2\rangle'$ and longitudinal $\langle\Delta V_z^2\rangle'$ fluctuations, and chemical potential μ. In particular, when $\langle\Delta V_\alpha^2\rangle' = 0$, Eqs. (10.33) and (10.34) turn into the Stoner mean-field theory equations. This gives an opportunity to obtain the effective constant u by solving the system of Eqs. (10.33) and (10.34) at $T = 0$ given the magnetic moment m_0.

In other approximations of SFT, only Eq. (10.24) for the spin fluctuations is modified [10]. In the static local approximation (SLA) we have

$$\langle\Delta V_\alpha^2\rangle_\mathrm{SLA} = \frac{uT}{2\lambda_\mathrm{L}^\alpha}, \tag{10.39}$$

in the static nonlocal approximation (SNA) we have

$$\begin{aligned}\langle\Delta V_\alpha^2\rangle_\mathrm{SNA} &= \frac{uT}{2\lambda_\mathrm{L}^\alpha}\int_0^1 \frac{1}{a_\alpha^2 + b_\alpha^2 k^2}3k^2\,\mathrm{d}k\\ &= \frac{uT}{2\lambda_\mathrm{L}^\alpha}\frac{3}{b_\alpha^3}\left(b_\alpha - a_\alpha \arctan\frac{b_\alpha}{a_\alpha}\right),\end{aligned} \tag{10.40}$$

and in the dynamic local approximation (DLA) we have

$$\langle\Delta V_\alpha^2\rangle'_\mathrm{DLA} = \frac{uT}{2\lambda_\mathrm{L}^\alpha}\frac{2}{\pi}\arctan c_\alpha. \tag{10.41}$$

10.4 Basic Magnetic Characteristics

As we showed in Chap. 8, the functional integral formalism reduces the canonical average to the average over configurations of a fluctuating field (8.37). In particular, the mean spin $\bar{s}_z = \langle s_j^z\rangle$ is related to the mean field by (8.54) and the *single-site*

[2] In practice, sufficient accuracy is achieved if one uses formulae (10.12) or (10.37) for $\Delta\Sigma_\sigma(\varepsilon)$ with the single-site Green function g_σ^S calculated by (10.35) without $\Delta\Sigma_\sigma(\varepsilon)$.

spin fluctuation $\langle \Delta s_\alpha^2 \rangle = \langle (\Delta s_j^\alpha)^2 \rangle$ is related to the mean-square fluctuation of the field by (8.63). For magnetization we immediately obtain $m = g\mu_B \bar{V}_z/u$.

The local magnetic moment is defined as $m_L = g\mu_B s_L$, where $s_L^2 \equiv \langle s_j^2 \rangle$. Rewriting relation (8.63) as

$$\langle \Delta s_\alpha^2 \rangle = \frac{1}{u^2} \langle \Delta V_\alpha^2 \rangle' . \tag{10.42}$$

and using $\langle s_j^2 \rangle = \bar{s}_z^2 + \langle \Delta s_j^2 \rangle$, we have

$$s_L^2 = u^{-2} (\bar{V}_z^2 + \langle \Delta V^2 \rangle'), \tag{10.43}$$

where $\langle \Delta V^2 \rangle' = \sum_\alpha \langle \Delta V_\alpha^2 \rangle'$ is the total mean-square fluctuation given by (10.15). Then the final computational formula is

$$s_L = u^{-1} (\bar{V}_z^2 + \langle \Delta V_x^2 \rangle' + \langle \Delta V_y^2 \rangle' + \langle \Delta V_z^2 \rangle')^{1/2}. \tag{10.44}$$

In the static approximation, we come to [9]

$$s_L = u^{-1} \left(\bar{V}_z^2 + \langle \Delta V_x^2 \rangle + \langle \Delta V_y^2 \rangle + \langle \Delta V_z^2 \rangle - \frac{3uT}{2} \right)^{1/2} .$$

In the presence of a *static* magnetic field $\mathbf{H}_j$, magnetic susceptibility is easily obtained as follows. The original Hamiltonian $\mathcal{H}$ gets the additional magnetic term $\mathcal{H}_M = -\sum_j \boldsymbol{\mathcal{M}}_j \mathbf{H}_j = 2\sum_j \mathbf{s}_j \mathbf{h}_j$, where $\boldsymbol{\mathcal{M}}_j = -g\mu_B \mathbf{s}_j$ is the magnetic moment operator, and $\mathbf{h}_j = \frac{1}{2} g\mu_B \mathbf{H}_j$ is the magnetic field in energy units. If we now carry out the Stratonovich-Hubbard transformation in the static approximation, the magnetic field $\mathbf{h}_j$ adds up to the exchange field $\mathbf{V}_j$, and we write the Hamiltonian of the noninteracting electrons in the field V as

$$\mathcal{H}(V) = \mathcal{H}_0 + 2\sum_j (\mathbf{V}_j + \mathbf{h}_j) \mathbf{s}_j . \tag{10.45}$$

The mean magnetic moment can be obtained by differentiating the total free energy (9.26) at the vanishing field $\mathbf{h} = 0$:

$$\langle \boldsymbol{\mathcal{M}}_j \rangle = -\frac{\partial \mathcal{F}}{\partial \mathbf{h}_j} = -\left\langle \frac{\partial F_1(V)}{\partial \mathbf{V}_j} \right\rangle = -2\langle \mathbf{s}_j \rangle,$$

in units of μ_B. Similarly, the enhanced susceptibility $\chi_{jj'}^{\alpha\alpha'} = \partial \langle \mathcal{M}_j^\alpha \rangle / \partial H_{j'}^{\alpha'}$ can be written as

$$\chi_{jj'}^{\alpha\alpha'} = -\frac{1}{2} \frac{\partial^2 \mathcal{F}}{\partial h_j^\alpha \partial h_{j'}^{\alpha'}} = -\frac{1}{2} \left\langle \frac{\partial^2 F_1(V)}{\partial V_j^\alpha \partial V_{j'}^{\alpha'}} \right\rangle = \frac{2}{T} \langle \Delta s_j^\alpha \Delta s_{j'}^{\alpha'} \rangle, \tag{10.46}$$

in units of $\frac{1}{2} g^2 \mu_B^2$. This choice of units allows to treat the exchange field V and magnetic field h on the equal footing in formulae (10.45) and (10.46).

In the presence of a *dynamic* magnetic field, the susceptibility $\chi_{\mathbf{q}m}^\alpha$ is given by formula (2.38). Transforming the relation between the spin and field fluctuations (8.63), in the momentum-"frequency" representation we have

$$\langle |\Delta s_{\mathbf{q}m}^\alpha|^2 \rangle = \frac{1}{\tilde{u}^2} \langle |\Delta V_{\mathbf{q}m}^\alpha|^2 \rangle - \frac{T}{2\tilde{u}}$$

(for details, see Appendix C.4). Using relations (6.65) and (9.35), in units of $\frac{1}{2} g^2 \mu_B^2$ we obtain

$$\chi_{\mathbf{q}m}^\alpha = \frac{\chi_{\mathbf{q}m}^{0\alpha}}{1 - \tilde{u}\chi_{\mathbf{q}m}^{0\alpha}}, \tag{10.47}$$

with the same enhancement factor as in the RPA. The critical difference between DSFT and RPA lies in the way the unenhanced susceptibility $\chi^{0\alpha}_{\mathbf{q}m}$ is calculated. In terms of the functional integral method, the unenhanced susceptibility in the RPA (9.21) is completely determined by the mean field $\bar{V}$:

$$\chi^{0\alpha}_{\mathbf{q}m}\Big|_{\text{RPA}} = -\frac{1}{2}\frac{\partial^2 F_1(\bar{V})}{\partial V^{\alpha}_{\mathbf{q}m}\,\partial V^{\alpha}_{-\mathbf{q}-m}} .$$

In the optimal Gaussian approximation of the DSFT, the right-hand side is averaged over all field configurations:

$$\chi^{0\alpha}_{\mathbf{q}m} = -\frac{1}{2}\left\langle \frac{\partial^2 F_1(V)}{\partial V^{\alpha}_{\mathbf{q}m}\,\partial V^{\alpha}_{-\mathbf{q}-m}} \right\rangle .$$

In the paramagnetic state, linear interpolation of the uniform static susceptibility $\chi^{\alpha}_0(0)$ yields the effective magnetic moment m_{eff} and paramagnetic Curie temperature Θ_{C} in the Curie-Weiss law (2.59). In the DSFT the static uniform susceptibility $\chi^{\alpha}_0(0)$ is measured in units of $\frac{1}{2}g^2\mu^2_{\text{B}}$, therefore its relation to the Curie-Weiss susceptibility (2.62) becomes $\chi^{\alpha}_{\text{CW}} = N^{-1}\chi^{\alpha}_0(0)\,\frac{1}{2}g^2\mu^2_{\text{B}}$.

10.5 Application to Ferromagnetic Metals

10.5.1 Iron

As the initial DOS, we take the one of nonmagnetic iron, calculated in the local-density approximation (LDA) by the Korringa-Kohn-Rostoker (KKR) method with a self-consistent potential [17]. The extended "tails" and the constant sp background were eliminated from this DOS, so that the area under the curve was equal to $2N_{\text{d}} = 10$ (the number of d states per atom). This yields the d bandwidth $W = 7.42\,\text{eV}$. The DOS is slightly smoothed out by convolution with the Lorentzian function of halfwidth $\Gamma = 0.01\,W$ to remove nonphysical sharp peaks (for details, see Appendix A.2.3). The peaks always appear in energy-band calculation, because it entirely ignores single-particle state damping due to electron–electron scattering. The smoothed DOS is then normalized to one state (per atom, band and spin), see Fig. 10.1. The number of d electrons per atom is $n_{\text{e}} = 7.43$. The effective interaction constant u determined from an experimental value of the magnetic moment $m_0 = 2.217\,\mu_{\text{B}}$ [18] is equal to 1.08 eV.

The results of the calculation [10] of the basic magnetic characteristics of iron in the Stoner mean-field theory and in various approximations of the SFT are represented in Table 10.1 and Figs. 10.2, 10.3 and 10.4. All characteristics are expressed in units of their experimental values given in Table 10.2.

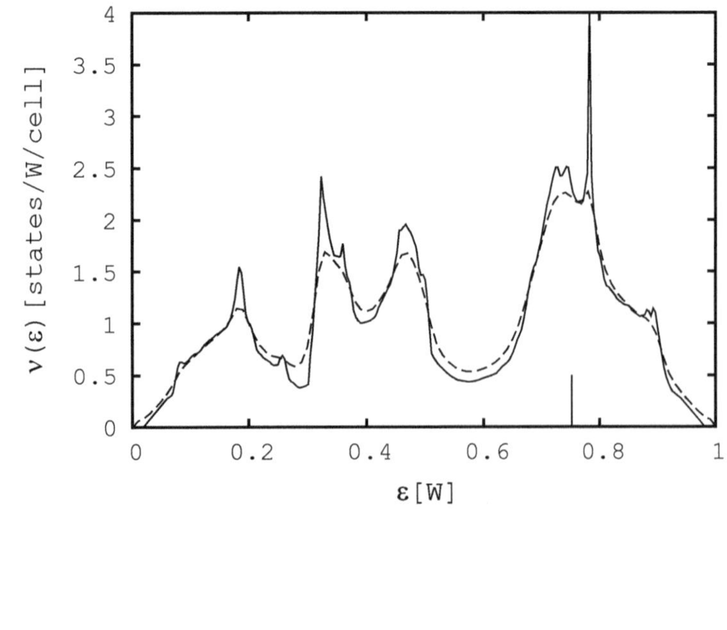

Fig. 10.1 The DOS of the d band of nonmagnetic iron, calculated by the KKR method with a self-consistent potential (solid line) and that smoothed out by convolution with Lorentzian function of halfwidth $\Gamma = 0.01$ (dashed line). The energy ε and halfwidth Γ are in units of the bandwidth $W = 7.42\,\text{eV}$. The vertical line indicates the position of the Fermi level ε_{F}

Table 10.1 The ferromagnetic T_C and paramagnetic Θ_C Curie temperatures, and the effective magnetic moment m_{eff} in the mean-field theory and in various approximations of the spin fluctuation theory

Metal	Magnetic characteristic	Stoner theory	Spin fluctuation theory			
			SLA	SNA	DLA	DNA
Fe	T_C/T_C^{exp}	5.84	1.28	0.76	2.45	1.49
	Θ_C/T_C^{exp}	5.82	1.24	0.77	2.29	1.45
	m_{eff}/m_{eff}^{exp}	0.86	0.66	0.98	0.80	1.30
Co	T_C/T_C^{exp}	3.61	0.55	0.32	1.23	0.63
	Θ_C/T_C^{exp}	3.60	0.54	0.34	1.22	0.68
	m_{eff}/m_{eff}^{exp}	0.84	0.41	0.56	0.58	0.92
Ni	T_C/T_C^{exp}	4.04	1.35	0.86	2.80	1.54
	Θ_C/T_C^{exp}	4.03	1.34	0.90	2.78	1.60
	m_{eff}/m_{eff}^{exp}	0.85	0.64	0.86	0.81	1.51

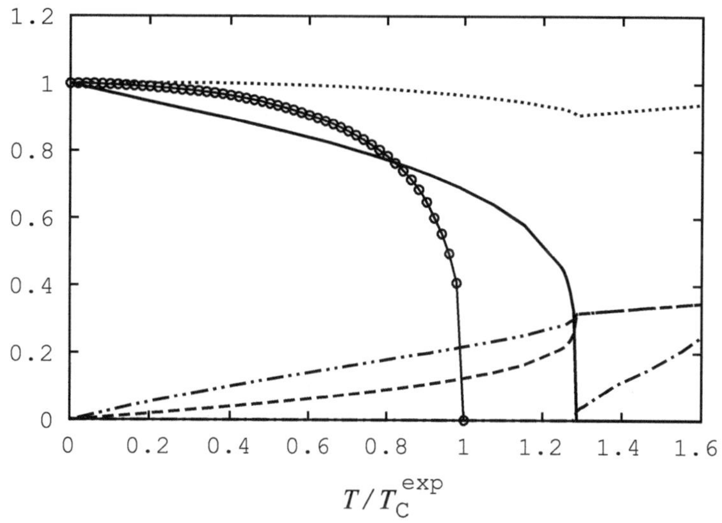

Fig. 10.2 Magnetization m/m_0 (solid line: calculation, circled line: experiment [18]), the mean-square fluctuations $\langle \Delta V_x^2 \rangle$ (dash dot dot dashed line) and $\langle \Delta V_z^2 \rangle$ (dashed line) in units of the square of the mean field $\bar{V}_z^2$ at $T = 0$, the reciprocal paramagnetic susceptibility χ^{-1} (dash dot dashed line) in units of $k_B T_C^{exp}/\mu_B^2$, and the local magnetic moment m_L/m_0 (dotted line) of iron, calculated in the SLA as functions of the reduced temperature T/T_C^{exp}

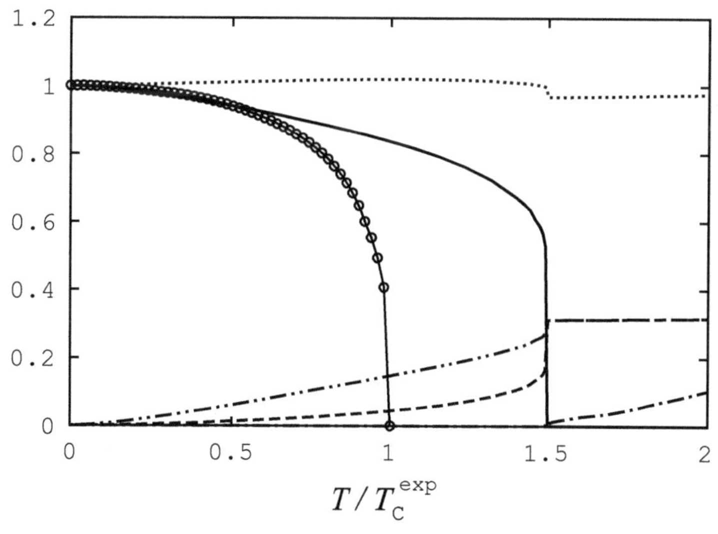

Fig. 10.3 As Fig. 10.2, but calculated in the DNA

As can be seen from Table 10.1, in the Stoner mean-field theory the temperature dependence of the magnetization is very weak and the calculated Curie temperature is almost six times greater than the observed one.

When the spin fluctuations are taken into account, the situation is substantially different. Let us start from the static local approximation (SLA), in which the fluctuations are calculated from formula (10.39). In the SLA, as can be seen from Fig. 10.2, the Curie temperature is close to the experimentally observed one, $T_C = 1.28 T_C^{exp}$. However, a noticeable decrease of the magnetization, $\sim T$, is seen over a wide temperature interval, because the spin fluctuations increase linearly with temperature. In general, the paramagnetic susceptibility follows the Curie-Weiss law, but the effective magnetic moment

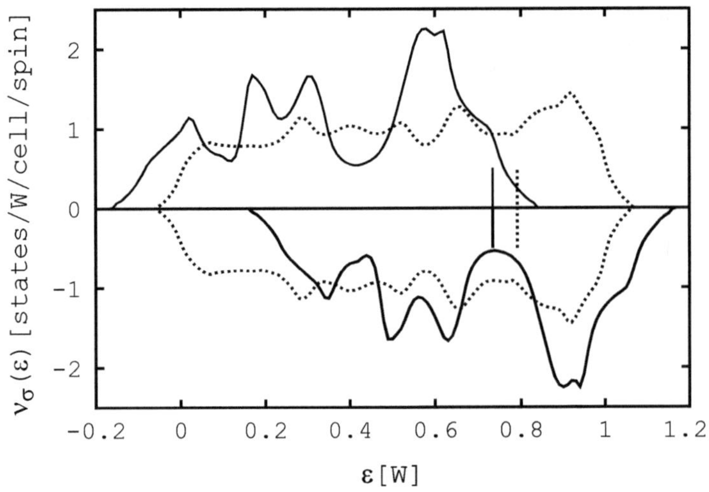

Fig. 10.4 Spin-polarized DOSs of iron in the ferromagnetic ($T = 0$, solid curves) and paramagnetic ($T = 1.5T_C^{exp}$, dotted curves) states calculated in the DNA. The vertical line indicates the position of the chemical potential μ

Table 10.2 Experimental values of fundamental magnetic characteristics of iron, cobalt and nickel

	m_0^{exp} (μ_B) [18]	T_C^{exp} (K) [18]	m_{eff}^{exp} (μ_B) [19]
Fe	2.217	1044.0	3.13
Co	1.753	1390.0	3.13
Ni	0.616	631.0	1.616

m_{eff} is only 0.66 of its experimental value. The paramagnetic Curie point Θ_C, obtained by the linear extrapolation of $\chi^{-1}(T)$ to zero, is nearly coincident with the ferromagnetic one.

In the static nonlocal approximation (SNA) the magnetization should decrease faster than in the SLA. The reason is that the fluctuations $\langle \Delta V_\alpha^2 \rangle_{SNA} \equiv \zeta_{SNA}^\alpha$ are greater than the fluctuations ζ_{SLA}^α at any temperature. Indeed, taking into account that $0 < a^2 < 1$, $b^2 = (1 - a^2)/0.6$ and $0 \le \arctan(b/a)$, from formula (10.40) we have

$$\zeta_{SNA}/\zeta_{SLA} = \frac{3}{b^3}\left(b - a \arctan\frac{b}{a}\right) \ge \frac{3}{b^2} = \frac{3 \times 0.6}{1 - a^2} \ge 1.8\,.$$

The calculations show that indeed in the SNA the decrease of the magnetization $m(T)$ is too fast.

In the dynamic local approximation (DLA) the situation is the opposite. The fluctuations ζ_{DLA}^α calculated by formula (10.41) are smaller than the fluctuations ζ_{SLA}^α at any temperature ($\zeta_{DLA}/\zeta_{SLA} = (2/\pi)\arctan c \le 1$). The calculations confirm that in the DLA the magnetization $m(T)$ decreases too slowly with temperature.

Only in the dynamic nonlocal approximation (DNA), as the temperature increases, the fluctuations ζ_{DNA}^α first increase slowly ($\sim T^2$), just as in DLA, and then increase quickly, as in the SNA. As a result, at low temperatures, the magnetization is proportional to T^2 in good agreement with the experimental curve (Fig. 10.3). At the same time, the calculated Curie temperature T_C is equal to $1.49T_C^{exp}$. The paramagnetic susceptibility follows the Curie-Weiss law, but the effective magnetic moment is a little greater than the experimental one: $m_{eff} = 1.3m_{eff}^{exp}$. Figure 10.4 shows the mean DOSs $\nu_\sigma(\varepsilon) = \pi^{-1}\text{Im}g_\sigma(\varepsilon)$ in the ferromagnetic ($T = 0$) and paramagnetic ($T = 1.5T_C^{exp}$) states. In contrast to the results of the Stoner mean-field theory, as the temperature increases, the functions $\nu_\uparrow(\varepsilon, T)$ and $\nu_\downarrow(\varepsilon, T)$ become noticeably smoothed when they shift towards each other.

10.5.2 Cobalt

As the initial DOS, we take the one of nonmagnetic fcc cobalt from [17]. The constant sp background is eliminated from it, just as in iron. After convolution with the Lorentzian function of halfwidth $\Gamma = 0.01\,W$ and normalization to one d band of unit width, we obtain the DOS represented in Fig. 10.5 by a solid curve. The bandwidth is $W = 7.50$ eV. The number of d electrons per atom is equal to 8.47. The electron–electron interaction constant $u = 1.25$ eV is determined at $T = 0$ from the magnetic moment $m_0 = 1.45\mu_B$. We are not able to find the constant u from the experimental value $m_0^{exp} = 1.753\mu_B$ because for cobalt the magnetic moment m_0^{exp} is comparatively large and the Fermi energy is close to the band edge. To

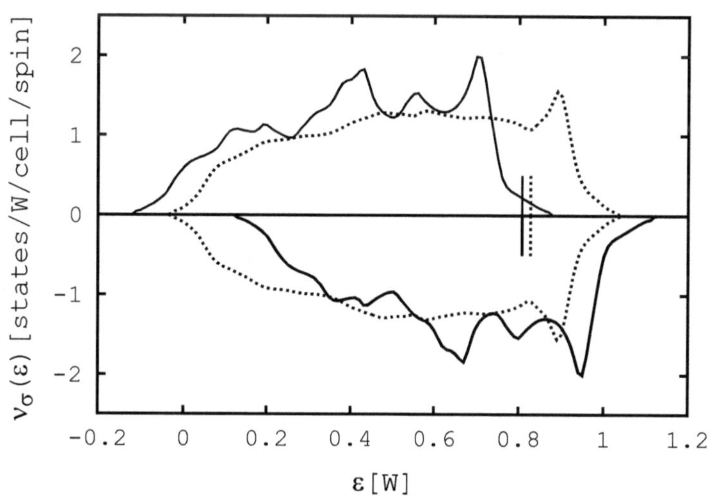

Fig. 10.5 Spin-polarized DOSs of cobalt in the ferromagnetic ($T = 0$) and paramagnetic ($T = 0.64T_C^{exp}$) states, calculated in the DNA (the notation is as for Fig. 10.4)

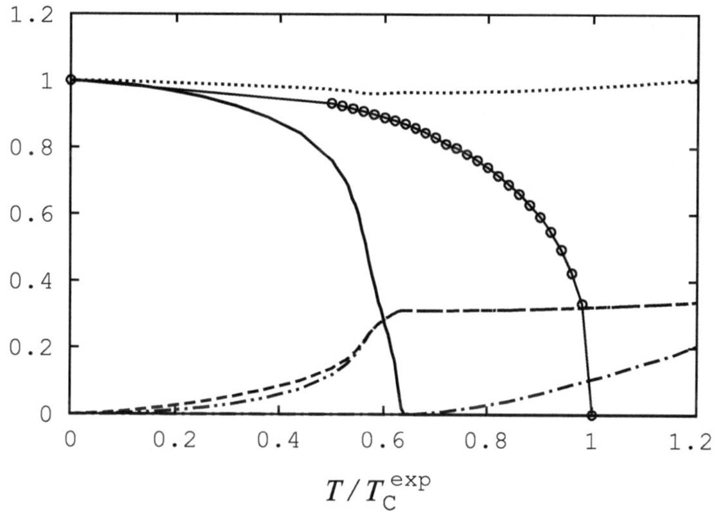

Fig. 10.6 Magnetization m/m_0, the mean-square fluctuations $\langle \Delta V_x^2 \rangle$ and $\langle \Delta V_z^2 \rangle$, the reciprocal paramagnetic susceptibility χ^{-1} and the local magnetic moment m_L/m_0 of cobalt, calculated in the DNA (the notation and units are as for Fig. 10.2)

obtain m_0^{exp} it is necessary to move the DOSs $\nu_\uparrow(\varepsilon)$ and $\nu_\downarrow(\varepsilon)$ so far apart that the sum $n_\uparrow + n_\downarrow$ becomes less than n_e. In our opinion, the discrepancy between $\nu(\varepsilon)$ and m_0^{exp} for cobalt is connected not with the fact that at $T = 0$ cobalt has an hcp structure, because the DOS curves of hcp and fcc cobalt are very similar (see Fig. 5 in [19]), but rather with the fact that, in addition to the spin magnetic moment of the d electrons, m_0^{exp} includes the spin magnetic moment of the s electrons and the orbital magnetic moment, which, generally speaking, do not compensate each other (see, e.g. [19–21]). The results of ab initio spin-polarized calculations of Fe, Co and Ni presented in [17, 22] also give evidence in favour of the chosen value $m_0 = 1.45\mu_B$. For iron and nickel the calculated spin magnetic moments m_0^{cal} are close to the experimental ones. If this is also the case in fcc cobalt, then we should have $m_0 = 1.54 \div 1.56\mu_B$. The value $m_0^{cal} = 1.56\mu_B$ for fcc Co was also obtained in [23].

The results of the calculations of the basic magnetic characteristics of cobalt within different approximations are listed in Table 10.1. Just as in iron, the best one is the DNA. As we see from Fig. 10.6, at low temperatures the calculated magnetization $m(T)$ is in good agreement with the experimental one. However, at high temperatures, $m(T)$ decreases too quickly and, as a consequence, we have $T_C = 0.63T_C^{exp}$. The paramagnetic susceptibility satisfies the Curie-Weiss law with $m_{eff} = 0.92m_{eff}^{exp}$. Thus, in spite of some disagreement with experiment, the theoretical description of the magnetic properties of cobalt may be considered quite satisfactory.

The mean DOSs $\nu_\sigma(\varepsilon, T)$ in the ferromagnetic ($T = 0$) and paramagnetic ($T = 0.64T_C^{exp}$) states are shown in Fig. 10.5. As the temperature increases, the curves $\nu_\sigma(\varepsilon, T)$ behave just as for iron, but are smoothed more noticeably.

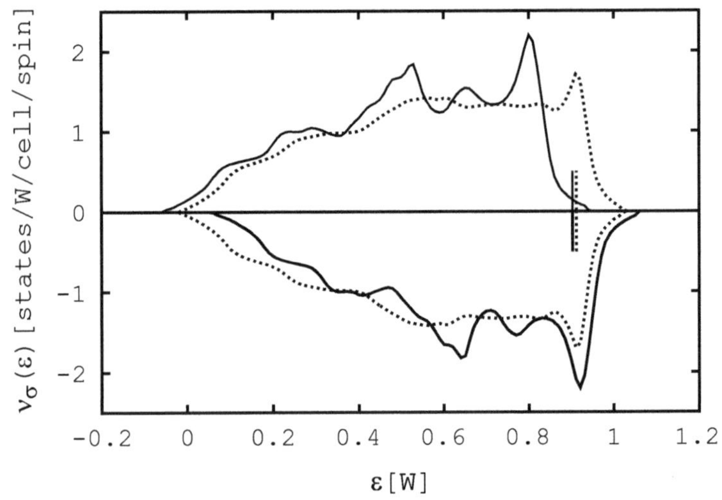

Fig. 10.7 Spin-polarized DOSs of nickel in the ferromagnetic ($T = 0$) and paramagnetic ($T = 1.55T_C^{exp}$) states, calculated in the DNA (the notation is as for Fig. 10.4)

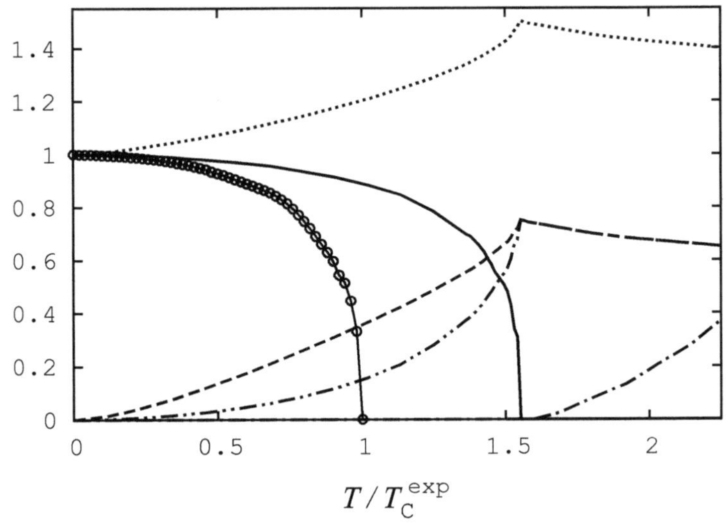

Fig. 10.8 Magnetization m/m_0, the mean-square fluctuations $\langle \Delta V_x^2 \rangle$ and $\langle \Delta V_z^2 \rangle$, the reciprocal paramagnetic susceptibility χ^{-1} and the local magnetic moment m_L/m_0 of nickel, calculated in the DNA (the notation and units are as for Fig. 10.2)

10.5.3 Nickel

Just as for cobalt, the initial nonmagnetic DOS was taken from [17]. After elimination of the sp background, convolution with the Lorentzian function of halfwidth $\Gamma = 0.01W$ and normalization to one d band of unit width, we obtain the DOS represented in Fig. 10.7 by a solid curve. The bandwidth is $W = 6.13$ eV, and the number of d electrons per atom is $n_e = 9.35$. The electron–electron interaction constant $u = 1.16$ eV is determined from the experimental value of the magnetic moment of nickel $m_0^{exp} = 0.616\,\mu_B$ [18].

The results of the calculations of the magnetic properties of nickel in various approaches are listed in Table 10.1. The quantitative characteristics are similar to those for iron. On the whole, the best results are obtained in the DNA. As can be seen from Fig. 10.8, in the DNA the spin fluctuations have the proper temperature behaviour and, as a consequence, the shape of the magnetization curve is in agreement with the experimental one. Note that in the ferromagnetic region, in iron the transverse fluctuations dominate, while in nickel the longitudinal fluctuations dominate and in cobalt the intermediate situation is realized: the transverse and longitudinal fluctuations are close in magnitude (see Figs. 10.3, 10.6 and 10.8).

10.5.4 Comparison with Other Studies

Now let us compare our results with the results of other calculations of the magnetic properties in Fe, Co and Ni [24–27] that also use the real (not model) band structure. Some results of these works are in better agreement with experiment than our results. However, good agreement, as a rule, is achieved only for several magnetic characteristics. For instance, in the paper

[24], for all three metals good agreement with experiment for the Curie temperature T_C is obtained. In the paper [25], for Fe good agreement with experiment is obtained for the Curie temperature: $T_C = 1015\,\text{K}$, but not for the effective magnetic moment: $m_{\text{eff}} = 1.96\,\mu_B$; for Ni, in contrast, the Curie temperature is almost one third smaller than the experimental one: $T_C = 450\,\text{K}$, but the effective moment is fairly close to the experimental one: $m_{\text{eff}} = 1.21\mu_B$. A similar situation is observed in the paper [26]: the Curie temperature for Fe is in good agreement with experiment, whereas for Co and Ni there is a discrepancy of about 30%. Moreover, for all three metals the slope of the reciprocal paramagnetic susceptibility is almost twice as large as the experimental one. Finally, in the paper [27] a reasonable value of the Curie temperature for Ni has been obtained in the Stoner approximation but with the use of an additional experimental information. At the same time, the temperature behaviour of the calculated magnetization curve remained static-like. As for our results for the local magnetic moment in the paramagnetic region, they agree with the results of all above-mentioned spin fluctuation calculations, except for those of [25], where the local moment of Ni above T_C appeared to be zero. Naturally, the same value of the local moment is obtained in the Stoner-like calculation in [27]. However, this result is so far debatable.

References

1. J. Hubbard, Phys. Rev. B **19**, 2626 (1979)
2. J. Hubbard, Phys. Rev. B **20**, 4584 (1979)
3. H. Hasegawa, J. Phys. Soc. Jpn. **46**, 1504 (1979)
4. H. Hasegawa, J. Phys. Soc. Jpn. **49**, 178 (1980)
5. H. Hasegawa, J. Phys. Soc. Jpn. **49**, 963 (1980)
6. J.A. Hertz, M.A. Klenin, Phys. Rev. B **10**(3), 1084 (1974)
7. J.A. Hertz, M.A. Klenin, Physica B&C **91B**, 49 (1977)
8. V.I. Grebennikov, Phys. Solid State **40**, 79 (1998)
9. B.I. Reser, V.I. Grebennikov, Phys. Met. Metallogr. **85**, 20 (1998)
10. B.I. Reser, J. Phys. Condens. Matter **11**, 4871 (1999)
11. N.B. Melnikov, B.I. Reser, V.I. Grebennikov, J. Phys. Condens. Matter **23**, 276003 (2011)
12. V.I. Grebennikov, J. Magn. Magn. Mater. **84**, 59 (1990)
13. N.B. Melnikov, B.I. Reser, Procs. Steklov Inst. Math. **271**, 149 (2010)
14. S. Hirooka, Physica B **149**, 156 (1988)
15. S. Hirooka, M. Shimizu, J. Phys. F: Met. Phys. **18**, 2237 (1988)
16. B.I. Reser, J. Phys. Condens. Matter **8**, 3151 (1996)
17. V.L. Moruzzi, J.F. Janak, A.R. Williams, *Calculated Electronic Properties of Metals* (Pergamon, New York, 1978)
18. J. Crangle, G.M. Goodman, Proc. R. Soc. Lond. A **321**, 477 (1971)
19. E.P. Wohlfarth, in *Ferromagnetic Materials*, vol. 1, ed. by E.P. Wohlfarth (North-Holland, Amsterdam, 1980), pp. 1–70
20. H.P.J. Wijn (ed.). *Magnetic Properties of Metals. 3d, 4d and 5d Elements, Alloys and Compounds. Landolt-Börnstein New Series*, vol. III/19a (Springer, Berlin, 1987)
21. O. Eriksson, B. Johansson, R.C. Albers, A.M. Boring, Phys. Rev. B **42**, 2707 (1990)
22. L.M. Sandratskii, J. Kübler, J. Phys. Condens. Matter **4**, 6927 (1992)
23. V.L. Moruzzi, P.M. Markus, K. Schwarz, P. Mohn, J. Magn. Magn. Mater. **54–57**, 955 (1986)
24. P. Mohn, E.P. Wohlfarth, J. Phys. F: Met. Phys. **17**, 2421 (1987)
25. J. Staunton, B.L. Gyorffy, Phys. Rev. Lett. **69**, 371 (1992)
26. M. Uhl, J. Kübler, Phys. Rev. Lett. **77**, 334 (1996)
27. N.C. Basalis, N. Theodorakopoulos, D.A. Papaconstantopoulos, Phys. Rev. B **55**, 11391 (1997)

11 High-Temperature Theory

Here must all distrust be left behind; all cowardice must be ended. (Dante Alighieri, The Divine Comedy, Inferno, Canto III; transl. by J.D. Sinclair)

As we showed in the previous chapter, the DSFT gives good agreement with experiment over a wide temperature range, which includes room temperatures. In this chapter, we demonstrate that at high temperatures, the temperature dependence of magnetic characteristics in the DSFT can become unstable and even discontinuous (first-order-like). We show that this type of behaviour is not an artefact of a numerical method. Same as in the simple Ising model (Chap. 7), the discontinuous jump and hysteresis behaviour in temperature dependency appear as a result of the Gaussian approximation when spin fluctuations become large. We show that proper temperature dependence of magnetization and second-order phase transition can be restored if we take into account higher-order terms of the free energy of electrons in the fluctuating exchange field.

11.1 Problem of Temperature Dependence

11.1.1 Discontinuous Jump of Magnetization

Calculations [1–4] showed that temperature behaviour of the DSFT solution can be unstable at high temperatures, well below the Curie temperature T_C. It is especially pronounced in Fe and Fe-Ni Invar, where spin fluctuations increase sharply in this region. Most often the instability takes the form of a discontinuous change of magnetic characteristics with temperature (the first-order-like transition). However, the discontinuous jump is usually too far from T_C to be interpreted in the framework of the critical phenomena.[1]

First-order magnetic phase transition in the *critical region* was observed in various versions [7–10] of the self-consistent renormalization (SCR) theory developed for weak ferromagnetic metals. In the SCR calculations, the jump in magnetization was usually explained by the character of the approximations. In the phenomenological theory [7] it was related to not taking into account the critical fluctuations, in [8] to the approximative character of the theory [7] itself, even after its improvement by going from a *scalar* field for the spin density to a *vector* field, which is more realistic. In the theory [9] the first-order transition at T_C appeared in a simplified model for the longitudinal susceptibility $\chi_\parallel$. When the $\chi_\parallel$ was treated properly, this discontinuity was eliminated. Finally, in [10, 11] the problem of the fictitious first-order phase transition at T_C has been solved by taking into account the zero-point spin fluctuations in the SCR theory.

In the DSFT the origin of the instability was discussed in [12–14]. Unlike the SCR theory, the DSFT is developed for the *strong* ferromagnets and uses the functional integral method [15, 16]. What is more important, the DSFT and SCR theory consider instability in different temperature ranges: over a wide range below T_C and in a narrow range near T_C, respectively. As is known [9], in the phase transition region the long-wave fluctuations play a dominant role. The nonlocal approximation [1] takes this fact into account, but insufficiently. Therefore, here we do not discuss the temperature dependence in the close neighbourhood of T_C.

[1]For the theory of the critical region, see, e.g. [5, 6].

N. B. Melnikov, B. I. Reser, *Dynamic Spin Fluctuation Theory of Metallic Magnetism*,
https://doi.org/10.1007/978-3-319-92974-3_11

In this section, we study the discontinuous jump in the temperature dependence of magnetic characteristics in the DSFT by the example of Fe and Invar alloy $Fe_{0.65}Ni_{0.35}$ (for more details, see [12]). First, we demonstrate that the instability is not an artefact of the calculation method. To this end, we show that the coordinate bisection method used in [1–4, 17–21] gives good agreement with the multidimensional minimization over a wide range of temperatures. Second, through the calculation with the forward and backward changes of temperature, we demonstrate the hysteresis behaviour of magnetic characteristics. Using methods of the catastrophe theory [22], we examine the effect of parameter changes on the temperature dependence. We show that the discontinuous temperature dependence can be smoothed out by small changes of the input data. Possible improvements of the DSFT approximations [1] are discussed in the next sections.

Finally, the choice of the disordered Invar alloy $Fe_{0.65}Ni_{0.35}$ is not accidental. First, this alloy has a large content of Fe, and thus exhibits all the problems connected to a sharp increase of the fluctuations at high temperatures. Second, and most important, $Fe_{0.65}Ni_{0.35}$ is the alloy that, in our opinion, can explain the Invar problem *per se* (for a review see [23–26]).

11.1.2 Instability Through Multiple Solutions

As we explained in the previous chapter, the DSFT calculation of magnetic properties in metals at finite temperatures is reduced to solution of the system of four nonlinear equations with respect to four unknowns: the chemical potential μ, mean single-site spin $\bar{s}^z$, and mean-square fluctuations $\zeta^x \equiv \langle \Delta V_x^2 \rangle'$ and $\zeta^z \equiv \langle \Delta V_z^2 \rangle'$:

$$\varphi_1(\mu, \bar{s}^z, \zeta^x, \zeta^z) \equiv n_\uparrow + n_\downarrow - n_e = 0, \tag{11.1}$$

$$\varphi_2(\mu, \bar{s}^z, \zeta^x, \zeta^z) \equiv (n_\uparrow - n_\downarrow)/2 - \bar{s}^z = 0, \tag{11.2}$$

$$\varphi_3(\mu, \bar{s}^z, \zeta^x, \zeta^z) \equiv uT/(2\lambda_L^x)I^x - \zeta^x = 0, \tag{11.3}$$

$$\varphi_4(\mu, \bar{s}^z, \zeta^x, \zeta^z) \equiv uT/(2\lambda_L^z)I^z - \zeta^z = 0. \tag{11.4}$$

Here

$$n_\sigma = \frac{N_d}{\pi} \int \mathrm{Im} g_\sigma(\varepsilon) f(\varepsilon)\, d\varepsilon \tag{11.5}$$

is the number of electrons with the spin projection $\sigma = \uparrow, \downarrow$ or ± 1, and the function I^α is calculated by the formula

$$I^\alpha = \int_0^1 \frac{1}{a_\alpha^2 + b_\alpha^2 k^2} \frac{2}{\pi} \arctan \frac{c_\alpha}{a_\alpha^2 + b_\alpha^2 k^2} 3k^2\, dk, \tag{11.6}$$

where a_α, b_α and c_α are given by (10.25).

Temperature dependence of a solution to the system of equations (11.1)–(11.4) is usually obtained by increasing temperature from $T = 0$, where the fluctuations vanish and solution is the easiest to calculate. Changing temperature in small steps one finds the solution at each T by a numerical method taking the solution from the previous step as the initial guess. Following this algorithm, one assumes that the solution changes continuously with temperature. Close to the point, where the derivative with respect to T becomes infinite, the solution can become unstable, i.e. the method convergence can slow down and the result can differ considerably from the initial guess.

Instability of the solution method appears at the point where the determinant of the Jacobi matrix vanishes:

$$\det \left\| \frac{\partial(\varphi_1, \varphi_2, \varphi_3, \varphi_4)}{\partial(\mu, \bar{s}^z, \zeta^x, \zeta^z)} \right\| = 0. \tag{11.7}$$

An explicit check of condition (11.7) in the DSFT is difficult, because finite-difference approximation of the gradients leads to considerable loss of precision. An alternative approach is to demonstrate that the solution to the system of equations (11.1)–(11.4) is nonunique (for details, see [12]).

We can show that the system of equations (11.1)–(11.4) has multiple solutions by obtaining a temperature hysteresis loop. The hysteresis in one of the variables, say $\bar{s}^z$, manifests itself in a jump down at a certain T_2 as T increases, and a jump back up at a smaller value T_1 as T is changed in the reverse direction (Fig. 11.1). Hence the curves $\bar{s}^z(T)$, obtained through the forward and backward change of temperature over $T_1 < T < T_2$, form a closed contour. In this case, we have three solutions at the interval $T_1 < T < T_2$: two stable and one unstable in between of them. At $T = T_1$ and $T = T_2$, two adjacent solutions

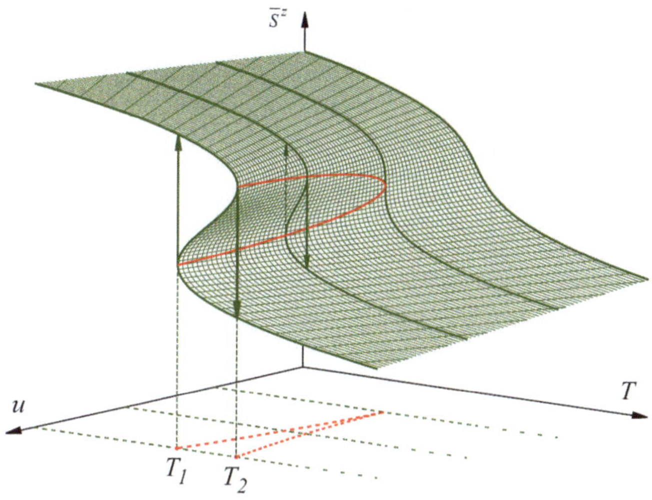

Fig. 11.1 Sketch of the instability through the multiple solution. The projection of the set of all degenerate solutions (red dotted line) separates the domains with three solutions and one solution on the parameter plane

merge into a degenerate one, i.e. condition (11.7) is satisfied. For $T < T_1$ and $T > T_2$, the system has only one solution, which is stable. At $T = T_1$ and $T = T_2$ we have first-order-like transition points, i.e. discontinuous jumps between two branches of solutions.

Consider now the effective interaction constant u as an additional parameter to the temperature T (Fig. 11.1). Changing u over an interval of its admissible values, we can make the hysteresis loop smaller. If, at a particular u value, the hysteresis loop eventually collapses to a point, we obtain the temperature dependence with the second-order-like transition. Decreasing u yet further, we obtain a temperature dependence without degenerate solutions. The projection of the set of all degenerate solutions (11.7) separates the domains with three solutions and one solution on the parameter plane (for more on bifurcations, see, e.g. [22]).

11.1.3 Temperature Hysteresis

Temperature dependence of the solution to the DSFT system of nonlinear equations is investigated first by the example of the Invar alloy $Fe_{0.65}Ni_{0.35}$. The initial nonmagnetic DOS $\nu(\varepsilon)$ (Fig. 11.2) is formed from the two spin-polarized DOSs obtained from the self-consistent calculation for the disordered $Fe_{0.65}Ni_{0.35}$ [27]. A detailed description of $\nu(\varepsilon)$ formation is given in [4]. The experimental value of the spin magnetic moment per atom $m_0^{\text{exp}} = 2\bar{s}^z(0)\mu_B = 1.75\mu_B$, used to determine the effective interaction constant u, is taken from [28].[2]

The results of the calculations of the basic magnetic characteristics of the $Fe_{0.65}Ni_{0.35}$ Invar obtained by the coordinate bisection method are represented in Fig. 11.3. Since the most important physical parameter in the Invar problem is not the magnetization $m(T) = 2\mu_B\bar{s}^z(T)$, but the *local* magnetic moment $m_L(T)$, calculated in the DNA by the formula (10.44):

$$m_L(T)/m(0) = [((u\bar{s}^z(T))^2 + 2\zeta^x + \zeta^z)/(u\bar{s}^z(0))^2]^{1/2},$$

it is also shown in the figure.

The investigation was carried out with the temperature T_W ranging from 0 to 0.01 (T_W is the temperature in units of the bandwidth $W = 9.70\,\text{eV} \approx 1.1 \times 10^5\,\text{K}$). First the temperature step is selected to be relatively large ($\Delta T_W = 0.001$), but as the values of $\bar{s}^z$, ζ^z and ζ^x start changing considerably, the program switches to a smaller step size ($\Delta T_W/20 = 0.00005$). Calculation accuracies in the variables μ, $\bar{s}^z$, ζ^x and ζ^z for one-dimensional bisections are selected so that the relative errors are approximately of the same order, 10^{-4}.

As can be seen from Fig. 11.3, at $T \leq 0.68\,T_C^{\text{exp}}$ a gradual increase of fluctuations and a smooth decrease of the magnetic moment are observed: during one small step in temperature the reduced fluctuations $\zeta^x(T)/\langle V_z(0)\rangle^2$ and $\zeta^z(T)/\langle V_z(0)\rangle^2$ increase by 0.001–0.01, and the reduced magnetic moment $\bar{s}^z(T)/\bar{s}^z(0)$ decreases by 0.002–0.02. The picture changes at

[2] Note that the experimental value m_0^{exp} cited in [28] is equal to $1.77\mu_B$, but it includes a small ($<0.1\mu_B$) positive contribution of the orbital magnetic moment [29].

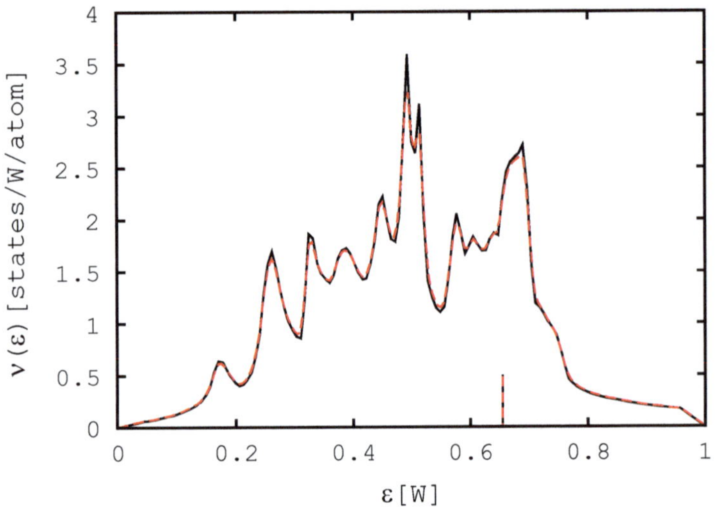

Fig. 11.2 The DOS of the d band of nonmagnetic $Fe_{0.65}Ni_{0.35}$, obtained from [27] (solid line), and the one smoothed out by convolution with the Lorentzian function of the halfwidth $\Gamma = 0.001$ (dashed line). The energy ε and halfwidth Γ are in units of the bandwidth $W = 9.70$ eV. The vertical line indicates the position of the Fermi level ε_F

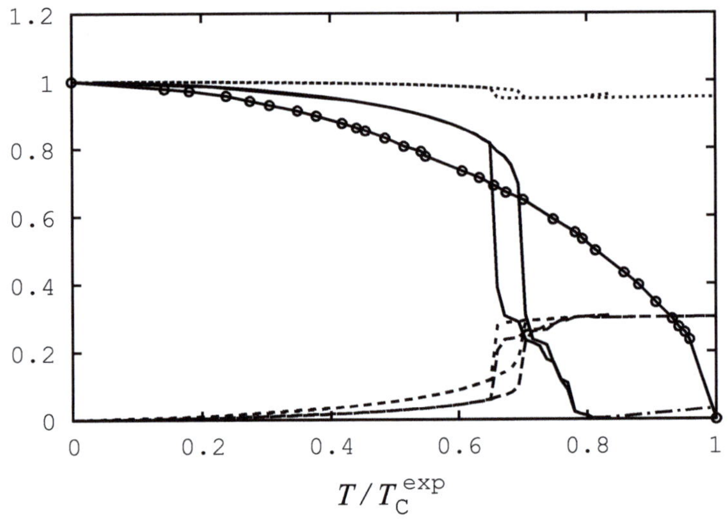

Fig. 11.3 The magnetization $m(T)/m(0)$ (solid line: calculation, circled line: experiment [28]), the mean square of spin fluctuations ζ^x (dashed line) and ζ^z (long dashed line) in units of the square of the mean exchange field at $T = 0$, the reciprocal paramagnetic susceptibility $\chi^{-1}(T)$ (dash dot dashed line) in units of $k_B T_C^{exp}/\mu_B^2$, and the local magnetic moment $m_L(T)/m(0)$ (dotted line) of the $Fe_{0.65}Ni_{0.35}$ Invar calculated in the DSFT with $m_0^{exp} = 1.75\mu_B$ through forward and backward changes of temperature

temperatures $T \geq 0.69\, T_C^{exp}$. First an abrupt change in fluctuations and a jump down in the magnetic moment are observed, then the change in this quantities fails to be smooth. However, the preset accuracy of the solution to the system of equations (11.1)–(11.4) is retained at all temperatures.

It was the problem of instability of the solution to the system of equations (a jump down in the magnetic moment and abrupt change in fluctuations) that urged us to use the multidimensional minimization (coordinate descent method) as a test for the results of the coordinate bisection method (for details, see Appendix I).

The calculation in the coordinate descent method was performed with the same initial data and at the same temperatures as in the coordinate bisection method. In the coordinate descent method the accuracy was chosen such that it guarantees the relative errors in $\bar{s}^z$, ζ^x and ζ^z to be nearly the same as calculated by the coordinate bisection method, i.e. within $10^{-3} - 10^{-4}$. At temperatures $T \leq 0.68\, T_C^{exp}$, the results obtained by the coordinate descent are in close agreement with those obtained by coordinate bisection (to the prescribed precision of 3–4 significant digits). However, starting from $T = 0.69\, T_C^{exp}$, the desired accuracy of the minimized function cannot be reached, although a monotonic decrease of the magnetic moment and a monotonic increase of the fluctuations still persist.

It is important to note that the values of $\bar{s}^z$, ζ^x and ζ^z obtained by the coordinate descent practically coincide with those obtained by coordinate bisection over a wide temperature range (see Table 1 in [12]). This is a strong indication that the instability at high temperatures is related to the system itself rather than the solution method.

To demonstrate that the numerical instability at $T \approx 0.69 T_C^{exp}$ is caused by the degeneracy of the system of equations (11.1)–(11.4), we investigate a possibility of the hysteresis behaviour of the solution. The calculation with the forward and backward changes of temperature shows that, contrary to the experiment, the curves $\bar{s}^z$, ζ^x and ζ^z have a small hysteresis loop

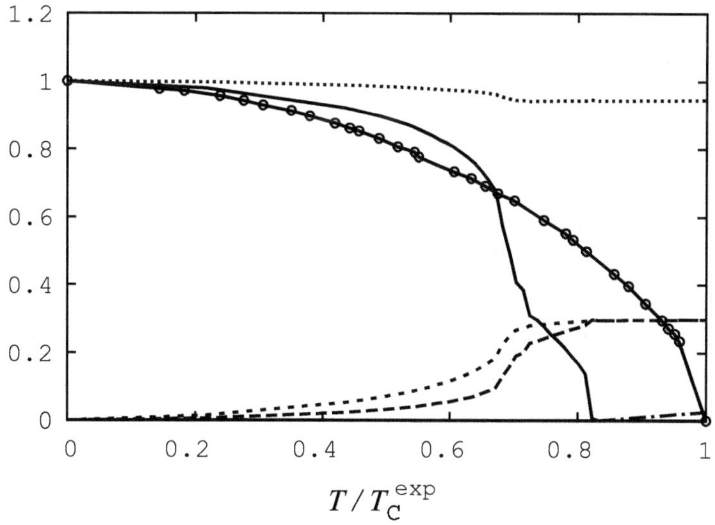

Fig. 11.4 As Fig. 11.3, but calculated with $m_0^{\rm exp} = 1.70\mu_{\rm B}$

(Fig. 11.3). However, with the admissible initial magnetic moment $m_0^{\rm exp} = 1.70\mu_{\rm B}$ (see footnote 2), the effective interaction constant u is equal to 1.10 eV (instead of $u = 1.13$ eV for $m_0^{\rm exp} = 1.75\mu_{\rm B}$), and the hysteresis loop disappears (Fig. 11.4). The curves $\bar{s}^z$, ζ^x and ζ^z obtained with the backward change of temperature almost replicate those obtained with the forward change of temperature, except for the instability region, where we have small deviations but without strictly vertical sections.

In order to reduce the fluctuations ζ^α one should take into account the higher-order terms in the expansion of the free energy $F_1(V)$. Instead, we can reduce the fluctuations ζ^α implicitly by decreasing the effective interaction constant u in formula (10.22) to $u_2 = cu$, where c is a parameter (slightly less than one) estimated from experience. With $c = 0.985$ it is possible to obtain full agreement with experiment for the Curie temperature: $T_{\rm C} = 1.01T_{\rm C}^{\rm exp}$, for the paramagnetic Curie point: $\Theta_{\rm C} = 1.06T_{\rm C}^{\rm exp}$ and for the effective magnetic moment: $m_{\rm eff} = 0.97m_{\rm eff}^{\rm exp}$. But overall the curve for the magnetization does not fit the experiment well enough. Qualitatively the calculated curve $m(T)$ stays similar to the one obtained with $c = 1$ (Fig. 11.4), i.e. it still has a snake-like form. The reason is that the above change of the constant u yields a uniform change of the fluctuations, while to remove the "snake" one has to account for the interaction between the fluctuations that becomes more intense with an increase of temperature.[3]

In the SCR-calculations [9], the first-order transition at $T_{\rm C}$ was eliminated entirely when the $\chi_\parallel^{-1}$ was approximated by $\chi_\perp^{-1}$ (the case $\eta = 0$ in (3.18) in [9]). However, in our calculation with $\zeta^z = \zeta^x$, the temperature behaviour represented in Fig. 11.4 remains. In [10, 11], to avoid the fictitious first-order transition in the SCR theory, a single equation for the longitudinal $\chi_\parallel$ and transverse $\chi_\perp$ susceptibilities was suggested. The relation that couples $\chi_\parallel$ and $\chi_\perp$ is based on the assumption that the total local spin fluctuation, i.e. the sum of the zero-point and thermal spin fluctuations, is conserved. This is the case in the Heisenberg local moment theory, and may be somehow justified for weak ferromagnets. In our case, the two equations for the fluctuations ζ^x and ζ^z are coupled to each other, as well as to the other two equations of system (11.1)–(11.4). However, system (11.1)–(11.4) is obtained using the quadratic approximation of the free energy, which does not account for the higher order interactions (the anharmonicity of the fluctuations). Apparently, the coupling of the ζ^x and ζ^z in the system (11.1)–(11.4) is insufficient.

Note that the temperature hysteresis is observed not only in the Fe-Ni Invar, but also in the elemental Fe, i.e. it is a general problem for strong ferromagnets with sharply increasing spin fluctuations. In Fe, as the initial DOS we use the same nonmagnetic DOS as in Sect. 10.5, which is calculated in the LDA by the KKR method with a self-consistent potential [30]. Then the DOS is smoothed out by convolution with the Lorentzian function of the halfwidth $\Gamma = 0.001W$ ($W = 7.16$ eV is the bandwidth) and normalized to one d band of unit width. The smoothed DOS of the d band $\nu(\varepsilon)$ used for calculation is represented in Fig. 11.5. The number of d electrons per atom is $n_{\rm e} = 7.43$, just as before. The effective interaction constant u determined from $m_0^{\rm exp} = 2.217\mu_{\rm B}$ [31] is 1.06 eV.

As can be seen from Fig. 11.6, which shows the results of our calculation for the magnetic characteristics with the forward and backward changes of temperature, in Fe the hysteresis loop is even larger than in the Fe-Ni Invar. However, the

[3]The reduction of the fluctuations with the help of u_2 is formally equivalent to replacing the $\lambda(\mathbf{q}, \omega)$ in the expression for the dynamic susceptibility (4.19) in [8] by a constant.

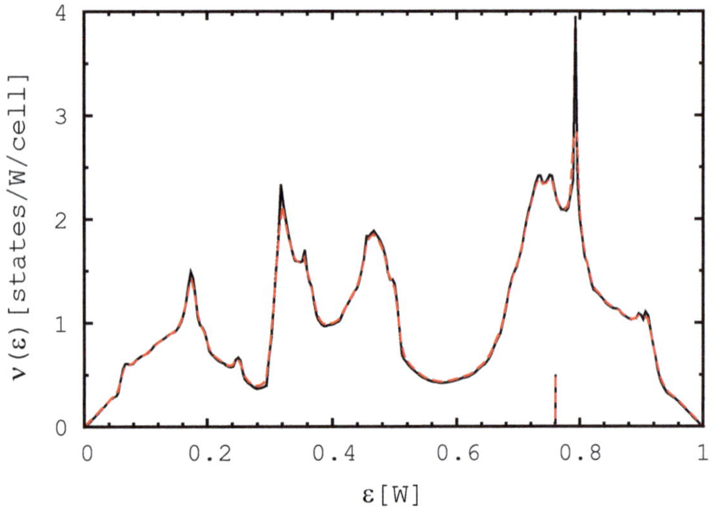

Fig. 11.5 The DOS of the d band of nonmagnetic Fe, calculated by the KKR method with a self-consistent potential (solid line), and the one smoothed out by convolution with the Lorentzian function of halfwidth $\Gamma = 0.001$ (dashed line). The energy ε and halfwidth Γ are in units of the bandwidth $W = 7.16\,\text{eV}$. The vertical line indicates the position of the Fermi level ε_F

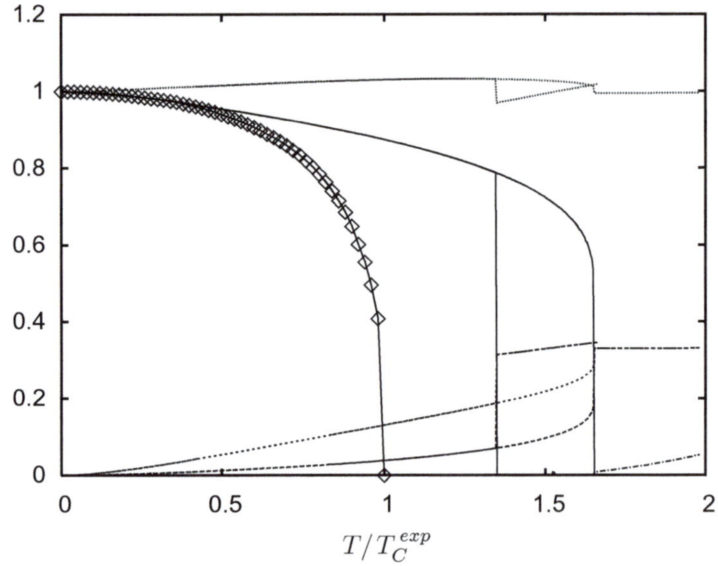

Fig. 11.6 The magnetization $m(T)/m(0)$ (solid line: calculation, diamond: experiment [31]), the mean square of spin fluctuations ζ^x (dashed line) and ζ^z (long dashed line) in units of the square of the mean exchange field at $T = 0$, the reciprocal paramagnetic susceptibility $\chi^{-1}(T)$ (dash dot dashed line) in units of $k_B T_C^{\text{exp}}/\mu_B^2$, and the local magnetic moment $m_L(T)/m(0)$ (dotted line) of Fe calculated in the DSFT with $m_0^{\text{exp}} = 2.217\mu_B$ through forward and backward changes of temperature

temperature behaviour of magnetization near the jump in the Fe-Ni Invar and Fe is qualitatively different (compare Figs. 11.3 and 11.6). In the former, the magnetization does not jump down to zero, i.e. the switching-over to the paramagnetic state does not occur. Therefore, in Fe-Ni Invar, a small change of the initial data only smears the discontinuous jump into a smooth curve with an inflexion. In contrast, in the case of Fe, the smoothing could transform the first-order transition to the second-order one.

Finally, as an alternative attempt to eliminate the discontinuous jump, we carried out calculations for Fe and Fe-Ni Invar with only the longitudinal or only transverse fluctuations. As should be expected, the switching-off of the transverse or longitudinal fluctuations leads to a Curie temperature almost twice as large as the experimental one. Worse agreement with experiment is achieved in this case for other magnetic characteristics as well. Most importantly, the use of models with a one-dimensional fluctuating field does not solve the problem of temperature dependence for Fe and Fe-Ni Invar. In particular, in the model that takes only longitudinal fluctuations into account, the magnetization decreases too slowly with the increase in temperature, and the discontinuous jump (first-order transition) turns out to be too sharp. In the model that takes only transverse fluctuations into account, the magnetization decreases faster and agreement with experiment at low temperatures is achieved, but singularities of the magnetization curve remain.

Thus, calculations of the magnetic properties of Fe and Fe-Ni Invar in the DSFT showed that the coordinate bisection method is quite applicable to practical calculations at finite temperatures. The instability at high temperatures is connected not with the solution method—two different numerical methods gave well-agreed results—but with the system of nonlinear equations itself.

The problem is that in Fe and in alloys with a considerable content of Fe, like Fe-Ni Invar, the spin fluctuations at high temperatures increase sharply, which means that $\lambda_{\mathrm{L}}^{\alpha} = 1 - u\chi_{\mathrm{L}}^{\alpha}(0)$ tends to zero. Consequently, there is a strong dependence of the solution to the system of nonlinear equations (11.1)–(11.4) on the accuracy of the effective constant u and the static local susceptibility $\chi_{\mathrm{L}}^{\alpha}(0)$.[4]

Note that the problem of temperature dependence is connected not only with the system of equations, but also with the initial parameters of the real ferromagnet (the DOS $\nu(\varepsilon)$, the number of d electrons n_{e}, the effective interaction constant u, etc.). Each particular set of data yields its own solution to the system of nonlinear equations (11.1)–(11.4). For example, in Co and Ni no instabilities were found at all (see, e.g. [2]). Even in the case of Fe and Fe-Ni Invar, reasonable changes of the initial data and/or the system of equations can either remove or at least reduce considerably the effect of the instability. Finally, as our calculations show, if only longitudinal fluctuations are taken into account the discontinuous jump near T_{C} only increases, i.e. taking into account both transverse and longitudinal fluctuations does not worsen the situation, as stated in [33], but improves it.

Apparently, for transition metals and alloys with strong spin fluctuations and large magnetization, the approximations [1] of the DSFT should be improved to be equally applicable at high temperatures. It is necessary to go beyond the quadratic approximations for the fluctuation contribution to the self-energy $\Delta\Sigma$ and for the free energy of electrons in the fluctuating exchange field $F_1(V)$. Quite possible, a more consistent account—not only in the derivation of formula (11.6)—of the short-range magnetic order is necessary. As we will show in the next section, damping of the sharp increase in the fluctuations allows not only to eliminate the instability of the solution but also, what is more important, to improve the agreement with experiment at high temperatures.

11.2 Beyond the Gaussian Approximation

As we have shown, for spin fluctuations with large amplitudes, the Gaussian approximation (GA) of the DSFT becomes insufficient to properly take the interactions into account and yields a jump phase transition to the paramagnetic state [1, 34]. In papers [35–37] the jump in temperature dependencies was eliminated and a proper second-order phase transition was obtained in the DSFT by taking into account higher-order terms of the free energy of electrons in the fluctuating exchange field. In the final computational formulae, the third-order term renormalizes the mean field, and fourth-order term renormalizes the susceptibility. The main novelty in the treatment of the higher-order terms by Reser et al. [35] and Melnikov et al. [36] is that the ferromagnetic state is changed self-consistently and an interpolation between local and long-wave limits is used (for treatments of the fourth-order term in the paramagnetic state, see [8, 38–40] and references therein). The renormalized Gaussian approximation (RGA) was enhanced further by taking into account uniform fluctuations (UF) in the single-site Green function [37].

11.2.1 Renormalized Gaussian Approximation

We briefly recall the main steps of the renormalization, suggested in [35] and developed in [36, 37]. We start with the fourth-order series expansion of the free energy (9.1) in powers of $\Delta V = V - \tilde{V}$ (for details, see Appendix A.1.2):

$$\begin{aligned} F(V) = {} & T\mathrm{Tr}\Big(\frac{\tilde{V}\Delta V}{uN_{\mathrm{d}}T} + G(\tilde{V})\Delta V\Big) + \frac{1}{2}T\mathrm{Tr}\Big(\frac{\Delta V^2}{uN_{\mathrm{d}}T} + \big(G(\tilde{V})\Delta V\big)^2\Big) \\ & + \frac{1}{3}T\mathrm{Tr}\big(G(\tilde{V})\Delta V\big)^3 + \frac{1}{4}T\mathrm{Tr}\big(G(\tilde{V})\Delta V\big)^4, \end{aligned} \tag{11.8}$$

where $\tilde{V}$ is an *arbitrary* value of the exchange field. First, we carry out the partial averaging of the cubic and quartic terms:

$$\Delta V^3 \approx 3\underbrace{\Delta V \Delta V}\Delta V, \qquad \Delta V^4 \approx 6\underbrace{\Delta V \Delta V}\Delta V \Delta V, \tag{11.9}$$

[4] High sensitivity of the local magnetic characteristics of iron to the value of u at high temperatures is well illustrated in [32].

where the prefactor is the number of all possible pair combinations, and the underbrace denotes the averaging with a Gaussian probability density $p^{(2)}(V)$, which we have to construct. Applying (11.9) to the right-hand side of (11.8), we obtain

$$\begin{aligned} F(V) = T\mathrm{Tr}\Big(\frac{\tilde{V}\Delta V}{uN_{\mathrm{d}}T} + G(\tilde{V})\Delta V + G(\tilde{V})\Delta \underbracket{V G(\tilde{V})\Delta V} G(\tilde{V})\Delta V\Big) \\ + \frac{1}{2}T\mathrm{Tr}\Big(\frac{\Delta V^2}{uN_{\mathrm{d}}T} + G(\tilde{V})\Delta V G(\tilde{V})\Delta V + 3G(\tilde{V})\Delta \underbracket{V G(\tilde{V})\Delta V} G(\tilde{V})\Delta V G(\tilde{V})\Delta V\Big). \end{aligned} \tag{11.10}$$

Next, we apply the Gaussian approximation (9.27) to the modified free energy (11.10). Henceforth we use the average $\langle\cdots\rangle_{(2)}$ and omit the subscript. Replacing averages of products of the Green functions $G(\tilde{V})$ by the products of the averages $\bar{G} \equiv \langle G(\tilde{V})\rangle$, we come to

$$F^{(2)}(V) = \frac{1}{2}T\mathrm{Tr}\Big(\frac{\Delta V^2}{uN_{\mathrm{d}}T} + \bar{G}\Delta V\bar{G}\Delta V + 3\bar{G}\Delta \underbracket{V\bar{G}\Delta V}\bar{G}\Delta V\bar{G}\Delta V\Big). \tag{11.11}$$

Here $\Delta V = V - \bar{V}$, and the mean field $\bar{V} \equiv \langle\tilde{V}\rangle$ is derived from the condition that the linear term annihilates:

$$T\mathrm{Tr}\Big(\frac{\bar{V}\Delta V}{uN_{\mathrm{d}}T} + \bar{G}\Delta V + \bar{G}\Delta \underbracket{V\bar{G}\Delta V}\bar{G}\Delta V\Big) = 0. \tag{11.12}$$

Further simplification of (11.11) and (11.12) is carried out through the splitting formula

$$\mathrm{Tr}(\bar{G}\Delta \underbracket{V\bar{G}\Delta V}\bar{G}\Delta V) \approx \frac{\pi T}{2N_{\mathrm{d}}NW}\mathrm{Tr}(\bar{G}\Delta \underbracket{V\bar{G}\Delta V})\mathrm{Tr}(\bar{G}\Delta V) \equiv \eta\mathrm{Tr}(\bar{G}\Delta V), \tag{11.13}$$

where the factors N_{d}^{-1}, N^{-1}, $\pi T/W$ and 1/2 appear due to partial averaging in band, site, "time" and spin, respectively (for details, see Appendix E.3). Applying (11.13) to the linear term (11.12), we have

$$T\mathrm{Tr}\Big(\frac{\bar{V}\Delta V}{uN_{\mathrm{d}}T} + (1+\eta)\bar{G}\Delta V\Big) = 0.$$

Splitting the correction due to the quartic term of the free energy in a similar fashion, we rewrite (11.11) as

$$F^{(2)}(V) = \frac{1}{2}T\mathrm{Tr}\Big(\frac{\Delta V^2}{uN_{\mathrm{d}}T} + (1+3\eta)\bar{G}\Delta V\bar{G}\Delta V\Big).$$

In the momentum-"frequency" representation, we obtain the same mean-field equation (9.29) as in the GA but $\bar{s}_z$ is now given by

$$\bar{s}_z = (1+\eta)\,N^{-1}N_{\mathrm{d}}T\sum_{\mathbf{q}n}\bar{G}^z_{\mathbf{q}n}. \tag{11.14}$$

Similarly, the fluctuation $\langle|\Delta V^\alpha_{\mathbf{q}m}|^2\rangle$ is calculated by the same formula (9.35) as in the GA but the *unenhanced* dynamic susceptibility $\chi^{0\alpha}_{\mathbf{q}m}$ becomes

$$\begin{aligned} \chi^{0\alpha}_{\mathbf{q}m} &\equiv -\frac{1}{2}\left\langle\frac{\partial^2 F^{(2)}_1(V)}{\partial V^\alpha_{\mathbf{q}m}\partial V^\alpha_{-\mathbf{q}-m}}\right\rangle \\ &= -\frac{1}{2}(1+3\eta)\,N_{\mathrm{d}}T\sum_{\mathbf{k}n}\sum_{\gamma_1\gamma_2}\bar{G}^{\gamma_1}_{\mathbf{k}n}\bar{G}^{\gamma_2}_{\mathbf{k}-\mathbf{q},\,n-m}\mathrm{Sp}(\sigma^{\gamma_1}\sigma^\alpha\sigma^{\gamma_2}\sigma^\alpha). \end{aligned} \tag{11.15}$$

The condition that the total number of electrons is conserved leads to the same equation on the chemical potential (9.36) as before.

Thus, the RGA yields the same system of nonlinear equations (9.29), (9.35) and (9.36) as in the GA but with mean spin (11.14) and susceptibility (11.15) renormalized self-consistently.

11.2.2 Local and Uniform Fluctuations

Just as in the GA, we introduce the single-site fluctuation $\langle \Delta V_\alpha^2 \rangle'$ using formula (10.15). Calculations of the sum over $\mathbf{q}$ and m follow closely those in the GA and give essentially the same result (10.24) but with the renormalization prefactor $1 + 3\eta$ in the local susceptibility.

The higher-order correction coefficient η, which appears in (11.14) and (11.15), is calculated as

$$\eta = -\frac{\pi}{W N_\mathrm{d}} \sum_\alpha \grave{\chi}_\mathrm{L}^{0\alpha}(0) \langle \Delta V_\alpha^2 \rangle', \tag{11.16}$$

where $\grave{\chi}_\mathrm{L}^{0\alpha}(0)$ is the static local susceptibility in the GA ($\eta = 0$). At low temperatures, the effect of the higher-order renormalization is small. At high temperatures, we have $\grave{\chi}_\mathrm{L}^{0\alpha}(0) \approx u^{-1}$, and hence (11.16) is proportional to the sum of the mean-square fluctuations: $\eta \approx -\pi (u N_\mathrm{d} W)^{-1} \sum_\alpha \langle \Delta V_\alpha^2 \rangle'$.

Note that the mean-square fluctuations (10.15) take into account both spatial correlations and interaction of different $\mathbf{q}$-harmonics due to the interpolation of the susceptibility $\chi_\mathbf{q}(\varepsilon)$ between the local $\chi_\mathrm{L}(\varepsilon)$ and uniform $\chi_0(\varepsilon)$ susceptibilities. However, the local susceptibility $\chi_\mathrm{L}(\varepsilon)$ is determined only by the site-diagonal part of the mean Green function: $\bar{G}_{jj'n} = g(\mathrm{i}\omega_n)\delta_{jj'}$.

We improve the effect of nonlocality by taking into account uniform fluctuations (UF) of the mean field in (10.35) as follows:

$$\tilde{g}(\varepsilon) \equiv \left\langle g^0(\varepsilon - \tilde{V} - \Delta\Sigma(\varepsilon, \tilde{V})) \right\rangle, \tag{11.17}$$

where the average is taken over the single-site Gaussian field $\tilde{V}$. Here

$$g^0(\varepsilon) = \int \frac{\nu(\varepsilon')}{\varepsilon - \varepsilon'} \mathrm{d}\varepsilon',$$

where $\nu(\varepsilon)$ is the nonmagnetic DOS at $T = 0$. To simplify the calculations, we approximate the fluctuating field $\tilde{V}$ by a field that takes only two values $\pm \bar{V}_z$, so that (11.17) is reduced to [37, 41]

$$\tilde{g}_\sigma(\varepsilon) = \sum_{\sigma'} P_{\sigma\sigma'} g^0(\varepsilon - \sigma' v - \Delta\Sigma_\sigma(\varepsilon, \sigma' v)), \tag{11.18}$$

$$P_{\sigma\sigma'} = \frac{1}{2}\left(1 + \frac{\sigma \bar{V}_z}{\sigma' v}\right),$$

where $v = \left(\bar{V}_z^2 + \langle \Delta V_x^2 \rangle' + \langle \Delta V_y^2 \rangle' + \langle \Delta V_z^2 \rangle'\right)^{1/2}$ (for details, see Appendix E.4).

In the RGA+UF, we still solve a system of four equations in four unknowns: the mean field $\bar{V}_z$, mean-square fluctuations $\langle \Delta V_x^2 \rangle' = \langle \Delta V_y^2 \rangle'$ and $\langle \Delta V_z^2 \rangle'$, and chemical potential μ, with temperature as a parameter. But the difference of the RGA+UF system of equation from the one in the GA is that the mean field (11.14) and unenhanced susceptibility (11.15) are renormalized self-consistently, and the uniform fluctuations are explicitly taken into account in the single-site mean Green function (11.18).

Similarly, the final expressions for most of the magnetic characteristics, such as magnetization $m_z = g\mu_\mathrm{B}\bar{s}_z$ and enhanced susceptibility (10.47), stay the same but the way they are calculated in the RGA or RGA+UF is different from the one in the GA. In particular, the local magnetic moment $m_\mathrm{L} = g\mu_\mathrm{B}s_\mathrm{L}$ is calculated in the RGA+UF as follows. We improve the effect of the uniform fluctuations on the square of the local spin moment (10.43) by averaging over the single-site field

$\tilde{V}$: $\tilde{s}_{\mathrm{L}}^2 = u^{-2}\langle \tilde{V}^2 + \langle \Delta V^2\rangle'\rangle$, just as we did for the single-site mean Green function $g(\varepsilon)$ in formula (11.17). Using the approximate fluctuating field $\tilde{V}$ that takes two values $\pm\bar{V}_z$, we come to the final expression

$$\tilde{s}_{\mathrm{L}} = u^{-1}\left(\bar{V}_z^2 + 2(\langle \Delta V_x^2\rangle' + \langle \Delta V_y^2\rangle' + \langle \Delta V_z^2\rangle')\right)^{1/2}. \tag{11.19}$$

Thus, in the RGA+UF formula (11.19) the total mean-square fluctuation is multiplied by the factor two as compared to the DSFT expression (10.44).

11.2.3 Application to Fe and Fe–Ni Invar

Iron

The extended DSFT allows to eliminate the first-order phase transition, which is observed in the GA of the DSFT. To demonstrate that, we consider bcc iron and take the same initial data as in Sect. 11.1.3 (the DOS is shown in Fig. 11.5). The number of d electrons per atom n_{e} is equal to 7.43. The effective interaction constant u, determined from $m_0^{\mathrm{exp}} = 2.217\,\mu_{\mathrm{B}}$ [31], is 1.06 eV.

The RGA calculation gives good agreement with experiment for the effective magnetic moment $m_{\mathrm{eff}} = 1.08 m_{\mathrm{eff}}^{\mathrm{exp}}$ ($m_{\mathrm{eff}}^{\mathrm{exp}} = 3.13\,\mu_{\mathrm{B}}$ [42]) and for temperature dependence of the local magnetic moment $m_{\mathrm{L}}(T)$. However, for the Curie temperature, the RGA yields $T_{\mathrm{C}} = 2.23 T_{\mathrm{C}}^{\mathrm{exp}}$ ($T_{\mathrm{C}}^{\mathrm{exp}} = 1044$ K [31]).

In the RGA+UF the fluctuations increase and this yields $T_{\mathrm{C}} = 1.56 T_{\mathrm{C}}^{\mathrm{exp}}$. A sharp increase of the fluctuations and sharp decrease in magnetization at high temperatures, which occurred in the GA of the DSFT (Fig. 11.7), disappear in the RGA+UF (Fig. 11.8). The temperature dependence of magnetization calculated in the RGA+UF gives a fairly good fit to the experimental one. The local magnetic moment calculated in the RGA+UF by formula (11.19) does not strongly depend on temperature as it must be (for details, see [43]).

Calculations in the GA+UF for Fe, Co and Ni were carried out and discussed in [34]. They give a smaller Curie temperature than the GA calculations but have the same first-order-like phase transition in Fe as the GA calculations [1, 2].

Fe-Ni Invar

Now the extended DSFT is applied to the calculation of the magnetic properties of the $Fe_{0.65}Ni_{0.35}$ Invar alloy at finite temperatures [37, 44]. The choice of the Fe-Ni Invar for demonstrating the possibilities of the extended DSFT is motivated by the problems of temperature dependence that were found within the quantum-statistical treatment of this Invar [3, 4, 12, 45].

It is known that the Fe-Ni Invar is a complex disordered system (see, e.g. [46, 47] and references therein). However, comparison of the calculation results for the disordered alloy $Fe_{0.65}Ni_{0.35}$ [4, 12] and the ordered compound Fe_3Ni [3, 45] showed that the effect of disorder in the filling of sites with Fe and Ni atoms on the magnetic properties of the Fe-Ni Invar is not critical. This conclusion agrees with earlier results for the ordered and disordered $Fe_{0.72}Pt_{0.28}$ Invar (see, e.g.

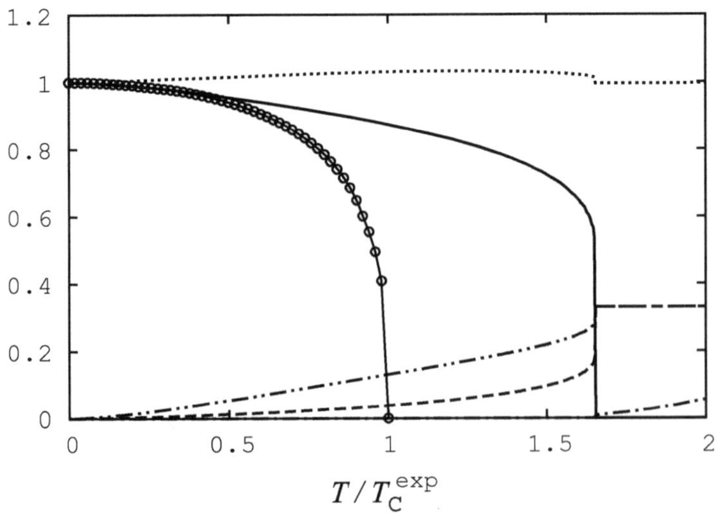

Fig. 11.7 The magnetization $m(T)/m(0)$ (solid line: calculation, circled line: experiment [31]), the mean-square fluctuations $\langle \Delta V_x^2\rangle'(T)$ (dash dot dot dashed line) and $\langle \Delta V_z^2\rangle'(T)$ (dashed line) in units of the square of the mean exchange field $\bar{V}_z^2(0)$, the reciprocal paramagnetic susceptibility $\chi^{-1}(T)$ (dash dot dashed line) in units of $k_{\mathrm{B}} T_{\mathrm{C}}^{\mathrm{exp}}/\mu_{\mathrm{B}}^2$, and the local magnetic moment $m_{\mathrm{L}}(T)/m(0)$ (dotted line) of the bcc iron calculated in the GA of the DSFT as functions of the reduced temperature $T/T_{\mathrm{C}}^{\mathrm{exp}}$

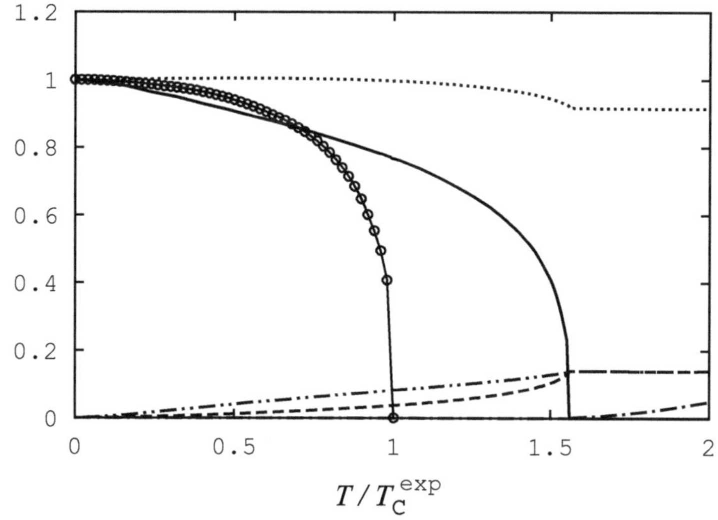

Fig. 11.8 As Fig. 11.7, but calculated in the RGA+UF of the DSFT

Table 11.1 The ferromagnetic (T_C) and paramagnetic (Θ_C) Curie temperatures, effective (m_{eff}) and local ($m_L(T_C)$) magnetic moments of $Fe_{0.65}Ni_{0.35}$ Invar, calculated in three approximations of the DSFT

DSFT	T_C/T_C^{exp}	Θ_C/T_C^{exp}	m_{eff}/m_{eff}^{exp}	$m_L(T_C)/m_L(0)$
GA	0.83	0.90	0.89	0.94
RGA	1.47	1.48	1.18	0.85
RGA+UF	1.07	1.15	1.01	0.65

[25, table 10-1]). The weak influence of the atomic disorder on the magnetic properties of the Fe-Ni Invar at finite temperatures is explained by the *integral* dependence of the SFT equations on the electronic energy structure. The details of the initial DOS do not exert the decisive effect on the results of the calculations.

We use the same initial nonmagnetic DOS $\nu(\varepsilon)$ as before (see Fig. 11.2). The number of d electrons per atom $n_e = 2N_d \int_0^{\varepsilon_F} \nu(\varepsilon)\,d\varepsilon$ (ε_F is the Fermi energy) is equal to 7.66. The effective interaction constant $u = 1.1$ eV is determined from (10.33) to (10.35) at $T = 0$ with the experimental value of the spin magnetic moment per atom $m_0^{exp} = 1.7\,\mu_B$ [28].

Note that here we neglect the fine effects of the atomic and/or magnetic short-range order (see, e.g. [46,47] and references therein). Moreover, the magnetic moment m_0^{exp} and the DOS $\nu(\varepsilon)$ represent the values per *averaged* atom. However, as we have mentioned, even with these initial data one can calculate the temperature dependence of the magnetic properties of an alloy in the DSFT.

Figure 11.4 and Table 11.1 present basic magnetic characteristics for the $Fe_{0.65}Ni_{0.35}$ Invar calculated within the GA of the DSFT. All the characteristics are represented in units of their experimental values $T_C^{exp} = 520$ K [28] and $m_{eff}^{exp} = 3.3\,\mu_B$ [48]. Clearly, at high temperatures, the calculated magnetization curve $m(T) = g\mu_B\bar{s}_z(T)$ does not fit the experimental one well enough. For the Curie temperature, we obtain $T_C = 0.83\,T_C^{exp}$. But most importantly, the calculated curve $m(T)$ has an inflection (see the discussion in [12]).

The situation improves when we take into account higher-order terms in the expansion of the free energy $F(V)$ using expression (11.16) for the correction coefficient η. The calculation results are represented in Fig. 11.9 and Table 11.1. As can be seen from Fig. 11.9, a sharp increase of the fluctuations and sharp decrease in magnetization at high temperatures, which occurred in the GA of the DSFT (Fig. 11.4), disappear in the RGA of the DSFT. On the whole, the curve for the magnetization fits the experimental one. However, there is no full quantitative agreement.

Finally, the RGA of the DSFT with $\tilde{g}_\sigma(\varepsilon)$ calculated by (11.18) gives good agreement with experiment at all temperatures (Fig. 11.10). In particular, the RGA+UF gives nearly full agreement for the Curie temperature: $T_C = 1.07\,T_C^{exp}$; for the paramagnetic Curie point: $\Theta_C = 1.15\,T_C^{exp}$; and for the effective magnetic moment: $m_{eff} = 1.01 m_{eff}^{exp}$ (Table 11.1).

As can be seen from Fig. 11.10, the local magnetic moment $m_L(T)/m_L(0)$ calculated by formula (10.44) strongly depends on temperature: with temperature increasing from zero to T_C^{exp}, the local moment decreases by 35% (Table 11.1). This change is considerable and quite sufficient for an explanation of the *Invar effect*. The assumption that the volume change of the Fe-Ni Invars is connected with the temperature variation of the local moment but not the magnetization was first made in [49] from an analysis of the experimental data. The subsequent SFT calculations [50,51] confirmed this assumption.

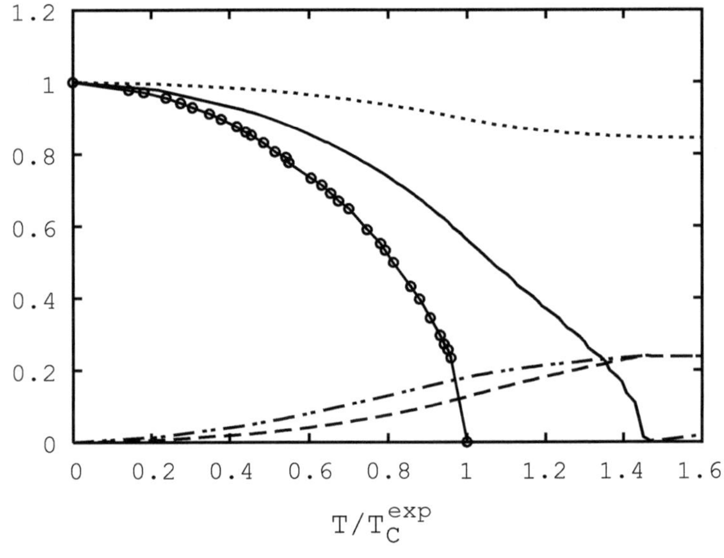

Fig. 11.9 As Fig. 11.3, but calculated in the RGA of the DSFT

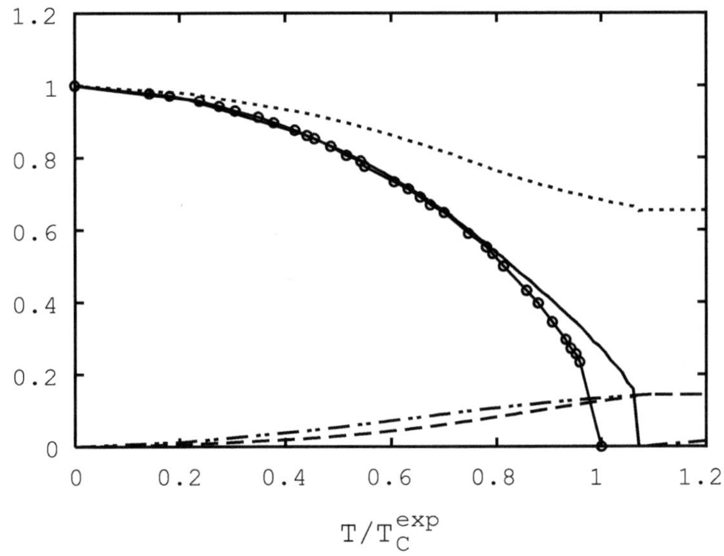

Fig. 11.10 As Fig. 11.3, but calculated in the RGA+UF of the DSFT

We use our calculated value $m_L(T_C)/m_L(0) = 0.65$ and experimentally estimated value $D_0/B \simeq 10^{-6}(\text{emu/g})^{-2}$ in the approximate formula (5) of [52] for the magnetic volume change: $\omega_m(T) = (D_0/B)M_0^2[(m_L(T)/m_L(0))^2 - 1]$, where D_0 is the magneto-volume coupling constant for $q = 0$, B is the bulk modulus and M_0 is the uniform magnetization at $T = 0\,\text{K}$. For $Fe_{0.65}Ni_{0.35}$, which has $M_0 \simeq 170$ emu/g [25], we obtain $\omega_m(T_C) = -0.017$, in good agreement with the experimentally estimated value -0.019 [53, 54].

References

1. B.I. Reser, V.I. Grebennikov, Phys. Met. Metallogr. **85**, 20 (1998)
2. B.I. Reser, J. Phys. Condens. Matter **11**, 4871 (1999)
3. B.I. Reser, J. Phys. Condens. Matter **16**, 361 (2004)
4. B.I. Reser, Phys. Met. Metallogr. **103**, 373 (2007)
5. J.A. Hertz, Int. J. Magn. **1**, 253 (1971)
6. R.M. White, *Quantum Theory of Magnetism*, 3rd edn. (Springer, Berlin, 2007)
7. K. Murata, S. Doniach, Phys. Rev. Lett. **29**(5), 285 (1972)
8. T. Moriya, *Spin Fluctuations in Itinerant Electron Magnetism* (Springer, Berlin, 1985)
9. G. Lonzarich, L. Taillefer, J. Phys. C Solid State Phys. **18**(22), 4339 (1985)

10. Y. Takahashi, J. Phys. Soc. Jpn. **55**, 3553 (1986)
11. Y. Takahashi, J. Phys. Condens. Matter **13**(29), 6323 (2001)
12. B.I. Reser, N.B. Melnikov, J. Phys. Condens. Matter **20**, 285205 (2008)
13. N.B. Melnikov, B.I. Reser, in *Proceedings of Dynamic Systems and Applications*, vol. 5 (Dynamic Publ., Atlanta, 2008), pp. 312–316
14. B.I. Reser, V.I. Grebennikov, N.B. Melnikov, Solid State Phenom. **152–153**, 579 (2009)
15. R.L. Stratonovich, Sov. Phys. Dokl. **2**, 416 (1958)
16. J. Hubbard, Phys. Rev. Lett. **3**(2), 77 (1959)
17. B.I. Reser, V.I. Grebennikov, Phys. Met. Metallogr. **83**, 127 (1997)
18. B.I. Reser, J. Phys. Condens. Matter **12**, 9323 (2000)
19. B.I. Reser, Phys. Met. Metallogr. **92**(Suppl. 1), S123 (2001)
20. B.I. Reser, J. Phys. Condens. Matter **14**, 1285 (2002)
21. B.I. Reser, J. Magn. Magn. Mater. **258–259**, 51 (2003)
22. T. Poston, I. Stewart, *Catastrophe Theory and Its Applications* (Dover, New York, 1998)
23. E.F. Wassermann, in *Ferromagnetic Materials*, vol. 5, ed. by K. Buschow, E. Wohlfarth (Elsevier, Amsterdam, 1990), pp. 237–322
24. E.F. Wassermann, J. Magn. Magn. Mater. **100**, 346 (1991)
25. M. Shiga, in *Materials Science and Technology*, vol. 3B, Part II, ed. by R. Cahn, P. Haasen, E. Kramer (VCH, Weinheim, 1994), p. 159
26. J. Wittenauer (ed.). *The Invar Effect: A Centennial Symposium* (The Minerals, Metals & Materials Society, Warrendale, 1997)
27. D.D. Johnson, F.J. Pinski, G.M. Stocks, J. Appl. Phys. **57**, 3018 (1985)
28. J. Crangle, G.C. Hallam, Proc. R. Soc. Lond. A **272**, 119 (1963)
29. G.G. Scott, J. Phys. Soc. Jpn. **17**(Suppl. B–I), 372 (1962)
30. V.L. Moruzzi, J.F. Janak, A.R. Williams, *Calculated Electronic Properties of Metals* (Pergamon, New York, 1978)
31. J. Crangle, G.M. Goodman, Proc. R. Soc. Lond. A **321**, 477 (1971)
32. Y.I. Prokopjev, B.I. Reser, J. Phys. Condens. Matter **3**, 6055 (1991)
33. Y. Takahashi, H. Nakano, J. Phys. Condens. Matter **18**, 521 (2006)
34. V.I. Grebennikov, S.A. Gudin, Phys. Met. Metallogr. **85**, 258 (1998)
35. B.I. Reser, N.B. Melnikov, V.I. Grebennikov, J. Phys. Confer. Ser. **200**, 012163 (2010)
36. N.B. Melnikov, B.I. Reser, V.I. Grebennikov, J. Phys. A Math. Theor. **43**, 195004 (2010)
37. N.B. Melnikov, B.I. Reser, V.I. Grebennikov, J. Phys. Condens. Matter **23**, 276003 (2011)
38. J.R. Schrieffer, W.E. Evanson, S.Q. Wang, J. Phys. Colloques **32**, C1 (1971)
39. J.A. Hertz, Phys. Rev. B **14**, 1165 (1976)
40. P.C.E. Stamp, J. Phys. F: Met. Phys. **15**, 1829 (1985)
41. V.I. Grebennikov, Phys. Solid State **40**, 79 (1998)
42. E.P. Wohlfarth, in *Ferromagnetic Materials*, vol. 1, ed. by E.P. Wohlfarth (North-Holland, Amsterdam, 1980), pp. 1–70
43. B.I. Reser, N.B. Melnikov, V.I. Grebennikov, Solid State Phenom. **190**, 55 (2012)
44. N.B. Melnikov, B.I. Reser, V.I. Grebennikov, Solid State Phenom. **190**, 43 (2012)
45. B.I. Reser, Phys. Met. Metallogr. **97**, 448 (2004)
46. V. Crisan, P. Entel, H. Ebert, H. Akai, D.D. Johnson, J.B. Staunton, Phys. Rev. B **66**, 014416 (2002)
47. A.V. Ruban, S. Khmelevskyi, P. Mohn, B. Johansson, Phys. Rev. B **76**, 014420 (2007)
48. M. Matsui, K. Adachi, S. Chikazumi, J. Appl. Phys. **51**, 6319 (1980)
49. M. Shiga, Y. Nakamura, J. Phys. Soc. Jpn. **26**, 24 (1969)
50. H. Hasegawa, J. Phys. C Solid State Phys. **14**, 2793 (1981)
51. Y. Kakehashi, J. Phys. Soc. Jpn. **50**(7), 2236 (1981)
52. T. Moriya, K. Usami, Solid State Commun. **34**, 95 (1980)
53. M. Shiga, in *Physics of Transition Metals 1980 (Inst. Phys. Conf. Ser. No. 55)*, ed. by P. Rhodes (IOP, Bristol, 1981), p. 241
54. M. Shiga, J. Phys. Soc. Jpn. **50**, 2573 (1981)

12 Low-Temperature Theory

Everything should be made as simple as possible, but not simpler. (A. Einstein)

At low temperatures ($T \ll T_C$), magnetization in metals follows the $T^{3/2}$ law. In Chap. 5 this was shown theoretically using the RPA. At finite temperature, the RPA does not give good agreement with experiment, because it neglects the feedback of the spin waves on the thermal equilibrium state. The spin fluctuation theory [1–6] allows to calculate the magnetization of the transition metals and alloys at all temperatures. However, at low temperatures, none of the approximations to the SFT fits well the experimental data. The static approximation [1–5] gives a too rapid decrease of the magnetization, $m(T) \propto T$, and the dynamic approximation [6] gives a too slow decrease, $m(T) \propto T^2$. The reasons for the disagreement with experiment are different [7]. The static SFT does not take into account the quantum nature of the spin fluctuations. In the DSFT the T^2 law results from the diagonal approximation of the dynamic susceptibility tensor [6].

In Sect. 12.1, we show that the transverse susceptibility reduces to the RPA form and magnetization follows the $T^{3/2}$ law if we use the DSFT without the diagonal approximation for the dynamic susceptibility tensor [8, 9]. Then we study the temperature interval where the $T^{3/2}$ law agrees with experiment. In Sect. 12.2 we present a simple low-temperature version of the DSFT that has only local transverse fluctuation (for details, see [10]), and compare its results with the ones of the RPA. The results are demonstrated by the example of the elemental Fe and disordered $Fe_{0.65}Ni_{0.35}$ Invar, where a proper explanation of the low-temperature excitations is still missing [11, 12].

12.1 Low-Temperature Region

12.1.1 Transverse Dynamic Susceptibility

In the linear response theory, the transverse dynamic susceptibility (2.41), expressed in units of $\frac{1}{2}g^2\mu_B^2$, is given by the formula

$$\chi_{\mathbf{q}}^{-+}(\omega) = 2\mathrm{i}\int_0^{\infty} \langle [s_{\mathbf{q}}^{-}(t), s_{-\mathbf{q}}^{+}]\rangle\, \mathrm{e}^{\mathrm{i}\omega t}\, \mathrm{d}t,$$

where $s^{\pm} = s^x \pm \mathrm{i}s^y$. By means of the analytic continuation $\omega + \mathrm{i}0^+ \to \mathrm{i}\omega_m$ the dynamic susceptibility is related to the thermodynamic susceptibility (6.65). For the transverse thermodynamic susceptibility, we write

$$\chi_{\mathbf{q}}^{-+}(\mathrm{i}\omega_m) = \frac{2}{T}\langle s_{\mathbf{q}m}^{-} s_{-\mathbf{q}-m}^{+}\rangle \tag{12.1}$$

(also in units of $\frac{1}{2}g^2\mu_B^2$), where the spin correlator is given by (6.64). Comparing formulae (12.1) and (6.65), it is easy to check that

$$\chi_{\mathbf{q}m}^{-+} = 2(\chi_{\mathbf{q}m}^{xx} + \mathrm{i}\chi_{\mathbf{q}m}^{xy}), \qquad \chi_{\mathbf{q}m}^{+-} = 2(\chi_{\mathbf{q}m}^{xx} - \mathrm{i}\chi_{\mathbf{q}m}^{xy}), \tag{12.2}$$

in full agreement with relations (2.16).

In the functional integral formalism, the spin correlator $\langle s_{\mathbf{q}m}^{\alpha} s_{-\mathbf{q}-m}^{\beta}\rangle$ is related to the field correlator $\langle V_{\mathbf{q}m}^{\alpha} V_{-\mathbf{q}-m}^{\beta}\rangle$ by formula (C.44). For the circular components we obtain

N. B. Melnikov, B. I. Reser, *Dynamic Spin Fluctuation Theory of Metallic Magnetism*,
https://doi.org/10.1007/978-3-319-92974-3_12

$$\langle s^{-}_{\mathbf{q}m} s^{+}_{-\mathbf{q}-m}\rangle = \frac{1}{\tilde{u}^2}\langle V^{-}_{\mathbf{q}m} V^{+}_{-\mathbf{q}-m}\rangle - \frac{T}{\tilde{u}},$$

where $V^{\pm}_{\mathbf{q}m} = V^{x}_{\mathbf{q}m} \pm \mathrm{i} V^{y}_{\mathbf{q}m}$. As a result, the susceptibility (12.1) becomes

$$\chi^{-+}_{\mathbf{q}m} = 2\left(\frac{1}{\tilde{u}^2 T}\langle V^{-}_{\mathbf{q}m} V^{+}_{-\mathbf{q}-m}\rangle - \frac{1}{\tilde{u}}\right). \tag{12.3}$$

Further calculation of the transverse susceptibility (12.3) is carried out in the optimal Gaussian approximation (see Sect. 9.4.2) but now *without the diagonal approximation*. Namely, the function $F(V) = F_0(V) + F_1(V)$ is replaced by the quadratic form (9.31):

$$F^{(2)}(V) = \sum_{\mathbf{q}m\alpha\beta} \Delta V^{\alpha}_{\mathbf{q}m} A^{\alpha\beta}_{\mathbf{q}m} \Delta V^{\beta}_{-\mathbf{q}-m}, \tag{12.4}$$

where $\Delta V^{\alpha}_{\mathbf{q}m} = V^{\alpha}_{\mathbf{q}m} - \bar{V}_z \delta_{\mathbf{q}0}\delta_{m0}\delta_{\alpha z}$ is the deviation from the mean field $\bar{V}_z$, and

$$A^{\alpha\beta}_{\mathbf{q}m} = \frac{1}{2}\left\langle \frac{\partial^2 F(V)}{\partial V^{\alpha}_{\mathbf{q}m}\partial V^{\beta}_{-\mathbf{q}-m}}\right\rangle.$$

Since $F_0(V)$ is the quadratic form

$$F_0(V) = \frac{1}{\tilde{u}}\sum_{\mathbf{q}m\alpha} |V^{\alpha}_{\mathbf{q}m}|^2,$$

we obtain

$$A^{\alpha\beta}_{\mathbf{q}m} = \frac{\delta_{\alpha\beta}}{\tilde{u}} - \chi^{0\alpha\beta}_{\mathbf{q}m},$$

where

$$\chi^{0\alpha\beta}_{\mathbf{q}m} = -\frac{1}{2}\left\langle \frac{\partial^2 F_1(V)}{\partial V^{\alpha}_{\mathbf{q}m}\partial V^{\beta}_{-\mathbf{q}-m}}\right\rangle$$

is the unenhanced susceptibility. Taking axial symmetry (9.24) into account:

$$\chi^{0}_{\mathbf{q}m} = \begin{pmatrix} \chi^{0xx}_{\mathbf{q}m} & \chi^{0xy}_{\mathbf{q}m} & 0 \\ -\chi^{0xy}_{\mathbf{q}m} & \chi^{0xx}_{\mathbf{q}m} & 0 \\ 0 & 0 & \chi^{0zz}_{\mathbf{q}m} \end{pmatrix},$$

we have

$$A_{\mathbf{q}m} = \begin{pmatrix} A^{xx}_{\mathbf{q}m} & A^{xy}_{\mathbf{q}m} & A^{xz}_{\mathbf{q}m} \\ A^{yx}_{\mathbf{q}m} & A^{yy}_{\mathbf{q}m} & A^{yz}_{\mathbf{q}m} \\ A^{zx}_{\mathbf{q}m} & A^{zy}_{\mathbf{q}m} & A^{zz}_{\mathbf{q}m} \end{pmatrix} = \begin{pmatrix} \dfrac{1}{\tilde{u}} - \chi^{0xx}_{\mathbf{q}m} & -\chi^{0xy}_{\mathbf{q}m} & 0 \\ \chi^{0xy}_{\mathbf{q}m} & \dfrac{1}{\tilde{u}} - \chi^{0xx}_{\mathbf{q}m} & 0 \\ 0 & 0 & \dfrac{1}{\tilde{u}} - \chi^{0zz}_{\mathbf{q}m} \end{pmatrix}.$$

Then the quadratic form (12.4) is written as

$$\begin{aligned} F^{(2)}(V) = \sum_{\mathbf{q}m}\Big[& \Delta V^{x}_{\mathbf{q}m}\left(\frac{1}{\tilde{u}} - \chi^{0xx}_{\mathbf{q}m}\right)\Delta V^{x}_{-\mathbf{q}-m} - \Delta V^{x}_{\mathbf{q}m}\chi^{0xy}_{\mathbf{q}m}\Delta V^{y}_{-\mathbf{q}-m} \\ & + \Delta V^{y}_{\mathbf{q}m}\chi^{0xy}_{\mathbf{q}m}\Delta V^{x}_{-\mathbf{q}-m} + \Delta V^{y}_{\mathbf{q}m}\left(\frac{1}{\tilde{u}} - \chi^{0xx}_{\mathbf{q}m}\right)\Delta V^{y}_{-\mathbf{q}-m} \\ & + \Delta V^{z}_{\mathbf{q}m}\left(\frac{1}{\tilde{u}} - \chi^{0zz}_{\mathbf{q}m}\right)\Delta V^{z}_{-\mathbf{q}-m}\Big]. \end{aligned}$$

Changing the variables by the formulae

$$V^x_{\mathbf{q}m} = \frac{1}{2}(V^+_{\mathbf{q}m} + V^-_{\mathbf{q}m}), \qquad V^y_{\mathbf{q}m} = -\frac{\mathrm{i}}{2}(V^+_{\mathbf{q}m} - V^-_{\mathbf{q}m}),$$

we obtain

$$F^{(2)}(V) = \sum_{\mathbf{q}m}\Bigg[\frac{1}{2}V^-_{\mathbf{q}m}\Big(\frac{1}{\tilde{u}} - \frac{1}{2}\chi^{0-+}_{\mathbf{q}m}\Big)V^+_{-\mathbf{q}-m} + \frac{1}{2}V^+_{\mathbf{q}m}\Big(\frac{1}{\tilde{u}} - \frac{1}{2}\chi^{0+-}_{\mathbf{q}m}\Big)V^-_{-\mathbf{q}-m}$$
$$+ \Delta V^z_{\mathbf{q}m}\Big(\frac{1}{\tilde{u}} - \chi^{0zz}_{\mathbf{q}m}\Big)\Delta V^z_{-\mathbf{q}-m}\Bigg], \tag{12.5}$$

where

$$\chi^{0-+}_{\mathbf{q}m} = 2(\chi^{0xx}_{\mathbf{q}m} - \mathrm{i}\chi^{0xy}_{\mathbf{q}m}), \qquad \chi^{0+-}_{\mathbf{q}m} = 2(\chi^{0xx}_{\mathbf{q}m} + \mathrm{i}\chi^{0xy}_{\mathbf{q}m}) \tag{12.6}$$

are the transverse unenhanced susceptibilities.[1]

To obtain a formula for the enhanced susceptibility, we need to calculate the field correlator on the right-hand side of (12.3). The quadratic form of the optimal Gaussian approximation (12.5) can be written as

$$F^{(2)}(V) = \sum_{\mathbf{q}m} \Delta\mathbf{V}_{\mathbf{q}m} A_{\mathbf{q}m} \Delta\mathbf{V}_{-\mathbf{q}-m},$$

where $\Delta\mathbf{V}_{\mathbf{q}m} = (V^-_{\mathbf{q}m}, V^+_{\mathbf{q}m}, \Delta V^z_{\mathbf{q}m})$ and

$$A_{\mathbf{q}m} = \begin{pmatrix} A^{--}_{\mathbf{q}m} & A^{-+}_{\mathbf{q}m} & A^{-z}_{\mathbf{q}m} \\ A^{+-}_{\mathbf{q}m} & A^{++}_{\mathbf{q}m} & A^{+z}_{\mathbf{q}m} \\ A^{z-}_{\mathbf{q}m} & A^{z+}_{\mathbf{q}m} & A^{zz}_{\mathbf{q}m} \end{pmatrix} = \begin{pmatrix} 0 & \frac{1}{2}\Big(\frac{1}{\tilde{u}} - \frac{1}{2}\chi^{0-+}_{\mathbf{q}m}\Big) & 0 \\ \frac{1}{2}\Big(\frac{1}{\tilde{u}} - \frac{1}{2}\chi^{0+-}_{\mathbf{q}m}\Big) & 0 & 0 \\ 0 & 0 & \frac{1}{\tilde{u}} - \chi^{0zz}_{\mathbf{q}m} \end{pmatrix}.$$

The correlator of the Gaussian field is given by the formula (see Appendix A.3.3)

$$\langle \Delta V^\alpha_{\mathbf{q}m} \Delta V^\beta_{-\mathbf{q}-m}\rangle = \frac{T}{2}\left(A^{-1}_{\mathbf{q}m}\right)^{\beta\alpha}, \tag{12.7}$$

where $\alpha, \beta = -, +, z$. Calculating the inverse matrix

$$A^{-1}_{\mathbf{q}m} = \begin{pmatrix} (A^{-1}_{\mathbf{q}m})^{--} & (A^{-1}_{\mathbf{q}m})^{-+} & (A^{-1}_{\mathbf{q}m})^{-z} \\ (A^{-1}_{\mathbf{q}m})^{+-} & (A^{-1}_{\mathbf{q}m})^{++} & (A^{-1}_{\mathbf{q}m})^{+z} \\ (A^{-1}_{\mathbf{q}m})^{z-} & (A^{-1}_{\mathbf{q}m})^{z+} & (A^{-1}_{\mathbf{q}m})^{zz} \end{pmatrix} = \begin{pmatrix} 0 & \dfrac{2}{\frac{1}{\tilde{u}} - \frac{1}{2}\chi^{0+-}_{\mathbf{q}m}} & 0 \\ \dfrac{2}{\frac{1}{\tilde{u}} - \frac{1}{2}\chi^{0-+}_{\mathbf{q}m}} & 0 & 0 \\ 0 & 0 & \dfrac{1}{\frac{1}{\tilde{u}} - \chi^{0zz}_{\mathbf{q}m}} \end{pmatrix},$$

by formula (12.7) we obtain

[1]One could notice that the sign in the formulae for the *unenhanced* susceptibility (12.6) differs from the one in the formulae for the *enhanced* susceptibility (12.2). This fact is merely a matter of defining $A_{\mathbf{q}m}$ and $\chi^0_{\mathbf{q}m}$ in the DSFT. They are defined in such a way that the second equation of the optimal Gaussian approximation looks as in (9.27) and unenhanced susceptibility is given by (9.32).

$$\langle V^-_{\mathbf{q}m} V^+_{-\mathbf{q}-m}\rangle = \frac{T}{2}\left(A^{-1}_{\mathbf{q}m}\right)^{+-} = \frac{T}{\frac{1}{\tilde{u}} - \frac{1}{2}\chi^{0-+}_{\mathbf{q}m}} = \frac{\tilde{u}T}{1 - \frac{\tilde{u}}{2}\chi^{0-+}_{\mathbf{q}m}}.$$

Then the enhanced susceptibility (12.3) becomes

$$\chi^{-+}_{\mathbf{q}m} = \frac{\chi^{0-+}_{\mathbf{q}m}}{1 - \frac{1}{2}\tilde{u}\chi^{0-+}_{\mathbf{q}m}}. \tag{12.8}$$

12.1.2 Spin Waves and $T^{3/2}$ Law

To study the spin waves in the DSFT, we need explicit formulae for the transverse susceptibilities (12.6). From expression (9.34), using formulae (A.22) for trace of the product of the Pauli matrices, we obtain

$$\begin{aligned}
\chi^{0xx}_{\mathbf{q}m} &= -\frac{N_\mathrm{d}}{2}T\sum_{\mathbf{k}n\gamma_1\gamma_2}\bar{G}^{\gamma_1}_{\mathbf{k}n}\bar{G}^{\gamma_2}_{\mathbf{k}-\mathbf{q},n-m}\mathrm{Sp}(\sigma^{\gamma_1}\sigma^x\sigma^{\gamma_2}\sigma^x)\\
&= -\frac{N_\mathrm{d}}{2}T\sum_{\mathbf{k}n}(\bar{G}^0_{\mathbf{k}n}\bar{G}^0_{\mathbf{k}-\mathbf{q},n-m}\mathrm{Sp}(\sigma^x\sigma^x) - \bar{G}^z_{\mathbf{k}n}\bar{G}^z_{\mathbf{k}-\mathbf{q},n-m}\mathrm{Sp}(\sigma^x\sigma^z\sigma^x\sigma^z))\\
&= -N_\mathrm{d}T\sum_{\mathbf{k}n}(\bar{G}^0_{\mathbf{k}n}\bar{G}^0_{\mathbf{k}-\mathbf{q},n-m} - \bar{G}^z_{\mathbf{k}n}\bar{G}^z_{\mathbf{k}-\mathbf{q},n-m}),
\end{aligned} \tag{12.9}$$

$$\begin{aligned}
\chi^{0xy}_{\mathbf{q}m} &= -\frac{N_\mathrm{d}}{2}T\sum_{\mathbf{k}n\gamma_1\gamma_2}\bar{G}^{\gamma_1}_{\mathbf{k}n}\bar{G}^{\gamma_2}_{\mathbf{k}-\mathbf{q},n-m}\mathrm{Sp}(\sigma^{\gamma_1}\sigma^x\sigma^{\gamma_2}\sigma^y)\\
&= -\frac{N_\mathrm{d}}{2}T\sum_{\mathbf{k}n}(\bar{G}^0_{\mathbf{k}n}\bar{G}^z_{\mathbf{k}-\mathbf{q},n-m}\mathrm{Sp}(\sigma^x\sigma^z\sigma^y) - \bar{G}^z_{\mathbf{k}n}\bar{G}^0_{\mathbf{k}-\mathbf{q},n-m}\mathrm{Sp}(\sigma^z\sigma^x\sigma^y))\\
&= \mathrm{i}N_\mathrm{d}T\sum_{\mathbf{k}n}(\bar{G}^0_{\mathbf{k}n}\bar{G}^z_{\mathbf{k}-\mathbf{q},n-m} - \bar{G}^z_{\mathbf{k}n}\bar{G}^0_{\mathbf{k}-\mathbf{q},n-m}).
\end{aligned} \tag{12.10}$$

Hence by (12.6) the transverse susceptibility is equal to

$$\begin{aligned}
\chi^{0-+}_{\mathbf{q}m} &= 2(\chi^{0xx}_{\mathbf{q}m} - \mathrm{i}\chi^{0xy}_{\mathbf{q}m})\\
&= -2N_\mathrm{d}T\sum_{\mathbf{k}n}(\bar{G}^0_{\mathbf{k}n} + \bar{G}^z_{\mathbf{k}n})(\bar{G}^0_{\mathbf{k}-\mathbf{q},n-m} - \bar{G}^z_{\mathbf{k}-\mathbf{q},n-m})\\
&= -2N_\mathrm{d}T\sum_{\mathbf{k}n}\bar{G}_{\mathbf{k}\uparrow}(\mathrm{i}\omega_n)\bar{G}_{\mathbf{k}-\mathbf{q},\downarrow}(\mathrm{i}\omega_n - \mathrm{i}\omega_m),
\end{aligned} \tag{12.11}$$

where $\bar{G}_\uparrow = \bar{G}^0 + \bar{G}^z$ and $\bar{G}_\downarrow = \bar{G}^0 - \bar{G}^z$. Here the mean Green function is given by

$$\bar{G}(z) = (z + \mu - \mathcal{H}_0 - \Sigma(z))^{-1},$$

where $\Sigma(z)$ is the self-energy and μ is the chemical potential. The mean Green function $\bar{G}$ is calculated self-consistently and takes into account thermal excitations with arbitrary wavevectors and "frequencies". At low temperatures, the fluctuational contribution $\Delta\Sigma = \Sigma - \bar{V}$ becomes small and can be neglected. Then

$$\bar{G}_{\mathbf{k}\sigma}(z) = \frac{1}{z + \mu - \varepsilon_\mathbf{k} - \sigma\bar{V}_z} = \frac{1}{z + \mu - \varepsilon_{\mathbf{k}\sigma}}.$$

Replacing the sum over the odd "frequencies" ω_n in formula (12.11) analogously to (A.63), we obtain

$$\chi_{\mathbf{q}}^{0-+}(\mathrm{i}\omega_m) = -2N_{\mathrm{d}} \sum_{\mathbf{k}} \frac{f(\varepsilon_{\mathbf{k}\uparrow}) - f(\varepsilon_{\mathbf{k}-\mathbf{q},\downarrow})}{\varepsilon_{\mathbf{k}\uparrow} - \varepsilon_{\mathbf{k}-\mathbf{q},\downarrow} - \mathrm{i}\omega_m} . \quad (12.12)$$

Switching to $\mathbf{k}' = \mathbf{k} - \mathbf{q}$ and making the analytic continuation $\mathrm{i}\omega_m \to \omega + \mathrm{i}0^+$, we have

$$\chi_{\mathbf{q}}^{0-+}(\omega) = 2N_{\mathrm{d}} \sum_{\mathbf{k}'} \frac{f(\varepsilon_{\mathbf{k}'\downarrow}) - f(\varepsilon_{\mathbf{k}'+\mathbf{q},\uparrow})}{\varepsilon_{\mathbf{k}'+\mathbf{q},\uparrow} - \varepsilon_{\mathbf{k}'\downarrow} - \omega} = 2N_{\mathrm{d}} F_{-+}(\mathbf{q}, \omega), \quad (12.13)$$

where $F_{-+}(\mathbf{q}, \omega)$ is the transverse Lindhard function (5.21). Using formula (12.8), we write the enhanced susceptibility in the form

$$\chi_{\mathbf{q}}^{-+}(\omega) = \frac{2N_{\mathrm{d}} F_{-+}(\mathbf{q}, \omega)}{1 - \tilde{u} N_{\mathrm{d}} F_{-+}(\mathbf{q}, \omega)} . \quad (12.14)$$

Recalling that DSFT susceptibilities are expressed in units of $g^2\mu_{\mathrm{B}}^2/2$, we see that the low-temperature susceptibility (12.14) is equal to the single-band RPA susceptibility (5.20) times the number of bands.

Now we introduce the spin-wave dispersion relation by (5.37). Deriving the low-temperature asymptotic expressions, similar to (5.38), we have

$$\chi_{\mathbf{q}}^{0-+}(\varepsilon) = \frac{2N}{u}\Big(1 - \frac{\varepsilon + Dq^2}{2u\bar{s}_z}\Big), \quad (12.15)$$

where D is the stiffness constant. Hence the low-temperature expression for the enhanced susceptibility (12.8) is given by

$$\chi_{\mathbf{q}}^{-+}(\varepsilon) = \frac{2N}{u}\Big(\frac{2u\bar{s}_z}{\varepsilon + Dq^2} - 1\Big), \quad (12.16)$$

where $\varepsilon = \varepsilon + \mathrm{i}0^+$. Using the Sokhotsky formula (A.44), we obtain

$$\mathrm{Im}\chi_{\mathbf{q}}^{-+}(\varepsilon) = -4N\bar{s}_z\pi\delta(\hbar\omega + Dq^2). \quad (12.17)$$

Following the argument of Sect. 5.2.2, we obtain the local transverse fluctuation (5.49) and the $T^{3/2}$ law for magnetization in metals:

$$\frac{m_z(T)}{m_z(0)} = 1 - a_{3/2}T^{3/2}, \qquad T \ll T_{\mathrm{C}},$$

where $a_{3/2}$ is given by (5.51).

First, the above low-temperature limit of the DSFT is illustrated by the example of iron. The lattice constant a of the bcc Fe is equal to 2.866 A [13], the magnetization at $T = 0$ is $m(0) = 2.217\,\mu_{\mathrm{B}}$ [14], and the spin-wave stiffness constant is $D_0 = 311$ meV A^2 [15]. Substituting these values to expression (5.51), we obtain $a_{3/2} = 2.87 \times 10^{-6}\,\mathrm{K}^{-\frac{3}{2}}$. The numerical results for the spin-wave approximation (SWA) of the magnetization (5.52) are presented in Fig. 12.1. The temperature in Fig. 12.1 is given in units of the Curie temperature $T_{\mathrm{C}}^{\mathrm{exp}} = 1044$ K [14].

For comparison, Fig. 12.1 shows the low-temperature magnetization $m(T)$ calculated in the Stoner approximation (STA) and two approximations of the SFT: SLA and DNA. (Recall that the approximations SLA and DNA differ in the DSFT only by expressions (10.39) and (10.24) for the fluctuations $\langle \Delta V_\alpha^2 \rangle'$, $\alpha = x, z$.) The initial data of the calculation, the value of the magnetic moment $m(0)$ and the first-principles DOS $\nu(\varepsilon)$ at $T = 0$, are the same as in Sect. 11.1.3.

As can be seen from Fig. 12.1, the decrease of $m(T)$ in the STA is negligible. In the SLA the magnetization decreases too fast ($\propto T$). Only in the DNA good agreement with the experimental data is obtained. As for the SWA, it gives a better agreement with experiment, as compared to the DNA, in the interval from zero to $0.2T_{\mathrm{C}}^{\mathrm{exp}} \approx 200$ K. However, the SWA is not sufficient with increasing temperature. Indeed, the DNA gives a better agreement at finite temperatures, including room temperatures. This is quite reasonable since in the DNA both transverse and longitudinal fluctuations are taken into account but not only transverse ones, as in the SWA. At low temperatures, the DNA fluctuation (10.24) has the asymptotic behaviour $\sim T^2$, while in the SLA the fluctuation (10.39) increases only linearly. Note that, near $T = 0$, the DNA curve $m(T)$ should go higher than the SWA curve, which has the asymptotic behaviour $\sim T^{3/2}$. In Fig. 12.1 this temperature interval is too small so that the DNA and SWA curves merge there.

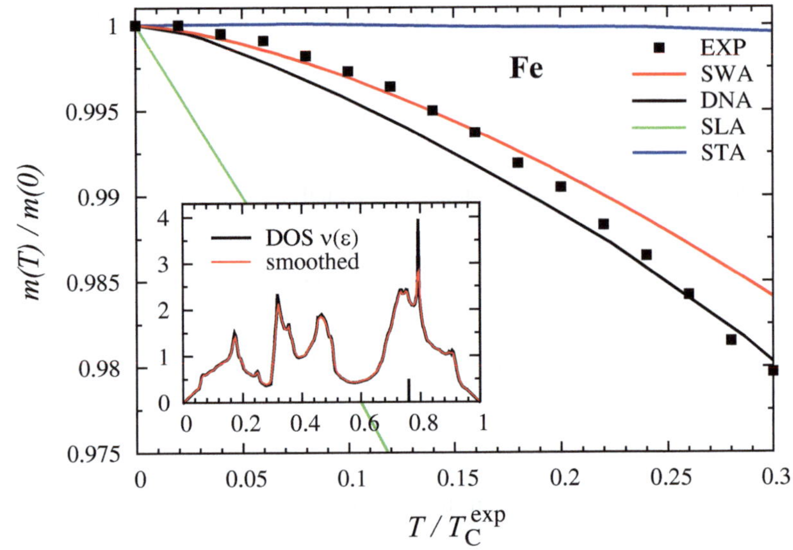

Fig. 12.1 Magnetization of Fe calculated in the SWA (red), STA (blue), SLA (green) and DNA (black). Filled squares are the experimental data of Crangle and Goodman [14]. The inset shows the initial DOS at T=0; the energy is in units of the bandwidth W = 7.16 eV, and the vertical line indicates the position of the Fermi level

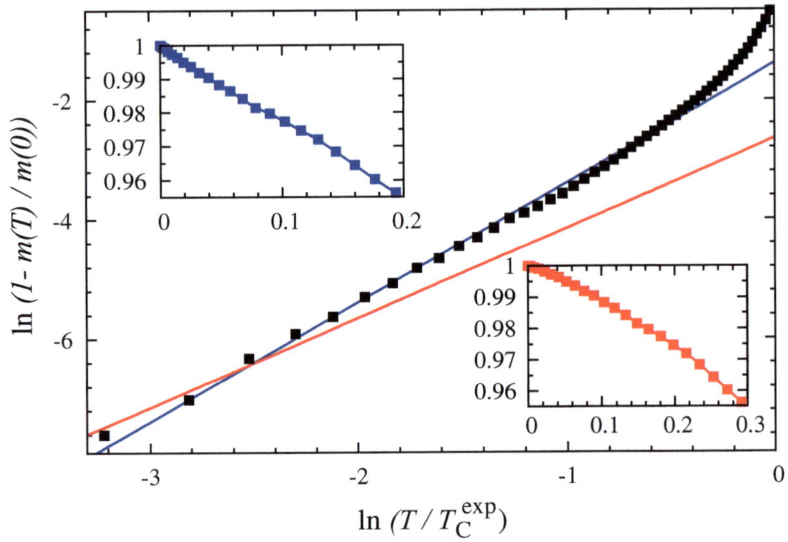

Fig. 12.2 The experimental data [14] for $1 - m(T)/m(0)$ of Fe in the logarithmic scale, at the interval $0 < T < T_C^{exp}$ (filled squares) interpolated by the line with the slope 2 (blue) and 3/2 (red). The insets show the experimental data [14] for magnetization $m(T)/m(0)$ vs. T^2 (blue) and $T^{3/2}$ (red) at the interval $0 < T < 0.45T_C^{exp}$

In Fig. 12.2, we plot the relative change of magnetization $1-m(T)/m(0)$ versus temperature T in the logarithmic scale. As can be seen, at low temperatures, the (red) line with the slope 3/2 gives the best agreement with experiment. This conclusion is confirmed by a number of measurements [14, 16, 17]. Figure 12.2 also shows that, in a large range of higher temperatures, the (blue) line with the slope 2 gives a better fit to the data. The crossover point from the $T^{3/2}$ to T^2 law is ambiguous (see, e.g. [18]). As can be seen from the insets of Fig. 12.2, both $T^{3/2}$ and T^2 dependencies of the experimental magnetization $m(T)/m(0)$ produce almost a straight line over the temperature interval $0 < T < 0.45T_C^{exp}$. The $T^{3/2}$ law has an asymptotic character and holds at least up to the temperature 32 K [16], which is equivalent to 0.03 in the reduced temperatures T/T_C^{exp}.

Next, we apply the low-temperature limit of the DSFT to the disordered Fe-Ni Invar, which has been studied in detail experimentally [19–22]. The lattice constant of the fcc $Fe_{0.65}Ni_{0.35}$ Invar is $a = 3.59$ Å, the magnetization at $T = 0$ is $m(0) = 1.7\,\mu_B$, and the spin-wave stiffness constant is $D_0 = 140.0$ meV Å^2 [20]. Substituting these values to expression (5.51), we obtain $a_{3/2} = 12.18 \times 10^{-6}\,K^{-\frac{3}{2}}$. The results for the magnetization in the spin-wave approximation (SWA) are presented in Fig. 12.3. As in iron, the results of the SWA are compared with those of the STA, SLA and DNA. The temperature is given in units of the Curie temperature $T_C^{exp} = 520$ K [23]. As the initial DOS for the SFT calculations, we use the DOS of the d band of nonmagnetic $Fe_{0.65}Ni_{0.35}$, which is obtained from two spin-polarized DOSs calculated in [24].

As can be seen from Fig. 12.3, the decrease of $m(T)$ in the STA is negligible and in the SLA it is too fast ($\propto T$), just as in pure iron. The SWA gives good agreement with experiment over the temperature interval from zero to $0.1\,T_C^{exp} \approx 50$ K, which is approximately four times smaller than in iron. This fact agrees with experiment and is usually attributed to the Invar

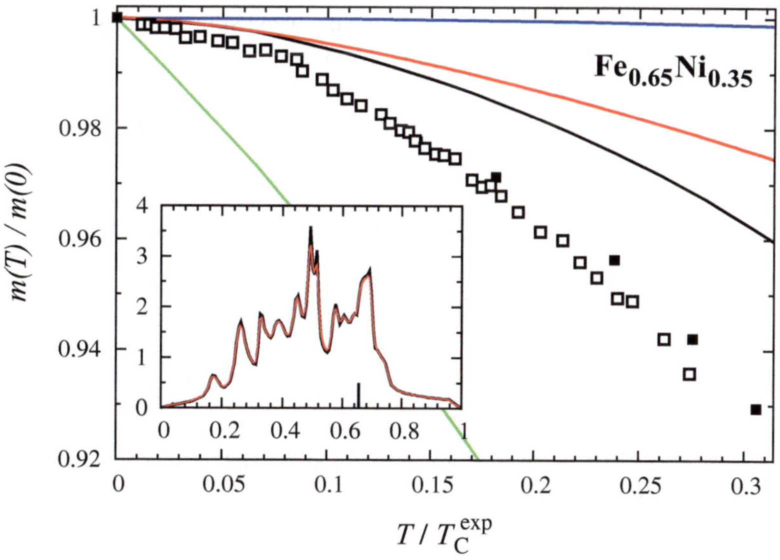

Fig. 12.3 Magnetization of $Fe_{0.65}Ni_{0.35}$ calculated in the SWA (red), STA (blue), SLA (green) and DNA (black). Filled squares are the experimental data of Crangle and Hallam [23] and open squares are the experimental data of Ishikawa et al. [25] The inset shows the initial DOS at $T = 0$ [24]; the energy is in units of the bandwidth W = 9.70 eV, and the vertical line indicates the position of the Fermi level

anomaly [19, 22]. Note that, in alloys with atoms that are different by their chemical nature (and local magnetic moment), our DSFT should be modified. Namely, atoms of each sort s should be characterized by their own intraatomic interaction constant u_s, which is determined by $m_s^{\text{exp}}(0)$, and their own initial DOS $\nu_s(\varepsilon)$. In other words, atoms in the alloy should not be averaged, as they are now.

The calculated value of the coefficient $a_{3/2}$ in the $Fe_{0.65}Ni_{0.35}$ Invar is approximately four times larger than in iron. The resulting decrease of magnetization in the $Fe_{0.65}Ni_{0.35}$ Invar is substantial and is in good agreement with the Invar effect. However, the magnetization in the SWA decreases slower than in experiment. The DNA results are in better agreement with experiment since the DNA takes into account longitudinal fluctuations, which are neglected in the SWA. Just as in iron, the DNA curve ($\propto T^2$) merges with the SWA curve ($\propto T^{3/2}$) near zero.

The DNA calculations at the interval from zero to 0.3 T_C^{exp} show that in Fe the longitudinal fluctuations are about ten times smaller than the transverse fluctuations, and in the $Fe_{0.65}Ni_{0.35}$ Invar the longitudinal fluctuations are about five times smaller than the transverse ones. However, this is not always the case in metals at finite temperatures. For instance, the longitudinal fluctuations are comparable with the transverse ones in Co and are larger than the transverse ones in Ni (see Sect. 10.5).

12.2 Beyond the Spin Waves

12.2.1 Low-Temperature DSFT

The simplest low-temperature version of the DSFT is the one with only transverse spin fluctuations (TDSFT), $\langle \Delta V_z^2 \rangle' = 0$. An important difference of this theory from the spin-wave approximation and some versions of the SCR theory for weakly ferromagnetic metals [26] is that conservation of the local spin is not assumed. For better comparison with previous results, the transverse fluctuation is calculated in the diagonal approximation: $\langle \Delta V_\perp^2 \rangle' = 2\langle \Delta V_x^2 \rangle'$. Thus, in the TDSFT, the problem is reduced to solving the system of three equations (10.24), (10.33) and (10.34) with respect to three variables $\langle \Delta V_x^2 \rangle'$, $\bar{V}_z$ and μ, for each temperature T.

Calculation of the local fluctuation $\langle \Delta V_\perp^2 \rangle'$ by formula (10.24) even in the TDSFT is based on a fairly complicated chain of approximations. We obtain a simplified formula for the fluctuation $\langle \Delta V_\perp^2 \rangle'$ from expression (5.49), which was derived from the low-temperature approximation of the transverse susceptibility (12.15) in the previous subsection. Namely, discarding the zero-point fluctuations in formula (5.49) and using $\langle \Delta V_\perp^2 \rangle' = u^2 \langle \Delta s_\perp^2 \rangle$, we have

$$\langle \Delta V_\perp^2 \rangle' = \frac{2u^2 \bar{s}_z}{\Omega_{\text{BZ}}}\, \zeta(3/2) \left(\frac{\pi}{D}\right)^{\frac{3}{2}} T^{3/2}. \qquad (12.18)$$

In this analytic expression, we calculate the mean spin $\bar{s}_z$ and spin-wave stiffness D self-consistently.[2] Averaging expression (12.15) over the Brillouin zone, we obtain

$$D = \frac{2u\bar{s}_z}{\overline{q^2}} \left(1 - \frac{u}{2}\chi_{\mathrm{L}}^{0-+}(0)\right),$$

where $\chi_{\mathrm{L}}^{0-+}(\varepsilon) = N^{-2}\sum_{\mathbf{q}} \chi_{\mathbf{q}}^{0-+}(\varepsilon)$ is the local susceptibility and $\overline{q^2} = N^{-1}\sum_{\mathbf{q}} q^2$. Approximating the Brillouin zone by the equal-volume sphere, we have $\overline{q^2} = 0.6q_{\mathrm{B}}^2$, just as before. The local susceptibility $\chi_{\mathrm{L}}^{0-+}(0)$ is expressed in terms of the single-site mean Green function $g(\varepsilon)$. Using formula (12.11), we obtain

$$\chi_{\mathrm{L}}^{0-+}(0) = -\frac{2N_{\mathrm{d}}}{\pi} \int \mathrm{Im}\big(g_{\downarrow}(\varepsilon)g_{\uparrow}(\varepsilon)\big) f(\varepsilon)\, \mathrm{d}\varepsilon,$$

where $g_{\sigma}(\varepsilon)$ is given by relations (10.35) and (10.37). The mean spin $\bar{s}_z$ is calculated by the formula $\bar{s}_z = -\bar{V}_z/u$, where $\bar{V}_z$ is given by Eq. (10.33).

Thus, in the low-temperature DSFT (LDSFT), we solve the system of three equations (10.33), (10.34) and (12.18) with respect to three variables $\bar{V}_z$, μ and $\langle \Delta V_{\perp}^2 \rangle'$, for each temperature T. Both in LDSFT and TDSFT, the initial data of the calculation are the first-principles DOS $\nu(\varepsilon)$ at $T = 0$ and the value of the magnetic moment $m(0) = g\mu_{\mathrm{B}}\bar{s}_z(0)$, which is used to obtain the interaction constant u.

12.2.2 Application to Fe and Fe-Ni Invar

We apply the above theory to calculate the temperature dependence of magnetization, Curie temperature and spin-wave stiffness. First we test the theory on a clean system such as elemental Fe, and then we consider the disordered $Fe_{0.65}Ni_{0.35}$ Invar.

In Fe the magnetization at $T = 0$ is $m(0) = 2.217\,\mu_{\mathrm{B}}$ [14] and nonmagnetic DOS of Fe is presented in Fig. 11.5. The lattice constant a of the bcc Fe is equal to 2.866 A [13] and hence the volume of the Wigner-Seitz cell is $\Omega_{\mathrm{WS}} = a^3/2 = 11.78\,\mathrm{A}^3$.

The numerical results for magnetization calculated in the LDSFT are presented in Fig. 12.4. The temperature is given in units of the Curie temperature $T_{\mathrm{C}}^{\mathrm{exp}} = 1044\,\mathrm{K}$ [14]. For comparison, we present the temperature dependence of magnetization $m(T)$ calculated in Stoner approximation (STA), two approximations of the DSFT: static local (SLA) and dynamic nonlocal (DNA), and TDSFT.

As we have already mentioned, the decrease of $m(T)$ in the STA is negligible, and in the SLA magnetization decreases too fast ($\propto T$). The DNA gives excellent agreement with experiment at the interval from zero to about $0.45T_{\mathrm{C}}^{\mathrm{exp}} \approx 450\,\mathrm{K}$. The calculation without the longitudinal fluctuation (TDSFT) changes the results of the DNA very little. This is quite reasonable since the transversal fluctuations in Fe are almost 10 times larger than the longitudinal fluctuations at this interval. The LDSFT results are different from those of DNA but are still within a 1.5% error, compared to experiment, over the interval from zero to $0.6T_{\mathrm{C}}^{\mathrm{exp}} \approx 600\,\mathrm{K}$. Note that, this interval is almost 20 times larger than the interval 32 K [16], where the experimental curve for the magnetization $m(T)$ follows the $T^{3/2}$ law.

Numerical results for magnetization of the $Fe_{0.65}Ni_{0.35}$ Invar calculated in the LDSFT are presented in Fig. 12.5. The temperature is given in units of the Curie temperature $T_{\mathrm{C}}^{\mathrm{exp}} = 520\,\mathrm{K}$ [23]. The magnetization at $T = 0$ is $m(0) = 1.7\,\mu_{\mathrm{B}}$ [7, 23]. As the initial DOS for the DSFT calculations, we use the DOS of the d band of nonmagnetic $Fe_{0.65}Ni_{0.35}$, which is obtained from two spin-polarized DOSs calculated in [24]. The lattice constant of the fcc $Fe_{0.65}Ni_{0.35}$ Invar is $a = 3.59\,\mathrm{A}$ [20], so that the volume of the Wigner-Seitz unit cell is $\Omega_{\mathrm{WS}} = a^3/4 = 11.56\,\mathrm{A}^3$. The results of the LDSFT are compared with those of the STA, SLA, DNA and TDSFT. Magnetization in the DNA is in good agreement with experiment over the interval from zero to $0.6T_{\mathrm{C}}^{\mathrm{exp}} \approx 300\,\mathrm{K}$ (which includes the room temperatures). The neglect of the longitudinal fluctuation (TDSFT) has a larger effect on the results, as compared to Fe, since in the $Fe_{0.65}Ni_{0.35}$ Invar the longitudinal fluctuations are only five times smaller than the transverse ones. The results of the LDSFT are close to those of the DNA at

[2] This method is similar to the use of a simple *analytic* expression for the Hartree-Fock exchange potential in the band calculations [27,28].

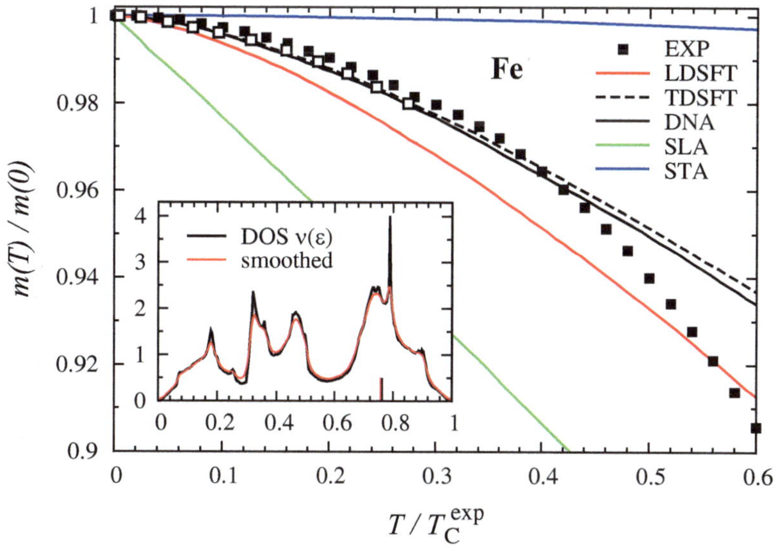

Fig. 12.4 Magnetization $m(T)$ of Fe calculated in the STA, SLA, DNA, TDSFT and LDSFT. Filled squares are the experimental data of Crangle and Goodman [14] and open squares are the experimental data of Pauthenet [17]. The inset shows the initial DOS at $T = 0$; the energy is in units of the bandwidth $W = 7.16\,\text{eV}$, and the vertical line indicates the position of the Fermi level

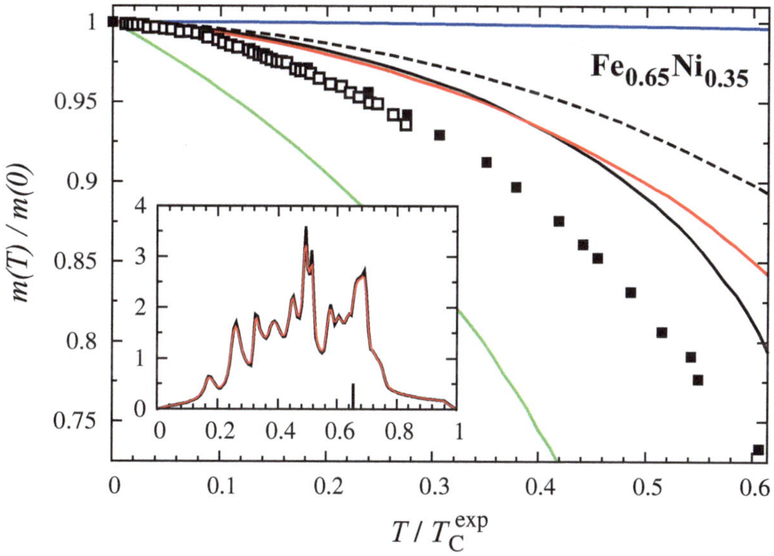

Fig. 12.5 Magnetization $m(T)$ of the $Fe_{0.65}Ni_{0.35}$ Invar calculated as in Fig. 12.4. Filled squares are the experimental data of Crangle and Hallam [23] and open squares are the experimental data of Ishikawa et al. [25]. The inset shows the initial DOS at $T = 0$ [24]; the energy is in units of the bandwidth $W = 9.70\,\text{eV}$

the interval from zero to $0.4T_C^{exp} \approx 200\,\text{K}$. At this interval both LDSFT and DNA agree with experiment within a 5% error. This difference in the $Fe_{0.65}Ni_{0.35}$ Invar should be attributed to the "hidden excitations", whose nature is still unknown.[3]

For the spin-wave stiffness at $T = 0$, our calculation in Fe yields $D_{m0}^{cal} = 200\,\text{meV A}^2$, which is in reasonable agreement with the experimental values $D_{m0}^{exp} = 270$ and $D_{n0}^{exp} = 280$ obtained from magnetic [17] and neutron-scattering [30] measurements, respectively. The calculations in the adiabatic approximation of the LDA+DMFT [31, 32] give the value $D_{n0}^{cal} = 250\,\text{meV A}^2$. In the $Fe_{0.65}Ni_{0.35}$ Invar, our calculated value $D_{m0}^{cal} = 110\,\text{meV A}^2$ is in good agreement with the experimental values $D_{m0}^{exp} = 110$ and $D_{n0}^{exp} = 140$ obtained from magnetic [33] and neutron-scattering [25] measurements, respectively. (As is known, the experimental values D_{m0}^{exp} and D_{n0}^{exp}, which coincide for most of ferromagnetic materials, are different for Invar alloys [22].)

Note that estimation of the spin-wave stiffness constant D_0 is a complex problem both experimentally and theoretically [34, 35]. The experimental values D_{m0}^{exp} obtained from the magnetization measurements [17, 33] depend on the temperature interval where one fits the data on magnetization with the $T^{3/2}$ law. Since the $T^{3/2}$ law has only an asymptotic character, the exact temperature interval is unknown. The neutron-scattering measurements [25,30] and numerical calculations [31, 32] have a similar problem of choosing the energy interval to fit the spin-wave spectrum $\varepsilon = Dq^2$. The

[3]For an explanation based on phonons, see [29, p. 218].

Table 12.1 Spin-wave stiffness D_0 and Curie temperature T_C of Fe and $Fe_{0.65}Ni_{0.35}$ Invar

	D_0^{cal} (meV Å^2)	D_0^{exp} (meV Å^2)	T_C^{cal} (K)	T_C^{exp} (K)
Fe	200	270[a], 280[b]	1654	1044[c]
$Fe_{0.65}Ni_{0.35}$	110	110[d], 140[e]	546	520[f]

[a]Magnetization measurements of Pauthenet [17]
[b]Neutron-scattering measurements of Mook and Nicklow [30]
[c]Crangle and Goodman [14]
[d]Magnetization measurements of Nakai [33]
[e]Neutron-scattering measurements of Ishikawa et al. [25]
[f]Crangle and Hallam [23]

results of the numerical calculations [31, 32] of D_{n0}^{cal} also depend on the estimation procedure for the exchange parameters J_{ij} in the Heisenberg Hamiltonian.

Finally, the LDSFT calculations give reasonable values for the Curie temperatures: $T_C^{cal} = 1.58\, T_C^{exp}$ for Fe and $T_C^{cal} = 1.05\, T_C^{exp}$ for $Fe_{0.65}Ni_{0.35}$ Invar. The calculated and experimental values of the spin-wave stiffness D_0 and Curie temperature T_C are summarized in Table 12.1.

Thus, we have shown analytically that the LDSFT becomes accurate in the low-temperature limit, exactly reproducing the $T^{3/2}$ law for the magnetization. At the same time, the calculated magnetization in Fe is in good agreement with experiment over the interval from zero to $0.6T_C^{exp} \approx 600\,K$. This interval is about 20 times larger than the interval where the $T^{3/2}$ law is valid. The calculated spin-wave stiffness D_0 and Curie temperature T_C for Fe are found in reasonable agreement with experimental data and results of previous calculations. For the $Fe_{0.65}Ni_{0.35}$ Invar, we obtained $D_0 = 110\,\mathrm{meV\,A}^2$ and $T_C = 546\,K$, which are in excellent agreement with experiment. Reliable calculations of these characteristics for the $Fe_{0.65}Ni_{0.35}$ Invar are not available. We argue that the present theory can be successfully applied to other ferromagnetic metals and alloys, where the longitudinal fluctuations are small compared to the transverse ones.

References

1. J. Hubbard, Phys. Rev. B **19**, 2626 (1979)
2. J. Hubbard, Phys. Rev. B **20**, 4584 (1979)
3. H. Hasegawa, J. Phys. Soc. Jpn. **46**, 1504 (1979)
4. H. Hasegawa, J. Phys. Soc. Jpn. **49**, 178 (1980)
5. H. Hasegawa, J. Phys. Soc. Jpn. **49**, 963 (1980)
6. N.B. Melnikov, B.I. Reser, V.I. Grebennikov, J. Phys. Condens. Matter **23**, 276003 (2011)
7. B.I. Reser, N.B. Melnikov, J. Phys. Condens. Matter **20**, 285205 (2008)
8. N.B. Melnikov, B.I. Reser, Theor. Math. Phys. **181**(2), 1435 (2014)
9. N.B. Melnikov, B.I. Reser, J. Supercond. Nov. Magn. **28**(3), 797 (2015)
10. N.B. Melnikov, B.I. Reser, Solid State Phenom. **233–234**, 20 (2015)
11. V. Crisan, P. Entel, H. Ebert, H. Akai, D.D. Johnson, J.B. Staunton, Phys. Rev. B **66**, 014416 (2002)
12. A.V. Ruban, S. Khmelevskyi, P. Mohn, B. Johansson, Phys. Rev. B **76**, 014420 (2007)
13. J. Donohue, *The Structure of the Elements* (Wiley, New York, 1974)
14. J. Crangle, G.M. Goodman, Proc. R. Soc. Lond. A **321**, 477 (1971)
15. A.T. Aldred, P.H. Froehle, Int. J. Magn. **2**, 195 (1972)
16. P.C. Riedi, Phys. Rev. B **8**, 5243 (1973)
17. R. Pauthenet, J. Appl. Phys. **53**, 8187 (1982)
18. U. Köbler, J. Phys. Condens. Matter **14**(38), 8861 (2002)
19. J.W. Lynn, N. Rosov, M. Acet, H. Bach, J. Appl. Phys. **75**, 6069 (1994)
20. M. Shiga, in *Materials Science and Technology*, vol. 3B, Part II, ed. by R. Cahn, P. Haasen, E. Kramer (VCH, Weinheim, 1994), p. 159
21. E.F. Wassermann, M. Acet, in *Magnetism and structure in functional materials*, ed. by A. Planes, L. Manosa, A. Saxena (Springer, Berlin, 2005), p. 177
22. A. Okorokov, S. Grigor'ev, V. Runov, G. Gordeev, Y. Chetverikov, G. Kopitsa, J. Surf. Invest. **1**, 542 (2007)
23. J. Crangle, G.C. Hallam, Proc. R. Soc. Lond. A **272**, 119 (1963)
24. D.D. Johnson, F.J. Pinski, G.M. Stocks, J. Appl. Phys. **57**, 3018 (1985)
25. Y. Ishikawa, S. Onodera, K. Tajima, J. Magn. Magn. Mater. **10**, 183 (1979)
26. Y. Takahashi, J. Phys. Condens. Matter **13**(29), 6323 (2001)
27. J.C. Slater, Phys. Rev. **81**, 385 (1951)

28. J.C. Slater, *Quantum Theory of Molecules and Solids, Vol. 4: The Self-Consistent Field for Molecules and Solids* (McGraw-Hill, New York, 1974)
29. D.J. Kim, *New Perspectives in Magnetism of Metals* (Kluwer/Plenum, New York, 1999)
30. H.A. Mook, R.M. Nicklow, Phys. Rev. B **7**, 336 (1973)
31. M. Pajda, J. Kudrnovsky, I. Turek, V. Drchal, P. Bruno, Phys. Rev. B **64**, 174402 (2001)
32. P. Buczek, A. Ernst, L.M. Sandratskii, Phys. Rev. B **84**, 174418 (2011)
33. I. Nakai, J. Phys. Soc. Jpn. **59**, 2211 (1990)
34. E.P. Wohlfarth, in *Ferromagnetic Materials*, vol. 1, ed. by E.P. Wohlfarth (North-Holland, Amsterdam, 1980), pp. 1–70
35. H.A. Mook, in *Spin Waves and Magnetic Excitations 1*, ed. by A.S. Borovik-Romanov, S.K. Sinha (Elsevier, Amsterdam, 1988), pp. 425–478

Temperature Dependence of Magnetic Characteristics

13

Don't look for the meaning; look for the use. (Ludwig Wittgenstein)

In this chapter we study spin-correlation effects in metals at finite temperatures. We start with qualitative estimates of the correlation effects on the magnitude and relaxation time of a single-site spin [1]. Then we apply the DSFT to calculate local magnetic characteristics such as the transverse and longitudinal susceptibilities, dynamic spin correlation function and local magnetic moment [2]. Finally, we calculate temperature dependencies of nuclear spin-relaxation rates [3].

13.1 Temporal Correlation Function

In Chap. 3 we introduced the single-site spin $\mathbf{s}_j$ in an itinerant electron ferromagnet as the integral of the spin-density over the Wigner-Seitz cell (3.30). The dynamics of the single-site spin is described by the Heisenberg representation

$$s_j^\alpha(t) = \mathrm{e}^{\mathrm{i}\mathcal{H}t/\hbar} s_j^\alpha \mathrm{e}^{-\mathrm{i}\mathcal{H}t/\hbar},$$

where $\alpha = x, y, z$. To quantify the coherence of spins at different sites and time moments, we introduce the spin correlation function

$$F_{jj'}^{\alpha\beta}(t) = \frac{1}{2}\langle\{s_j^\alpha(t), s_{j'}^\beta(0)\}\rangle,$$

where the braces denote the anticommutator of operators and angle brackets denote the canonical average. Using the relation

$$\langle s_j^\alpha(t) s_{j'}^\beta(0)\rangle = \langle s_j^\alpha(t)\rangle\langle s_{j'}^\beta(0)\rangle + \langle \Delta s_j^\alpha(t)\Delta s_{j'}^\beta(0)\rangle,$$

we write

$$F_{jj'}^{\alpha\beta}(t) = \langle s_j^\alpha(t)\rangle\langle s_{j'}^\beta(0)\rangle + A_{jj'}^{\alpha\beta}(t), \tag{13.1}$$

where the first term is time-independent and is equal to $\langle s_j^\alpha\rangle\langle s_j^\beta\rangle$, and

$$A_{jj'}^{\alpha\beta}(t) = \frac{1}{2}\langle\{\Delta s_j^\alpha(t), \Delta s_{j'}^\beta(0)\}\rangle$$

is the oscillating part. Since at large times the spin fluctuations $\Delta\mathbf{s}_j(t)$ and $\Delta\mathbf{s}_{j'}(0)$ become uncorrelated, $A_{jj'}^{\alpha\beta}(t)$ tends to zero as time goes to infinity. The frequency spin correlation function is given by the Fourier transformation,

$$F_{jj'}^{\alpha\beta}(\omega) = \int F_{jj'}^{\alpha\beta}(t)\mathrm{e}^{\mathrm{i}\omega t}\,\mathrm{d}t. \tag{13.2}$$

N. B. Melnikov, B. I. Reser, *Dynamic Spin Fluctuation Theory of Metallic Magnetism*,
https://doi.org/10.1007/978-3-319-92974-3_13

By formula (13.1) we have

$$F^{\alpha\beta}_{jj'}(\omega) = 2\pi \langle s^\alpha_j\rangle\langle s^\beta_{j'}\rangle\delta(\omega) + A^{\alpha\beta}_{jj'}(\omega), \tag{13.3}$$

where $\delta(\omega)$ is the delta function and $A^{\alpha\beta}_{jj'}(\omega)$ is the Fourier transform of $A^{\alpha\beta}_{jj'}(t)$.

We define the susceptibility in the site representation by

$$\chi^{\alpha\beta}_{jj'}(\omega) = \frac{1}{N}\sum_{\mathbf{q}} \chi^{\alpha\beta}_{\mathbf{q}}(\omega)\, \mathrm{e}^{\mathrm{i}\mathbf{q}(\mathbf{R}_j-\mathbf{R}_{j'})}.$$

Then using (2.38) and (3.32), we obtain

$$\chi^{\alpha\beta}_{jj'}(\omega) = \frac{\mathrm{i}}{\hbar}\int_0^\infty \langle [s^\alpha_j(t), s^\beta_{j'}]\rangle\, \mathrm{e}^{\mathrm{i}\omega t}\, \mathrm{d}t.$$

The fluctuation-dissipation theorem relates the function $A^{\alpha\beta}_{jj'}(\omega)$ to the susceptibility. Namely, the Fourier transformation of (2.53) gives

$$A^{\alpha\beta}_{jj'}(\omega) = -\frac{\mathrm{i}\hbar}{2}\coth\left(\frac{\hbar\omega}{2T}\right)\left[\chi^{\alpha\beta}_{jj'}(\omega) - \left(\chi^{\beta\alpha}_{jj'}(\omega)\right)^*\right].$$

In particular,

$$A^{\alpha\alpha}_{jj'}(\omega) = \hbar\coth\left(\frac{\hbar\omega}{2T}\right)\mathrm{Im}\chi^{\alpha\alpha}_{jj'}(\omega).$$

The *spatial* correlations will be considered in Chap. 15. In this chapter, we consider the local spin moment and *temporal* correlations of the single-site spin.

We define the single-site temporal correlation function as

$$F(t) = \sum_\alpha F^{\alpha\alpha}_{jj}(t) = \frac{1}{2}\langle\{\mathbf{s}_j(t), \mathbf{s}_j(0)\}\rangle. \tag{13.4}$$

Then from (13.1) it follows that

$$F(t) = \langle\mathbf{s}_j\rangle^2 + A(t),$$

where

$$A(t) = \sum_\alpha A^{\alpha\alpha}_{jj}(t) = \frac{1}{2}\langle\{\Delta\mathbf{s}_j(t), \Delta\mathbf{s}_j(0)\}\rangle.$$

Since $A(t)$ tends to zero at large t, we have

$$F(t=\infty) = \langle\mathbf{s}_j\rangle^2 = \bar{s}_z^2,$$

and the function $A(t)$ describes the characteristic time of the spin fluctuations.

The Fourier transform of the temporal correlation function is obtained by formula (13.3). Recalling that $s^z_j = \frac{1}{2}(n_{j\uparrow} - n_{j\downarrow})$, we come to

$$F(\omega) = 2\pi\left(\frac{n_\uparrow - n_\downarrow}{2}\right)^2\delta(\omega) + A(\omega), \tag{13.5}$$

where n_σ is the average number of electrons with spin σ and

$$A(\omega) = \hbar\coth\left(\frac{\hbar\omega}{2T}\right)\sum_\alpha \mathrm{Im}\chi^{\alpha\alpha}_{jj}(\omega). \tag{13.6}$$

Since $\mathrm{Im}\chi^{\alpha\alpha}_{jj}(\omega)$ is an odd function of ω, it follows that $A(\omega)$ is an even function. Recalling the relations (3.22) and (3.23), we have

$$\chi^{xx}_{jj}(\omega) + \chi^{yy}_{jj}(\omega) = \frac{1}{2}[\chi^{+-}_{jj}(\omega) + \chi^{-+}_{jj}(\omega)].$$

Using $\chi^{-+}_{jj}(\omega) = (\chi^{+-}_{jj}(-\omega))^*$, we write formula (13.6) as

$$A(\omega) = \hbar \coth\left(\frac{\hbar\omega}{2T}\right) \mathrm{Im}\left[\frac{1}{2}\left(\chi_{jj}^{+-}(\omega) + (\chi_{jj}^{+-}(-\omega))^*\right) + \chi_{jj}^{zz}(\omega)\right]. \qquad (13.7)$$

Taking the inverse Fourier transform of (13.5), we obtain the temporal correlation function

$$F(t) = \left(\frac{n_\uparrow - n_\downarrow}{2}\right)^2 + \frac{1}{\pi}\int_0^\infty A(\omega)\cos(\omega t)\,\mathrm{d}\omega. \qquad (13.8)$$

To compare the calculation results with the ones of the polarized neutron scattering experiment, we introduce the effective local spin in a frequency interval $[-\omega, \omega]$ by the formula

$$s_{\mathrm{L}}(\omega) = \left(\frac{1}{\pi}\int_0^\omega F(\omega')\,\mathrm{d}\omega'\right)^{1/2} = \left[\left(\frac{n_\uparrow - n_\downarrow}{2}\right)^2 + \frac{1}{\pi}\int_0^\omega A(\omega')\,\mathrm{d}\omega'\right]^{1/2}. \qquad (13.9)$$

Clearly, the effective local spin in the *infinite* frequency interval is equal to the square root of the temporal correlation function (13.8) at zero:

$$s_{\mathrm{L}}(\omega = \infty) = (F(t = 0))^{1/2}.$$

By the definition of the temporal correlation function (13.4), the value $(F(t = 0))^{1/2}$ coincides with a root mean square of the spin operator, i.e. the local spin:

$$s_{\mathrm{L}}(\omega = \infty) = \langle s_j^2 \rangle^{1/2} = s_{\mathrm{L}}.$$

Note that the second term on the right-hand side in formula (13.9) does not vanish at $T = 0$ due to zero-point spin fluctuations.

13.2 Qualitative Analysis of Spin Correlations

Spin-density correlation in metals at finite temperatures can be described by SFT but the results in different modifications of SFT can differ quite substantially (for a review, see, e.g. [4,5]). The question naturally arises: when does the correlation make a major contribution and when does it merely lead to more or less important corrections to the band calculations? Indeed, the self-consistent potential, used in the band calculation, already takes some of the "Hund interaction" into account, and so the question is to what extent the remaining part is important in a particular case.

For the mean square spin at the jth site $\langle \mathbf{s}_j^2 \rangle$, we can take results of the one-electron calculations in the paramagnetic state as a lower bound of any SFT result. This estimate implies that the correlation leads to a Hund's rule, which leads to an increase of the local moment, because each site tends to collect electrons with the same spin (see the discussion at the end of Sect. 8.1.1). On the other hand, if we consider only correlations at nonzero temperatures, and the Stoner ground state is a good approximation for the true ground state of a metal, then the one-electron calculation of local moment in the ferromagnetic state at $T = 0$ gives an upper bound for a SFT calculation of $\langle \mathbf{s}^2 \rangle$ (here and hereafter, the site index j is omitted for brevity).

Notice that the value of $\langle \mathbf{s}^2 \rangle$ is not necessarily small even in the absence of correlations. In fact, even in the simplest case of *one* nondegenerate band with the mean number of electrons per site equal to n, the value of $\langle \mathbf{s}^2 \rangle$ will be equal to the product of the square of the electron spin by the probability that there is one and only one electron at a site (for details, see Appendix A.3):

$$\langle \mathbf{s}^2 \rangle = \frac{3}{4}\, 2\frac{n}{2}\left(1 - \frac{n}{2}\right) = \frac{3}{8}\, n(2 - n). \qquad (13.10)$$

So at the half-filling ($n = 1$) the mean square spin in a *metal* will be only half the one in a *dielectric* with one electron at a site. In the case of *several* independent bands, it is easy to show that, instead of (13.10), we have the expression

$$\langle \mathbf{s}^2 \rangle = \frac{3}{8}\sum_{\nu=1}^{N_{\mathrm{b}}} n_\nu(2 - n_\nu), \qquad (13.11)$$

where N_b is the number of bands; n_ν is the number of electrons per site in the ν-th band; $\sum_\nu n_\nu = n$. If all the bands are identically occupied, i.e. $n_\nu = n/N_b$, the mean square spin reaches its maximum value

$$\langle \mathbf{s}^2 \rangle = \frac{3}{8} \frac{n(2N_b - n)}{N_b}. \tag{13.12}$$

Let us examine another limiting case—a dielectric with n electrons (holes) per site, which occupy with equal probability any of the N_b ($N_b \geq n$) orbital states that are doubly degenerate with respect to the spin projection. This case is obtained from the previous one by totally suppressing charge fluctuations, i.e. if there is no intersite electron hopping. It is easy to show that if the n electrons occupy n *different* orbital states and both of their spin states are equally probable, then the mean square spin of the system will be equal to $\frac{3}{4}n$.[1] Now taking into account the possibility that an orbital state is occupied by a pair of electrons (total spin is zero), for n electrons in N_b orbital states, it is easy to show that the mean square spin is $2N_b/(2N_b - 1)$ times larger than the value (13.12). Thus, it is only in the s band that the suppression of charge fluctuations alters the mean square spin substantially (by the factor of two). Already in the d band this effect is no more than $\sim$10% and falls as the orbital moment increases.[2] Finally, recalling that the maximum possible value of the square of the spin in a system of n electrons is equal to $\frac{n}{2}(\frac{n}{2} + 1)$, we conclude that, with a small number of electrons (holes) per site, the mean square of its spin can easily reach values close to half the maximum, even in the absence of correlations. Thus, inclusion of the electron–electron interactions should mainly result in slowing down of the spin "rotation" at each site, rather than in growth of its mean magnitude.

Next we analyse the single-site spin relaxation time t_0, i.e. the time that is necessary for a spin at a site to "forget" its initial magnitude and direction. In the absence of the electron correlation, the spin relaxation time is about the electron lifetime $t_0 \simeq t_W \simeq \hbar/W$ s (W is the bandwidth). Indeed, the spin direction of an electron arriving at a site is not correlated in any way with that of the electrons that were already present at this site. As a result, in the absence of correlation, the characteristic spin relaxation time of a 3d transition metals with d bandwidth $W \simeq 5$ eV would be $t_0 \simeq 10^{-16}$ s (for details, see [6]).

The above estimate of the spin relaxation time in the absence of the electron correlation is a qualitative one. The result could be affected by large groups of electrons with energies close to the Fermi level ε_F. The existence of such groups is associated with flattening of $\varepsilon(\mathbf{k})$ in certain regions of $\mathbf{k}$-space (small $\nabla_\mathbf{k}\varepsilon$), which produces sharp singularities of the density of states. The characteristic energy width of such singularities is one order less than the width of the d band, and that could have a strong effect on the estimate of [6] and similar ones.

The electron correlation leads to a Hund's rule. On the one hand, this leads to growth in the magnitude of the spin at a site, and on the other hand, the arrival of electrons with a spin direction different from the spin of the site becomes energetically unfavourable and, therefore, less likely. So the spin relaxation time t_0 grows. The single-site spin relaxation time is determined by the intersite exchange energy. If the correlations are very strong, the situation becomes similar to Heisenberg magnets with localized spins, where the characteristic energy of the exchange interaction is about the Curie temperature: $t_0 \simeq \hbar/(k_B T_C)$. For 3d metals with $T_C \simeq 1000$ K this leads to the estimate, $t_0 \simeq 10^{-14}$ s. The difference between the latter and band calculation estimate can be attributed to correlation effects.

13.3 Spin Correlations in the One-Electron Approximation

13.3.1 Unenhanced Susceptibilities

To calculate the temporal correlation function (13.8) and its Fourier transform (13.5), we calculate the transverse and longitudinal susceptibilities of the noninteracting electrons:

$$\chi_{jj}^{0+-}(\omega) = \frac{\mathrm{i}}{\hbar} \int_0^\infty \langle [s_j^+(t), s_j^-(0)] \rangle \, \mathrm{e}^{\mathrm{i}\omega t} \, \mathrm{d}t, \tag{13.13}$$

[1] This is simply the sum of squares of electronic spins in different orbital states, since in these conditions the mean value of the scalar product of spins of different electrons is zero. The same argument applies in the derivation of (13.11).

[2] These results are valid only in the case of non-interacting electrons.

$$\chi_{jj}^{0zz}(\omega) = \frac{\mathrm{i}}{\hbar} \int_0^\infty \langle [s_j^z(t), s_j^z(0)] \rangle \, \mathrm{e}^{\mathrm{i}\omega t} \, \mathrm{d}t, \tag{13.14}$$

expressed in units of $g^2\mu_{\mathrm{B}}^2$. Similar to (3.31) the single-site spin-flip operators are defined by

$$s_j^+ = \int_{\Omega_j} \psi_\uparrow^\dagger(\mathbf{r}) \, \psi_\downarrow(\mathbf{r}) \, \mathrm{d}\mathbf{r}, \qquad s_j^- = \int_{\Omega_j} \psi_\downarrow^\dagger(\mathbf{r}) \, \psi_\uparrow(\mathbf{r}) \, \mathrm{d}\mathbf{r},$$

where Ω_j is the Wigner-Seitz unit cell centred at the jth site, and $\psi_\sigma^\dagger(\mathbf{r})$ and $\psi_\sigma(\mathbf{r})$ are the field operators. Using formulae (3.26), we write

$$s_j^+ = \frac{1}{N} \sum_{\nu\nu'\mathbf{k}\mathbf{k}'} M_{\nu\mathbf{k}\downarrow}^{\nu'\mathbf{k}'\uparrow} a_{\nu'\mathbf{k}'\uparrow}^\dagger a_{\nu\mathbf{k}\downarrow}, \qquad s_j^- = \frac{1}{N} \sum_{\nu\nu'\mathbf{k}\mathbf{k}'} M_{\nu\mathbf{k}\uparrow}^{\nu'\mathbf{k}'\downarrow} a_{\nu'\mathbf{k}'\downarrow}^\dagger a_{\nu\mathbf{k}\uparrow}. \tag{13.15}$$

Here

$$M_{\nu\mathbf{k}\sigma}^{\nu'\mathbf{k}'\sigma'} = \int_{\Omega_j} \varphi_{\nu'\mathbf{k}'\sigma'}^*(\mathbf{r}) \, \varphi_{\nu\mathbf{k}\sigma}(\mathbf{r}) \, \mathrm{d}\mathbf{r}, \tag{13.16}$$

and the eigenfunctions $\varphi_{\nu\mathbf{k}\sigma}(\mathbf{r})$ of the Hamiltonian

$$\mathcal{H}_0 = \sum_{\nu\mathbf{k}\sigma} \varepsilon_{\nu\mathbf{k}\sigma} a_{\nu\mathbf{k}\sigma}^\dagger a_{\nu\mathbf{k}\sigma} \tag{13.17}$$

are assumed to be normalized *per unit cell*. Clearly the absolute value of the matrix element (13.16) is site-independent. Using expressions (13.15) and calculating the commutator in (13.13), we obtain

$$\chi_{jj}^{0+-}(\omega) = \frac{1}{N^2} \sum_{\nu\nu'\mathbf{k}\mathbf{k}'} \frac{|M_{\nu\mathbf{k}\downarrow}^{\nu'\mathbf{k}'\uparrow}|^2 (f(\varepsilon_{\nu\mathbf{k}\downarrow}) - f(\varepsilon_{\nu'\mathbf{k}'\uparrow}))}{\varepsilon_{\nu'\mathbf{k}'\uparrow} - \varepsilon_{\nu\mathbf{k}\downarrow} + \hbar\omega}, \tag{13.18}$$

where $f(\varepsilon) = [\exp((\varepsilon - \mu)/T) + 1]^{-1}$ is the Fermi function.

Similarly, we introduce

$$s_j^\sigma = \int_{\Omega_j} \psi_\sigma^\dagger(\mathbf{r}) \, \psi_\sigma(\mathbf{r}) \, \mathrm{d}\mathbf{r},$$

so that $s_j^z = \frac{1}{2}(s_j^\uparrow - s_j^\downarrow)$. Then we write the longitudinal susceptibility (13.14) as

$$\chi_{jj}^{0zz}(\omega) = \frac{1}{4} \left[\chi_{jj}^{\uparrow\uparrow}(\omega) - \chi_{jj}^{\downarrow\uparrow}(\omega) - \chi_{jj}^{\uparrow\downarrow}(\omega) + \chi_{jj}^{\downarrow\downarrow}(\omega) \right], \tag{13.19}$$

where

$$\chi_{jj}^{0\sigma\sigma'}(\omega) = \frac{\mathrm{i}}{\hbar} \int_0^\infty \langle [s_j^\sigma(t), s_j^{\sigma'}(0)] \rangle \, \mathrm{e}^{\mathrm{i}\omega t} \, \mathrm{d}t.$$

Calculation yields

$$\chi_{jj}^{0\sigma\sigma'}(\omega) = \frac{1}{N^2} \sum_{\nu\nu'\mathbf{k}\mathbf{k}'} \frac{|M_{\nu\mathbf{k}\sigma}^{\nu'\mathbf{k}'\sigma'}|^2 (f(\varepsilon_{\nu\mathbf{k}\sigma}) - f(\varepsilon_{\nu'\mathbf{k}'\sigma'}))}{\varepsilon_{\nu'\mathbf{k}'\sigma'} - \varepsilon_{\nu\mathbf{k}\sigma} + \hbar\omega} \delta_{\sigma\sigma'}. \tag{13.20}$$

Thus, $\chi_{jj}^{0\uparrow\downarrow}(\omega) = \chi_{jj}^{0\downarrow\uparrow}(\omega) = 0$, and

$$\chi_{jj}^{0zz}(\omega) = \frac{1}{4} \left[\chi_{jj}^{0\uparrow\uparrow}(\omega) + \chi_{jj}^{0\downarrow\downarrow}(\omega) \right].$$

13.3.2 Computational Formulae

We start with the Fourier transform of the temporal correlation function (13.5). We substitute the unenhanced susceptibilities (13.18)–(13.20) in formula (13.7). Then taking into account that $\coth(\hbar\omega/(2T)) = 1$ at $T = 0$ and applying the Sokhotsky formula (A.44), we obtain[3]

$$F(\omega) = 2\pi \left(\frac{n_\uparrow - n_\downarrow}{2}\right)^2 \delta(\omega) + \frac{2\pi}{8N^2} \sum_{\substack{\nu\mathbf{k}\sigma \\ \nu'\mathbf{k}'\sigma'}} 2^{|\sigma-\sigma'|/2} \left|M_{\nu\mathbf{k}\sigma}^{\nu'\mathbf{k}'\sigma'}\right|^2 \tag{13.21}$$

$$\times \left[f(\varepsilon_{\nu\mathbf{k}\sigma})(1 - f(\varepsilon_{\nu'\mathbf{k}'\sigma'})) - f(\varepsilon_{\nu'\mathbf{k}'\sigma'})(1 - f(\varepsilon_{\nu\mathbf{k}\sigma}))\right] \delta\left(\omega - \frac{\varepsilon_{\nu'\mathbf{k}'\sigma'} - \varepsilon_{\nu\mathbf{k}\sigma}}{\hbar}\right).$$

There are two special cases when (13.21) can be largely simplified. Firstly, when the overlap of Wannier functions of neighbouring sites can be neglected ("tight binding"), we have the already mentioned case of independent bands: $\left|M_{\nu\mathbf{k}\sigma}^{\nu'\mathbf{k}'\sigma'}\right| = \delta_{\nu\nu'}$. In that case, using (3.55) and (4.24), we have

$$F(\omega) = 2\pi \left(\frac{n_\uparrow - n_\downarrow}{2}\right)^2 \delta(\omega) + \frac{2\pi\hbar}{8} \sum_{\nu\sigma\sigma'} 2^{|\sigma-\sigma'|/2} \int n_{\nu\sigma}(\varepsilon)\, n_{\nu\sigma'}(\varepsilon + \hbar\omega)$$

$$\times \left[f(\varepsilon)(1 - f(\varepsilon + \hbar\omega)) - f(\varepsilon + \hbar\omega)(1 - f(\varepsilon))\right] \mathrm{d}\varepsilon, \tag{13.22}$$

where $n_{\nu\sigma}(\varepsilon)$ is the density of states in the ν-th subband with spin projection σ. Then, the mean square single-site spin

$$\langle \mathbf{s}^2 \rangle = F(t = 0) = \frac{1}{2\pi} \int_{-\infty}^{\infty} F(\omega)\, \mathrm{d}\omega$$

is written as

$$\langle \mathbf{s}^2 \rangle = \left(\frac{n_\uparrow - n_\downarrow}{2}\right)^2 + \frac{3}{4} n - \frac{3}{8} \sum_\nu n_\nu^2 + \frac{1}{8} \sum_\nu (n_{\nu\uparrow} - n_{\nu\downarrow})^2. \tag{13.23}$$

Here $n_{\nu\sigma}$ is the mean number of electrons with the spin σ in the ν-th band (pet site) and $n = \sum_{\nu\sigma} n_{\nu\sigma} = n_\uparrow + n_\downarrow$. In the nonmagnetic case ($n_{\nu\uparrow} = n_{\nu\downarrow}$) expression (13.23) naturally coincides with (13.11).

The second case is the constant matrix element approximation: $|M_{\nu\mathbf{k}\sigma}^{\nu'\mathbf{k}'\sigma'}|^2 = C$. It is clear that the latter can hold approximately only when a limited number of bands N_b, genetically related to the site states of the same type (such as five d bands, for example), is used in the calculation. Then the value of C can be determined from the completeness condition:

$$\sum_{\nu\mathbf{k}} \varphi_{\nu\mathbf{k}\sigma}^*(\mathbf{r})\, \varphi_{\nu\mathbf{k}\sigma}(\mathbf{r}') = N\, \delta(\mathbf{r} - \mathbf{r}'). \tag{13.24}$$

Calculating $\sum_{\nu\mathbf{k}} |M_{\nu\mathbf{k}\sigma}^{\nu'\mathbf{k}'\sigma'}|^2$ with the help of that condition, we obtain $C = 1/N_\mathrm{b}$. So in this case (13.21) can be written in the form

$$F(\omega) = 2\pi \left(\frac{n_\uparrow - n_\downarrow}{2}\right)^2 \delta(\omega) + \frac{2\pi\hbar}{8N_\mathrm{b}} \sum_{\sigma\sigma'} 2^{|\sigma-\sigma'|/2} \int n_\sigma(\varepsilon)\, n_{\sigma'}(\varepsilon + \hbar\omega)$$

$$\times \left[f(\varepsilon)(1 - f(\varepsilon + \hbar\omega)) - f(\varepsilon + \hbar\omega)(1 - f(\varepsilon))\right] \mathrm{d}\varepsilon, \tag{13.25}$$

[3] If detailed calculations of the generalized susceptibility $\chi(\mathbf{q}, \omega)$ were available, the Fourier-transform of correlator (13.21) could be obtained by only triple (rather than sixfold) integration over the Brillouin zone. Indeed, by the fluctuation-dissipation theorem (2.56), we have

$$F(\omega) \propto \coth(\beta\hbar\omega/2) \int \mathrm{Im}\chi(\mathbf{q}, \omega)\, \mathrm{d}\mathbf{q}.$$

However, calculations of $\chi(\mathbf{q}, \omega)$ are usually given only in a few symmetrical directions of $\mathbf{q}$ (see, e.g. [7]), and so this formula cannot be used.

where $n_\sigma(\varepsilon)$ is the total density of states for electrons with the spin σ. For $\langle \mathbf{s}^2 \rangle$ from (13.25) we find

$$\langle \mathbf{s}^2 \rangle = \left(1 + \frac{1}{2N_\mathrm{b}}\right)\left(\frac{n_\uparrow - n_\downarrow}{2}\right)^2 + \frac{3}{8}\frac{n(2N_\mathrm{b} - n)}{N_\mathrm{b}}. \tag{13.26}$$

Clearly, (13.26) coincides with (13.12) in the paramagnetic state and is temperature independent. Moreover, if all the bands are identically occupied at any temperature, so that $n_\nu = n/N_\mathrm{b}$, (13.26) will also coincide with (13.23).

In numerical calculations with a specific spectrum and wave functions, we ignore the thermal broadening of the Fermi function, using the approximation

$$f(\varepsilon) = \theta(\varepsilon_\mathrm{F} - \varepsilon), \tag{13.27}$$

where $\theta(x)$ is the step-function, equal to zero for $x < 0$ and unity for $x > 0$. Then, for $\omega > 0$,[4] formula (13.21) is easily transformed to

$$F(\omega) = 2\pi\left(\frac{n_\uparrow - n_\downarrow}{2}\right)^2 \delta(\omega) + \frac{2\pi\hbar}{8}\sum_{\substack{\nu\sigma \\ \nu'\sigma'}} 2^{|\sigma-\sigma'|/2}\, I_{\nu\sigma}^{\nu'\sigma'}(\omega), \tag{13.28}$$

$$I_{\nu\sigma}^{\nu'\sigma'}(\omega) = \frac{1}{\Omega_\mathrm{BZ}^2}\frac{\mathrm{d}}{\mathrm{d}(\hbar\omega)} \iint\limits_{\substack{\varepsilon_{\nu\mathbf{k}\sigma} \le \varepsilon_\mathrm{F} \\ \varepsilon_{\nu'\mathbf{k}'\sigma'} > \varepsilon_\mathrm{F} \\ \varepsilon_{\nu'\mathbf{k}'\sigma'} - \varepsilon_{\nu\mathbf{k}\sigma} \le \hbar\omega}} \left|M_{\nu\mathbf{k}\sigma}^{\nu'\mathbf{k}'\sigma'}\right|^2 \mathrm{d}\mathbf{k}'\,\mathrm{d}\mathbf{k}, \tag{13.29}$$

where the integration with respect to $\mathbf{k}$ and $\mathbf{k}'$ is performed in the Brillouin zone of the volume Ω_BZ.

The calculation of spin correlators by formulae (13.28), (13.29) and

$$F(t) = \frac{1}{\pi}\int_0^\infty F(\omega)\cos(\omega t)\,\mathrm{d}\omega \tag{13.30}$$

involves four computational problems. The first one consists in calculating one-electron states and is not considered here. We only mention that the energy spectrum $\varepsilon_{\nu\mathbf{k}\sigma}$ and wave functions $\varphi_{\nu\mathbf{k}\sigma}(\mathbf{r})$ are calculated by the KKR method [8].

The second problem consists in calculating the matrix elements $M_{\nu\mathbf{k}\sigma}^{\nu'\mathbf{k}'\sigma'}$. We represent (13.16) as the sum of two integrals: over the Slater sphere [9] and over the remaining part of the Wigner-Seitz cell. Since the contribution from the second integral is small for d wave functions, it can be neglected, thereby the wave functions is normalized in the Slater sphere. Putting the expression for the wave function in the KKR method [10] into the first integral and integrating in spherical coordinates, we have

$$M_{\nu\mathbf{k}\sigma}^{\nu'\mathbf{k}'\sigma'} = \sum_l \int_0^{r_\mathrm{S}} R_l^{\sigma'}(r, \varepsilon_{\nu'\mathbf{k}'\sigma'}) R_l^{\sigma}(r, \varepsilon_{\nu\mathbf{k}\sigma})\, r^2\,\mathrm{d}r \sum_m (c_{lm}^{\nu'\mathbf{k}'\sigma'})^* c_{lm}^{\nu\mathbf{k}\sigma}, \tag{13.31}$$

where $c_{lm}^{\nu\mathbf{k}\sigma}$ are the coefficients of the expansion, R_l^σ are the bounded at zero solutions of the radial Schrödinger equation; r_S is the radius of the Slater sphere.

The third problem is to calculate the integrals (13.29). The difficulty of numerical integration over the Wigner-Seitz cell is that it is necessary to calculate many sixfold integrals with a complex domain of integration, while the number of points $\mathbf{k}$ and $\mathbf{k}'$ at which the integrand is known is relatively small. We represent the sixfold integral (13.29) in the form of two triple integrals (in $\mathbf{k}$ and $\mathbf{k}'$) and apply the tetrahedral method [11] to each of them.

Finally, the last computational problem involves the approximate Fourier transformation [12]. Without going into detail, we note that the computational formula for the numerical cosine Fourier transform of the function $F(\omega)$ is obtained as follows. Using the uniform grid with the step h, we interpolate the function $F(\omega)$ by a second-order polynomial at each segment of the length $2h$, then analytically integrate by parts (see also Appendix H.5).

In the approximation of a constant matrix element, we calculate $F(\omega)$ by the formula

$$F(\omega) = 2\pi\left(\frac{n_\uparrow - n_\downarrow}{2}\right)^2 \delta(\omega) + \frac{2\pi\hbar}{8N_\mathrm{b}}\sum_{\sigma\sigma'} 2^{|\sigma-\sigma'|/2}\int_{\varepsilon_\mathrm{F}-\hbar\omega}^{\varepsilon_\mathrm{F}} n_\sigma(\varepsilon)\, n_{\sigma'}(\varepsilon + \hbar\omega)\,\mathrm{d}\varepsilon, \tag{13.32}$$

[4] The function $F(\omega)$ is even.

which can be obtained from (13.25) using (13.27). In this case the calculation of the spin correlators simplifies substantially, because the calculation of the wave functions, matrix elements and sixfold integrals is no longer necessary.

13.3.3 Band and Model Calculations

The energy spectrum $\varepsilon_{\nu\mathbf{k}\sigma}$ and wave functions $\varphi_{\nu\mathbf{k}\sigma}(\mathbf{r})$ of ferromagnetic (F) and nonmagnetic (N) bcc iron are calculated by the KKR method with the self-consistent potentials given by Moruzzi et al. [13]. The number of states $N_\sigma(\varepsilon)$ and density of states $n_\sigma(\varepsilon)$ were calculated by the hybrid tetrahedron method examined in detail in [14]. The results for $N_\sigma(\varepsilon)$ and $n_\sigma(\varepsilon)$ are in good agreement with [13]. The values n_σ are obtained as values of $N_\sigma(\varepsilon)$ at $\varepsilon = \varepsilon_F$.

The Fourier transform $A(\omega)$ of the oscillating part $A(t)$ is calculated for the F- and N-iron in two modifications: (1) with a $\nu\mathbf{k}\sigma$-dependent matrix element and (2) in the approximation of constant matrix element with $N_b = 6$. The results for $A(\omega)$ and its integral over the interval $[0, \omega]$:

$$\Phi(\omega) = \frac{1}{\pi}\int_0^{\omega} A(\omega')\,\mathrm{d}\omega', \tag{13.33}$$

are shown in Fig. 13.1. The character of $A(\omega)$ is mainly governed by the form of the density of states $n(\varepsilon)$ (see, e.g. [15]). So in the ferromagnetic case the curve $A^F(\omega)$ is formed of three substantial contributions at $\hbar\omega \simeq 2.4$, 4.6 and 5.5 eV due to transitions from the three occupied peaks $n^F(\varepsilon)$ to the fourth one situated above ε_F. In the nonmagnetic case the curve $n^N(\varepsilon)$ contains three peaks, one of which is close to ε_F, and so there are three sharp peaks of $A^N(\omega)$, at $\hbar\omega \simeq 0.6$, 3.2 and 4.6 eV. On the whole, the $A(\omega)$ curve shifts towards lower energies and slightly contracts when iron changes from the ferromagnetic to nonmagnetic state.

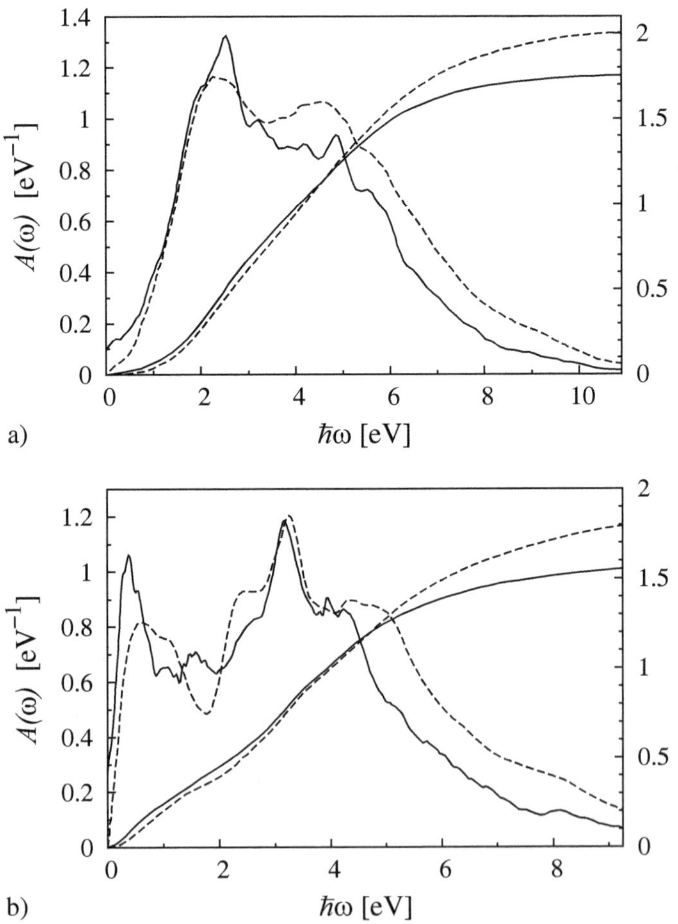

Fig. 13.1 The Fourier transform $A(\omega)$ and integral Fourier transform $\Phi(\omega)$ of the spin correlator of ferromagnetic (**a**) and nonmagnetic (**b**) bcc iron: with $\nu\mathbf{k}\sigma$-dependent matrix element (solid line) and with constant matrix element (dashed line)

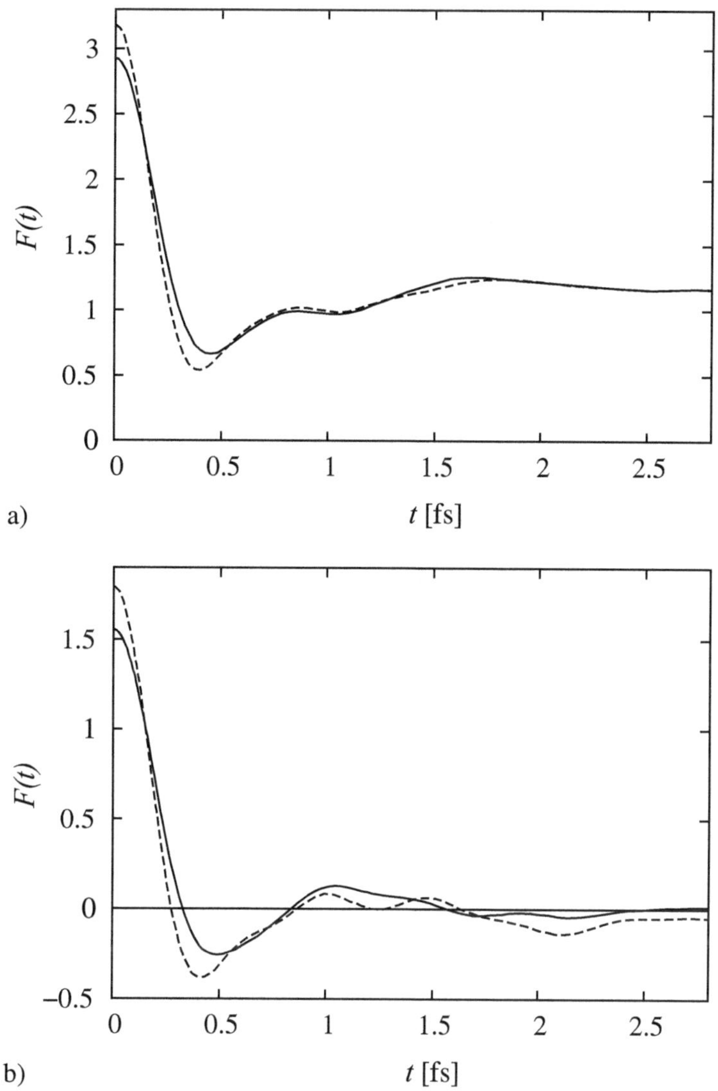

Fig. 13.2 Spin correlator $F(t)$ of ferromagnetic (**a**) and nonmagnetic (**b**) bcc iron: with $\nu\mathbf{k}\sigma$-dependent matrix element (solid line) and with constant matrix element (dashed line)

As for the spin correlator $F(t)$ presented in Fig. 13.2, its constant part is equal to 1.18 in the ferromagnetic case and is equal to zero in the nonmagnetic case. The oscillating part behaves similarly in both cases: it is equal to 1.75 and 1.55 at $t = 0$ and declines to zero during $t_0 \simeq 2.7 \times 10^{-16}$ and 3.3×10^{-16} s for F and N-iron, respectively.[5] The difference between the t_0 values for F and N-iron is due to the difference in the behaviour of their densities of states near ε_F: when the peak of $n^N(\varepsilon)$ gets close to ε_F the number of transitions with small $\hbar\omega$ increases. It is clear from Figs. 13.1 and 13.2 that calculation with $\nu\mathbf{k}\sigma$-dependent matrix element results in the growth of $A(\omega)$ for small values of ω and to its decline for large ω values. As a consequence, the decay of $F(t)$ is delayed, i.e. the relaxation time of spin fluctuations increases. However, this effect is not large enough in iron, because the electronic states at all the peaks of the density of states are genetically related mainly with the atomic 3d states, and hence the matrix elements decrease slowly as $\hbar\omega = \varepsilon_{\nu'\mathbf{k}'\sigma'} - \varepsilon_{\nu\mathbf{k}\sigma}$ increase.

Results of a band calculation of the mean square single-site spin for F and N-iron are given in Table 13.1. For comparison, we also present the values of $\langle \mathbf{s}^2 \rangle$ obtained in the model with independent equally occupied bands and in the localized-states model.

Experimental estimates of the mean square single-site spin are obtained as follows. In the ferromagnetic state, by applying formula (5.53), we have $\langle \mathbf{s}^2 \rangle = s_z(s_z + 1)$ at $T = 0$. Using the experimental value $m_0^{\exp} = 2.2\mu_B$, we obtain $s_z = 1.1$ and $\langle \mathbf{s}^2 \rangle = 2.3$. In the paramagnetic state, if we consider iron as a purely localized-spin magnet with a completely frozen orbital

[5]Since the calculations with $\nu\mathbf{k}\sigma$-dependent matrix element and with constant matrix element yield similar results, for brevity, we give numerical results only for the former case.

Table 13.1 Mean square single-site spin $\langle \mathbf{s}^2 \rangle$ in the band and model calculations of ferromagnetic (F) and nonmagnetic (N) iron

Calculation method	Approximation		F-Fe	N-Fe
Band calculation	$\nu \mathbf{k}\sigma$-dependent matrix element		2.93	1.55
	Constant matrix element		3.18	1.79
Independent bands model	d bands	$n = 3$, $N_b = 5$	2.91	1.58
	d and s bands	$n = 4$, $N_b = 6$	3.31	2.00
Localized states model	d states	$n = 3$, $N_b = 5$	2.55*	1.75
	d and s states	$n = 4$, $N_b = 6$	2.40*	2.18

*It is assumed that the values $s = s_{max}$ and $s = s_{max} - 1$ have different probabilities a and b that correspond to the experimental value $m_0 = 2.2\mu_B$, i.e. they are obtained from the system of equations $as_{max} + b(s_{max} - 1) = 1.1$ and $a + b = 1$

moment, we have $m_{eff} = 2\mu_B\sqrt{\langle \mathbf{s}^2 \rangle}$ (see, e.g. [16]). Using the value $m_{eff}^{exp} = 3.12\mu_B$, obtained from the experimental susceptibility, we find $\langle \mathbf{s}^2 \rangle = 2.4$. This is, of course, a rough estimate, but a better one is probably unavailable.

The calculation for the F and N-iron gives $\langle \mathbf{s}^2 \rangle \simeq 2.93$ and 1.55, respectively. As we explained in the previous section, these values are the upper and lower bounds of $\langle \mathbf{s}^2 \rangle$. The calculated values differ from the experimental estimates by about 30%. The values calculated by Hasegawa [17] using the two-field method $\langle \mathbf{s}^2 \rangle \simeq 3.6$ and four-field method $\langle \mathbf{s}^2 \rangle \simeq 1.12$ are too high and too low, respectively. The underestimated second value can be associated with the use of the approximate formula (4.20) in [17],[6] which underestimates the contribution to $\langle \mathbf{s}^2 \rangle$ at high energies. The overestimated first value of [17] is due to other approximations.

As for the lifetime t_0 of the spin fluctuations, the value obtained in the band calculation is somewhat larger than the lower estimate $\hbar/W \simeq 10^{-16}$ s but considerably smaller than the upper estimate of $\hbar/(k_B T_C) \simeq 10^{-14}$ s. In order that t_0 may increase without any substantial loss in $\langle \mathbf{s}^2 \rangle$, the $A(\omega)$ curve must increase for low ω and decrease for large ω, but in such a way that the area under the curve remains roughly the same. In the one-electron approximation, this can only happen if the wave functions of states close to the Fermi surface—or, equivalently, corresponding matrix elements (13.16)—differ substantially from those of other states.

The correlation effects on the function $A(\omega)$ can be estimated from the neutron scattering experiment [18]. Comparing (13.9) and (13.33), in the paramagnetic state, we have $\Phi(\omega) = s_L^2(\omega)$. In the experiment with the energy window up to $\simeq 0.1$ eV, the atomic magnetic moment $m_L = 2\mu_B s_L$ is about 1.3 μ_B. Therefore,

$$\Phi(\hbar\omega = 0.1) = s_L^2(\hbar\omega = 0.1) = 0.4.$$

It is clear from Fig. 13.1b that, in the absence of correlation, this value can be achieved in the N-iron only when integration is performed over an energy range $\simeq 2$ eV. This value is one order greater than the value 0.1 eV used in the experiment.

13.3.4 Interim Conclusions

The values of mean square single-site spin $\langle \mathbf{s}^2 \rangle$ obtained in the band calculations and model calculations differ from the experimental value $\langle \mathbf{s}^2 \rangle_{exp} \simeq 2.3$–2.4 by not more than 30%; the calculated results are lower than the experimental one in N-Fe and higher in F-Fe (Table 13.1).

Calculation of the spin correlator with constant matrix element gives a quite satisfactory agreement with the results of the calculation with $\nu \mathbf{k}\sigma$-dependent matrix element (see Figs. 13.1 and 13.2). During the transition from F to N-iron, the change of $\langle \mathbf{s}^2 \rangle$ is mainly due to disappearance of the long-range order term $(\bar{s}^z)^2\delta(\omega)$ in (13.21). If that term is excluded the values $\langle \mathbf{s}^2 \rangle$ in F and N-Fe are almost equal, being $\simeq 1.7$ and $\simeq 1.6$, respectively.

As long as the correlation is taken into account only at $T \neq 0$ and the ground state is well approximated by the Stoner state, the value $\langle \mathbf{s}^2 \rangle_F \simeq 2.9$ is the upper bound and $\langle \mathbf{s}^2 \rangle_N \simeq 1.6$ is the lower bound for $\langle \mathbf{s}^2 \rangle$ in iron at all temperatures (Table 13.1).

[6]A similar expression can be obtained from the formula given in the footnote 3 using the Kramers-Kronig relation for $\hbar\omega \ll k_B T$.

In the paramagnetic state, correlation should lead to a very sharp and narrow peak of $A(\omega)$ near $\omega = 0$ (of width $\simeq 0.1$ eV and with area under the peak close to 0.4). Explanation of that peak in the Hubbard [19, 20] or Hasegawa [17] models is unclear. We will demonstrate the appearance of such a peak in the following sections using the DSFT.

13.4 Magnetic Properties in the DSFT

13.4.1 Local Magnetic Characteristics

As we discussed in Chap. 4, magnetic properties of ferromagnetic metals at *zero* temperature are fairly successfully described [13, 21–25] in the framework of the local spin-density approximation (LSDA) to the density functional theory [26–29]. However, attempts to describe *temperature* dependence of magnetic properties within the LSDA do not lead to satisfactory results. Here we apply the DSFT to calculations of the *local* magnetic characteristics such as the transverse and longitudinal susceptibilities, dynamic spin correlation function and local magnetic moment of Fe, Co and Ni (for details, see [2, 30]).

In Sect. 13.3 we showed that calculations with constant matrix element (13.16) give a good approximation to the ones with $\nu\mathbf{k}\sigma$-dependent matrix element. Substituting $|M_{\nu\mathbf{k}\sigma}^{\nu'\mathbf{k}'\sigma'}|^2 = 1/N_\mathrm{d}$ in (13.18), where N_d is the number of (degenerate) d bands, we see that $\chi_{jj}^{0+-}(\omega)$ is equal to the unenhanced local susceptibility

$$\chi_\mathrm{L}^{0+-}(\omega) = \frac{1}{N^2}\sum_{\mathbf{q}} \chi_{\mathbf{q}}^{0+-}(\omega),$$

where $\chi_{\mathbf{q}}^{0+-}(\omega)$ is the transverse Lindhard function (5.23) times N_d.[7] Analogously, $\chi_{jj}^{0\sigma}(\omega)$ is equal to the unenhanced local susceptibility

$$\chi_\mathrm{L}^{0\sigma}(\omega) = \frac{1}{N^2}\sum_{\mathbf{q}} \chi_{\mathbf{q}}^{0\sigma}(\omega),$$

where $\chi_{\mathbf{q}}^{0\sigma}(\omega)$ is the function (5.12) times N_d. In this case, the imaginary parts of the unenhanced local susceptibilities can be written as

$$\mathrm{Im}\chi_\mathrm{L}^{0+-}(\omega) = N_\mathrm{d}\pi \int f(\varepsilon)[\nu_\uparrow(\varepsilon)\nu_\downarrow(\varepsilon+\hbar\omega) - \nu_\downarrow(\varepsilon)\nu_\uparrow(\varepsilon-\hbar\omega)]\,\mathrm{d}\varepsilon, \tag{13.34}$$

$$\mathrm{Im}\chi_\mathrm{L}^{0\sigma}(\omega) = N_\mathrm{d}\pi \int f(\varepsilon)\nu_\sigma(\varepsilon)[\nu_\sigma(\varepsilon+\hbar\omega) - \nu_\sigma(\varepsilon-\hbar\omega)]\,\mathrm{d}\varepsilon. \tag{13.35}$$

Here

$$\nu_\sigma(\varepsilon) = \frac{1}{\pi}\mathrm{Im}g_\sigma(\varepsilon)$$

is the spin-polarized DOS (per band, site and spin), where $g_\sigma(\varepsilon)$ is the mean single-site Green function (10.35). The latter is calculated self-consistently in the DSFT.

We obtain the enhanced susceptibilities at low energies $\hbar\omega$ using the RPA. Replacing the susceptibilities $\chi_{\mathbf{q}}^{0\alpha\beta}(\omega)$ by $\chi_\mathrm{L}^{0\alpha\beta}(\omega)$ in the RPA formulae (5.11) and (5.20), we come to

$$\chi_\mathrm{L}^{+-}(\omega) = \frac{\chi_\mathrm{L}^{0+-}(\omega)}{1 - u\chi_\mathrm{L}^{0+-}(\omega)}, \tag{13.36}$$

$$\chi_\mathrm{L}^{z}(\omega) = \frac{1}{4}\frac{\chi_\mathrm{L}^{0\uparrow}(\omega) + \chi_\mathrm{L}^{0\downarrow}(\omega) + 2u\chi_\mathrm{L}^{0\uparrow}(\omega)\chi_\mathrm{L}^{0\downarrow}(\omega)}{1 - u^2\chi_\mathrm{L}^{0\uparrow}(\omega)\chi_\mathrm{L}^{0\downarrow}(\omega)}. \tag{13.37}$$

In calculations, we take the same initial DOSs for Fe, Co and Ni as described in Sect. 10.5. The results of the calculation of the effective constant u for Fe, Co and Ni are given in Table 13.2. In the same table, for comparison, we present the values

[7] Use of the constant matrix element approximation is the same as setting the form-factor (3.28) to unity.

Table 13.2 Results for the effective interaction constant u of Fe, Co and Ni, compared to the corresponding results for the exchange-correlation (effective Stoner) parameter I obtained in LSDA (all values are in units of eV)

	Fe	Co	Ni
Reser [2]	1.08	1.25	1.16
Gunnarsson [21,22]	0.92	0.99	1.01
Andersen et al. [24]	0.91	–	0.99
Janak [25]	0.92	0.98	1.01
Ma and Dudarev [31]	1.0	–	–

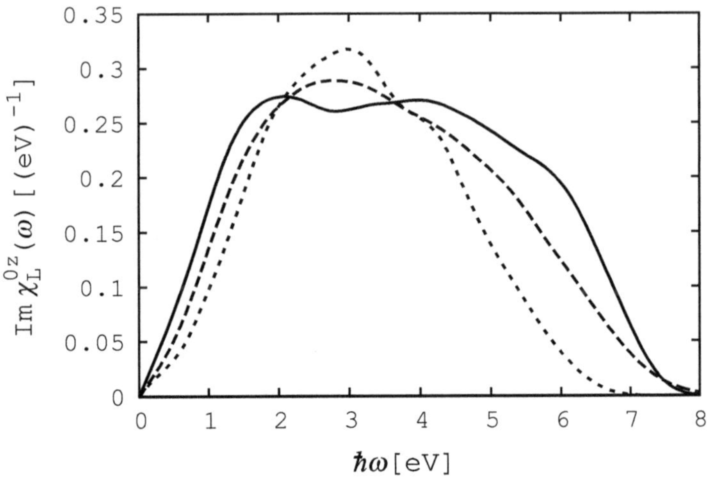

Fig. 13.3 The imaginary part of the *zeroth* longitudinal local susceptibility of iron at $T/T_C = 0.0$ (dotted line), 0.93 (dashed line), and 1.0 (solid line)

of the effective Stoner parameter I, obtained in the LSDA calculations [21, 22, 24, 25, 31]. The deviations of our values of u from the corresponding ones of I are not considerable.[8]

Let us illustrate the behaviour of the imaginary part of the local susceptibilities by the example of the zeroth and enhanced longitudinal susceptibilities of iron represented in Figs. 13.3 and 13.4. As can be seen from Fig. 13.3, the temperature dependence of $\mathrm{Im}\chi_L^{0z}(\omega) = \frac{1}{4}\sum_\sigma \mathrm{Im}\chi_L^{0\sigma}(\omega)$ is weak. (Since $\mathrm{Im}\chi_L^{0\sigma}(\omega)$ is an odd function, the calculations of $\mathrm{Im}\chi_L^{0z}(\omega)$ and $\mathrm{Im}\chi_L^{z}(\omega)$ are performed only for $\omega \geq 0$.) According to the formula

$$\mathrm{Im}\chi_L^{0\sigma}(\omega) \approx \pi N_d \nu_\sigma^2(\mu)\hbar\omega$$

resulting from (13.35) at low $\hbar\omega$ with neglect of the temperature broadening of the Fermi function, the linear dependence of $\mathrm{Im}\chi_L^{0z}(\omega)$ is retained in a sufficiently large energy interval and, in full accord with the behaviour of the DOS at the chemical-potential level (see Fig. 10.4), $\mathrm{Im}\chi_L^{0z}(\omega)$ increases with increasing temperature. From comparison of Fig. 13.4 with Fig. 13.3, it is seen that the enhancement of the susceptibility due to electron–electron correlations occurs mainly in the region $\hbar\omega < 2\,\mathrm{eV}$. As for the temperature variations, they are qualitatively similar to those of the zeroth susceptibility, but quantitatively they are considerable. Besides, with the increase of temperature the enhancement increases. This is due to the fact that, with the increase of temperature, the *real* part of the zeroth susceptibility in the energy region $\hbar\omega < 2\,\mathrm{eV}$ increases considerably. Due to the formula

$$\mathrm{Im}\chi_L^{z}(\omega) = \pi N_d \hbar\omega \frac{1}{4}\sum_\sigma \left(\frac{1 + u\mathrm{Re}\chi_L^{0\bar{\sigma}}(0)}{1 - u^2\mathrm{Re}\chi_L^{0\sigma}(0)\mathrm{Re}\chi_L^{0\bar{\sigma}}(0)}\right)^2 \nu_\sigma^2(\mu),$$

which comes from (13.37) at small $\hbar\omega$, the growth of $\mathrm{Re}\chi_L^{0\bar{\sigma}}(0)$ leads to the sharp increase of the enhancement factor.

[8]In the DSFT we use $g^2\mu_B^2/2$ units. If we come back to $g^2\mu_B^2$ units, one can see that

$$\chi = \frac{\chi^0}{1 - 2u\chi^0},$$

and the Stoner constant I must be compared with u (see (4.32)).

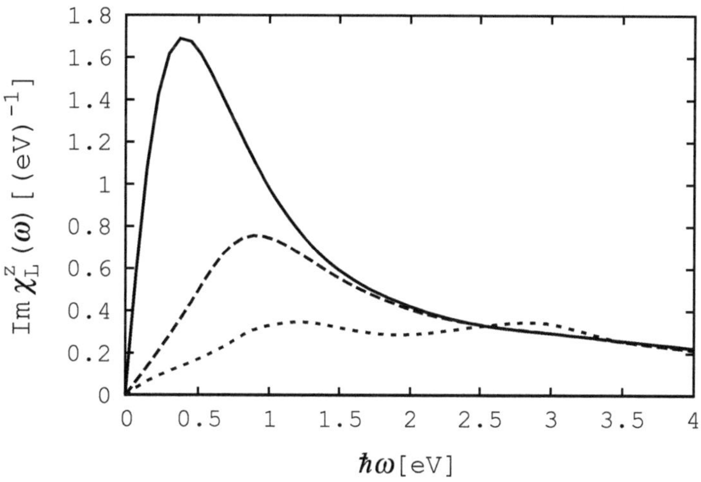

Fig. 13.4 The imaginary part of the *enhanced* longitudinal local susceptibility of iron (the notation is as for Fig. 13.3)

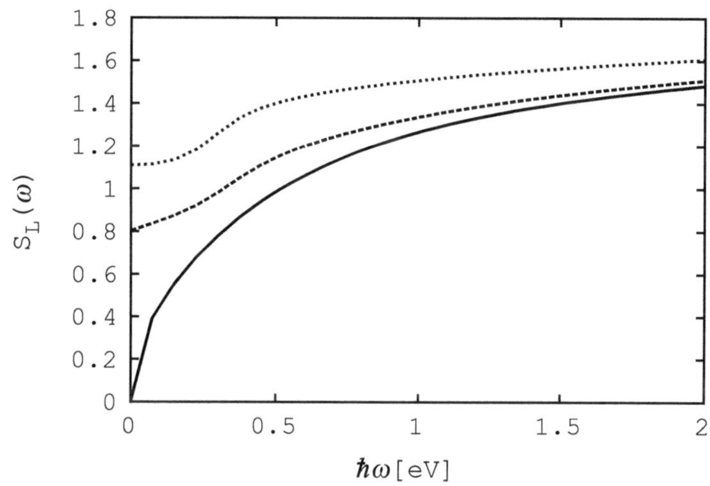

Fig. 13.5 The effective local spin $s_L(\omega)$ in a frequency interval $[-\omega, \omega]$ for iron (the notation is as for Fig. 13.3)

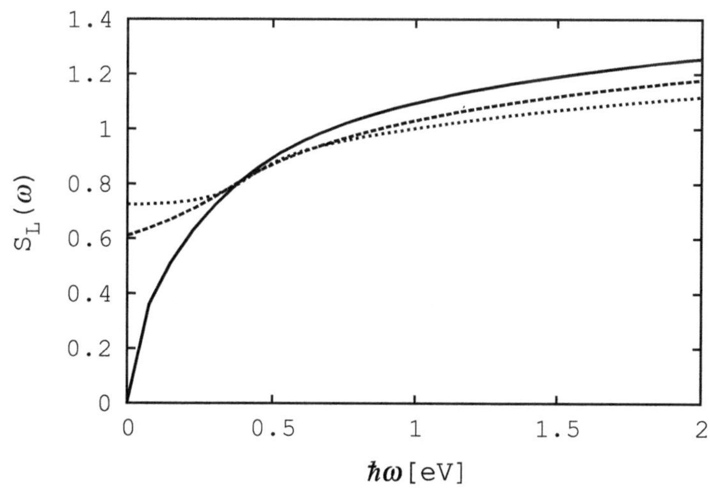

Fig. 13.6 The effective local spin $s_L(\omega)$ in a frequency interval $[-\omega, \omega]$ for cobalt at $T/T_C = 0.0$ (dotted line), 0.78 (dashed line) and 1.0 (solid line)

The results of the calculation of the effective local spin $s_L(\omega)$ are represented in Figs. 13.5, 13.6 and 13.7. As can be seen from the figures, the local spin in a wide energy interval varies only slightly with the increase of temperature. This is due to the fact that a decrease in the first term in (13.9) is compensated by an increase in the second one. However, in a small energy interval such a compensation does not occur. Here the local spin depends on the temperature strongly, and the smaller the interval, the sharper the dependence. For Fe at $T = T_C = 1.49T_C^{\mathrm{exp}}$, the computed values $1.0\,\mu_B$ and $1.3\,\mu_B$ of the local magnetic moment $m_L = g\mu_B s_L$ in the energy intervals $\hbar\omega = 0.12$ and 0.2 eV are fairly well agree

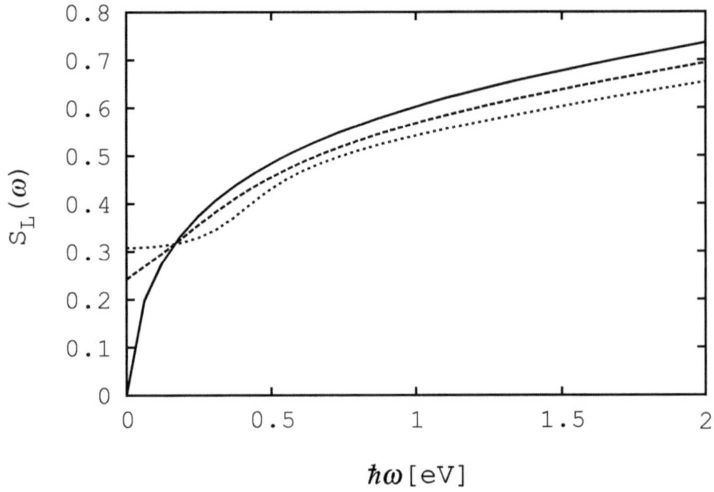

Fig. 13.7 The effective local spin $s_L(\omega)$ in a frequency interval $[-\omega, \omega]$ for nickel at $T/T_C = 0.0$ (dotted line), 0.88 (dashed line) and 1.0 (solid line)

Table 13.3 The local magnetic moment m_L for Fe, Co and Ni at zero, near critical and at the Curie temperatures (T_C is the Curie temperature calculated in the DNA and given in the last column of Table 10.1)

Fe		Co		Ni	
$\frac{T}{T_C}$	$\frac{m_L(T)}{m_L(0)}$	$\frac{T}{T_C}$	$\frac{m_L(T)}{m_L(0)}$	$\frac{T}{T_C}$	$\frac{m_L(T)}{m_L(0)}$
0.00	1.00	0.00	1.00	0.00	1.00
0.93	0.98	0.78	1.06	0.88	1.06
1.00	0.99	1.00	1.14	1.00	1.12

with the experimental values 1.3 μ_B and 1.55 μ_B, obtained at $T = 1.25T_C^{exp}$ in [18, 32]. Note that the value of the local magnetic moment in the energy window 0.1 eV in the one-electron calculations [33] was only 0.36 μ_B, which is much less than the experimental value. Good agreement with the experimental data on the polarized neutron scattering is also obtained for nickel (the experimental data for Co are absent from the literature): the values of $m_L(\hbar\omega = 0.12\text{eV}) = 0.55\,\mu_B$ and $m_L(\hbar\omega = 0.2\text{eV}) = 0.7\,\mu_B$, computed at $T = T_C = 1.54T_C^{exp}$, are in good agreement with the values 0.6 μ_B and 0.9 μ_B, respectively, measured at $T = 2T_C^{exp}$ [5].

The results of the calculation of the local magnetic moment in the *infinite* energy interval, $m_L(T)/m_L(0)$, of Fe, Co and Ni by formula (13.9) are given in Table 13.3. As can be seen from Table 13.3, the local moment in the infinite energy interval depends on temperature only slightly, being almost constant for Fe, and slightly increasing for Co and Ni. The fact that $\langle m_L^2 \rangle^{1/2}$ only slightly changes with temperature is supported by experiments on magnetovolume effect (see, e.g. [34–37] and references therein). Note that for Fe the weak change of the local magnetic moment with increasing temperature was obtained in other theoretical calculations as well. However, for the values of $m_L(T_C)/m_L(0)$ in Ni, none of the calculation results cited in [37] are close to unity.

The results of the calculation of the temporal correlation function $F(t)$ for Fe, Co and Ni are represented in Figs. 13.8, 13.9 and 13.10. As can be seen from Fig. 13.8, the square of the local spin $s_L^2 = F(t = 0)$ for iron at $T = 0$ equals 3.22, which almost coincides with the value 3.18, obtained in the one-electron calculation [1] using the constant-matrix-element approximation. At $T = T_C$, the square of the local spin of iron varies insignificantly and reaches the value 3.16, which is close to the value $s_L^2 \simeq 3.6$, obtained by Hasegawa [17] in the two-field method at $T = 1.1T_C^{exp}$. For Co and Ni, the computed values of $s_L^2(T_C)$ are equal to 2.20 and 0.81, respectively. The latter substantially differs from the value $s_L^2 \simeq 0.2$, obtained by Hasegawa [38] for Ni at $T = 1.1T_C^{exp}$. However, as can be seen from Table 13.4, our values for $s_L^2(T_C)$ are in better agreement with the experimental values, obtained from the relation $m_{eff} = g\mu_B\sqrt{s(s+1)}$ for the effective magnetic moment than those of Hasegawa [17, 38]. Clearly, estimating the square of the local spin in the paramagnetic region by the formula $s_L^2 = s(s+1) = m_{eff}^2/(g\mu_B)^2$ for ferromagnetic metals is fairly rough, but a better estimate is unavailable.

As for the temporal correlation function $F(t)$ at $t \neq 0$, its oscillating part $A(t)$ is large enough also beyond the main maximum, determined by the bandwidth. For example, in iron at $T = 0$, $A(t)$ calculated in the one-electron approximation vanishes during $t_0 \simeq 0.3$ fs, but $A(t)$ calculated with the account of spin fluctuations vanishes during 4 fs. With an increase of the temperature, the damping time of the local spin increases. This is due to the increase of the Fourier transform $A(\omega)$ in the low-frequency region. According to (13.6), the temperature dependence of the function $A(\omega)$ is determined by two factors: the hyperbolic cotangent and the sum of the imaginary parts of the susceptibilities. At the temperatures under consideration,

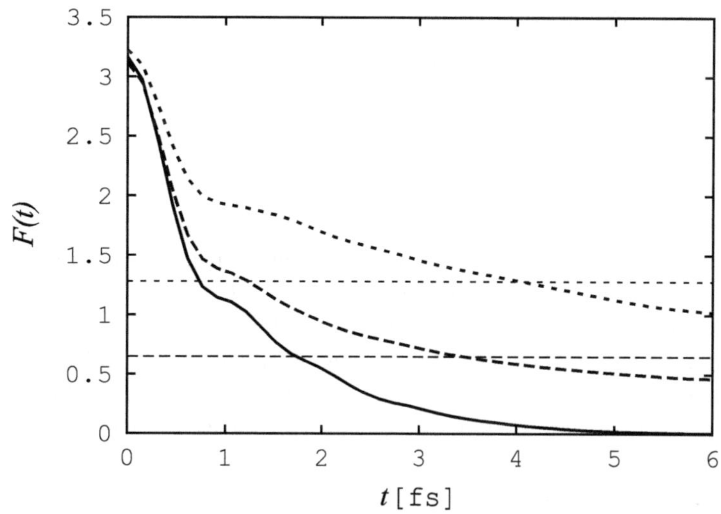

Fig. 13.8 Time-dependent spin correlation function $F(t)$ for iron. The horizontal line indicates the asymptotic value of $[(n_\uparrow - n_\downarrow)/2]^2$ (the notation is as for Fig. 13.3)

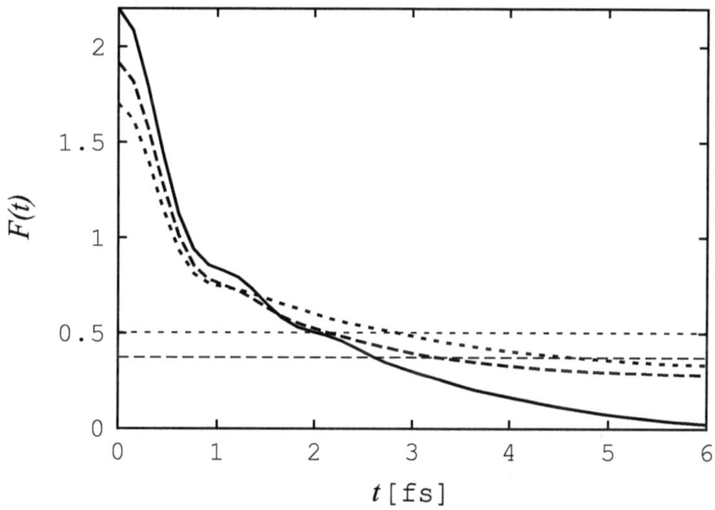

Fig. 13.9 Time-dependent spin correlation function $F(t)$ for cobalt. The horizontal line indicates the asymptotic value $[(n_\uparrow - n_\downarrow)/2]^2$ (the notation is as for Fig. 13.6)

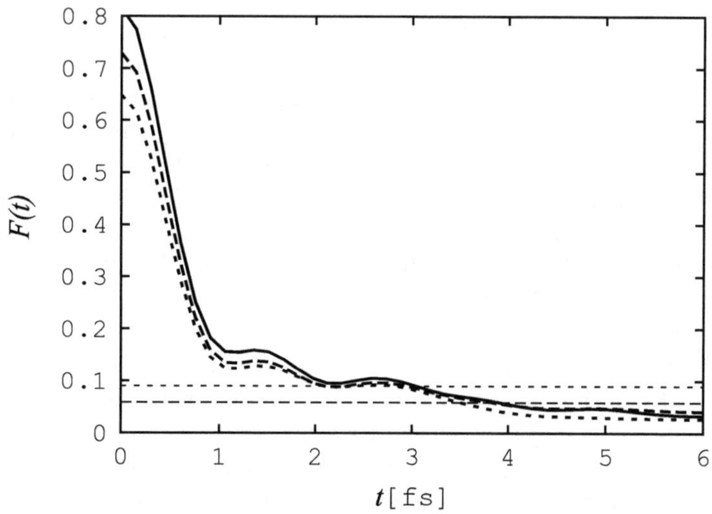

Fig. 13.10 Time-dependent spin correlation function $F(t)$ for nickel. The horizontal line indicates the asymptotic value $[(n_\uparrow - n_\downarrow)/2]^2$ (the notation is as for Fig. 13.7)

Table 13.4 Calculated and experimental values of the square of the local spin for Fe, Co and Ni at the Curie temperature

	Fe	Co	Ni
Hasegawa [17, 38]	3.6	–	0.2
Reser [2]	3.16	2.20	0.81
Experiment [39]	2.45	2.45	0.65

the hyperbolic cotangent differs from unity only in the low-energy region: $\hbar\omega < 2T \sim 0.1$ eV. Outside of this energy region, the temperature dependence of $A(\omega)$ is determined by the sum of the imaginary parts of the susceptibilities, whose behaviour is similar to $\mathrm{Im}\chi_L^z(\omega)$ represented in Fig. 13.4.

To conclude, the numerical investigation of local magnetic characteristics of Fe, Co and Ni at finite temperatures using the DSFT has shown that the correlation effects significantly change the local characteristics, especially at low energies. The values 1.0 and 1.3 μ_B of the local magnetic moment m_L in Fe at $T = T_C$ obtained in the energy windows 0.12 and 0.2 eV, respectively, are in good agreement with values 1.3 and 1.55 μ_B, obtained in polarized neutron scattering experiments at $T = 1.25T_C$ [18, 32]. Thus, the results of the calculations have predictive character for future experiments on neutron scattering for Fe, Co and Ni in a sufficiently large energy window.

The value of $m_L(T_C)/m_L(0)$ for Ni (as for Fe and Co) is near unity in full agreement with experimental data on the magnetovolume effect, whereas in previous treatments it was substantially smaller (see Table 1 of [37]).

The calculated damping time for the spin correlation function is one order greater than the one-electron hopping time, but is still less than the characteristic value, determined by the Curie temperature: $t \sim \hbar/T_C \sim 10^{-14}$ s.

The papers [40–42] present calculations of temperature dependence of the local magnetic moment in Fe-Ni Invar alloys, which are of great interest (for discussion see, e.g. [43–48] and references therein).

13.4.2 Nuclear Spin Relaxation Rates

It is well known that in the ferromagnetic metals the contact hyperfine interaction gives the main contribution to the nuclear spin relaxation (see, e.g. [49, 50]). While a theoretical explanation of the temperature dependence of the relaxation rates for simple and nonmagnetic transition metals is available [51–53], for ferromagnetic ones it is absent even at a qualitative level. For example, according to Moriya's estimates [54] (see also [55–57]), the largest contribution to T_1^{-1} for Fe, Co and Ni comes from the orbital (not contact) interaction. In our opinion, this result is a consequence of the fact that the spin fluctuations increasing sharply with increasing temperature were not taken into account in [54]. We believe that only correct account of the spin fluctuations permits one to explain the temperature behaviour of the relaxation rates in ferromagnetic metals properly. We emphasize that by ferromagnetic metals we always mean the *strongly* ferromagnetic metals Fe, Co and Ni. As for the *weakly* ferromagnetic metals, a quite satisfactory treatment of the temperature dependence of the relaxation rates based on the SFT is available (see [5] and references therein).

Note that an attempt to take into account the effect of the electron–electron interaction on the nuclear spin relaxation in metals was undertaken in [58], but within a very simplified model: the wavelength- and frequency-dependent magnetic susceptibilities were calculated within a model of a free-electron gas with interaction of a delta function type. As follows from the arguments in [58], the relaxation rate T_1^{-1} is enhanced by the electron–electron interaction. However, detailed comparison between the theory and the experimental data is absent from [58]. Moreover, at that time, experimental data were available only for multidomain Fe, Co and Ni [59, 60], since they were obtained in zero applied field. (The high-field rates which are two-to-three times less than the low-field ones [61] are intrinsic relaxation rates for ferromagnetic metals.)

Finally, calculations of the relaxation rates T_1^{-1} for most elements as impurities in ferromagnetic iron using the KKR-Green's function method [62] revealed that almost all of the calculated rates are smaller than the experimental ones. Observing a systematic tendency for theory to underestimate the experimental rates, including those of Fe*Fe*, the authors of [63–65] suggested that an important relaxation mechanism was missing in the one-electron theory [62]. The contribution of spin fluctuations is a reasonable candidate for providing the missing rate as we demonstrate below (for details, see [3, 66]).

The usual quantum-mechanical consideration of the contact hyperfine interaction (see, e.g. [49]) gives the following expressions for the longitudinal (T_1^{-1}) and transverse (T_2^{-1}) nuclear spin relaxation rates:

$$\frac{1}{T_1} = \frac{B}{2}\int_{-\infty}^{\infty} \langle\{\Delta s_j^+(t), \Delta s_j^-(0)\}\rangle\, \mathrm{e}^{\mathrm{i}\omega_0 t}\, \mathrm{d}t, \tag{13.38}$$

$$\frac{1}{T_2} = \frac{1}{2T_1} + \frac{B}{2}\int_{-\infty}^{\infty} \langle\{\Delta s_j^z(t), \Delta s_j^z(0)\}\rangle\, \mathrm{e}^{\mathrm{i}\omega_0 t}\, \mathrm{d}t, \tag{13.39}$$

where $\langle\{\Delta s_j^\alpha(t), \Delta s_j^\beta(0)\}\rangle$ is a single-site electron spin correlator and ω_0 is the nuclear magnetic resonance frequency. The expressions (13.38) and (13.39) are derived under the assumption that the hyperfine magnetic field at a nucleus is proportional to the total spin at a site, and that the constant B effectively depends on the magnitude of the nuclear spin (not necessarily equal to 1/2). Applying the fluctuation-dissipation theorem (2.55), we obtain [67, 68]

$$\frac{1}{T_1} = B\hbar \coth\frac{\hbar\omega_0}{2T}\mathrm{Im}\chi_{\mathrm{L}}^{+-}(\omega_0, T),$$

$$\frac{1}{T_2} = \frac{1}{2T_1} + B\hbar \coth\frac{\hbar\omega_0}{2T}\mathrm{Im}\chi_{\mathrm{L}}^{zz}(\omega_0, T),$$

where χ_{L}^{+-} and χ_{L}^{zz} are the transverse and longitudinal local susceptibilities expressed in units of $g^2\mu_B^2$, and T is the temperature in energy units. Since the energy $\hbar\omega_0$ is close to zero ($\hbar\omega_0 \sim 10^{-4}$–$10^{-5}$ eV), we use $\coth(\hbar\omega/2T) \simeq 2T/\hbar\omega$. Then

$$\frac{1}{T_1 T} = 2B\hbar\frac{\mathrm{Im}\chi_{\mathrm{L}}^{+-}(\omega_0, T)}{\hbar\omega_0}, \tag{13.40}$$

$$\frac{1}{T_2 T} = \frac{1}{2}\frac{1}{T_1 T} + 2B\hbar\frac{\mathrm{Im}\chi_{\mathrm{L}}^{zz}(\omega_0, T)}{\hbar\omega_0}. \tag{13.41}$$

As we see, at a fixed temperature, the relaxation rates are defined by the slopes of the imaginary parts of the local susceptibilities near zero energy.

At small energies, taking into account the expansion $\chi_{\mathrm{L}}^{0\alpha}(\varepsilon) = \chi_{\mathrm{L}}^{0\alpha}(0) + \mathrm{i}\varphi_{\mathrm{L}}^{\alpha}\varepsilon$ and keeping only linear terms in $\varepsilon = \hbar\omega$, from (13.36) and (13.37) we obtain [3]

$$\mathrm{Im}\chi_{\mathrm{L}}^{+-}(\varepsilon) = \frac{2\varphi_{\mathrm{L}}^{x}}{(1 - 2u\chi_{\mathrm{L}}^{0x}(0))^2}\varepsilon, \tag{13.42}$$

$$\mathrm{Im}\chi_{\mathrm{L}}^{zz}(\varepsilon) = \varepsilon\frac{1}{4}\sum_{\sigma}\left(\frac{1 + u\chi_{\mathrm{L}}^{0\bar\sigma}(0)}{1 - u^2\chi_{\mathrm{L}}^{0\sigma}(0)\chi_{\mathrm{L}}^{0\bar\sigma}(0)}\right)^2 \varphi_{\mathrm{L}}^{\sigma} \tag{13.43}$$

($\sigma = \uparrow, \downarrow$ or ± 1, and $\bar\sigma$ denotes $-\sigma$). Substituting (13.42) and (13.43) in (13.40) and (13.41), respectively, and expressing all susceptibilities in units of $\frac{1}{2}g^2\mu_B^2$, we finally obtain

$$\frac{1}{T_1 T} = c\frac{2\varphi_{\mathrm{L}}^{x}}{(1 - u\chi_{\mathrm{L}}^{0x}(0))^2}, \qquad c \equiv B\hbar, \tag{13.44}$$

$$\frac{1}{T_2 T} = \frac{1}{2}\frac{1}{T_1 T} + c\frac{1}{4}\sum_{\sigma}\left(\frac{1 + u\chi_{\mathrm{L}}^{0\bar\sigma}(0)/2}{1 - u^2\chi_{\mathrm{L}}^{0\sigma}(0)\chi_{\mathrm{L}}^{0\bar\sigma}(0)/4}\right)^2 \varphi_{\mathrm{L}}^{\sigma}. \tag{13.45}$$

In particular, for $u \to 0$, i.e. without the enhancement, we have

$$\frac{1}{T_1 T} = c2\varphi_{\mathrm{L}}^{x}, \qquad \frac{1}{T_2 T} = \frac{1}{2}\frac{1}{T_1 T} + c\varphi_{\mathrm{L}}^{z}, \tag{13.46}$$

where $\varphi_{\mathrm{L}}^{z} = \frac{1}{4}(\varphi_{\mathrm{L}}^{\uparrow} + \varphi_{\mathrm{L}}^{\downarrow})$.

At first sight, the final expressions for relaxation rates (13.44) and (13.45) appear to have an approximate character, because they are based on the linear approximation of the local susceptibilities. However, in numerical calculations the formulae (13.44) and (13.45) are preferable to the initial formulae (13.40) and (13.41), because the calculation by the formulae (13.40) and (13.41) reduces, in fact, to numerical differentiation, which is known to be an ill-posed problem. In addition, it is very difficult to carry out the self-consistent calculation of local susceptibilities with a small ε-step.

Here we study the temperature dependence (not numerical values) of the relaxation rates. Therefore, for simplicity, the constant c in the formulae (13.44)–(13.46) is set to unity.

We calculate the relaxation rates of Fe, Co and Ni with the zeroth and enhanced susceptibilities in the SLA and DNA of the DSFT. The calculation results are presented in Figs. 13.11, 13.12, 13.13, 13.14, 13.15 and 13.16. As can be seen from the figures, the relaxation rates have qualitatively similar behaviour for all three metals. The quantities $(T_1T)^{-1}$ and $(T_2T)^{-1}$ calculated without the enhancement of the susceptibility depend only slightly on temperature, which is consistent with the well-known Korringa formula [51] for the nuclear spin-relaxation rate for simple metals. If the enhancement is taken into account, the quantities $(T_1T)^{-1}$ and $(T_2T)^{-1}$ increase considerably and manifest significant temperature dependence in the ferromagnetic region. The increase near T_C is monotone and similar for both rates, just as in the experiment [69]. For Fe and Ni near T_C, the longitudinal and transverse relaxation rates completely coincide: $T_1^{-1} = T_2^{-1}$; for Co they are close.

As can be seen from Figs. 13.11, 13.13 and 13.15, in the SLA the quantities $(T_1T)^{-1}$ and $(T_2T)^{-1}$ depend on temperature too strongly. This is explained by the fact that the spin fluctuations in the SLA increase linearly with temperature (see (10.39)) and hence the magnetization decreases too rapidly over a wide temperature interval.

Let us now analyse the results obtained in the DNA (Figs. 13.12, 13.14 and 13.16) in more detail. We begin with the ferromagnetic region. In agreement with experiment [69], at low temperatures the temperature dependence of the quantities $(T_1T)^{-1}$ and $(T_2T)^{-1}$ is weak, it is more pronounced at room and higher temperatures and sharply increases in the critical region. On the whole, the temperature dependence of the quantities $(T_1T)^{-1}$ and $(T_2T)^{-1}$ is similar for all three metals. However, for Fe the curve $(T_1T)^{-1}$ is well above the $(T_2T)^{-1}$ curve; for Co these curves get close, and for Ni they change

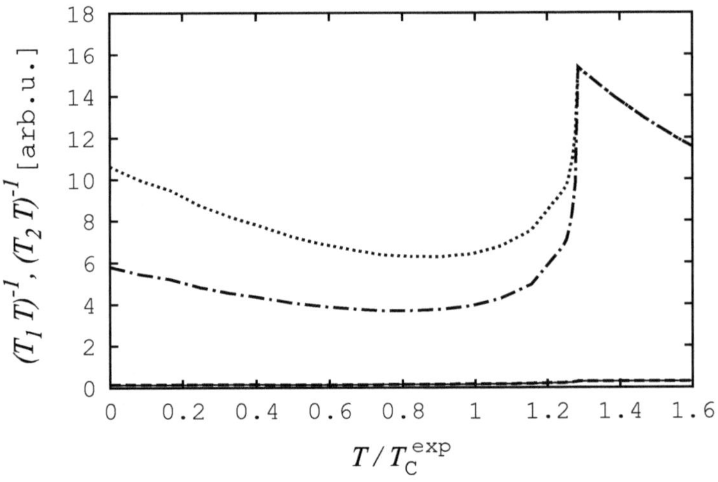

Fig. 13.11 The temperature dependence of the longitudinal (T_1^{-1}) and transverse (T_2^{-1}) nuclear spin relaxation rates (divided by T) of Fe, calculated in the SLA with the zero (solid line and dashed line) and enhanced (dotted line and dash dot dashed line) susceptibilities

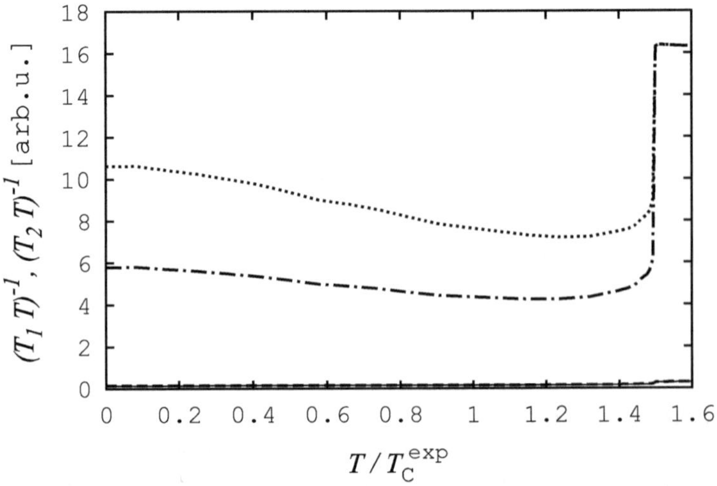

Fig. 13.12 The temperature dependence of the longitudinal (T_1^{-1}) and transverse (T_2^{-1}) nuclear spin relaxation rates (divided by T) of Fe, calculated in the DNA with the zero (solid line and dashed line) and enhanced (dotted line and dash dot dashed line) susceptibilities

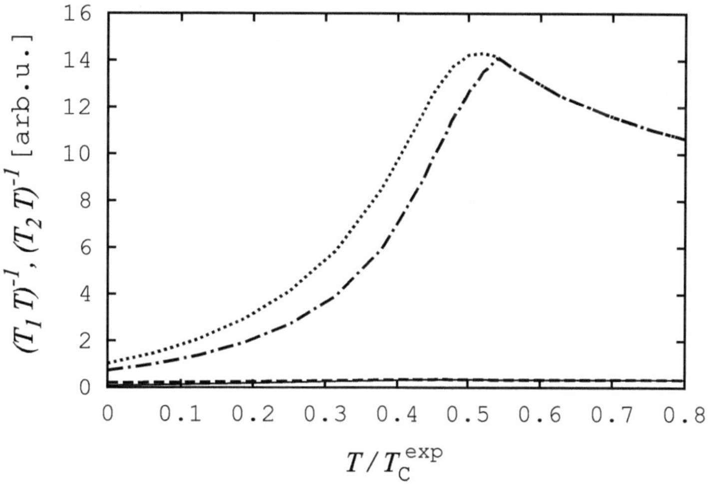

Fig. 13.13 The temperature dependence of the longitudinal (T_1^{-1}) and transverse (T_2^{-1}) nuclear spin relaxation rates (divided by T) of Co, calculated in the SLA with the zero (solid line and dashed line) and enhanced (dotted line and dash dot dashed line) susceptibilities

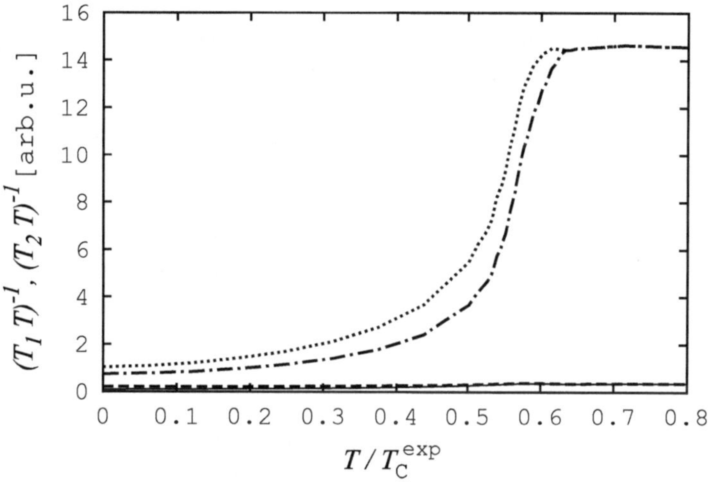

Fig. 13.14 The temperature dependence of the longitudinal (T_1^{-1}) and transverse (T_2^{-1}) nuclear spin relaxation rates (divided by T) of Co, calculated in the DNA with the zero (solid line and dashed line) and enhanced (dotted line and dash dot dashed line) susceptibilities

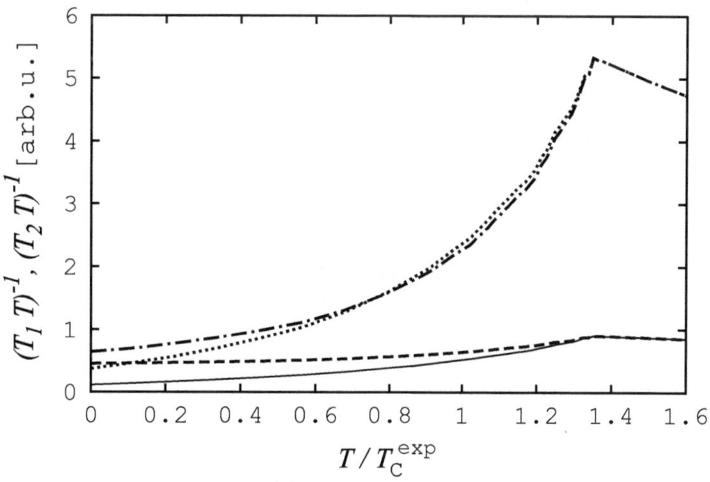

Fig. 13.15 The temperature dependence of the longitudinal (T_1^{-1}) and transverse (T_2^{-1}) nuclear spin relaxation rates (divided by T) of Ni, calculated in the SLA with the zero (solid line and dashed line) and enhanced (dotted line and dash dot dashed line) susceptibilities

places. This is connected with the fact that in Fe the transverse fluctuations dominate, while in Ni the longitudinal fluctuations dominate, and in Co the intermediate situation takes place: the transverse and longitudinal fluctuations are close.

The strong temperature dependence of the relaxation rates is principally due to resonance behaviour of the imaginary parts of the enhanced susceptibilities (13.42) and (13.43) at low energies. This behaviour was analysed in detail in [68] by the example of iron.

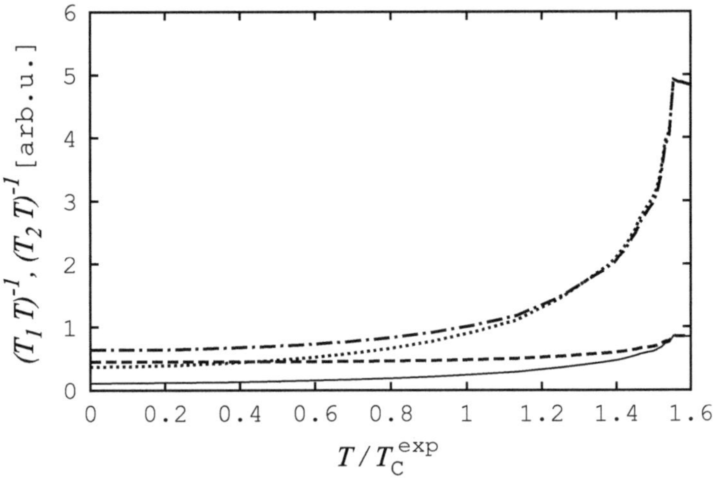

Fig. 13.16 The temperature dependence of the longitudinal (T_1^{-1}) and transverse (T_2^{-1}) nuclear spin relaxation rates (divided by T) of Ni, calculated in the DNA with the zero (solid line and dashed line) and enhanced (dotted line and dash dot dashed line) susceptibilities

It should be noted that near T_C the calculated relaxation rates, particularly for Co, do not increase as sharply as those in [69]. This is possibly due to the single-site approximation, which insufficiently takes into account the *spatial* spin-density correlations.[9]

In the paramagnetic region, the computed quantities $(T_1T)^{-1}$ and $(T_2T)^{-1}$ are equal to each other. They slowly decrease with the temperature increasing. The experimental data for Co and Ni [69, 70] confirm this temperature behaviour. (The data for paramagnetic iron are absent from the literature.) However, the fact that over a wide range of temperatures the computed quantities $(T_1T)^{-1}$ and $(T_2T)^{-1}$ remain on a level with the critical ones attracts our attention. This is connected with the fact that in the paramagnetic region the imaginary part of the local susceptibility $\mathrm{Im}\chi_L(\omega, T)$ decreases too slowly with temperature increasing. Such behaviour of $\mathrm{Im}\chi_L(\omega, T)$ is observed in all local models of the SFT (see, e.g. [5] and references therein).

Thus, the principal results here are as follows. The temperature behaviour of the nuclear spin relaxation rates in ferromagnetic metals is determined by the electron–electron correlations and these correlations can be adequately described within DSFT. The DNA gives a qualitative agreement with experimental data [69,70], over a wide range of temperatures.

Note that, in addition to relaxation due to the electron spin fluctuations, there are important nuclear relaxation mechanisms due to the crystal imperfections, e.g. the relaxation due to the magnetic impurities [71]. Moreover, the experimental data themselves differ in the ferromagnetic metals (see, e.g. the data for Ni in Fig. 5 of [70] and Fig. 8 of [69]).

References

1. B.I. Reser, E.V. Rosenfeld, E.V. Shipitsyn, Phys. Met. Metallogr. **69**(6), 48 (1990)
2. B.I. Reser, J. Phys. Condens. Matter **14**, 1285 (2002)
3. B.I. Reser, J. Phys. Condens. Matter **12**, 9323 (2000)
4. T. Moriya, Rep. Progr. Phys. **44**(4), 329 (1981)
5. T. Moriya, *Spin Fluctuations in Itinerant Electron Magnetism* (Springer, Berlin, 1985)
6. D.M. Edwards, J. Magn. Magn. Mater. **36**(3), 213 (1983)
7. J.F. Cooke, J.W. Lynn, H.L. Davis, Phys. Rev. B **21**, 4118 (1980)
8. W. Kohn, N. Rostoker, Phys. Rev. **94**(5), 1111 (1954)
9. J.C. Slater, Phys. Rev. **51**, 846 (1937)
10. B. Segall, Phys. Rev. **105**, 108 (1957)
11. O. Jepson, O.K. Andersen, Solid State Comm. **9**(20), 1763 (1971)
12. V.I. Krylov, N.S. Skoblya, *A Handbook of Methods of Approximate Fourier Transformation and Inversion of the Laplace Transformation* (Mir Publ., Moscow, 1977)
13. V.L. Moruzzi, J.F. Janak, A.R. Williams, *Calculated Electronic Properties of Metals* (Pergamon, New York, 1978)
14. B.I. Reser, Phys. Status Solidi B **116**, 31 (1983)
15. B.I. Reser, L.M. Sandratskii, Phys. Met. Metallogr. **62**(4), 66 (1986)
16. S. Chikazumi, *Physics of Ferromagnetism*, 2nd edn. (Clarendon, Oxford, 1997)
17. H. Hasegawa, J. Phys. F: Met. Phys. **13**, 2655 (1983)

[9]Compare the results obtained in the DNA with those obtained in the SLA, where the intersite correlations are not taken into account at all.

18. P.J. Brown, H. Capellmann, J. Déportes, D. Givord, K.R.A. Ziebeck, J. Magn. Magn. Mater. **30**, 243 (1982)
19. J. Hubbard, Phys. Rev. B **19**, 2626 (1979)
20. J. Hubbard, Phys. Rev. B **20**, 4584 (1979)
21. O. Gunnarsson, J. Phys. F: Met. Phys. **6**, 587 (1976)
22. O. Gunnarsson, Physica **91B**, 329 (1977)
23. U.K. Poulsen, J. Kollár, O.K. Andersen, J. Phys. F: Met. Phys. **6**, L241 (1976)
24. O.K. Andersen, J. Madsen, U.K. Poulsen, O. Jepsen, J. Kollár, Physica **86–88B**, 249 (1977)
25. J.F. Janak, Phys. Rev. B **16**, 255 (1977)
26. U. von Barth, L. Hedin, J. Phys. C Solid State Phys. **5**, 1629 (1972)
27. O. Gunnarsson, B.I. Lundqvist, Phys. Rev. B **13**, 4274 (1976)
28. W. Kohn, P. Vashishta, in *Theory of the Inhomogeneous Electron Gas*, ed. by S. Lundqvist, N.H. March (Plenum, New York, 1983), pp. 79–147
29. R.O. Jones, O. Gunnarsson, Rev. Mod. Phys. **61**, 689 (1989)
30. B.I. Reser, J. Magn. Magn. Mater. **258–259**, 51 (2003)
31. P.-W. Ma, S.L. Dudarev, Phys. Rev. B **86**, 054416 (2012)
32. P.J. Brown, H. Capellmann, J. Deportes, D. Givord, S.M. Johnson, J.W. Lynn, K.R.A. Ziebeck, J. Phys. (Paris) **46**, 827 (1985)
33. Y.I. Prokopjev, B.I. Reser, J. Phys. Condens. Matter **3**, 6055 (1991)
34. H. Capellmann (ed.), *Metallic Magnetism* (Springer, Berlin, 1987)
35. M. Shiga, in *Physics of Transition Metals 1980 (Inst. Phys. Conf. Ser. No. 55)*, ed. by P. Rhodes (IOP, Bristol, 1981), p. 241
36. M. Shiga, J. Phys. Soc. Jpn. **50**, 2573 (1981)
37. A.J. Holden, V. Heine, J.H. Samson, J. Phys. F: Met. Phys. **14**, 1005 (1984)
38. H. Hasegawa, J. Phys. F: Met. Phys. **14**, 1235 (1984)
39. E.P. Wohlfarth, in *Ferromagnetic Materials*, vol. 1, ed. by E.P. Wohlfarth (North-Holland, Amsterdam, 1980), pp. 1–70
40. B.I. Reser, J. Phys. Condens. Matter **16**, 361 (2004)
41. B.I. Reser, Phys. Met. Metallogr. **97**, 448 (2004)
42. B.I. Reser, Phys. Met. Metallogr. **103**, 373 (2007)
43. P. James, O. Eriksson, B. Johansson, I.A. Abrikosov, Phys. Rev. B **59**, 419 (1999)
44. P. Mohn, Nature **400**, 18 (1999)
45. M. van Schilfgaarde, I.A. Abrikosov, B. Johansson, Nature **400**, 46 (1999)
46. P. Weinberger, L. Szunyogh, C. Blaas, C. Sommers, P. Entel, Phys. Rev. B **63**, 094417 (2001)
47. P. Brown, K.U. Neumann, K.A. Ziebeck, J. Phys. Condens. Matter **13**(7), 1563 (2001)
48. J.P. Rueff, A. Shukla, A. Kaprolat, M. Krisch, M. Lorenzen, F. Sette, R. Verbeni, Phys. Rev. B **63**, 132409 (2001)
49. R.M. White, *Quantum Theory of Magnetism*, 3rd edn. (Springer, Berlin, 2007)
50. C.P. Slichter, *Principles of Magnetic Resonance*, 3rd edn. (Springer, Berlin, 1990)
51. J. Korringa, Physica **16**, 601 (1950)
52. Y. Obata, J. Phys. Soc. Jpn. **18**, 1020 (1963)
53. Y. Yafet, V. Jaccarino, Phys. Rev. **133**, A1630 (1964)
54. T. Moriya, J. Phys. Soc. Jpn. **19**, 681 (1964)
55. V.J. N. Kaplan, J.H. Wernick, Phys. Rev. Lett. **16**, 1142 (1966)
56. R.E. Walstedt, V. Jaccarino, N. Kaplan, J. Phys. Soc. Jpn. **21**, 1843 (1966)
57. M.B. Salamon, J. Phys. Soc. Jpn. **21**, 2746 (1966)
58. T. Moriya, J. Phys. Soc. Jpn. **18**, 516 (1964)
59. M. Weger, E.L. Hahn, A.M. Portis, J. Appl. Phys. **32**, 124S (1961)
60. M. Weger, Phys. Rev. **128**, 1505 (1962)
61. V. Jaccarino, N. Kaplan, R.E. Walstedt, J.H. Wernick, Phys. Lett. **23**, 514 (1966)
62. H. Akai, Hyperfine Interact. **43**, 255 (1988)
63. H. Akai, M. Akai, S. Blügel, B. Drittler, H. Ebert, K. Terakura, R. Zeller, P.H. Dederichs, Progr. Theor. Phys. Suppl. **101**, 11 (1990)
64. G. Seewald, E. Hagn, E. Zech, Phys. Rev. Lett. **78**, 5002 (1997)
65. T. Funk, E. Beck, W.D. Brewer, C. Bobek, E. Klein, J. Magn. Magn. Mater. **195**, 406 (1999)
66. B.I. Reser, Phys. Met. Metallogr. **92**(Suppl. 1), S123 (2001)
67. B.I. Reser, Y.I. Prokopjev, Phys. Met. Metallogr. **74**, 123 (1992)
68. B.I. Reser, Phys. Met. Metallogr. **77**, 451 (1994)
69. M. Shaham, J. Barak, U. El-Hanany, W.W. Warren Jr., Phys. Rev. B **22**, 5400 (1980)
70. P.J. Ségransan, Y. Chabre, W.G. Clark, J. Phys. F: Met. Phys. **8**, 1513 (1978)
71. S.V. Ivanov, M.I. Kurkin, in *Dynamic and Kinetic Properties of Magnets*, ed. by S.V. Vonsovskii, E.A. Turov (Nauka, Moscow, 1986), p. 223 [in Russian]

Neutron Scattering in Metals

14

On the pragmatic view the only thing that matters is that the theory is efficacious, that it "works" and that the necessary preliminaries and side issues do not cost too much in time and effort. (B.N. Brockhouse, Nobel Lecture, 1994)

Neutron scattering is one of the main methods to study magnetic properties of metals and alloys. Firstly, neutron is an uncharged particle with spin and, secondly, slow neutrons have energies (<1 eV) of the same order as most of magnetic excitations. So, inelastic neutron scattering allows to analyse fluctuations of the spin-density.

Results of neutron scattering experiments are expressed in terms of cross-sections (Sect. 14.1). The basic problem, with which this chapter is concerned, is to derive expressions for these quantities. An authoritative and comprehensive treatment of neutron scattering can be found in [1–5]; for a readable introduction, see [6].

In magnets with *localized* spins, an expression for the nuclear scattering cross-section was obtained in [7] and expression for the nonpolarized magnetic scattering cross-section was obtained in [8]. In the papers [7, 8] the magnetic scattering cross-section was related to the spin correlator. Calculating the spin correlator for the electron gas in the RPA, the paper [9] explained the possibility of magnetic neutron scattering in metals above the Curie temperature.

In this chapter we consider neutron scattering in magnets with *itinerant* electrons [10, 11] (see also [12]). In Sect. 14.2, we obtain an expression for the cross-section of neutron scattering by an arbitrary potential, which can describe the interaction with nuclei, orbital currents and spins. In Sect. 14.3, we study the correction factor in the scattering cross-section formula due to the lattice vibrations. The coefficient that corresponds to the elastic phonon scattering is called the Debye-Waller factor (DWF). We present a concise and simple method for calculating the DWF in the harmonic approximation and estimate the effect of lattice vibrations in ferromagnetic metals above the Curie temperature.

14.1 Scattering Cross-Section

A neutron scattering experiment is set up as follows. A collimated beam of monochromatic neutrons is incident on a scattering system (in our case, a crystal). As a result of the interaction, the neutrons are scattered with a certain probability in each direction. A quantitative characteristic that describes this process is the scattering cross-section (see, e.g. [6, 13–16]).

We assume that the incident neutrons are all in the same state with the wavevector $\mathbf{k}$. Then the (doubly) *differential scattering cross-section* is defined as

$$\frac{\mathrm{d}^2\sigma}{\mathrm{d}\Omega'\mathrm{d}E_{k'}} = \frac{1}{\Phi\,\mathrm{d}\Omega'\mathrm{d}E_{k'}}\begin{pmatrix}\text{number of neutrons scattered per unit time}\\ \text{into a solid angle } \mathrm{d}\Omega' \text{ with the final energy}\\ \text{between } E_{k'} \text{ and } E_{k'}+\mathrm{d}E_{k'}\end{pmatrix}, \tag{14.1}$$

where Φ is the flux density of the incident neutrons defined as the number of neutrons through unit area per unit time, the area being perpendicular to the direction of the neutron beam (here and hereafter the prime denotes a scattered quantity).

Let $w_{\mathbf{k}\to\mathbf{k}'}$ be the transition probability per unit time from the state $\mathbf{k}$ to the state $\mathbf{k}'$ and $N(E_{k'})$ is the number of states $\mathbf{k}'$ scattered into the solid angle $\mathrm{d}\Omega'$ with the energy less or equal to $E_{k'}$. Then $w_{\mathbf{k}\to\mathbf{k}'}\mathrm{d}N(E_{k'})$ is the numerator on the right-hand side of (14.1). Defining the neutron density of states as

$$n(E_{k'}) = \frac{\mathrm{d}N(E_{k'})}{\mathrm{d}E_{k'}}, \tag{14.2}$$

N. B. Melnikov, B. I. Reser, *Dynamic Spin Fluctuation Theory of Metallic Magnetism*,
https://doi.org/10.1007/978-3-319-92974-3_14

we write the scattering cross-section (14.1) in the form

$$\frac{\mathrm{d}^2\sigma}{\mathrm{d}\Omega'\mathrm{d}E_{k'}} = \frac{n(E_{k'})}{\Phi\,\mathrm{d}\Omega'}\,w_{\mathbf{k}\to\mathbf{k}'}. \tag{14.3}$$

To include the magnetic interaction it is necessary to specify not only the wavevector $\mathbf{k}$ of the neutron but also its spin state σ. First, we consider the cross-section for a process in which the crystal changes from the state λ to λ' and the neutron changes from the state $(\mathbf{k}, \sigma)$ to $(\mathbf{k}', \sigma')$,

$$\left(\frac{\mathrm{d}^2\sigma}{\mathrm{d}\Omega'\mathrm{d}E_{k'}}\right)_{\sigma\lambda\to\sigma'\lambda'} = \frac{n(E_{k'})}{\Phi\,\mathrm{d}\Omega'}\,w_{\mathbf{k}\sigma\lambda\to\mathbf{k}'\sigma'\lambda'}. \tag{14.4}$$

In the first-order perturbation theory, the quantity $w_{\mathbf{k}\sigma\lambda\to\mathbf{k}'\sigma'\lambda'}$ is given by *Fermi's golden rule* (see Appendix F.1)

$$w_{\mathbf{k}\sigma\lambda\to\mathbf{k}'\sigma'\lambda'} = \frac{2\pi}{\hbar}|\langle\mathbf{k}'\sigma'\lambda'|\,V\,|\mathbf{k}\sigma\lambda\rangle|^2\delta(E_\lambda - E_{\lambda'} + E_k - E_{k'}), \tag{14.5}$$

where V is the scattering potential, and E_λ and E_k are the energies of the crystal and neutron, respectively. Then (14.4) becomes

$$\left(\frac{\mathrm{d}^2\sigma}{\mathrm{d}\Omega'\mathrm{d}E_{k'}}\right)_{\sigma\lambda\to\sigma'\lambda'} = \frac{2\pi}{\hbar}\frac{n(E_{k'})}{\Phi\,\mathrm{d}\Omega'}|\langle\mathbf{k}'\sigma'\lambda'|\,V\,|\mathbf{k}\sigma\lambda\rangle|^2\delta(E_\lambda - E_{\lambda'} + E_k - E_{k'}). \tag{14.6}$$

The next step is to calculate the matrix element $\langle\mathbf{k}'|\,V\,|\mathbf{k}\rangle$, neutron density of states $n(E_{k'})$ and flux density Φ in formula (14.6). To do this we need to determine the neutron wave functions. In the *Born approximation* it is sufficient to use plane waves $|\mathbf{k}\rangle = \mathcal{V}^{-1/2}\mathrm{e}^{\mathrm{i}\mathbf{kr}}$ (see Appendix F.2), where $\mathcal{V}$ denotes the volume of the crystal to distinguish from the scattering potential V. Then the matrix element $\langle\mathbf{k}'|\,V\,|\mathbf{k}\rangle$ is given by

$$\langle\mathbf{k}'|\,V\,|\mathbf{k}\rangle = \frac{1}{\mathcal{V}}\int \mathrm{e}^{-\mathrm{i}\mathbf{k}'\mathbf{r}}\,V(\mathbf{r})\,\mathrm{e}^{\mathrm{i}\mathbf{kr}}\,\mathrm{d}\mathbf{r} = \frac{1}{\mathcal{V}}V_{-\boldsymbol{\kappa}}, \tag{14.7}$$

where

$$V_{\boldsymbol{\kappa}} = \int V(\mathbf{r})\,\mathrm{e}^{-\mathrm{i}\boldsymbol{\kappa}\mathbf{r}}\,\mathrm{d}\mathbf{r} \tag{14.8}$$

is the Fourier transform of the potential and $\boldsymbol{\kappa} = \mathbf{k} - \mathbf{k}'$ is the *scattering vector*.

For the neutron density of states (14.2) we have

$$n(E_{k'}) = \frac{\mathrm{d}N(E_{k'})}{\mathrm{d}E_{k'}} = \frac{\mathrm{d}N(k')}{\mathrm{d}k'}\frac{\mathrm{d}k'}{\mathrm{d}E_{k'}}. \tag{14.9}$$

Here $\mathrm{d}N(k')$ is the number of the wavevectors in $\mathrm{d}\Omega'$ with the magnitude between k' and $k' + \mathrm{d}k'$, which is the number of wavevector points in the element of volume $k'^2\,\mathrm{d}k'\mathrm{d}\Omega'$,

$$\mathrm{d}N(k') = \frac{\mathcal{V}k'^2\,\mathrm{d}k'\mathrm{d}\Omega'}{(2\pi)^3}. \tag{14.10}$$

Differentiating the final energy $E_{k'} = \hbar^2k'^2/(2m)$, where m is the mass of a neutron, we have

$$\frac{\mathrm{d}E_{k'}}{\mathrm{d}k'} = \frac{\hbar^2k'}{m}. \tag{14.11}$$

Substituting (14.10) and (14.11) in (14.9), we come to

$$n(E_{k'}) = \frac{\mathcal{V}k'm\mathrm{d}\Omega'}{\hbar^2(2\pi)^3}. \tag{14.12}$$

Finally, we obtain the flux density of the incident neutrons:

$$\boldsymbol{\Phi} = \frac{\mathrm{i}\hbar}{2m}\left(\psi_{\mathbf{k}}\nabla\psi_{\mathbf{k}}^{*} - \psi_{\mathbf{k}}^{*}\nabla\psi_{\mathbf{k}}\right) = \frac{\hbar\mathbf{k}}{\mathcal{V}m}, \tag{14.13}$$

where $\psi_{\mathbf{k}}$ is the wave function of a neutron (plane wave). Substituting (14.7), (14.12) and (14.13) into (14.6), we obtain the scattering cross-section in the Born approximation:

$$\left.\frac{\mathrm{d}^2\sigma}{\mathrm{d}\Omega'\mathrm{d}E_{k'}}\right|_{\sigma\lambda\to\sigma'\lambda'} = \frac{k'}{k}\left(\frac{m}{2\pi\hbar^2}\right)^2 |\langle\sigma'\lambda'|\, V_{-\kappa}\, |\sigma\lambda\rangle|^2\delta(E_\lambda - E_{\lambda'} + E_k - E_{k'}). \tag{14.14}$$

14.2 Scattering Potential Correlator

The cross-section (14.14) corresponds to the scattering from one specific state $(\mathbf{k}, \sigma, \lambda)$ to another one $(\mathbf{k}', \sigma', \lambda')$. Next we obtain the cross-section $\mathrm{d}^2\sigma/\mathrm{d}\Omega'\mathrm{d}E_{k'}$ defined in (14.1). By the law of total probability (for details, see Appendix A.3.2), we have

$$P_{\mathbf{k}\to\mathbf{k}'\sigma'\lambda'} = \sum_{\sigma\lambda} P_{\mathbf{k}\sigma\lambda\to\mathbf{k}'\sigma'\lambda'}P_{\sigma\lambda}, \tag{14.15}$$

where $P_{\sigma\lambda}$ is the probability of the initial state (σ, λ). Summing (14.15) over the final states (σ', λ'), we obtain the transition probability from $\mathbf{k}$ to $\mathbf{k}'$:

$$P_{\mathbf{k}\to\mathbf{k}'} = \sum_{\sigma'\lambda'} P_{\mathbf{k}\to\mathbf{k}'\sigma'\lambda'} = \sum_{\sigma'\lambda'}\sum_{\sigma\lambda} P_{\mathbf{k}\sigma\lambda\to\mathbf{k}'\sigma'\lambda'}P_{\sigma\lambda}.$$

Differentiation with respect to time leads to

$$w_{\mathbf{k}\to\mathbf{k}'} = \sum_{\sigma'\lambda'}\sum_{\sigma\lambda} w_{\mathbf{k}\sigma\lambda\to\mathbf{k}'\sigma'\lambda'}P_{\sigma\lambda}. \tag{14.16}$$

Substituting (14.16) in (14.3), we have

$$\frac{\mathrm{d}^2\sigma}{\mathrm{d}\Omega'\mathrm{d}E_{k'}} = \sum_{\sigma\lambda} P_{\sigma\lambda}\sum_{\sigma'\lambda'}\left.\frac{\mathrm{d}^2\sigma}{\mathrm{d}\Omega'\mathrm{d}E_{k'}}\right|_{\sigma\lambda\to\sigma'\lambda'}. \tag{14.17}$$

Finally, substituting (14.14) into (14.17), we obtain the general expression for the differential cross-section:

$$\frac{\mathrm{d}^2\sigma}{\mathrm{d}\Omega'\mathrm{d}E_{k'}} = \frac{k'}{k}\left(\frac{m}{2\pi\hbar^2}\right)^2 \sum_{\sigma\lambda} P_{\sigma\lambda}\sum_{\sigma'\lambda'} |\langle\sigma'\lambda'|\, V_{-\kappa}\, |\sigma\lambda\rangle|^2\delta(E_\lambda - E_{\lambda'} + E_k - E_{k'}). \tag{14.18}$$

We now calculate the average over initial states σ, λ and sum over the final states σ', λ' in formula (14.18) and thus relate the nonpolarized scattering cross-section to the scattering potential correlator. We assume that the potential $V(\mathbf{r})$ can be expressed as

$$V(\mathbf{r}) = W(\mathbf{r})\, U, \tag{14.19}$$

where $W(\mathbf{r})$ is an operator acting on the neutron $\mathbf{k}$ states and crystal λ states, and U is an operator acting on the neutron spin states σ. Using the expression for the Hermitian 2×2 matrix U (see Appendix A.1.4)

$$U = \sum_{\mu} U^{\mu}\sigma^{\mu}, \qquad U^{\mu} = \frac{1}{2}\mathrm{Sp}(U\sigma^{\mu}), \qquad \mu = 0, x, y, z,$$

where σ^0 is the identity 2×2 matrix and $\sigma^x, \sigma^y, \sigma^z$ are the Pauli matrices, we write (14.19) as

$$V(\mathbf{r}) = \sum_\mu V^\mu(\mathbf{r})\,\sigma^\mu, \tag{14.20}$$

where $V^\mu(\mathbf{r}) = U^\mu\, W(\mathbf{r})$. A potential of this form describes an arbitrary neutron-crystal interaction including the interaction of the neutron with nucleus and interactions of the neutron spin with electron spin and orbital current.

Independence of the σ and λ states leads to the following simplifications. First, the matrix element of the potential (14.20) is factored as

$$\langle\sigma'\lambda'|\, V_{-\kappa}\, |\sigma\lambda\rangle = \sum_\mu \langle\sigma'\lambda'|\, V^\mu_{-\kappa}\,\sigma^\mu\, |\sigma\lambda\rangle = \sum_\mu \langle\lambda'|\, V^\mu_{-\kappa}\, |\lambda\rangle\, \langle\sigma'|\,\sigma^\mu\, |\sigma\rangle\,.$$

Second, for the probability, we have $P_{\sigma\lambda} = P_\sigma P_\lambda$. As a result, formula (14.18) becomes

$$\frac{\mathrm{d}^2\sigma}{\mathrm{d}\Omega'\mathrm{d}E_{k'}} = \frac{k'}{k}\left(\frac{m}{2\pi\hbar^2}\right)^2 \sum_{\mu\nu}\sum_\lambda P_\lambda \sum_{\lambda'} \langle\lambda|\, V^\mu_\kappa\, |\lambda'\rangle\, \langle\lambda'|\, V^\nu_{-\kappa}\, |\lambda\rangle$$
$$\times\delta(E_\lambda - E_{\lambda'} + E_k - E_{k'}) \sum_\sigma P_\sigma \sum_{\sigma'} \langle\sigma|\,\sigma^\mu\, |\sigma'\rangle\, \langle\sigma'|\,\sigma^\nu\, |\sigma\rangle\,. \tag{14.21}$$

Next we calculate the average over the spin states σ and sum over the spin states σ' in expression (14.21). Using the formula for a matrix product element, we obtain

$$\sum_\sigma P_\sigma \sum_{\sigma'} \langle\sigma|\,\sigma^\mu\, |\sigma'\rangle\, \langle\sigma'|\,\sigma^\nu\, |\sigma\rangle = \sum_\sigma P_\sigma\, \langle\sigma|\,\sigma^\mu\sigma^\nu\, |\sigma\rangle\,.$$

For the *nonpolarized* neutron beam, from $P_\uparrow = P_\downarrow = 1/2$ we have

$$\sum_\sigma P_\sigma\, \langle\sigma|\,\sigma^\mu\sigma^\nu\, |\sigma\rangle = \delta_{\mu\nu}, \qquad \mu, \nu = 0, x, y, z.$$

Thus, (14.21) becomes

$$\frac{\mathrm{d}^2\sigma}{\mathrm{d}\Omega'\mathrm{d}E_{k'}} = \frac{k'}{k}\left(\frac{m}{2\pi\hbar^2}\right)^2 \sum_\mu\sum_\lambda P_\lambda \sum_{\lambda'} \langle\lambda|\, V^\mu_\kappa\, |\lambda'\rangle\, \langle\lambda'|\, V^\mu_{-\kappa}\, |\lambda\rangle\, \delta(E_\lambda - E_{\lambda'} + E_k - E_{k'}). \tag{14.22}$$

Finally, we calculate the sum over λ' and average over λ. Using the inverse Fourier transform of the delta-function (A.41), we write

$$\delta(E_\lambda - E_{\lambda'} + \hbar\omega) = \frac{1}{2\pi\hbar}\int \mathrm{e}^{\mathrm{i}(E_\lambda - E_{\lambda'})t/\hbar}\mathrm{e}^{\mathrm{i}\omega t}\,\mathrm{d}t,$$

where $\hbar\omega = E_k - E_{k'}$ is the energy transfer. Substituting the latter in (14.22), we have

$$\frac{\mathrm{d}^2\sigma}{\mathrm{d}\Omega'\mathrm{d}E_{k'}} = \frac{k'}{k}\left(\frac{m}{2\pi\hbar^2}\right)^2 \sum_\mu \frac{1}{2\pi\hbar}\int\sum_{\lambda\lambda'} P_\lambda\, \langle\lambda|\, V^\mu_\kappa\, \mathrm{e}^{-\mathrm{i}E_{\lambda'}t/\hbar}\, |\lambda'\rangle\, \langle\lambda'|\, V^\mu_{-\kappa}\, \mathrm{e}^{\mathrm{i}E_\lambda t/\hbar}\, |\lambda\rangle\, \mathrm{e}^{\mathrm{i}\omega t}\,\mathrm{d}t.$$

Choosing λ to be eigenstates of the crystal Hamiltonian,

$$\mathcal{H}\,|\lambda\rangle = E_\lambda\, |\lambda\rangle\,, \qquad \mathrm{e}^{-\mathrm{i}\mathcal{H}t/\hbar}\, |\lambda\rangle = \mathrm{e}^{-\mathrm{i}E_\lambda t/\hbar}\, |\lambda\rangle\,, \tag{14.23}$$

we obtain

$$\frac{\mathrm{d}^2\sigma}{\mathrm{d}\Omega'\mathrm{d}E_{k'}} = \frac{k'}{k}\left(\frac{m}{2\pi\hbar^2}\right)^2 \sum_{\mu} \frac{1}{2\pi\hbar} \int \sum_{\lambda} P_\lambda \,\langle\lambda|\, \mathrm{e}^{\mathrm{i}\mathcal{H}t/\hbar}\, V^{\mu}_{\kappa}\, \mathrm{e}^{-\mathrm{i}\mathcal{H}t/\hbar}\, V^{\mu}_{-\kappa}\, |\lambda\rangle\, \mathrm{e}^{\mathrm{i}\omega t}\, \mathrm{d}t. \tag{14.24}$$

Since $P_\lambda = Z^{-1}\mathrm{e}^{-E_\lambda/T}$, the sum over λ is nothing but the canonical average in the interacting electrons system:

$$\sum_{\lambda} P_\lambda \,\langle\lambda| \ldots |\lambda\rangle = Z^{-1}\mathrm{Tr}(\ldots \mathrm{e}^{-\mathcal{H}/T}) \equiv \langle \ldots \rangle.$$

Using the Heisenberg representation $V^{\mu}_{\kappa}(t) = \mathrm{e}^{\mathrm{i}\mathcal{H}t/\hbar}\, V^{\mu}_{\kappa}\, \mathrm{e}^{-\mathrm{i}\mathcal{H}t/\hbar}$, we write the scattering cross-section (14.24) as

$$\frac{\mathrm{d}^2\sigma}{\mathrm{d}\Omega'\mathrm{d}E_{k'}} = \frac{k'}{k}\left(\frac{m}{2\pi\hbar^2}\right)^2 \sum_{\mu} \frac{1}{2\pi\hbar} \int \langle V^{\mu}_{\kappa}(t)\, V^{\mu}_{-\kappa}\rangle\, \mathrm{e}^{\mathrm{i}\omega t}\, \mathrm{d}t, \tag{14.25}$$

where $\mu = 0, x, y, z$.

14.3 Neutron Scattering and Phonons

When we derived the scattering cross-section formula (14.25), we ignored the lattice vibrations (phonons). Taking lattice vibrations into account leads to a correction factor in the scattering cross-section formula. The correction factor that corresponds to the elastic phonon scattering is called the *Debye-Waller factor* (DWF).

In the harmonic approximation, an expression for the DWF was obtained separately for the nuclear scattering (see, e.g. [4, 6]) and magnetic scattering [10]. We present a concise, simple calculation of the DWF in the harmonic approximation for an arbitrary scattering process in an itinerant electron ferromagnet. The canonical average of exponentials of operators linear in the atomic displacements is calculated by the formula obtained in the paper [17]. This method allows avoiding the cumbersome calculation of the mean-square atomic displacement (see, e.g. [6, 10]). Using the Debye model, we then derive a computational formula that allows evaluating the Debye-Waller factor in real metals.

14.3.1 Lattice Vibrations

We now assume that each lattice site can be displaced by $\mathbf{u}_j$ from the equilibrium position $\mathbf{R}_j$. Replacing $\mathbf{R}_j$ by $\mathbf{R}_j + \mathbf{u}_j$, we write the Fourier transform of the potential as

$$V^{\mu}_{\kappa} = \sum_{j} V^{\mu}_{j}\, \mathrm{e}^{-\mathrm{i}\kappa(\mathbf{R}_j + \mathbf{u}_j)}, \qquad V^{\mu}_{j} = \int_{\Omega_j} V^{\mu}(\mathbf{r})\, \mathrm{d}\mathbf{r}, \tag{14.26}$$

where Ω_j is the Wigner-Seitz cell centred at $\mathbf{R}_j$. With (14.26) taken into account, the correlator of the potential becomes

$$\langle V^{\mu}_{\kappa}(t)\, V^{\mu}_{-\kappa}\rangle = \sum_{jj'} \mathrm{e}^{-\mathrm{i}\kappa(\mathbf{R}_j - \mathbf{R}_{j'})} \left\langle V^{\mu}_{j}(t)\, \mathrm{e}^{-\mathrm{i}\kappa \mathbf{u}_j(t)}\, V^{\mu}_{j'}\, \mathrm{e}^{\mathrm{i}\kappa \mathbf{u}_{j'}} \right\rangle. \tag{14.27}$$

Assuming that the lattice vibrations and interaction between the neutron and electron subsystem are independent, we factor the average as

$$\left\langle V^{\mu}_{j}(t)\, \mathrm{e}^{-\mathrm{i}\kappa \mathbf{u}_j(t)}\, V^{\mu}_{j'}\, \mathrm{e}^{\mathrm{i}\kappa \mathbf{u}_{j'}} \right\rangle = \langle V^{\mu}_{j}(t) V^{\mu}_{j'}\rangle \left\langle \mathrm{e}^{-\mathrm{i}\kappa \mathbf{u}_j(t)}\, \mathrm{e}^{\mathrm{i}\kappa \mathbf{u}_{j'}} \right\rangle. \tag{14.28}$$

For brevity, we introduce the notations $A = -\mathrm{i}\kappa \cdot \mathbf{u}_j(t)$ and $B = \mathrm{i}\kappa \cdot \mathbf{u}_{j'}$. Substituting (14.27) to formula (14.25) and taking (14.28) into account, we have

$$\frac{d^2\sigma}{d\Omega' dE_{k'}} = \frac{k'}{k}\left(\frac{m}{2\pi\hbar^2}\right)^2 \sum_{jj'} e^{-i\kappa(\mathbf{R}_j-\mathbf{R}_{j'})} \sum_{\mu} \frac{1}{2\pi\hbar} \int \langle V_j^{\mu}(t) V_{j'}^{\mu}\rangle \left\langle e^A e^B \right\rangle e^{i\omega t}\, dt. \tag{14.29}$$

Let us transform the expression $\langle e^A e^B\rangle$. The displacement operator at the lattice site $\hat{\mathbf{u}}_j(t)$ is given by formula (G.18):

$$\hat{\mathbf{u}}_j(t) = \left(\frac{\hbar}{2MN}\right)^{1/2} \sum_s \frac{\mathbf{e}_s}{\sqrt{\omega_s}} \left(b_s\, e^{i(\mathbf{q}\mathbf{R}_j-\omega_s t)} + b_s^{\dagger}\, e^{-i(\mathbf{q}\mathbf{R}_j-\omega_s t)}\right), \tag{14.30}$$

where M is the atomic mass, N is the number of crystal lattice sites, $b_s^{\dagger}$ and b_s are the phonon creation and annihilation operators, ω_s is the vibration frequency and $\mathbf{e}_s$ is the polarization vector of the normal mode $s = (\mathbf{q}, i)$ corresponding to the wavevector $\mathbf{q}$ and polarization $i = 1, 2, 3$. Taking (14.30) into account, we write the operators A and B as

$$A = -i\sum_s (\alpha_s b_s + \alpha_s^* b_s^{\dagger}), \qquad B = i\sum_s (\beta_s b_s + \beta_s^* b_s^{\dagger}), \tag{14.31}$$

where

$$\alpha_s = \left(\frac{\hbar}{2MN}\right)^{1/2} \frac{\kappa\cdot\mathbf{e}_s}{\sqrt{\omega_s}}\, e^{i(\mathbf{q}\mathbf{R}_j-\omega_s t)}, \qquad \beta_s = \left(\frac{\hbar}{2MN}\right)^{1/2} \frac{\kappa\cdot\mathbf{e}_s}{\sqrt{\omega_s}}\, e^{i\mathbf{q}\mathbf{R}_{j'}}. \tag{14.32}$$

Since the commutator $[A, B] = \sum_s \left(\alpha_s\beta_s^* - \alpha_s^*\beta_s\right)$ is proportional to the identity operator, the following relation holds (see, e.g. [6])

$$e^A e^B = e^{A+B} e^{\frac{1}{2}[A,B]}. \tag{14.33}$$

The displacement $\mathbf{u}_j$ has a Gaussian distribution (see Appendix G.2). Therefore, the quantities A, B and $A + B$ also have a Gaussian distribution. A Gaussian random variable Q satisfies the relation $\langle e^Q\rangle = e^{\frac{1}{2}\langle Q^2\rangle}$. Then the canonical average of expression (14.33) takes the form

$$\left\langle e^A e^B\right\rangle = e^{\frac{1}{2}\langle (A+B)^2\rangle} e^{\frac{1}{2}\langle [A,B]\rangle} = e^{\frac{1}{2}\langle A^2\rangle} e^{\frac{1}{2}\langle B^2\rangle} e^{\langle AB\rangle} = \left\langle e^A\right\rangle \left\langle e^B\right\rangle e^{\langle AB\rangle}. \tag{14.34}$$

Considering the contribution only from the *elastic* phonon scattering: $e^{\langle AB\rangle} = 1$ (for details, see [6]), we come to the relation

$$\left\langle e^A e^B\right\rangle = \left\langle e^A\right\rangle \left\langle e^B\right\rangle. \tag{14.35}$$

In the harmonic approximation, the canonical average $\left\langle e^Q\right\rangle$ with an arbitrary operator linear in b_s and $b_s^{\dagger}$: $Q = \sum_s (c_s b_s + d_s b_s^{\dagger})$, is given by the formula [17]

$$\left\langle e^Q\right\rangle = \exp\left[\frac{1}{2}\sum_s c_s d_s \coth\left(\frac{\hbar\omega_s}{2T}\right)\right]. \tag{14.36}$$

We apply the latter to calculate the averages $\left\langle e^A\right\rangle$ and $\left\langle e^B\right\rangle$ in expression (14.35). Taking (14.31) into account, we come to

$$\left\langle e^A e^B\right\rangle = \exp\left[-\frac{1}{2}\sum_s \left(|\alpha_s|^2 + |\beta_s|^2\right) \coth\left(\frac{\hbar\omega_s}{2T}\right)\right].$$

Substituting (14.32) for α_s and β_s, we have

$$\left\langle e^A e^B\right\rangle = e^{-2W(\kappa)}, \tag{14.37}$$

where

$$2W(\kappa) = \frac{\hbar}{2MN} \sum_s \frac{(\kappa\cdot\mathbf{e}_s)^2}{\omega_s} \coth\left(\frac{\hbar\omega_s}{2T}\right). \tag{14.38}$$

Finally, substituting (14.37) in (14.29) and going back to Fourier transform of the potential (14.26), we obtain the expression for the differential scattering cross-section

$$\frac{d^2\sigma}{d\Omega' dE_{k'}} = \frac{k'}{k}\left(\frac{m}{2\pi\hbar^2}\right)^2 e^{-2W(\kappa)} \sum_{\mu} \frac{1}{2\pi\hbar} \int \langle V_{\kappa}^{\mu}(t) V_{-\kappa}^{\mu} \rangle e^{i\omega t}\, dt. \tag{14.39}$$

The exponential $e^{-2W(\kappa)}$ is called the *Debye-Waller factor* (DWF). As can be seen from (14.38), the DWF depends on the scattering vector and temperature.

14.3.2 Debye-Waller Factor

To evaluate (14.38) we relate it to the phonon density of states and make use of a simple assumption about the phonon spectrum. First, we replace the multiplier $(\boldsymbol{\kappa} \cdot \mathbf{e}_s)^2$ by the average over the Brillouin zone ($s = (\mathbf{q}, i), i = 1, 2, 3$). Transforming the sum over $\mathbf{q}$ into an integral over the Brillouin zone,

$$\frac{1}{V}\sum_{\mathbf{q}} = \frac{1}{(2\pi)^3}\int d\mathbf{q},$$

we obtain

$$\frac{1}{3N}\sum_{\mathbf{q}i}(\boldsymbol{\kappa} \cdot \mathbf{e}_{\mathbf{q}i})^2 = \frac{1}{3}\frac{\Omega_{WS}}{(2\pi)^3}\sum_{i}\int (\boldsymbol{\kappa} \cdot \mathbf{e}_{\mathbf{q}i})^2\, d\mathbf{q}.$$

Replacing the Brillouin zone by the equal-volume sphere, we have

$$\frac{1}{3N}\sum_{\mathbf{q}i}(\boldsymbol{\kappa} \cdot \mathbf{e}_{\mathbf{q}i})^2 = \frac{1}{3}\kappa^2. \tag{14.40}$$

Substituting the average $\kappa^2/3$ for $(\boldsymbol{\kappa} \cdot \mathbf{e}_s)^2$ in (14.38), we obtain

$$2W(\kappa) = \frac{\hbar\kappa^2}{6MN}\sum_{s}\frac{1}{\omega_s}\coth\left(\frac{\hbar\omega_s}{2T}\right). \tag{14.41}$$

This result may be written in a compact way as (for details, see Appendix G.3)

$$2W(\kappa) = \frac{1}{3}\kappa^2\langle u^2\rangle.$$

Since the DWF in the cubic ferromagnets depends only on the frequencies of the normal modes and not on the polarization vectors, it can be expressed in terms of the phonon density of states $n(\omega)$. Transforming the sum over frequencies into the integral, we write (14.41) as

$$2W(\kappa) = \frac{\hbar\kappa^2}{6MN}\int\frac{1}{\omega}\coth\left(\frac{\hbar\omega}{2T}\right) n(\omega)\, d\omega, \tag{14.42}$$

where $\int n(\omega)\, d\omega = 3N$ is the number of the normal modes in the Brillouin zone. In the Debye model, the phonon density of states is given by $n(\omega) = 9N\omega^2/\omega_D^3$ for $0 \le \omega \le \omega_D$, where $\hbar\omega_D = \Theta_D$ is the Debye temperature in energy units. Using $\coth x \approx 1/x$ for $x \ll 1$, we have

$$2W(\kappa) = \frac{3\hbar^2\kappa^2}{M}\frac{T}{\Theta_D^2}, \qquad T \gg \Theta_D. \tag{14.43}$$

Table 14.1 Experimental Curie temperature T_C^{exp}, Debye temperature Θ_D and atomic mass M for ferromagnetic metals

	Fe	Co	Ni
T_C^{exp}, K [24]	1044	1390	631
Θ_D, K [23]	418	385	375
M, a.m.u. [25]	55.85	58.93	58.69

Table 14.2 The DWF $e^{-2W(\kappa)}$ for ferromagnetic metals calculated using formula (14.43) in the paramagnetic state at $\kappa = 3\,\text{Å}^{-1}$

T/T_C^{exp}	Fe	T/T_C^{exp}	Co	T/T_C^{exp}	Ni
1.2	0.845	1.05	0.803	1.25	0.882
1.3	0.833	1.1	0.795	1.5	0.860
1.4	0.822	1.15	0.787	2.0	0.818

In experiments studying magnetic properties of metals, the characteristic range of the scattering vector modulus is from 0.1 to $10\,\text{Å}^{-1}$ [18]. In this range, the DWF in metals gives a correction of up to 10% at $T = 0$ [19], and this correction substantially increases at finite temperatures. For instance, in Bragg neutron scattering, the DWF correction for Cu is about 2% at $T = 10$ K and is about 18% at $T = 1000$ K (see [6]). But theoretical estimates in most cases are available for only specific values of the scattering vector, and experimental measurements of the DWF are absent from the literature.

We calculate the DWF $e^{-2W(\kappa)}$ for ferromagnetic metals as a function of the scattering vector modulus and temperature [11]. The values of the Debye temperature Θ_D are determined as follows.[1] The lattice heat capacity is calculated as a function of temperature T, using the phonon density of states obtained from experimental data, and is fitted to the heat capacity in the Debye model (see, e.g. [21,22]). Then, assuming Θ_D an unknown parameter, one calculates the "experimental" Debye temperature $\Theta_D^{exp}(T)$ for each T. In solids the variation of $\Theta_D^{exp}(T)$ with temperature is up to about 10% [21]. In our calculations, Θ_D is the high temperatures value of $\Theta_D^{exp}(T)$ taken from [23]. The initial data are given in Table 14.1.

Table 14.2 shows the calculated values of the DWF in ferromagnetic metals above the Curie temperature. The DWF is calculated using formula (14.43) at $\kappa = 3\,\text{Å}^{-1}$, which corresponds to the maximal value of the scattering vector in the polarized magnetic scattering experiment [26,27]. The temperature values for Fe and Ni in the paramagnetic state correspond to the experimental ones in [26,27]. As can be seen from Table 14.2, taking lattice vibrations into account by means of the DWF, we obtain a correction of about 20% for all ferromagnetic metals. Our results are in good agreement with the correction of 18% for copper [6] obtained at the same value of the scattering vector at $T = 1000$ K.

In Fig. 14.1, we present the calculation results for the DWF (14.43) in ferromagnetic metals above T_C for a wider range of κ from zero to $4\,\text{Å}^{-1}$. This range corresponds to the polarized neutron scattering experiment described in the paper [28]. As can be seen from Fig. 14.1, the correction is up to 25–30% for all ferromagnetic metals.[2] At larger κ, it is necessary to take phonons into account in a more consistent way (see, e.g. [29]).

In the paper [11] we showed that the anharmonic contribution to the DWF can be appreciable above the Curie temperature at large values of the scattering vector, but the anharmonic contribution requires cumbersome calculations, and the results largely depend on the approximations used. The harmonic approximation gives satisfactory values of the DWF even outside of the Brillouin zone [11]. These results can be used for estimating the DWF in various neutron scattering experiments (e.g. Bragg scattering and magnetic form-factor measurements), and also in the X-ray and Mössbauer spectroscopy.

As can be seen from Fig. 14.1, the DWF gives an insignificant correction within the Brillouin zone. For example, for bcc Fe the correction is not more than 4%.[3] Therefore, the DWF can be ignored in calculation of the local magnetic moment using results of the neutron scattering experiment (see, e.g. [26]). For the same reason, the correction from lattice vibrations does not play an important role when calculating other magnetic characteristics that are determined by an integral over the Brillouin zone, such as magnetization, local susceptibility, single-site spin correlator and nuclear spin-relaxation rates. It is the thermal spin fluctuations that play a dominant role in describing magnetic characteristics in metals at finite temperatures.

[1] In the book [20] the calculation formula for $W(\kappa)$ is obtained from the Lindemann melting formula. This result can be used to estimate the DWF using the melting temperature instead of the Debye temperature. However, the Lindemann formula can only give a qualitative estimate (see [21]).

[2] Strictly speaking, calculating the DWF for Co is applicable only up to the melting temperature $T/T_C^{exp} \approx 1.27$.

[3] Indeed, substituting the atomic mass $M = 55.85$ a.m.u. and Debye temperature $\Theta_D = 464$ K of Fe, and $q_B = 1.71\,\text{Å}^{-1}$ into formula (14.43), at $\kappa = q_B$ and $T = T_C^{exp}$, we obtain $2W \approx 0.037$ and $e^{-2W} \approx 0.96$.

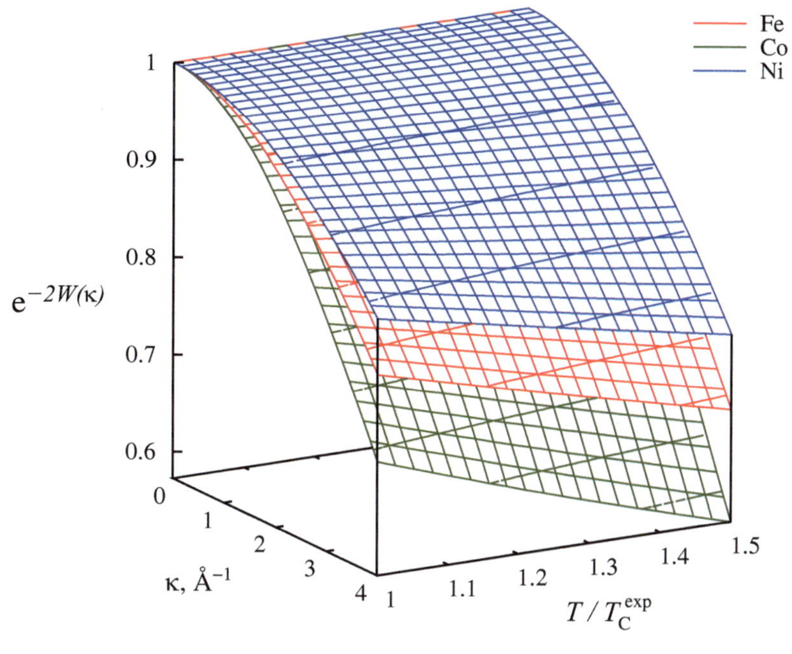

Fig. 14.1 The DWF $e^{-2W(\kappa)}$ for ferromagnetic metals calculated using formula (14.43)

References

1. G.E. Bacon, *Neutron Diffraction*, 2nd edn. (Oxford University Press, Oxford, 1962)
2. V.F. Turchin, *Slow Neutrons* (Atomizdat, Moscow, 1963) [in Russian]. Engl. transl.: Israel Program for Scientific Translations, Jerusalem, 1965
3. I.I. Gurevich, L. Tarasov, *Low-Energy Neutron Physics* (Elsevier, Amsterdam, 1968)
4. Y.A. Izyumov, R.P. Ozerov, *Magnetic Neutron Diffraction* (Plenum, New York, 1970)
5. S.W. Lovesey, *Theory of Neutron Scattering from Condensed Matter*, vols. 1 and 2 (Clarendon, Oxford, 1986)
6. G.L. Squires, *Introduction to the Theory of Thermal Neutron Scattering*, 2nd edn. (Dover, New York, 1996)
7. L. Van Hove, Phys. Rev. **95**, 249 (1954)
8. L. Van Hove, Phys. Rev. **95**, 1374 (1954)
9. T. Izuyama, D.J. Kim, R. Kubo, J. Phys. Soc. Jpn. **18**(7), 1025 (1963)
10. N.B. Melnikov, G.V. Paradezhenko, B.I. Reser, Theor. Math. Phys. **191**(1), 602 (2017)
11. G.V. Paradezhenko, N.B. Melnikov, B.I. Reser, Theor. Math. Phys. **195**(1), 572 (2018)
12. R.M. White, *Quantum Theory of Magnetism*, 3rd edn. (Springer, Berlin, 2007)
13. L. Schiff, *Quantum Mechanics*, 3rd edn. (McGraw-Hill, New York, 1968)
14. I.V. Savelyev, *Fundamentals of Theoretical Physics*, vol. 2. Quantum Mechanics (Mir Publ., Moscow, 1982)
15. C. Cohen-Tannoudji, B. Diu, F. Laloe, *Quantum Mechanics* (Wiley, New York, 1991)
16. E. Mertzbacher, *Quantum Mechanics*, 3rd edn. (Wiley, New York, 1997)
17. N.D. Mermin, J. Math. Phys. **7**, 1038 (1966)
18. C.G. Windsor, Physica B+C. **91**, 119 (1977)
19. C. Kittel, *Introduction to Solid State Physics*, 8th edn. (Wiley, New York, 2005)
20. J. Ziman, *Principles of the Theory of Solids* (Cambridge University Press, Cambridge, 1972)
21. J. Quinn, K. Yi, *Solid State Physics. Principles and Modern Applications* (Springer, Berlin, 2009)
22. A.A. Maradudin, E.W. Montroll, G.H. Weiss, *Theory of Lattice Dynamics in the Harmonic Approximation* (Academic, New York, 1963)
23. *Phonon States of Elements. Electron States and Fermi Surfaces of Alloys. Landolt-Börnstein New Series*, vol. III/13a (Springer, Berlin, 1981)
24. J. Crangle, G.M. Goodman, Proc. R. Soc. Lond. A **321**, 477 (1971)
25. W.B. Pearson, *Handbook of Lattice Spacings and Structures of Metals and Alloys* (Pergamon, New York, 1958)
26. P.J. Brown, H. Capellmann, J. Déportes, D. Givord, K.R.A. Ziebeck, J. Magn. Magn. Mater. **30**, 243 (1982)
27. P.J. Brown, H. Capellmann, J. Deportes, D. Givord, S.M. Johnson, K.R.A. Ziebeck, J. Phys. (Paris) **47**, 491 (1986)
28. P.J. Brown, H. Capellmann, J. Déportes, D. Givord, K.R.A. Ziebeck, J. Magn. Magn. Mater. **31–34**, 295 (1983)
29. D.J. Kim, *New Perspectives in Magnetism of Metals* (Kluwer/Plenum, New York, 1999)

Short-Range Order Above T_C 15

A mathematician may say anything he pleases, but a physicist must be at least partially sane. (J.W. Gibbs)

Inelastic neutron scattering measurements give the most direct experimental information on spin fluctuations (see, e.g. [1–4]). In particular, the *polarized* neutron scattering analysis is one of the most important methods of investigating the spin-density correlations and short-range order (SRO). Large SRO of about 20 Å was obtained in metals above the Curie temperature based on the spin-wave interpretation of the polarized neutron scattering experiments [5]. But this interpretation of the experiment is controversial [6, 7]. Static spin fluctuation theories all support the existence of SRO, but there is no agreement about the size of the SRO domain (see, e.g. [1]). The dynamic theories [8–10] are based on the local approximation and do not allow to estimate the SRO quantitatively. The dynamic spin fluctuation theory (DSFT) [11–13] takes into account both quantum nature and nonlocal character of the spin fluctuations. We use the DSFT to calculate the spin correlator and compare the results with experiment [14–17].

15.1 Magnetic Neutron Scattering

In magnets with *itinerant* electrons, magnetic neutron scattering has been considered in less detail than in magnets with *localized* spins.[1] We begin by deriving the formula for the *nonpolarized* magnetic scattering cross-section in the itinerant electron magnets following our papers [16, 19].

The scattering of *polarized* neutrons allows to separate the magnetic contribution from both instrumental background and nuclear scattering contributions. The integral of the magnetic scattering contribution over energies is explicitly measured in the experiment [20]. We derive the expression for the magnetic contribution to the neutron scattering cross-section following [16, 21] and relate it to the squared *effective moment* (the Fourier-transform of the spin-density correlator).

15.1.1 Magnetic Interaction Potential

We are interested in the *magnetic* scattering arising from the interaction between the magnetic moment of a neutron and magnetic moments of electrons in the crystal. In the ferromagnetic metals, the interaction between a neutron and orbital current gives a small contribution to the scattering cross-section (see, e.g. [22]). The interaction potential between an electron's spin magnetic moment $\boldsymbol{\mu}_i = -g\mu_B \mathbf{s}_i$ at a point $\mathbf{r}_i$ and neutron's magnetic moment $\boldsymbol{\mu}$ at a point $\mathbf{r}$ is given by

$$V_i(\mathbf{x}_i) = -\frac{e\hbar}{m_e c} \mathbf{s}_i \cdot \nabla_{\mathbf{x}_i} \times \mathbf{A}(\mathbf{x}_i). \tag{15.1}$$

Here m_e and $-e$ are the mass and charge of an electron, c is the velocity of light and $\mathbf{A} = \boldsymbol{\mu} \times \mathbf{x}_i / |\mathbf{x}_i|^3$ is the vector potential, where $\mathbf{x}_i = \mathbf{r}_i - \mathbf{r}$.

[1] For a brief introduction to neutron scattering in itinerant electron magnets, see [18].

N. B. Melnikov, B. I. Reser, *Dynamic Spin Fluctuation Theory of Metallic Magnetism*,
https://doi.org/10.1007/978-3-319-92974-3_15

We first calculate the Fourier transform of the potential $V_i(\mathbf{x}_i)$ by formula (14.8). The integration variable may be changed from $\mathbf{r}$ to $\mathbf{x}_i$, because $\mathbf{r}_i$ is constant. Taking (15.1) into account, we obtain

$$V_\kappa^i = -\frac{e\hbar}{m_e c}\, e^{-i\kappa\mathbf{r}_i}\mathbf{s}_i \cdot \int \nabla_{\mathbf{x}_i} \times \mathbf{A}(\mathbf{x}_i)\, e^{i\kappa\mathbf{x}_i}\, d\mathbf{x}_i .$$

The interaction potential $V(\mathbf{x})$ is the sum of potentials (15.1) over all electrons in the crystal: $V(\mathbf{x}) = \sum_i V_i(\mathbf{x}_i)$. Using the relation (see, e.g. [23, Appendix B])

$$\int \nabla_{\mathbf{x}} \times \mathbf{A}(\mathbf{x})\, e^{i\kappa\mathbf{x}}\, d\mathbf{x} = \frac{4\pi}{\kappa^2}\left(\kappa \times (\boldsymbol{\mu} \times \kappa)\right) = 4\pi\left(\boldsymbol{\mu} - (\hat{\kappa}\cdot\boldsymbol{\mu})\,\hat{\kappa}\right),$$

where $\hat{\kappa} = \kappa/\kappa$ is the unit vector, we have

$$V_\kappa = \sum_i V_\kappa^i = -\frac{4\pi e\hbar}{m_e c}\sum_i \mathbf{s}_i \cdot \left(\boldsymbol{\mu} - (\hat{\kappa}\cdot\boldsymbol{\mu})\,\hat{\kappa}\right) e^{-i\kappa\mathbf{r}_i}.$$

The multiplier $\sum_i \mathbf{s}_i e^{-i\kappa\mathbf{r}_i}$ in the scalar product is equal to the Fourier transform $\mathbf{s}_\kappa$ of the spin density $\mathbf{s}(\mathbf{r}) = \sum_i \mathbf{s}_i\delta(\mathbf{r}-\mathbf{r}_i)$. Therefore,

$$V_\kappa = -\frac{4\pi e\hbar}{m_e c}\,\mathbf{s}_\kappa \cdot \left(\boldsymbol{\mu} - (\hat{\kappa}\cdot\boldsymbol{\mu})\,\hat{\kappa}\right).$$

Rearranging, we obtain

$$V_\kappa = -\frac{4\pi e\hbar}{m_e c}\,\boldsymbol{\mu}\cdot\tilde{\mathbf{s}}_\kappa, \tag{15.2}$$

where

$$\tilde{\mathbf{s}}_\kappa = \mathbf{s}_\kappa - (\hat{\kappa}\cdot\mathbf{s}_\kappa)\hat{\kappa}$$

is the component of the vector $\mathbf{s}_\kappa$ perpendicular to the direction of κ. The neutron's magnetic moment $\boldsymbol{\mu}$ is related to its spin $\mathbf{S}$ by the formula $\boldsymbol{\mu} = -\gamma e\hbar/(mc)\mathbf{S}$, where $\gamma \approx 1.913$. Taking $\mathbf{S} = \boldsymbol{\sigma}/2$ into account, we have

$$V_\kappa^\alpha = \frac{2\pi\gamma e^2\hbar^2}{m_e m\, c^2}\tilde{s}_\kappa^\alpha . \tag{15.3}$$

15.1.2 Nonpolarized Magnetic Scattering

To obtain the scattering cross-section, we substitute (15.3) in (14.25) and recall that the term with V_κ^0 corresponds to the nuclear interaction, which we do not consider here. Then

$$\frac{d^2\sigma}{d\Omega' dE'} = \left(\frac{\gamma e^2}{m_e c^2}\right)^2 \frac{k'}{k}\sum_\alpha \frac{1}{2\pi\hbar}\int \langle \tilde{s}_\kappa^\alpha(t)\tilde{s}_{-\kappa}^\alpha\rangle\, e^{i\omega t}\, dt,$$

where $\alpha = x, y, z$. Passing from $\tilde{\mathbf{s}}_\kappa$ to $\mathbf{s}_\kappa$ by the formula

$$\tilde{\mathbf{s}}_\kappa \cdot \tilde{\mathbf{s}}_{-\kappa} = \sum_{\alpha\beta}(\delta_{\alpha\beta} - \hat{\kappa}^\alpha\hat{\kappa}^\beta)s_\kappa^\alpha s_{-\kappa}^\beta ,$$

we obtain

$$\frac{d^2\sigma}{d\Omega' dE'} = \left(\frac{\gamma e^2}{m_e c^2}\right)^2 \frac{k'}{k}\sum_{\alpha\beta}(\delta_{\alpha\beta} - \hat{\kappa}^\alpha\hat{\kappa}^\beta)\frac{1}{2\pi\hbar}\int \langle s_\kappa^\alpha(t)s_{-\kappa}^\beta\rangle\, e^{i\omega t}\, dt. \tag{15.4}$$

Introducing the *scattering function*

$$S^{\alpha\beta}(\kappa,\omega) = \frac{1}{2\pi\hbar N}\int \langle s^{\alpha}_{\kappa}(t)s^{\beta}_{-\kappa}\rangle\, e^{i\omega t}\, dt, \tag{15.5}$$

we write formula (15.4) as

$$\frac{d^2\sigma}{d\Omega' dE'} = N\left(\frac{\gamma e^2}{m_e c^2}\right)^2 \frac{k'}{k}\sum_{\alpha\beta}(\delta_{\alpha\beta} - \hat{\kappa}^{\alpha}\hat{\kappa}^{\beta})S^{\alpha\beta}(\kappa,\omega).$$

In the *paramagnetic* case, $S^{\alpha\beta}(\kappa,\omega) = S^z(\kappa,\omega)\delta_{\alpha\beta}$, denoting $S(\kappa,\omega) = 3S^z(\kappa,\omega)$ and taking $N = \mathcal{V}/\Omega_{WS}$ into account ($\mathcal{V}$ denotes the volume of the crystal), we finally obtain

$$\frac{d^2\sigma}{d\Omega' dE'} = 2\left(\frac{\gamma e^2}{m_e c^2}\right)^2 \frac{1}{3}\frac{\mathcal{V}}{\Omega_{WS}}\frac{k'}{k} S(\kappa,\omega). \tag{15.6}$$

In the experiment, a direct application of formula (15.6) is difficult, because it is necessary to separate the magnetic scattering from the nuclear scattering background.

15.1.3 Polarized Magnetic Scattering

In polarized neutron scattering, we consider the cross-section

$$\left.\frac{d^2\sigma}{d\Omega' dE_{k'}}\right|_{\sigma\to\sigma'} = \sum_{\lambda} P_{\lambda}\sum_{\lambda'}\left.\frac{d^2\sigma}{d\Omega' dE_{k'}}\right|_{\sigma\lambda\to\sigma'\lambda'},$$

where the state of the neutron beam changes from σ to σ'. Taking (14.14) into account, we obtain

$$\left.\frac{d^2\sigma}{d\Omega' dE_{k'}}\right|_{\sigma\to\sigma'} = \frac{k'}{k}\left(\frac{m}{2\pi\hbar^2}\right)^2 \sum_{\lambda} P_{\lambda}\sum_{\lambda'} |\langle\sigma'\lambda'|\, V_{-\kappa}\,|\sigma\lambda\rangle|^2 \delta(E_{\lambda} - E_{\lambda'} + E_k - E_{k'}).$$

Substituting expression (15.2) for the magnetic scattering potential $V_{-\kappa}$, we have

$$\left.\frac{d^2\sigma}{d\Omega' dE_{k'}}\right|_{\sigma\to\sigma'} = 4\left(\frac{\gamma e^2}{m_e c^2}\right)^2 \frac{k'}{k}\sum_{\lambda\lambda'} P_{\lambda}|\langle\sigma'\lambda'|\,\mathbf{S}\cdot\tilde{\mathbf{s}}_{-\kappa}\,|\sigma\lambda\rangle|^2 \delta(E_{\lambda} - E_{\lambda'} + E_k - E_{k'}). \tag{15.7}$$

We consider the uniaxial polarization analysis experiment, in which the spin of the incident neutron beam changes to the opposite one in the scattering process (see, e.g. [24]). The *polarized* neutron scattering experiment allows to eliminate both background and nuclear scattering contributions. To do this, one uses two polarization directions: along the scattering vector $\boldsymbol{\kappa}$ and perpendicular to it.

We choose the coordinate system such that the x-axis is oriented along the scattering vector $\boldsymbol{\kappa}$, i.e. $\hat{\boldsymbol{\kappa}} = \boldsymbol{\kappa}/\kappa = (1,0,0)$. First, we consider the beam that is polarized *along the vector* $\boldsymbol{\kappa}$ ($\mathbf{P} \parallel \boldsymbol{\kappa}$). Then all the incident neutrons are spin-up and scattered neutrons are spin-down *with respect to the x-axis*. The corresponding spin states have the form (see, e.g. [25,26])

$$|\sigma\rangle = |\uparrow_x\rangle = \frac{1}{\sqrt{2}}\begin{pmatrix}1\\1\end{pmatrix}, \qquad |\sigma'\rangle = |\downarrow_x\rangle = \frac{1}{\sqrt{2}}\begin{pmatrix}1\\-1\end{pmatrix}. \tag{15.8}$$

Making use of

$$\mathbf{S}\cdot\tilde{\mathbf{s}}_{-\boldsymbol{\kappa}} = \frac{1}{2}\sum_{\alpha}\sigma^{\alpha}\tilde{s}^{\alpha}_{-\boldsymbol{\kappa}},$$

where σ^{α} is the Pauli matrix, we obtain

$$\langle\downarrow_x|\,\mathbf{S}\cdot\tilde{\mathbf{s}}_{-\boldsymbol{\kappa}}\,|\uparrow_x\rangle = \frac{1}{2}\sum_{\alpha}\langle\downarrow_x|\,\sigma^{\alpha}\,|\uparrow_x\rangle\,\tilde{s}^{\alpha}_{-\boldsymbol{\kappa}} = \frac{1}{2}(\tilde{s}^{z}_{-\boldsymbol{\kappa}} - \mathrm{i}\tilde{s}^{y}_{-\boldsymbol{\kappa}}).$$

Hence formula (15.7) becomes

$$\left(\frac{\mathrm{d}^2\sigma^{\uparrow\downarrow}}{\mathrm{d}\Omega'\mathrm{d}E_{k'}}\right)_{\parallel}$$
$$= \left(\frac{\gamma e^2}{m_e c^2}\right)^2\frac{k'}{k}\sum_{\lambda\lambda'}P_{\lambda}\,\langle\lambda|\,\tilde{s}^{z}_{\boldsymbol{\kappa}} + \mathrm{i}\tilde{s}^{y}_{\boldsymbol{\kappa}}\,|\lambda'\rangle\,\langle\lambda'|\,\tilde{s}^{z}_{-\boldsymbol{\kappa}} - \mathrm{i}\tilde{s}^{y}_{-\boldsymbol{\kappa}}\,|\lambda\rangle\,\delta(E_{\lambda} - E_{\lambda'} + E - E').$$

Calculating the average over λ and sum over λ' as in Sect. 14.2, we have

$$\left(\frac{\mathrm{d}^2\sigma^{\uparrow\downarrow}}{\mathrm{d}\Omega'\mathrm{d}E_{k'}}\right)_{\parallel} = \left(\frac{\gamma e^2}{m_e c^2}\right)^2\frac{k'}{k}\frac{1}{2\pi\hbar}\int\big[\langle\tilde{s}^{z}_{\boldsymbol{\kappa}}(t)\tilde{s}^{z}_{-\boldsymbol{\kappa}}\rangle + \langle\tilde{s}^{y}_{\boldsymbol{\kappa}}(t)\tilde{s}^{y}_{-\boldsymbol{\kappa}}\rangle$$
$$-\mathrm{i}(\langle\tilde{s}^{z}_{\boldsymbol{\kappa}}(t)\tilde{s}^{y}_{-\boldsymbol{\kappa}}\rangle - \langle\tilde{s}^{y}_{\boldsymbol{\kappa}}(t)\tilde{s}^{z}_{-\boldsymbol{\kappa}}\rangle)\big]\,\mathrm{e}^{\mathrm{i}\omega t}\,\mathrm{d}t. \tag{15.9}$$

Now we consider the beam that is polarized *along the z-axis*, i.e. perpendicular to $\boldsymbol{\kappa}$ ($\mathbf{P}\perp\boldsymbol{\kappa}$). In this case the incident and scattered spin states are

$$|\sigma\rangle = |\uparrow_z\rangle = \begin{pmatrix}1\\0\end{pmatrix}, \qquad |\sigma'\rangle = |\downarrow_z\rangle = \begin{pmatrix}0\\1\end{pmatrix}. \tag{15.10}$$

Hence,

$$\langle\downarrow_z|\,\mathbf{S}\cdot\tilde{\mathbf{s}}_{-\boldsymbol{\kappa}}\,|\uparrow_z\rangle = \frac{1}{2}\sum_{\alpha}\langle\downarrow_z|\,\sigma^{\alpha}\,|\uparrow_z\rangle\,\tilde{s}^{\alpha}_{-\boldsymbol{\kappa}} = \frac{1}{2}(\tilde{s}^{x}_{-\boldsymbol{\kappa}} + \mathrm{i}\tilde{s}^{y}_{-\boldsymbol{\kappa}}).$$

Calculating the average over λ and sum over λ', we have

$$\left(\frac{\mathrm{d}^2\sigma^{\uparrow\downarrow}}{\mathrm{d}\Omega'\mathrm{d}E_{k'}}\right)_{\perp} = \left(\frac{\gamma e^2}{m_e c^2}\right)^2\frac{k'}{k}\frac{1}{2\pi\hbar}\int\big[\langle\tilde{s}^{x}_{\boldsymbol{\kappa}}(t)\tilde{s}^{x}_{-\boldsymbol{\kappa}}\rangle + \langle\tilde{s}^{y}_{\boldsymbol{\kappa}}(t)\tilde{s}^{y}_{-\boldsymbol{\kappa}}\rangle$$
$$+\mathrm{i}(\langle\tilde{s}^{x}_{\boldsymbol{\kappa}}(t)\tilde{s}^{y}_{-\boldsymbol{\kappa}}\rangle - \langle\tilde{s}^{y}_{\boldsymbol{\kappa}}(t)\tilde{s}^{x}_{-\boldsymbol{\kappa}}\rangle)\big]\,\mathrm{e}^{\mathrm{i}\omega t}\,\mathrm{d}t. \tag{15.11}$$

In the chosen coordinate system, where $\hat{\boldsymbol{\kappa}} = (1, 0, 0)$, we have $\tilde{\mathbf{s}}_{\boldsymbol{\kappa}} = (0, s^{y}_{\boldsymbol{\kappa}}, s^{z}_{\boldsymbol{\kappa}})$. Then the cross-section (15.9) takes the form

$$\left(\frac{\mathrm{d}^2\sigma^{\uparrow\downarrow}}{\mathrm{d}\Omega'\mathrm{d}E_{k'}}\right)_{\parallel} = \left(\frac{\gamma e^2}{m_e c^2}\right)^2\frac{k'}{k}\frac{1}{2\pi\hbar}\int\big[\langle s^{z}_{\boldsymbol{\kappa}}(t)s^{z}_{-\boldsymbol{\kappa}}\rangle + \langle s^{y}_{\boldsymbol{\kappa}}(t)s^{y}_{-\boldsymbol{\kappa}}\rangle$$
$$-\mathrm{i}(\langle s^{z}_{\boldsymbol{\kappa}}(t)s^{y}_{-\boldsymbol{\kappa}}\rangle - \langle s^{y}_{\boldsymbol{\kappa}}(t)s^{z}_{-\boldsymbol{\kappa}}\rangle)\big]\,\mathrm{e}^{\mathrm{i}\omega t}\,\mathrm{d}t \tag{15.12}$$

and the cross-section (15.11) takes the form

$$\left(\frac{\mathrm{d}^2\sigma^{\uparrow\downarrow}}{\mathrm{d}\Omega'\mathrm{d}E_{k'}}\right)_{\perp} = \left(\frac{\gamma e^2}{m_e c^2}\right)^2\frac{k'}{k}\frac{1}{2\pi\hbar}\int\langle s^{y}_{\boldsymbol{\kappa}}(t)s^{y}_{-\boldsymbol{\kappa}}\rangle\mathrm{e}^{\mathrm{i}\omega t}\,\mathrm{d}t. \tag{15.13}$$

In the paramagnetic region, for the isotropic crystal we have $\langle s^{\alpha}_{\boldsymbol{\kappa}}(t)s^{\beta}_{-\boldsymbol{\kappa}}\rangle = 0$, where $\alpha\neq\beta$ (see Sect. 2.1.2). As a result, the cross-section (15.12) is written as

$$\left(\frac{\mathrm{d}^2\sigma^{\uparrow\downarrow}}{\mathrm{d}\Omega'\mathrm{d}E_{k'}}\right)_{\parallel} = \left(\frac{\gamma e^2}{m_e c^2}\right)^2 \frac{k'}{k}\frac{1}{2\pi\hbar}\int \left[\langle s^y_{\kappa}(t)s^y_{-\kappa}\rangle + \langle s^z_{\kappa}(t)s^z_{-\kappa}\rangle\right] \mathrm{e}^{\mathrm{i}\omega t}\,\mathrm{d}t, \tag{15.14}$$

in full accord with the fact that the third term in expression (15.12) is responsible for the chiral magnetic scattering (see, e.g. [2]), which is absent in the isotropic crystal.

We now consider the difference between the scattering cross-sections $\mathbf{P} \parallel \kappa$ and $\mathbf{P} \perp \kappa$. Since nuclear scattering does not depend on polarization, we are left only with magnetic contributions (15.14) and (15.13):

$$\left(\frac{\mathrm{d}^2\sigma^{\uparrow\downarrow}}{\mathrm{d}\Omega'\mathrm{d}E_{k'}}\right)_{\parallel} - \left(\frac{\mathrm{d}^2\sigma^{\uparrow\downarrow}}{\mathrm{d}\Omega'\mathrm{d}E_{k'}}\right)_{\perp} = \left(\frac{\gamma e^2}{m_e c^2}\right)^2 \frac{k'}{k}\frac{1}{2\pi\hbar}\int \langle s^z_{\kappa}(t)s^z_{-\kappa}\rangle\,\mathrm{e}^{\mathrm{i}\omega t}\,\mathrm{d}t.$$

Transforming to the scattering function (15.5), and taking $S(\kappa,\omega) = 3S^z(\kappa,\omega)$ and $N = \mathcal{V}/\Omega_{\mathrm{WS}}$ into account, we obtain

$$\left(\frac{\mathrm{d}^2\sigma^{\uparrow\downarrow}}{\mathrm{d}\Omega'\mathrm{d}E_{k'}}\right)_{\parallel} - \left(\frac{\mathrm{d}^2\sigma^{\uparrow\downarrow}}{\mathrm{d}\Omega'\mathrm{d}E_{k'}}\right)_{\perp} = \left(\frac{\gamma e^2}{m_e c^2}\right)^2 \frac{1}{3}\frac{\mathcal{V}}{\Omega_{\mathrm{WS}}}\frac{k'}{k} S(\kappa,\omega). \tag{15.15}$$

In the neutron scattering experiment [5,27–29] the measured quantity is the paramagnetic spin-flip scattering cross-section integrated over energies. Integration of (15.15) over energies $E_{k'} = E_k - \hbar\omega$ gives

$$\left(\frac{\mathrm{d}\sigma^{\uparrow\downarrow}}{\mathrm{d}\Omega'}\right)_{\parallel} - \left(\frac{\mathrm{d}\sigma^{\uparrow\downarrow}}{\mathrm{d}\Omega'}\right)_{\perp} = \left(\frac{\gamma e^2}{m_e c^2}\right)^2 \frac{\hbar}{3}\frac{\mathcal{V}}{\Omega_{\mathrm{WS}}}\frac{k'}{k}\int S(\kappa,\omega)\,\mathrm{d}\omega. \tag{15.16}$$

Inverting the Fourier transform (15.5), we have

$$\frac{1}{\hbar N}\langle s^{\alpha}_{\kappa}(t)s^{\alpha}_{-\kappa}\rangle = \int S^{\alpha}(\kappa,\omega)\,\mathrm{e}^{-\mathrm{i}\omega t}\,\mathrm{d}\omega.$$

Using this expression at $t = 0$ and taking the Debye-Waller factor into account, we write formula (15.16) as

$$\left(\frac{\mathrm{d}\sigma^{\uparrow\downarrow}}{\mathrm{d}\Omega'}\right)_{\parallel} - \left(\frac{\mathrm{d}\sigma^{\uparrow\downarrow}}{\mathrm{d}\Omega'}\right)_{\perp} = \left(\frac{\gamma e^2}{m_e c^2}\right)^2 \frac{1}{3g^2\mu_{\mathrm{B}}^2}\frac{\mathcal{V}}{\Omega_{\mathrm{WS}}}\frac{k'}{k}\,\mathrm{e}^{-2W(\kappa)}\,M^2(\kappa), \tag{15.17}$$

where

$$M^2(\kappa) = \left(\hbar\int S(\kappa,\omega)\,\mathrm{d}\omega\right) g^2\mu_{\mathrm{B}}^2 = \frac{1}{N}\langle \mathbf{s}_{\kappa}\mathbf{s}_{-\kappa}\rangle\, g^2\mu_{\mathrm{B}}^2 \tag{15.18}$$

is the square of the effective moment.

15.2 Spin-Density Correlations

The magnetic neutron scattering experiments in ferromagnetic metals above the Curie temperature have been mainly interpreted using the spin-wave theory [5,27–33]. Analysing the peak of the scattering function, the short-range order (SRO) of about 15–20 Å was obtained. This interpretation was criticized, because the peak height and width are about equal [6,7].

Spin fluctuation theories of metallic magnetism all support the existence of SRO above the Curie temperature in the ferromagnetic metals but there is no agreement about the extent of the SRO (see, e.g. [1]). The fluctuating-local-band theory [34–38] is based upon the existence of very strong SRO well above T_{C} but it is unlikely to apply to any real material [39]. The static spin fluctuation theories [40–45] describe the paramagnetic phase as having no appreciable SRO outside the critical region. The situation is similar in the dynamic theories [8,10,46–48], all based on the local approximation.

In this section, we present theoretical results on the spin-density correlations in the DSFT and compare our calculations with polarized neutron scattering experiments, which play a major role in testing the theory. Our theoretical results are demonstrated by the example of bcc Fe. We calculate the spin-density correlator as a function of distance and temperature and calculate its Fourier transform (squared effective moment) as a function of wavevector and temperature in a systematic way.[2] A number of magnetic characteristics, such as effective moment and local moment, are compared with experiment over a large temperature range (for details see, e.g. [14–16, 51]).

15.2.1 Spatial Spin Correlator

We begin with the spin-density correlator $\langle \mathbf{s}(\mathbf{r}, t)\, \mathbf{s}(\mathbf{r}', t')\rangle$, where $\langle \dots \rangle$ is the canonical average and time dependence means the Heisenberg representation. For a system with a time-independent Hamiltonian, the spin correlator is a function of the time difference $t - t'$. If the system is translationally invariant, the spin correlator is a function of $\mathbf{r} - \mathbf{r}'$.

The equal-time correlation function or *spatial* spin correlator

$$\langle \mathbf{s}(\mathbf{r})\mathbf{s}(0)\rangle = \langle \Delta\mathbf{s}(\mathbf{r})\Delta\mathbf{s}(0)\rangle + \langle \mathbf{s}(\mathbf{r})\rangle\langle \mathbf{s}(0)\rangle$$

contains information about how much the spin densities at different positions are related (see, e.g. [52]). At small distances spins have the strongest influence on each other. Therefore, the fluctuation term $\langle \Delta\mathbf{s}(\mathbf{r})\Delta\mathbf{s}(0)\rangle$ is the largest at $r = 0$ and decays to zero as the distance increases to infinity. In the ferromagnetic metals, the decay with distance is usually monotone. The second term $\langle \mathbf{s}(\mathbf{r})\rangle\langle \mathbf{s}(0)\rangle$ is constant; it is equal to $\bar{s}_z^2/\Omega_{\mathrm{WS}}^2$ below T_C state and zero above T_C.

To analyse the short-range order it is convenient to define the *normalized* correlation function

$$C(\mathbf{r}) = \frac{\langle \mathbf{s}(\mathbf{r})\, \mathbf{s}(0)\rangle}{\langle \mathbf{s}(0)\, \mathbf{s}(0)\rangle}, \tag{15.19}$$

which varies between zero and unity.[3] Figure 15.1 sketches $C(r)$ as a function of distance r/a between the lattice sites (a is the lattice constant). Most important are the values of $C(r)$ at the nearest neighbours, next nearest neighbours, etc., of the lattice site at the origin (see arrows in Fig. 15.1).

We first consider the spin correlator $\langle s^\alpha(\mathbf{r}) s^\alpha(0)\rangle$, where $\alpha = x, y, z$. The spatial Fourier transformation of the spin-density operator (3.19) is defined by

$$s^\alpha_{\mathbf{q}} = \int s^\alpha(\mathbf{r})\, \mathrm{e}^{-\mathrm{i}\mathbf{q}\mathbf{r}}\, \mathrm{d}\mathbf{r}, \qquad s^\alpha(\mathbf{r}) = \frac{1}{\mathcal{V}} \sum_{\mathbf{q}} s^\alpha_{\mathbf{q}}\, \mathrm{e}^{\mathrm{i}\mathbf{q}\mathbf{r}},$$

where $\mathcal{V} = N\Omega_{\mathrm{WS}}$ is the volume of the crystal. Translational invariance of the system leads to

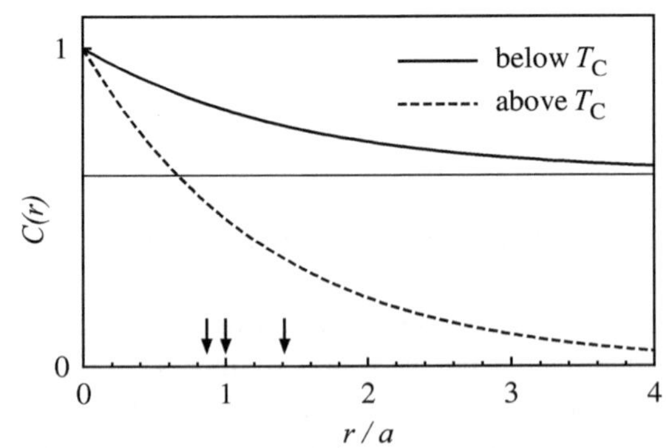

Fig. 15.1 Sketch of the normalized correlation function $C(r)$, where thin horizontal line shows the long-distance value of $C(r)$ below T_C and arrows point to the nearest neighbours, next nearest neighbours, etc., for the bcc lattice

[2]For selected temperatures, calculations of the spin-density correlator and its Fourier transform in static and dynamic approximations of spin fluctuation theory were carried out by Hasegawa [49] and Grebennikov [11, 50], respectively.

[3]*In the Heisenberg model*, this reduces to $C_j = \langle \mathbf{S}_j \mathbf{S}_0\rangle / \langle \mathbf{S}_0 \mathbf{S}_0\rangle$, where the jth spin $\mathbf{S}_j$ is a vector of the modulus S_0. Therefore, in the paramagnetic region, C_j is the mean cosine of the angle between the spins.

$$\langle s^\alpha(\mathbf{r}) s^\alpha(0)\rangle = \frac{1}{\mathcal{V}^2}\sum_{\mathbf{q}} \langle s^\alpha_{\mathbf{q}} s^\alpha_{-\mathbf{q}}\rangle\, \mathrm{e}^{\mathrm{i}\mathbf{q}\mathbf{r}} \tag{15.20}$$

(for details, see Appendix C.4). Transforming the sum into an integral over the Brillouin zone and replacing the latter by the equal-volume sphere with radius q_{B}, we have

$$\langle s^\alpha(\mathbf{r}) s^\alpha(0)\rangle = \frac{1}{2\pi^2 N \Omega_{\mathrm{WS}}} \int_0^{q_{\mathrm{B}}} \langle s^\alpha_{\mathbf{q}} s^\alpha_{-\mathbf{q}}\rangle \frac{q\sin(qr)}{r}\, \mathrm{d}q. \tag{15.21}$$

The local spin s_{L} is defined by the formula

$$s_{\mathrm{L}}^2 = \iint_{\mathrm{WS}} \langle \mathbf{s}(\mathbf{r})\mathbf{s}(\mathbf{r}')\rangle\, \mathbf{dr}\, \mathbf{dr}', \tag{15.22}$$

where both integrals are taken over the same Wigner-Seitz cell. Replacing the integral by the value of the integrand at the site multiplied by the cell volume, we write formula (15.22) as

$$s_{\mathrm{L}}^2 = \Omega^2_{\mathrm{WS}} \langle \mathbf{s}(\mathbf{R}_j)\mathbf{s}(\mathbf{R}_j)\rangle = \Omega^2_{\mathrm{WS}} \langle \mathbf{s}(0)\mathbf{s}(0)\rangle. \tag{15.23}$$

Passing to the limit in (15.21) as $r \to 0$ and using $\sin(rq) \approx rq$, we have

$$\langle s^\alpha(0) s^\alpha(0)\rangle = \frac{1}{2\pi^2 N \Omega_{\mathrm{WS}}} \int_0^{q_{\mathrm{B}}} \langle s^\alpha_{\mathbf{q}} s^\alpha_{-\mathbf{q}}\rangle\, q^2\, \mathrm{d}q. \tag{15.24}$$

Substituting the latter into formula for the local moment (15.23) and taking into account $\Omega_{\mathrm{WS}} = (2\pi)^3/\Omega_{\mathrm{BZ}}$, we obtain

$$s_{\mathrm{L}}^2 = \frac{4\pi}{N\Omega_{\mathrm{BZ}}} \int_0^{q_{\mathrm{B}}} \langle \mathbf{s}_{\mathbf{q}} \mathbf{s}_{-\mathbf{q}}\rangle\, q^2\, \mathrm{d}q. \tag{15.25}$$

In order to calculate the spatial correlator (15.21) and local moment (15.25), it is necessary to find the spin-density correlator in the momentum representation $\langle s^\alpha_{\mathbf{q}} s^\alpha_{-\mathbf{q}}\rangle$. By the fluctuation-dissipation theorem (2.51), the spin correlator $\langle \Delta s^\alpha_{\mathbf{q}} \Delta s^\alpha_{-\mathbf{q}}\rangle$ is related to the imaginary part of the susceptibility. In the *paramagnetic* state, we have

$$\langle s^\alpha_{\mathbf{q}} s^\alpha_{-\mathbf{q}}\rangle = \frac{1}{2\pi} \int B(\varepsilon)\, \mathrm{Im}\chi^\alpha_{\mathbf{q}}(\varepsilon)\, \mathrm{d}\varepsilon, \tag{15.26}$$

where $\chi^\alpha_{\mathbf{q}}(\varepsilon)$ is the enhanced susceptibility in units of $\frac{1}{2}g^2\mu_{\mathrm{B}}^2$ and $B(\varepsilon)$ is the Bose function.

15.2.2 Spin Correlator in the DSFT

In the DSFT, the enhanced susceptibility $\chi^\alpha_{\mathbf{q}}(\varepsilon)$ is expressed in terms of the unenhanced susceptibility $\chi^{0\alpha}_{\mathbf{q}}(\varepsilon)$ using formula (10.47):

$$\chi^\alpha_{\mathbf{q}}(\varepsilon) = \frac{\chi^{0\alpha}_{\mathbf{q}}(\varepsilon)}{1 - \tilde{u}\chi^{0\alpha}_{\mathbf{q}}(\varepsilon)}, \tag{15.27}$$

where $\tilde{u} = u/N$ is the Fourier transform of the effective interaction constant u. Due to strong localization of the Bose function at zero energy, we replace $B(\varepsilon)$ and susceptibility $\chi^{0\alpha}_{\mathbf{q}}(\varepsilon)$ by their Taylor series in ε (for details, see Sect. 10.2). As a result, formula (15.26) becomes

$$\langle s^\alpha_{\mathbf{q}} s^\alpha_{-\mathbf{q}}\rangle = \frac{T}{2\tilde{u}\lambda^\alpha_{\mathbf{q}}} \frac{2}{\pi} \arctan \frac{\tilde{u}\varphi^\alpha_{\mathbf{q}}\pi^2 T}{6\lambda^\alpha_{\mathbf{q}}},$$

where $\lambda^\alpha_{\mathbf{q}} = 1 - \tilde{u}\chi^{0\alpha}_{\mathbf{q}}(0)$ and $\varphi^\alpha_{\mathbf{q}} = \mathrm{d}\chi^{0\alpha}_{\mathbf{q}}(0)/\mathrm{d}\varepsilon$. The interaction of the modes is taken into account by interpolating the static susceptibility $\chi^{0\alpha}_{\mathbf{q}}$ between the uniform susceptibility $\chi^{0\alpha}_0$ and local susceptibility (10.23). Thus, $\lambda^\alpha_{\mathbf{q}}$ is calculated by formula (10.21). The function $\varphi_{\mathbf{q}}^{\alpha}$ is replaced, for simplicity, by its mean value $N\varphi_{\mathrm{L}}^{\alpha}$, where $\varphi_{\mathrm{L}}^{\alpha} = N^{-2}\sum_{\mathbf{q}} \varphi_{\mathbf{q}}^{\alpha}(0)$. The final expression for spin-density correlator (15.26) takes the form (for details, see [16])

$$\langle s_{\mathbf{q}}^{\alpha} s_{-\mathbf{q}}^{\alpha} \rangle = \frac{NT}{2u\lambda_{\mathrm{L}}^{\alpha}} \frac{1}{a_{\alpha}^2 + b_{\alpha}^2 (q/q_{\mathrm{B}})^2} \frac{2}{\pi} \arctan \frac{c_{\alpha}}{a_{\alpha}^2 + b_{\alpha}^2 (q/q_{\mathrm{B}})^2}, \tag{15.28}$$

where a_{α}, b_{α} and c_{α} are given by (10.25).

15.2.3 High-Temperature Approximation

At $T \gg T_{\mathrm{C}}$, we can use the high-temperature version of the fluctuation-dissipation theorem (2.57):

$$\langle s_{\mathbf{q}}^{\alpha} s_{-\mathbf{q}}^{\alpha} \rangle = \frac{T}{2} \chi_{\mathbf{q}}^{\alpha}(0), \tag{15.29}$$

where $\chi_{\mathbf{q}}^{\alpha}$ is the enhanced susceptibility in units of $\frac{1}{2}g^2\mu_{\mathrm{B}}^2$. Expressing the enhanced susceptibility in terms of the unenhanced one according to (15.27), we have

$$\langle s_{\mathbf{q}}^{\alpha} s_{-\mathbf{q}}^{\alpha} \rangle = \frac{T}{2} \frac{\chi_{\mathbf{q}}^{0\alpha}(0)}{1 - \tilde{u}\chi_{\mathbf{q}}^{0\alpha}(0)} = \frac{T}{2\tilde{u}} \frac{1}{1 - \tilde{u}\chi_{\mathbf{q}}^{0\alpha}(0)} - \frac{T}{2\tilde{u}}.$$

Neglecting the second term, which is responsible for the intrinsic fluctuations of the field (for details, see [13]), and using formula (10.21), we obtain the Lorentzian function

$$\langle s_{\mathbf{q}}^{\alpha} s_{-\mathbf{q}}^{\alpha} \rangle = \frac{NT}{2u\lambda_{\mathrm{L}}^{\alpha}} \frac{1}{a_{\alpha}^2 + b_{\alpha}^2 (q/q_{\mathrm{B}})^2}. \tag{15.30}$$

At distances $r > \pi/q_{\mathrm{B}}$, spatial correlator (15.21) is determined by small wavevectors q, thus the upper integration limit in expression (15.21) can be extended to infinity [14]. Substituting (15.30) in (15.21) and integrating over positive q, we obtain the Ornstein-Zernike correlator

$$\langle s^{\alpha}(\mathbf{r}) s^{\alpha}(0) \rangle = \frac{B_{\alpha}}{q_{\mathrm{B}} r} \mathrm{e}^{-r/r_{\mathrm{c}}^{\alpha}}, \tag{15.31}$$

which appears in magnets with *localized* spins.[4] Here $B_{\alpha} = 3\pi T/(4u\Omega_{\mathrm{WS}}^2 \lambda_{\mathrm{L}}^{\alpha} b_{\alpha}^2)$ and

$$r_{\mathrm{c}}^{\alpha} = |b_{\alpha}|/(|a_{\alpha}| q_{\mathrm{B}}) \tag{15.32}$$

is the *correlation radius*. In the DSFT, we have shown that r_{c} increases to infinity as T goes to T_{C} and decreases as $r_{\mathrm{c}} \sim 1/T$ at high temperatures $T \gg T_{\mathrm{C}}$ [14].

15.3 Application to Iron

15.3.1 Comparison of DSFT and Experiment

We demonstrate our theoretical results by the example of bcc Fe following [15,21]. The initial data of the DSFT calculations are the magnetic moment per atom $m(0) = 2.217\,\mu_{\mathrm{B}}$ [53] and the first-principles density of states at $T = 0\,\mathrm{K}$. Here we use the DOS from Fig. 11.5. The number of d electrons per atom is equal to 7.43. The effective interaction constant u, determined from the DSFT equations at $T = 0$, is equal to 1.06 eV. The Brillouin zone is approximated by an equal-volume

[4] For discussion of the Ornstein-Zernike correlator in the Ising model see, e.g. [18].

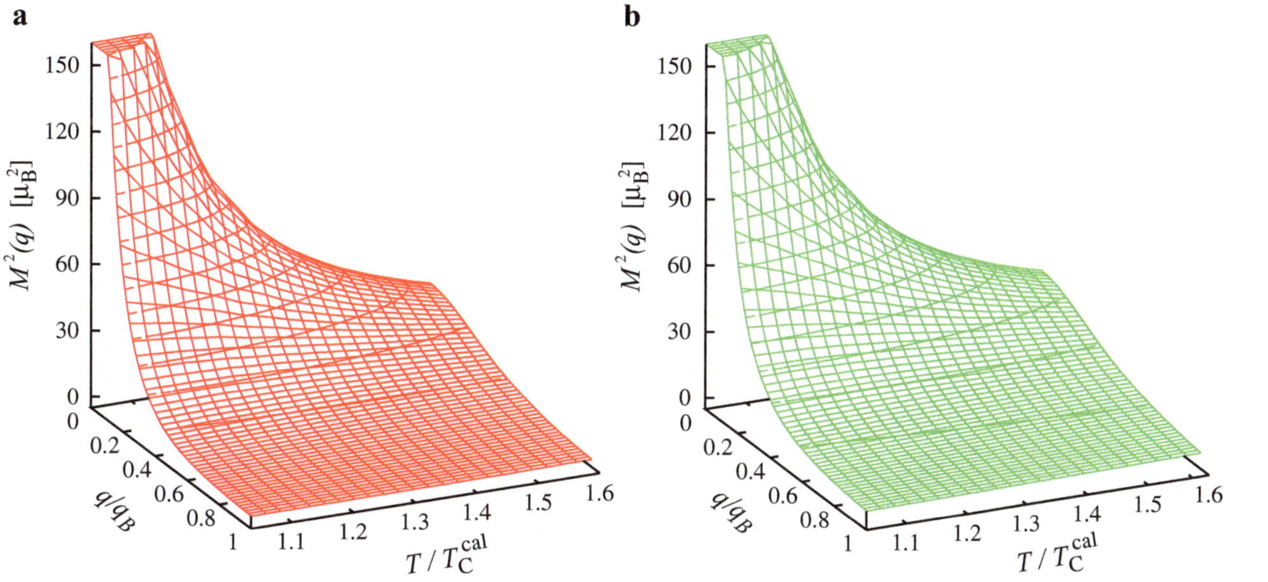

Fig. 15.2 Square of the effective moment $M^2(q)$ for bcc Fe calculated in the GA with the use of (**a**) exact formula (15.28) and (**b**) approximate high-temperature formula (15.30)

sphere of radius $q_B = (3\Omega_{BZ}/(4\pi))^{1/3} = 1.71\,\text{Å}^{-1}$. All calculations are carried out in the Gaussian approximation (GA) of the DSFT [13].[5]

Figure 15.2 shows the square of the effective moment (15.18). The temperature is given in units of the calculated Curie temperature $T_C^{cal} = 1.65\,T_C^{exp}$. The experimental Curie temperature for Fe is $T_C^{exp} = 1044\,\text{K}$ [53]. As can be seen from Fig. 15.2, $\langle s^\alpha_{\mathbf{q}} s^\alpha_{-\mathbf{q}}\rangle$ calculated by approximate high-temperature formula (15.30) (Fig. 15.2b) is in good agreement with $\langle s^\alpha_{\mathbf{q}} s^\alpha_{-\mathbf{q}}\rangle$ calculated by exact formula (15.28) (Fig. 15.2a) in a wide range of temperatures for all **q** from the Brillouin zone.

Comparing the effective moment $M(q)$ calculated in the DSFT with the experiment [5,27–29], it is necessary to take into account three circumstances. First, in the DSFT we consider the *reduced* spin density, which differs from the total one by the magnetic form-factor $F(q)$ (for details, see [16]). In papers [5,27,28] values of the total effective moment $M^2(q)|F(q)|^2$ are presented, where the value $|F(q)|^2 \approx 0.81$ is used for Fe (see [29]). The second circumstance is that taking into account the atom vibrations we have the Debye-Waller factor e^{-2W} in the formula for the magnetic scattering cross-section. But in papers [5,27–29], just as in the DSFT, the Debye-Waller factor is ignored, so we set it to unity.[6] Third, in experiment the transferred energy ε is finite and belongs to a certain interval $[0, \varepsilon_{max}]$. Therefore, in formula (15.26) we should also restrict the integration to the same energy interval. As a result, we obtain

$$\begin{aligned}\langle s^\alpha_{\mathbf{q}} s^\alpha_{-\mathbf{q}}\rangle &= \frac{1}{2\pi}\int_0^{\varepsilon_{max}} B(\varepsilon)\,\mathrm{Im}\chi^\alpha_{\mathbf{q}}(\varepsilon)\,d\varepsilon \\ &= \frac{NT}{2u\lambda^\alpha_L}\frac{1}{a^2_\alpha + b^2_\alpha(q/q_B)^2}\frac{2}{\pi}\arctan\frac{\varepsilon_{max}\tilde{c}_\alpha}{a^2_\alpha + b^2_\alpha(q/q_B)^2},\end{aligned} \tag{15.33}$$

where $\tilde{c}_\alpha = u\varphi^\alpha_L/\lambda^\alpha_L$.

In Fig. 15.3, we compare $M^2(q)$ calculated at T/T_C^{cal} with experimental $M^2(q)$ [5,27,28] obtained at the same values of T/T_C^{exp}.[7] As can be seen from Fig. 15.3, the calculation of the effective moment with the use of formula (15.33) with the energy cutoff $\varepsilon_{max} = 50\,\text{meV}$ gives a good quantitative agreement with the experiment (for details, see [21]). The curve $M^2(q)$ calculated with the use of exact formula (15.28) goes higher than the experimental values, because the energy interval

[5]Calculations in the renormalized Gaussian approximation with uniform fluctuations (RGA+UF) of the DSFT give close results [14,16].

[6]We consider only the *spin* contribution from the electron subsystem. The magnetic force arising from the *orbital* motion of electrons gives only a small cross-section for the neutron scattering and may be ignored [22]. As for the *phonon* contribution, it is not the main one in the ferromagnetic metals (see, e.g. [16,19]). In detail the phonon mechanism is discussed in [54].

[7]Experimental measurements up to the Brillouin zone boundary [29] confirm the results of [5,27,28].

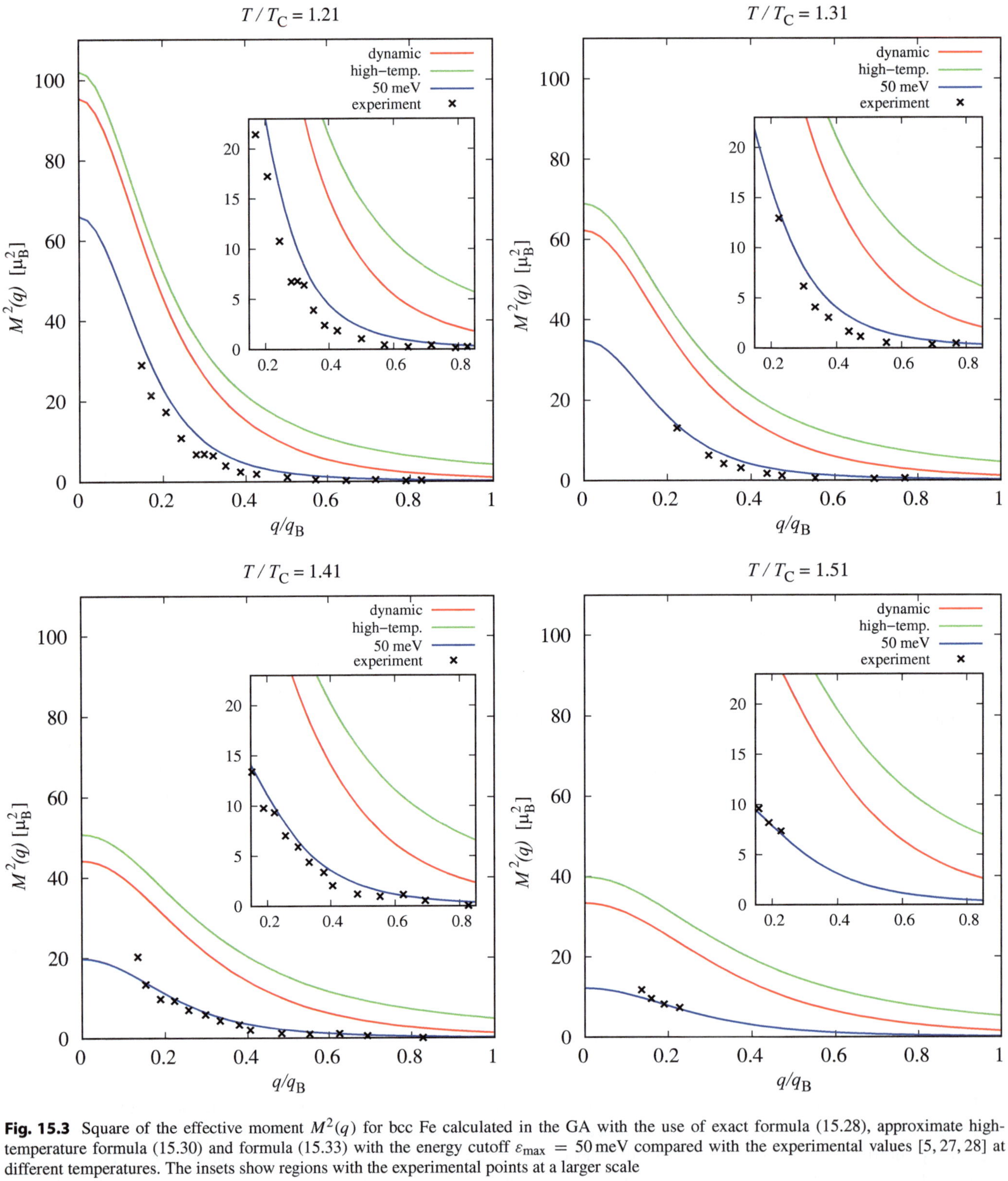

Fig. 15.3 Square of the effective moment $M^2(q)$ for bcc Fe calculated in the GA with the use of exact formula (15.28), approximate high-temperature formula (15.30) and formula (15.33) with the energy cutoff $\varepsilon_{\max} = 50\,\mathrm{meV}$ compared with the experimental values [5, 27, 28] at different temperatures. The insets show regions with the experimental points at a larger scale

Table 15.1 Local magnetic moment m_L for bcc Fe calculated in the GA with the use of exact formula (15.28), approximate high-temperature formula (15.30) and formula (15.33) with the energy cutoff $\varepsilon_{max} = 50$ meV

T/T_C^{cal}	Dynamic, μ_B	High-temp., μ_B	50 meV, μ_B
1.226	2.214	3.060	1.120
1.322	2.218	3.080	1.116
1.418	2.222	3.097	1.047
1.514	2.228	3.114	0.993

is wider than in the experiment. The approximate high-temperature formula gives good agreement with the exact formula uniformly at all temperatures in the paramagnetic state. From Fig. 15.3 we see that the value of $M(0)$ calculated with the use of exact formula (15.28) decreases and tends to the value of the effective moment in the Curie-Weiss law $m_{eff}^{exp} = 3.13\,\mu_B$ [55] as temperature increases, in agreement with the theoretical result (2.62). Indeed, we have $M(0) = 5.7\,\mu_B$ at $T = 1.51\,T_C^{cal}$ and $M(0) = 4.0\,\mu_B$ at $T = 1.92\,T_C^{cal}$. But the convergence is fairly slow, so one should be careful using m_{eff}^{exp} instead of $M(0)$.

Table 15.1 shows the local magnetic moment $m_L = g\mu_B s_L$ at the same temperatures as in [5, 27, 28]. Calculated values with the use of the exact formula are approximately equal to the experimental value of the magnetic moment $m_0 = 2.217\,\mu_B$ at $T = 0$ [53] and do not change much with temperature (second column). Calculations with classical Hamiltonians [56, 57] give comparable results. In [56] m_L decreases from 1.95 to 1.85 μ_B over the interval 1.1–1.5T_C^{exp} and in [57] m_L is about 2.2 μ_B and is almost constant over the same temperature interval. Approximate high-temperature formula (15.30) gives somewhat higher values of m_L (third column).

In order to compare the calculated local moment with the experimental one, we note that m_L is proportional to the area under the curve $q^2M^2(q)$. Indeed, substituting (15.18) in (15.25), we obtain

$$m_L^2 = \frac{4\pi}{\Omega_{BZ}} \int_0^{q_B} q^2 M^2(q)\, dq. \tag{15.34}$$

As can be seen from Fig. 15.4, the calculated curve $q^2M^2(q)$ with the use of formula (15.33) with the energy cutoff $\varepsilon_{max} = 50$ meV is in good agreement with the experimental points. Therefore, the numerical values of m_L, presented in the fourth column of Table 15.1, should also give a good approximation to the experimental ones.[8] Comparing the second and third columns in Table 15.1, we see that, despite qualitative similarities, the exact and high-temperature formulae give different quantitative results.

15.3.2 Short-Range Order Analysis

Next we calculate the correlation function (15.19). From formulae (15.20) and (15.18) we see that $\langle \mathbf{s}(\mathbf{r})\,\mathbf{s}(0)\rangle$ is obtained by the inverse Fourier transform of $M^2(\kappa)$. Taking (15.21), (15.24) and (15.18) into account, we write the normalized correlation function as

$$C(r, T) = \left(\int_0^{q_B} q^2 M^2(q)\, dq\right)^{-1} \int_0^{q_B} q^2 M^2(q)\, \frac{\sin(qr)}{qr}\, dq. \tag{15.35}$$

Note that the reconstruction of the spin correlator based on the experimental data alone can lead to ambiguous results (see, e.g. [58]). Indeed, $M^2(\kappa)$ cannot be measured at small κ (see, e.g. [5]), where it has a strong peak. The inverse Fourier transform of $M^2(\kappa)$, i.e. the spin correlator $\langle \mathbf{s}(\mathbf{r})\,\mathbf{s}(0)\rangle$, is unstable with respect to extrapolation of $M^2(\kappa)$ at small κ.[9] One way to overcome this difficulty is to obtain $M^2(0)$ from the static susceptibility by formula (15.29) and fit the experimental data using the Lorentzian function (see, e.g. [29]). But, as we showed, the result of the high-temperature approximation (15.29) for $M^2(0)$ can differ substantially from the experimental one because of the energy window used in the experiment.

[8] A slightly larger experimental value of $m_L = 1.3\,\mu_B$ is obtained in [5, 27, 28] with the use of another measurement, and the value $m_L = 1.55\,\mu_B$ presented in [1] corresponds to the energy cutoff 200 meV.

[9] Roughly speaking, the higher the peak of $M^2(\kappa)$, the wider the correlator, and hence the larger is the SRO domain (for details, see [21]).

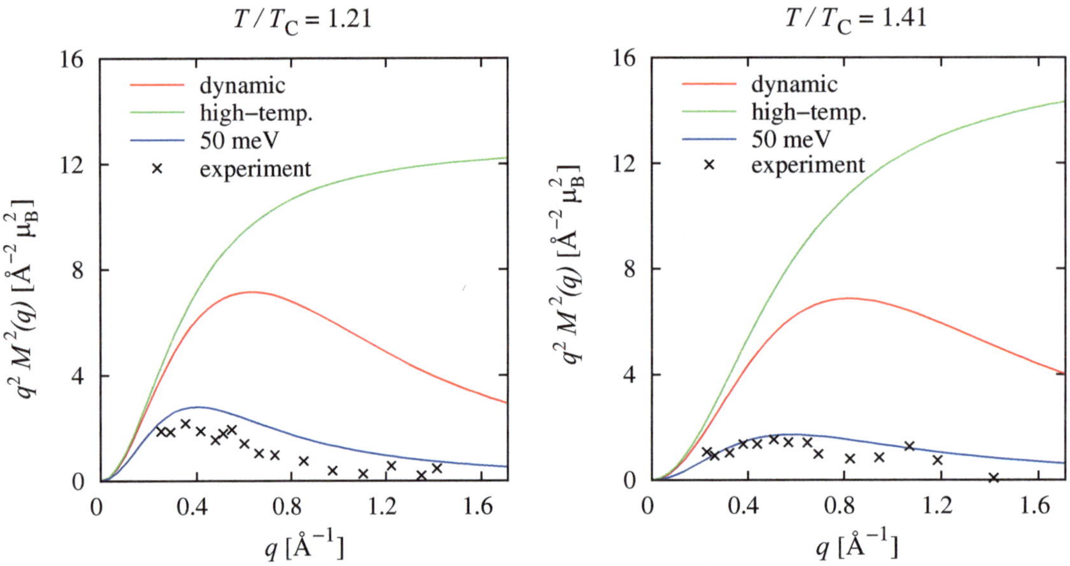

Fig. 15.4 Values of $q^2M^2(q)$ for bcc Fe calculated in the GA with the use of exact formula (15.28), approximate high-temperature formula (15.30) and formula (15.33) with the energy cutoff $\varepsilon_{\max} = 50\,\text{meV}$ compared with the experimental values [5, 27, 28] at different temperatures

The spin-wave interpretation of the experimental results can be explained as follows. Assuming that $q^2M^2(q) \propto \delta(q-q_0)$ in formula (15.35), we obtain

$$C(r,T) = \int_0^{q_B} \delta(q-q_0)\,\frac{\sin(qr)}{qr}\,\mathrm{d}q = \frac{\sin(q_0r)}{q_0r}.$$

This expression is associated with a (damped) spin wave of the wavelength $\lambda_0 = 2\pi/q_0$. From Fig. 15.4, at $T = 1.21T_C^{\text{exp}}$ we have $\lambda_0 \approx 18\,\text{Å}$. Arguments of this kind were used in [5, 27, 28] and other experimental papers to obtain an estimate for the SRO domain of about 15–20 Å. But the hypothesis that the spin waves persist above T_C is controversial. Indeed, as one can see from Fig. 15.4, the peak of the function $q^2M^2(q)$ is strongly spread out, its height and width being about equal at $T = 1.21T_C^{\text{exp}}$, and the width being larger than the height at $T = 1.41T_C^{\text{exp}}$.

In the DSFT we are able to calculate the correlation function explicitly in a wide range of temperatures above T_C. Good agreement of the DSFT calculation results for the effective moment with the experiment validates our results for the correlation function.

Figure 15.5 shows correlator (15.35) as a function of distance r and temperature T. As we see from Fig. 15.5, the approximate high-temperature formula gives good agreement with the exact one at temperatures $T/T_C^{\text{cal}} = 1.5$ and above. The calculation results with the use of Ornstein-Zernike formula (15.31) do not differ much from the results with the use of high-temperature formula (15.30) starting already from distances of 1–2 Å, and are not presented here. Moreover, the correlation radius in the Ornstein-Zernike formula (15.31) describes only the correlations at large distances. To describe the SRO of a specific metal it is necessary to use other characteristics. In the paper [14], we introduced the halfwidth at half maximum (HWHM) of the normalized spatial correlator $C(r, T)$, *correlation halfwidth* for short, as a measure of the SRO.

Figure 15.6 shows the halfwidth $r_{1/2}(T)$ of the correlation function (15.35) calculated in different approximations. As can be seen from Fig. 15.6, the SRO domain is small, even if we take into account only energies up to $\varepsilon_{\max} = 50\,\text{meV}$. The halfwidth of the correlation function is 2.0–3.5 Å at $T/T_C^{\text{cal}} = 1.05$ and about 1.8–2.0 Å at $T/T_C^{\text{cal}} = 1.6$. As we showed in paper [14], inverse correlation radius (15.32) varies linearly with temperature. From the inset in Fig. 15.6 we see that the temperature dependence of the inverse halfwidth is also close to linear.

At temperatures $1.1T_C^{\text{cal}}$–$1.2T_C^{\text{cal}}$, the correlation radius r_c is substantially larger than the halfwidth $r_{1/2}$ and cannot be used as a measure of the SRO [14]. For instance, at $T = 1.1T_C^{\text{cal}}$ we have $r_c \approx 8\,\text{Å}$. But as T increases the values of the correlation radius and halfwidth approach each other, and starting from $1.6T_C^{\text{cal}}$ they are almost equal.

Thus, the calculated correlation halfwidth in bcc Fe is within 5 Å.[10] The results suggest that the SRO in the ferromagnetic metals above the Curie temperature is small and slowly decreases with temperature.

[10]Correlation length of approximately $2a$ (a is the lattice constant) was reported by Tao et al. [59].

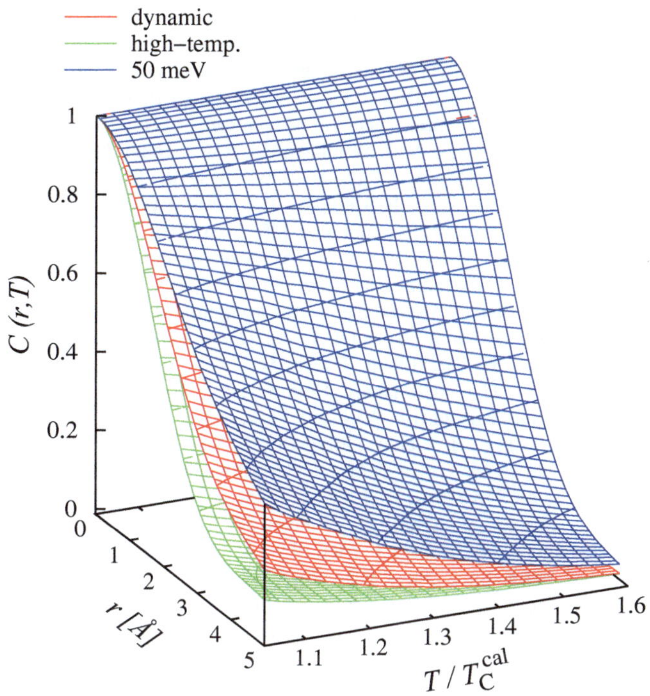

Fig. 15.5 Spatial correlation function $C(r, T)$ for bcc Fe calculated in the GA with the use of exact formula (15.28), approximate high-temperature formula (15.30) and formula (15.33) with the energy cutoff $\varepsilon_{\max} = 50\,\text{meV}$

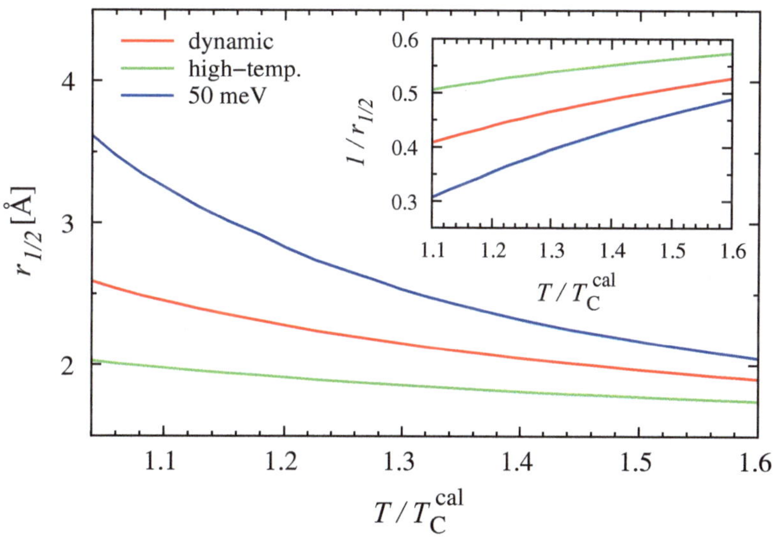

Fig. 15.6 Halfwidth $r_{1/2}(T)$ of correlation function (15.35) for bcc Fe calculated in the GA with the use of exact formula (15.28), approximate high-temperature formula (15.30) and formula (15.33) with the energy cutoff $\varepsilon_{\max} = 50\,\text{meV}$. The inset shows the inverse halfwidth $r_{1/2}^{-1}(T)$

References

1. T. Moriya, *Spin Fluctuations in Itinerant Electron Magnetism* (Springer, Berlin, 1985)
2. S.V. Maleyev, Physics-Uspekhi **45**, 569 (2002)
3. T. Chatterji (ed.), *Neutron Scattering from Magnetic Materials* (Elsevier, Amsterdam, 2006)
4. N. Plakida, *High-Temperature Cuprate Superconductors* (Springer, Berlin, 2010)
5. P.J. Brown, H. Capellmann, J. Déportes, D. Givord, K.R.A. Ziebeck, J. Magn. Magn. Mater. **30**, 243 (1982)
6. J. Hubbard, in *Physics of Transition Metals 1980 (Inst. Phys. Conf. Ser., No. 55)*, ed. by P. Rhodes (IOP, Bristol, 1981), pp. 669–687
7. J. Wicksted, G. Shirane, P. Böni, Phys. Rev. B **30**, 3655 (1984)
8. A.I. Lichtenstein, M.I. Katsnelson, G. Kotliar, Phys. Rev. Lett. **87**, 067205 (2001)
9. V.P. Antropov, Phys. Rev. B **72**, 140406 (2005)
10. Y. Kakehashi, M.A.R. Patoary, J. Phys. Soc. Jpn. **80**, 034706 (2011)
11. V.I. Grebennikov, Phys. Solid State **40**, 79 (1998)
12. B.I. Reser, V.I. Grebennikov, Phys. Met. Metallogr. **85**, 20 (1998)
13. N.B. Melnikov, B.I. Reser, V.I. Grebennikov, J. Phys. Condens. Matter **23**, 276003 (2011)

14. N.B. Melnikov, B.I. Reser, J. Magn. Magn. Mater. **397**, 347 (2016)
15. N.B. Melnikov, B.I. Reser, G.V. Paradezhenko, J. Magn. Magn. Mater. **411**, 133 (2016)
16. N.B. Melnikov, G.V. Paradezhenko, B.I. Reser, Theor. Math. Phys. **191**(1), 602 (2017)
17. N.B. Melnikov, B.I. Reser, G.V. Paradezhenko, EPJ Web of Conferences **185**, 11011 (2018)
18. R.M. White, *Quantum Theory of Magnetism*, 3rd edn. (Springer, Berlin, 2007)
19. G.V. Paradezhenko, N.B. Melnikov, B.I. Reser, Theor. Math. Phys. **195**(1), 572 (2018)
20. K. Ziebeck, P. Brown, J. Phys. F: Metal Phys. **10**, 2015 (1980)
21. G.V. Paradezhenko, N.B. Melnikov, B.I. Reser, Vychisl. Metody Programm. **17**, 474 (2016) [in Russian]
22. R.J. Elliott, Proc. R. Soc. Lond. A **235**, 289 (1956)
23. G.L. Squires, *Introduction to the Theory of Thermal Neutron Scattering*, 2nd edn. (Dover, New York, 1996)
24. J. Schweizer, in *Neutron Scattering from Magnetic Materials*, chap. 4, ed. by T. Chatterji (Elsevier, Amsterdam, 2006), pp. 153–213
25. K. Blum, *Density Matrix Theory and Applications* (Plenum, New York, 1981)
26. C. Cohen-Tannoudji, B. Diu, F. Laloe, *Quantum Mechanics* (Wiley, New York, 1991)
27. P.J. Brown, H. Capellmann, J. Déportes, D. Givord, K.R.A. Ziebeck, J. Magn. Magn. Mater. **30**, 335 (1983)
28. P.J. Brown, H. Capellmann, J. Déportes, D. Givord, K.R.A. Ziebeck, J. Magn. Magn. Mater. **31–34**, 295 (1983)
29. G. Shirane, P. Böni, J.P. Wicksted, Phys. Rev. B **33**, 1881 (1986)
30. H.A. Mook, R.M. Nicklow, Phys. Rev. B **7**, 336 (1973)
31. H.A. Mook, J.W. Lynn, R.M. Nicklow, Phys. Rev. Lett. **30**, 556 (1973)
32. J.W. Lynn, Phys. Rev. B **11**, 2624 (1975)
33. H.A. Mook, J.W. Lynn, J. Appl. Phys. **57**, 3006 (1985)
34. H. Capellmann, J. Phys. F: Met. Phys. **4**, 1466 (1974)
35. H. Capellmann, Z. Phys. B **34**, 29 (1979)
36. V. Korenman, J.L. Murray, R.E. Prange, Phys. Rev. B **16**, 4032 (1977)
37. V. Korenman, J.L. Murray, R.E. Prange, Phys. Rev. B **16**, 4048 (1977)
38. V. Korenman, J.L. Murray, R.E. Prange, Phys. Rev. B **16**, 4058 (1977)
39. D.M. Edwards, J. Magn. Magn. Mater. **15–18**, 262 (1980)
40. J. Hubbard, Phys. Rev. B **19**, 2626 (1979)
41. J. Hubbard, Phys. Rev. B **20**, 4584 (1979)
42. H. Hasegawa, J. Phys. Soc. Jpn. **46**, 1504 (1979)
43. H. Hasegawa, J. Phys. Soc. Jpn. **49**, 178 (1980)
44. H. Hasegawa, J. Phys. Soc. Jpn. **49**, 963 (1980)
45. V.I. Grebennikov, Y.I. Prokopjev, O.B. Sokolov, E.A. Turov, Phys. Met. Metallogr. **52**, 1 (1981)
46. B.L. Gyorffy, A.J. Pindor, J. Staunton, G.M. Stocks, H. Winter, J. Phys. F: Met. Phys. **15**, 1337 (1985)
47. J. Staunton, B.L. Gyorffy, A.J. Pindor, G.M. Stocks, H. Winter, J. Phys. F: Met. Phys. **15**, 1387 (1985)
48. J. Staunton, B.L. Gyorffy, G. Stocks, J. Wadsworth, J. Phys. F: Met. Phys. **16**, 1761 (1986)
49. H. Hasegawa, J. Phys. F: Met. Phys. **13**, 2655 (1983)
50. V.I. Grebennikov, J. Magn. Magn. Mater. **84**, 59 (1990)
51. N.B. Melnikov, B.I. Reser, Phys. Procedia **75**, 739 (2015)
52. J.P. Sethna, *Statistical Mechanics: Entropy, Order Parameters and Complexity* (Oxford University Press, New York, 2006)
53. J. Crangle, G.M. Goodman, Proc. R. Soc. Lond. A **321**, 477 (1971)
54. D.J. Kim, *New Perspectives in Magnetism of Metals* (Kluwer/Plenum, New York, 1999)
55. E.P. Wohlfarth, in *Ferromagnetic Materials*, vol. 1, ed. by E.P. Wohlfarth (North-Holland, Amsterdam, 1980), pp. 1–70
56. A. Ruban, S. Khmelevskyi, P. Mohn, B. Johansson, Phys. Rev. B **75**, 054402 (2007)
57. P.-W. Ma, S.L. Dudarev, Phys. Rev. B **86**, 054416 (2012)
58. D.M. Edwards, J. Magn. Magn. Mater. **36**(3), 213 (1983)
59. X. Tao, D.P. Landau, T.C. Schulthess, G.M. Stocks, Phys. Rev. Lett. **95**, 087207 (2005)

16 Conclusion

We have given a detailed introduction to the quantum theory of metallic magnetism. Magnetism is a cooperative phenomenon, and its study requires various many-body techniques such as linear response theory, fluctuation-dissipation theorem and Green functions. Application of these techniques to magnetic susceptibility is given in the introductory chapters and appendices. As a reference point, we presented two important early models: the Stoner mean-field theory and RPA. The rest of the book is devoted to the dynamic spin fluctuation theory (DSFT).

The DSFT describes thermal spin fluctuations taking their quantum nature and nonlocal character into account. The spin fluctuations are treated microscopically using the functional integral method. The essence of the method is to replace the electron–electron interaction by the interaction with a fluctuating exchange field. Magnetic characteristics in the functional integral method are obtained as averages over the fluctuating field configurations. Since the exchange interaction has a purely quantum character, the fluctuating field is both space- and *"time"*-dependent. When the fluctuations are not too large, calculation of the functional integral can be carried out using the Gaussian approximation. The simplest Gaussian approximation is obtained by the saddle-point method and leads to the RPA magnetic susceptibility. But the RPA gives reasonable results only at low temperatures, because it neglects the feedback of the spin fluctuations on the mean field.

The basis of the DSFT is the *optimal* Gaussian approximation of the fluctuating field. We have derived the formulae of the optimal Gaussian approximation for any magnetic ground state. In ferromagnets we use these formulae to calculate the mean field and mean-square fluctuations self-consistently. Application of the optimal Gaussian approximation to real metals requires further approximations. The DSFT employs a single-site Gaussian fluctuating field, which is completely described by the chemical potential, mean field and two fluctuations: longitudinal and transverse. Each fluctuation is a sum over the wavevectors and "frequencies".

For spin fluctuations with large amplitudes, the Gaussian approximation can be insufficient. The DSFT can lead to a first-order phase transition to the paramagnetic state, contrary to experiment in metals. To obtain a proper second-order phase transition in this case, we take into account higher-order terms of the free energy of electrons in the fluctuating exchange field. In the computational formulae of the extended DSFT, the third-order term renormalizes the mean field, and fourth-order term renormalizes the susceptibility.

In its final form the DSFT is not much more complicated than the Stoner mean-field theory. In the Stoner theory one has to solve two equations with two unknowns—the mean field and chemical potential—for each temperature. In the DSFT we have two extra equations and two extra variables: longitudinal and transverse fluctuations. In particular temperature regions, we can predict which of the two fluctuations dominate. Based on these predictions one can further simplify the theory.

Results of the DSFT for metals are a clear improvement of the ones of the Stoner model and static single-site SFT. The DSFT calculations are in good agreement with the neutron scattering, nuclear magnetic resonance and magneto-volume experiments. The main conclusion that can be drawn from our numerical calculations is as follows. In ferromagnetic metals the temperature behaviour of major magnetic characteristics, such as magnetization, local and effective magnetic moments and nuclear spin-relaxation rates, is determined mainly by the electron–electron correlations. These correlations can be adequately described within the DSFT over a wide temperature range including the intermediate temperature range. By estimating the Debye-Waller factor within the Brillouin zone as a function of temperature, we show that the effect of phonons on local magnetic characteristics is small.

N. B. Melnikov, B. I. Reser, *Dynamic Spin Fluctuation Theory of Metallic Magnetism*,
https://doi.org/10.1007/978-3-319-92974-3_16

Clearly the DSFT has its limitations. The most important one, in our opinion, is the use of the Hubbard-type Hamiltonian, which takes into account only local interaction of d electrons. Another serious simplification is the use of the nonpolarized DOS of d electrons and interaction constant as the input data. These and other approximations seem to be justified when we study local characteristics such as magnetization and local moment, but could become critical if we consider fine effects of electronic structure on magnetic properties. In our defence we can say that all modern theories that try to describe metallic magnetism based on detailed electronic structure have a largely simplified treatment of thermal spin fluctuations. A lot should be done to reconcile the two approaches.

There is a long-lasting controversy about whether the localized or itinerant models should be used to describe magnetism of metals. In this discussion the Heisenberg and Hubbard models are often contrasted to each other. The most important advantage of the Hubbard Hamiltonian and its generalizations is that they allow to explain noninteger single-site spin in metals. However, our results show that the assumption about the single-site interaction makes the results of the Hubbard model in many respects similar to the ones of the Heisenberg model. Indeed, in the DSFT we obtain the low-temperature $T^{3/2}$ law, Ornstein-Zernike form for the spin correlator at high temperatures and Curie-Weiss law for the paramagnetic susceptibility using the multiband Hubbard Hamiltonian.

The choice of the Hubbard-type Hamiltonian is decisive in SFT, e.g. it leads to small SRO and weak change of the local moment with temperature. Various approximations in SFT have a secondary effect on the results. Therefore, it seems that the progress in metallic magnetism requires two things. The first is reconciling the input data based on band calculations of the ground state with the model Hamiltonian. Indeed, the DFT already takes into account some of the electron–electron correlations which then appear in the interaction part of the Hamiltonian (the infamous "double counting" problem). The second, and most important, is improving the approximation of the electron–electron Hamiltonian beyond the localized interaction (such as the Hubbard or Heisenberg Hamiltonians) in a microscopic model.

One last remark. After carefully reading the text, we saw that some parts of the book could have been written more concisely and clearly. But the timeline is strict, and one can only exclaim as Cinderella:

> Goodness me, the clock has struck—
> Alackday, and fuck my luck.

(Kurt Vonnegut, *Slaughterhouse-Five*, 1969)

Appendices

Basic Mathematical Results

A

A mathematician proves, and a physicist convinces. (Author unknown)

A.1 Linear Operators and Matrices

Most of the texts on quantum mechanics, such as [1, 2], limit themselves to the basics of the operator theory (eigenvalues and eigenvectors, Hermitian and unitary operators, etc.), leaving aside important topics like functions of an operator and operator power series, which are necessary in quantum statistical physics. In this section, we collect some results about linear operators that are used in the main text. For a more systematic overview of matrix analysis, we refer to [3–5].

A.1.1 Inverse Operator

The operator A^{-1} is called the *inverse* of an operator A if

$$AA^{-1} = A^{-1}A = 1,$$

where 1 is the unity operator. The inverse of the product of operators A and B satisfies the equality

$$(AB)^{-1} = B^{-1}A^{-1}. \tag{A.1}$$

For an operator A we have a series expansion analogous to the geometric series[1]:

$$(1 - A)^{-1} = 1 + A + A^2 + \cdots \tag{A.2}$$

This can be verified as follows. Multiplying

$$S \equiv 1 + A + A^2 + \cdots$$

on the left by A, we obtain

$$AS = A + A^2 + A^3 + \cdots = S - 1.$$

Hence

$$(1 - A)S = 1.$$

This proves formula (A.2). An immediate corollary is

$$\begin{aligned} A(1 - BA)^{-1} &= (1 - AB)^{-1}A \\ &= A + ABA + ABABA + \cdots \end{aligned} \tag{A.3}$$

[1]Here and hereafter, we consciously adhere to the physical level of rigour. In particular, we consider formal series and do not discuss their convergence, as is typical of quantum mechanics.

N. B. Melnikov, B. I. Reser, *Dynamic Spin Fluctuation Theory of Metallic Magnetism*,
https://doi.org/10.1007/978-3-319-92974-3

In particular,

$$A^{-1}(1-BA^{-1})^{-1} = (1-A^{-1}B)^{-1}A^{-1}$$
$$= A^{-1} + A^{-1}BA^{-1} + A^{-1}BA^{-1}BA^{-1} + \cdots$$

As an example, we consider the zeroth Green function

$$G^0(z) = (z + \mu - \mathcal{H}_0)^{-1}$$

and the Green function in the presence of a static field V:

$$G(z) = (z + \mu - \mathcal{H}_0 - V)^{-1}.$$

Then clearly

$$\begin{aligned} G &= [(z+\mu-\mathcal{H}_0)(1-(z+\mu-\mathcal{H}_0)^{-1}V)]^{-1} \\ &= (1-(z+\mu-\mathcal{H}_0)^{-1}V)^{-1}(z+\mu-\mathcal{H}_0)^{-1} \\ &= (1-G^0V)^{-1}G^0. \end{aligned} \tag{A.4}$$

Using formula (A.3), we obtain

$$G = G^0 + G^0VG^0 + G^0VG^0VG^0 + \cdots$$

From this we come to the *Dyson equation*

$$G = G^0(1+VG) \quad \text{or} \quad G = (1+GV)G^0.$$

A.1.2 Functions of an Operator

Note that the operator expansion (A.2) can be obtained by substituting the operator A to the Taylor series $1/(1-x) = 1 + x + x^2 + \cdots$. Similarly, we can define $f(A)$ for any function $f(x)$ represented by its Taylor series

$$f(x) = f(\tilde{x}) + \mathrm{d}f(\tilde{x}) + \frac{1}{2!}\mathrm{d}^2 f(\tilde{x}) + \cdots + \frac{1}{n!}\mathrm{d}^n f(\tilde{x}) + \cdots \tag{A.5}$$

Here $\mathrm{d}^n f(\tilde{x}) = f^{(n)}(\tilde{x})(\mathrm{d}x)^n$ is the nth differential at the point $\tilde{x} = x + \Delta x$, where $\mathrm{d}x$ is just $\Delta x = x - \tilde{x}$. We need an analogue of the nth differential $\mathrm{d}^n f(\tilde{x})$ for an operator.

The (first) differential $\mathrm{d}f(A)$ is the linear part of $f(A+\Delta A) - f(A)$ with respect to ΔA. For example, the differential of the inverse operator is the linear part of $(A+\Delta A)^{-1} - A^{-1}$. The latter can be written as

$$(A+\Delta A)^{-1} - A^{-1} = A^{-1}(A(A+\Delta A)^{-1} - 1).$$

Using property (A.1), we have

$$(A+\Delta A)^{-1} - A^{-1} = A^{-1}((1+\Delta A\,A^{-1})^{-1} - 1).$$

Expanding $(1+\Delta A\,A^{-1})^{-1}$ into the power series by formula (A.2) and leaving only the first-order term in ΔA, we obtain

$$(A+\Delta A)^{-1} - A^{-1} = A^{-1}\Delta A\,A^{-1} + \cdots.$$

Thus,

$$\mathrm{d}(A^{-1}) = -A^{-1}\Delta A\,A^{-1}. \tag{A.6}$$

The nth differential $\mathrm{d}^n f(A)$ is the nth-order term of $f(A+\Delta A) - f(A)$ with respect to ΔA. Analogously to the Taylor series for ordinary functions (A.5), for the function of an operator $f(A)$ we can write the Taylor series expansion around $\tilde{A}$ as

$$f(A) = f(\tilde{A}) + \mathrm{d}f(\tilde{A}) + \frac{1}{2!}\mathrm{d}^2 f(\tilde{A}) + \cdots + \frac{1}{n!}\mathrm{d}^n f(\tilde{A}) + \cdots ,$$

where $A = \tilde{A} + \mathrm{d}A$ and $\mathrm{d}A = \Delta A$.

As an application of the above formulae, let us calculate the expansion of the logarithm the Green function $G = G(V)$ in the field V. By formula (A.4) the Green function is given by

$$G(V) = (1 - G^0 V)^{-1} G^0. \tag{A.7}$$

Then the differential is

$$\mathrm{d}G(V) = \mathrm{d}((1 - G^0 V)^{-1}) G^0. \tag{A.8}$$

Using formula (A.6), we obtain

$$\begin{aligned}\mathrm{d}((1 - G^0 V)^{-1}) &= -(1 - G^0 V)^{-1}(-G^0 \mathrm{d}V)(1 - G^0 V)^{-1} \\ &= (1 - G^0 V)^{-1} G^0 \mathrm{d}V (1 - G^0 V)^{-1}.\end{aligned}$$

Substituting this result in (A.8) and taking (A.7) into account, we have

$$\mathrm{d}G(V) = G(V)\mathrm{d}V G(V). \tag{A.9}$$

Similar to the well-known formula from calculus:

$$\frac{\mathrm{d}}{\mathrm{d}x} \ln f(x) = \frac{1}{f(x)} \frac{\mathrm{d}}{\mathrm{d}x} f(x),$$

we write the differential of the operator $\ln G(V)$ as

$$\mathrm{d} \ln G(V) = G^{-1}(V)\mathrm{d}G(V).$$

Taking (A.9) into account, we obtain

$$\mathrm{d} \ln G(V) = \mathrm{d}V G(V). \tag{A.10}$$

Next, we calculate the second differential

$$\mathrm{d}^2 \ln G(V) = \mathrm{d}(\mathrm{d} \ln G(V)) = \mathrm{d}(\mathrm{d}V G(V)). \tag{A.11}$$

The differential of the product of operators A and B is calculated as

$$\mathrm{d}(AB) = (\mathrm{d}A)B + A\mathrm{d}B.$$

Applying this formula to (A.11) and taking $\mathrm{d}^2 V = 0$ into account, we have

$$\mathrm{d}^2 \ln G(V) = \mathrm{d}V \mathrm{d}G(V).$$

Using (A.9) again, we obtain

$$\mathrm{d}^2 \ln G(V) = \mathrm{d}V G(V) \mathrm{d}V G(V) = (\mathrm{d}V G(V))^2. \tag{A.12}$$

Repeating this procedure, for the higher-order differential we have

$$\mathrm{d}^n \ln G(V) = (n-1)!(\mathrm{d}V G(V))^n. \tag{A.13}$$

The expansion of the operator $\ln G(V)$ is written as

$$\ln G(V) = \ln G(\tilde{V}) + \mathrm{d}\ln G(\tilde{V}) + \frac{1}{2!}\mathrm{d}^2 \ln G(\tilde{V}) + \cdots + \frac{1}{n!}\mathrm{d}^n \ln G(\tilde{V}) + \cdots$$

Using formula (A.13), we obtain

$$\ln G(V) = \ln G(\tilde{V}) + \mathrm{d}VG(\tilde{V}) + \frac{1}{2}\left(\mathrm{d}VG(\tilde{V})\right)^2 + \cdots + \frac{1}{n}\left(\mathrm{d}VG(\tilde{V})\right)^n + \cdots,$$

Now we recall that for the independent variable $\mathrm{d}V$ is equal to $\Delta V = V - \tilde{V}$. Therefore, the expansion is finally written as

$$\ln G(V) = \ln G(\tilde{V}) + \Delta VG(\tilde{V}) + \frac{1}{2}\left(\Delta VG(\tilde{V})\right)^2 + \cdots + \frac{1}{n}\left(\Delta VG(\tilde{V})\right)^n + \cdots \tag{A.14}$$

In statistical physics the increment ΔA of an operator A is often taken with respect to its mean value $\langle A \rangle$:

$$\Delta A = A - \langle A \rangle. \tag{A.15}$$

In this case the mean of the commutator $[A, B] = AB - BA$ satisfies the relation

$$\langle [\Delta A, \Delta B] \rangle = \langle [A, B] \rangle. \tag{A.16}$$

Indeed, using (A.15), we write the left-hand side of (A.16) as

$$\langle [A - \langle A \rangle, B - \langle B \rangle] \rangle = \langle [A, B] \rangle + \langle [\langle A \rangle, \langle B \rangle] \rangle - \langle [A, \langle B \rangle] \rangle - \langle [\langle A \rangle, B] \rangle.$$

The second term is equal to zero, because $\langle A \rangle$ and $\langle B \rangle$ are just numbers multiplied by the unity operator. Calculating the third terms, we obtain

$$\langle [A, \langle B \rangle] \rangle = \langle A \langle B \rangle - \langle B \rangle A \rangle = \langle A \rangle \langle B \rangle - \langle B \rangle \langle A \rangle = 0.$$

Similarly, the fourth term is also zero. Thus, we have proved (A.16).

A.1.3 Trace of an Operator

The trace of an operator A is defined as

$$\mathrm{tr}A = \sum_i A_{ii}, \qquad A_{ii} = (\psi_i, A\psi_i),$$

and does not depend on the basis ψ_i. The trace satisfies the following properties:

$$\mathrm{tr}(\alpha A) = \alpha \mathrm{tr}A, \qquad \mathrm{tr}(A + B) = \mathrm{tr}A + \mathrm{tr}B, \qquad \mathrm{tr}(BA) = \mathrm{tr}(AB).$$

The latter immediately follows if we write the trace of the product:

$$\mathrm{tr}(AB) = \sum_{ik} A_{ik} B_{ki} = \sum_{ik} B_{ki} A_{ik} = \mathrm{tr}(BA).$$

As a corollary, we obtain the cyclic property of trace

$$\mathrm{tr}(ABC...DF) = \mathrm{tr}(BC...DFA)$$

(one needs to swap A with the rest of the product and use the previous formula).

For a Hermitian matrix A the trace of $f(A)$ is calculated by the formula

$$\mathrm{tr} f(A) = \sum_i f(\lambda_i), \tag{A.17}$$

where λ_i are the eigenvalues of A. Indeed, for a Hermitian matrix A we have

$$A = UDU^\dagger,$$

where D is diagonal and U is a unitary transformation: $UU^\dagger = 1$. Then by the cyclic property of trace we obtain

$$\mathrm{tr}A = \mathrm{tr}(UDU^\dagger) = \mathrm{tr}D.$$

Since

$$A^m = \underbrace{UDU^\dagger UDU^\dagger \ldots}_{m} = UD^m U^\dagger,$$

we have

$$\mathrm{tr}A^m = \mathrm{tr}(UD^m U^\dagger) = \mathrm{tr}D^m.$$

For any analytic function

$$f(x) = \sum_{n=0}^{\infty} a_n x^n,$$

we obtain

$$f(A) = \sum_{n=0}^{\infty} a_n A^n = \sum_{n=0}^{\infty} a_n U D^n U^\dagger = Uf(D)U^\dagger,$$

and hence

$$\mathrm{tr} f(A) = \mathrm{tr}(Uf(D)U^\dagger) = \mathrm{tr} f(D).$$

The latter gives us formula (A.17). For example, if $f(x) = \ln x$, we come to

$$\mathrm{tr} \ln A = \sum_i \ln \lambda_i = \ln\Big(\prod_i \lambda_i\Big) = \ln \det A. \tag{A.18}$$

For noncommuting operators A and B, we have

$$\mathrm{e}^A \mathrm{e}^B \neq \mathrm{e}^{A+B}. \tag{A.19}$$

This can be seen directly (see [6]). Using the expansion of e^x, we have

$$\begin{aligned} \mathrm{e}^A \mathrm{e}^B &= \left(1 + A + \frac{1}{2!}A^2 + \cdots\right)\left(1 + B + \frac{1}{2!}B^2 + \cdots\right) \\ &= 1 + (A + B) + \frac{1}{2!}(A^2 + 2AB + B^2) + \cdots \end{aligned}$$

On the other hand,

$$\begin{aligned} \mathrm{e}^{A+B} &= 1 + (A + B) + \frac{1}{2!}(A + B)^2 + \cdots \\ &= 1 + (A + B) + \frac{1}{2!}(A^2 + AB + BA + B^2) + \cdots . \end{aligned}$$

The operators $e^A e^B$ and e^{A+B} are different, because $AB \neq BA$. A more specific result is as follows. If the commutator is proportional to the unity matrix: $[A, B] = c$ (here c is a number), then

$$e^A e^B = e^{A+B} e^{\frac{1}{2}[A,B]}$$

(for a proof, see, e.g. [7, Appendix I]).

Analogously to relation (A.19), for noncommuting operators A and B, we have $\ln(AB) \neq \ln A + \ln B$. However, for trace of the logarithm of Hermitian operators we obtain

$$\operatorname{tr}\ln(AB) = \operatorname{tr}\ln A + \operatorname{tr}\ln B. \tag{A.20}$$

Indeed, equality (A.18) leads to

$$\begin{aligned}\operatorname{tr}\ln(AB) &= \ln\det(AB) = \ln(\det A \cdot \det B)\\ &= \ln\det A + \ln\det B = \operatorname{tr}\ln A + \operatorname{tr}\ln B.\end{aligned}$$

A.1.4 Pauli Spin Matrices

The Pauli spin matrices are given by

$$\sigma^x = \begin{pmatrix} 0 & 1 \\ 1 & 0 \end{pmatrix}, \quad \sigma^y = \begin{pmatrix} 0 & -\mathrm{i} \\ \mathrm{i} & 0 \end{pmatrix}, \quad \sigma^z = \begin{pmatrix} 1 & 0 \\ 0 & -1 \end{pmatrix}. \tag{A.21}$$

Explicit calculation shows that

$$\begin{aligned}\operatorname{tr}(\sigma^i\sigma^k) &= 2\delta_{ik},\\ \operatorname{tr}(\sigma^i\sigma^k\sigma^l) &= \mathrm{i}2\varepsilon_{ikl},\\ \operatorname{tr}(\sigma^i\sigma^k\sigma^l\sigma^j) &= 2(\delta_{ik}\delta_{lj} - \delta_{il}\delta_{kj} + \delta_{ij}\delta_{kl}).\end{aligned} \tag{A.22}$$

Here $i, k, l, j = x, y, z$ and

$$\varepsilon_{ikl} = [\mathbf{e}_i \times \mathbf{e}_k]\mathbf{e}_l$$

is the Levi-Civita symbol, where $\mathbf{e}_x, \mathbf{e}_y, \mathbf{e}_z$ are the orthonormal basis vectors in the three-dimensional Euclidean space and times stands for the vector product (see, e.g. [4, 8]).

Any Hermitian 2×2 matrix

$$A = \begin{pmatrix} A_{11} & A_{12} \\ A_{21} & A_{22} \end{pmatrix}$$

can be represented as a linear combination of the 2×2 unity matrix σ^0 and Pauli matrices:

$$A = \sum_\mu A^\mu \sigma^\mu, \qquad \mu = 0, x, y, z. \tag{A.23}$$

Using the explicit form of the Pauli matrices (A.21), this can be written as

$$A = \begin{pmatrix} A^0 + A^z & A^x - \mathrm{i}A^y \\ A^x + \mathrm{i}A^y & A^0 - A^z \end{pmatrix}. \tag{A.24}$$

The scalar coefficients A^μ are obtained as follows. Multiplying (A.23) on the right by $\sigma^{\mu'}$ and taking the trace, we come to

$$\operatorname{tr}(A\sigma^{\mu'}) = \operatorname{tr}\left(\sum_\mu A^\mu \sigma^\mu \sigma^{\mu'}\right) = \sum_\mu A^\mu \operatorname{tr}(\sigma^\mu \sigma^{\mu'}).$$

Calculating the trace by formula (A.22), we have

$$\mathrm{tr}(A\sigma^{\mu'}) = \sum_{\mu} A^{\mu} 2\delta_{\mu\mu'} = 2A^{\mu'}.$$

Hence, dropping the prime, we finally write

$$A^{\mu} = \frac{1}{2}\mathrm{tr}(A\sigma^{\mu}). \tag{A.25}$$

A.1.5 Operator Tensor Product

The *tensor product* $A \otimes B$ of a $m \times n$-matrix $A \equiv [a_{ik}]$ and $m' \times n'$-matrix $B \equiv [b_{i'k'}]$ is the $mm' \times nn'$-matrix

$$A \otimes B \equiv [c_{jh}] \qquad (c_{jh} = a_{ik} b_{i'k'}), \tag{A.26}$$

where the index j denotes the running number of the pair (i, i') in the sequence $(1, 1), (1, 2), \dots, (1, m'), (2, 1), (2, 2), \dots, (2, m'), \dots, (m, m')$, and the index h is the running number of the pair (k, k') in the analogous sequence (see, e.g. [4]). Note that

$$(A \otimes B)(C \otimes D) = AC \otimes BD, \tag{A.27}$$

$$\mathrm{tr}(A \otimes B) = \mathrm{tr}(A)\mathrm{tr}(B). \tag{A.28}$$

In (A.27) it is assumed that the number of rows in the matrix C is equal to the number of columns of A and number of row in the matrix D is equal to the number of columns of B; and in (A.28) it is assumed that A and B are square matrices.

One-electron operators, such as the exchange field V, can be expressed as

$$V = W \otimes U, \tag{A.29}$$

where W is an operator acting on the *space* states (the Bloch functions $\psi_{\nu\mathbf{k}}(\mathbf{r})$ or Wannier functions $w_{\nu j}(\mathbf{r})$) and U is an operator acting on the *spin* states χ_{σ}. Using the expression for the Hermitian 2×2 matrix (A.23):

$$U = \sum_{\mu} U^{\mu}\sigma^{\mu}, \qquad U^{\mu} = \frac{1}{2}\mathrm{tr}(U\sigma^{\mu}), \qquad \mu = 0, x, y, z,$$

we write (A.29) as

$$V = \sum_{\mu} V^{\mu} \otimes \sigma^{\mu}, \tag{A.30}$$

where $V^{\mu} = U^{\mu}W$ is an *operator* acting on the space states. Similarly, the Green function is written as

$$G = \sum_{\mu} G^{\mu} \otimes \sigma^{\mu}, \tag{A.31}$$

where G^{μ} is an *operator* acting on the space states. Then, using formula (A.27), we can write

$$\mathrm{tr}(VG) = \sum_{\mu\mu'} \mathrm{tr}\big((V^{\mu} \otimes \sigma^{\mu})(G^{\mu'} \otimes \sigma^{\mu'})\big) = \sum_{\mu\mu'} \mathrm{tr}\big((V^{\mu}G^{\mu'}) \otimes (\sigma^{\mu}\sigma^{\mu'})\big)$$

and using formula (A.28) we come to

$$\mathrm{tr}(VG) = \sum_{\mu\mu'} \mathrm{tr}(V^{\mu}G^{\mu'})\mathrm{tr}(\sigma^{\mu}\sigma^{\mu'}).$$

Finally, calculating the trace of the product of the Pauli matrices by formula (A.22), we have

$$\mathrm{tr}(VG) = 2\sum_{\mu} \mathrm{tr}(V^{\mu}G^{\mu}). \tag{A.32}$$

For brevity, we omit the tensor multiplication sign $\otimes$ throughout the book as is typical of the solid-state literature.

A.2 Functions of Real and Complex Variable

In this section, we collect some formulae from calculus and complex analysis that are used in the main text (for a more systematic overview, see, e.g. [4, 7, 9]).

A.2.1 Dirac Delta Function

The function $\delta(x)$ can be thought of as the density function of the probability distribution localized at the origin (for details on random variables, see Appendix A.3). That means the average of a function of the random variable $f(x)$ is equal to its value at $x = 0$:

$$\int_{-\infty}^{\infty} f(x)\delta(x)\,\mathrm{d}x = f(0), \tag{A.33}$$

and, in particular,

$$\int_{-\infty}^{\infty} \delta(x)\,\mathrm{d}x = 1. \tag{A.34}$$

These relations imply that

$$\delta(x) = \begin{cases} 0, & x \neq 0, \\ \infty, & x = 0, \end{cases} \tag{A.35}$$

so that $\delta(x)$ cannot be a proper function. Nevertheless, the δ-function is a useful instrument when one deals with integrals.

From formula (A.33), we obtain the δ-function properties

$$\int_{-\infty}^{\infty} f(x)\delta(x - x_0)\,\mathrm{d}x = f(x_0), \tag{A.36}$$

and

$$\delta(ax) = \frac{1}{|a|}\delta(x), \qquad a \neq 0. \tag{A.37}$$

In particular, from (A.37) it follows that $\delta(x)$ is an even function: $\delta(-x) = \delta(x)$. From (A.36) it is easy to see that

$$f(x)\delta(x - x_0) = f(x_0)\delta(x - x_0).$$

Another way of representing the δ-function is the following. If $\theta(x)$ is the step function, which is equal to zero for $x < 0$ and unity for $x > 0$, then

$$\delta(x) = \frac{\mathrm{d}}{\mathrm{d}x}\theta(x). \tag{A.38}$$

Indeed, if we substitute the latter in the integral on the left-hand side of (A.33), then, integrating by parts and assuming that $f(x)$ vanishes at infinity, we obtain $f(0)$.

The δ-function can be obtained as a limit of the proper functions. This can be done in different ways. Let us consider several examples. First, we consider the Gaussian distribution with the probability density function

$$\rho_{\mathrm{G}}(x) = \frac{1}{\sqrt{2\pi}\,\sigma} \exp\left(-\frac{x^2}{2\sigma^2}\right). \tag{A.39}$$

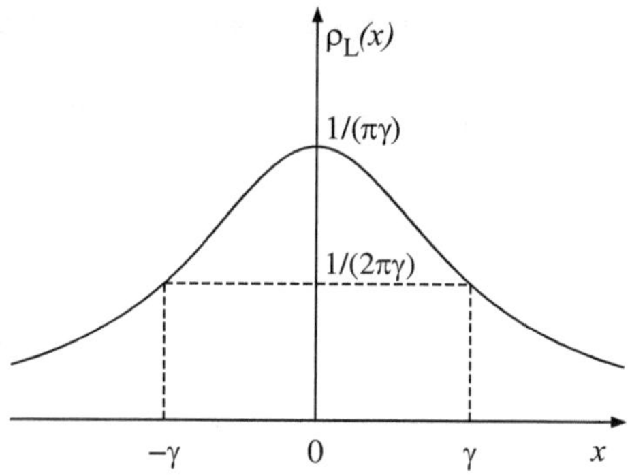

Fig. A.1 Sketch of the Lorenzian function $\rho_L(x)$ of the halfwidth 2γ

The width of the peak at $x = 0$ is characterized by the root-mean-square deviation σ. As $\sigma \to 0$, we have

$$\int_{-\infty}^{\infty} f(x)\rho_G(x - x_0)\,dx \to f(x_0).$$

Our second example is the density function of the Cauchy-Lorentz distribution (see, e.g. [10]), often called the *Lorentzian function*:

$$\rho_L(x) = \frac{1}{\pi\gamma}\frac{1}{1 + (x/\gamma)^2} = \frac{1}{\pi}\frac{\gamma}{x^2 + \gamma^2}. \tag{A.40}$$

Here the value γ is the *halfwidth* at half maximum (HWHM). Indeed, the maximum of the Lorentzian function is attained at $x = 0$ and is equal to $1/(\pi\gamma)$. The value $1/(2\pi\gamma)$ is taken by the Lorentzian function at $x = \pm\gamma$ (Fig. A.1). If γ tends to zero, then

$$\int_{-\infty}^{\infty} f(x)\rho_L(x - x_0)\,dx \to f(x_0).$$

Yet another example of a delta sequence is given by the integral (see also Appendix F.1)

$$\frac{1}{2\pi}\int_{-k_0}^{k_0} e^{ikx}\,dk = \frac{1}{\pi x}\frac{e^{ik_0x} - e^{-ik_0x}}{2i} = \frac{1}{\pi x}\sin(k_0x).$$

At large k_0 the right-hand side sharply peaks around $x = 0$, but the integral over x is equal to unity for all k_0. Therefore, in the limit $k_0 \to \infty$, we come to the well-known integral representation:

$$\delta(x) = \frac{1}{2\pi}\int_{-\infty}^{\infty} e^{ikx}\,dk, \tag{A.41}$$

which means that unity is the Fourier transform of the δ-function.

In the three-dimensional Euclidean space, we define the δ-function as

$$\delta(\mathbf{r}) = \delta(x)\delta(y)\delta(z),$$

so that (A.33) and (A.34) become

$$\int f(\mathbf{r})\delta(\mathbf{r})\,d\mathbf{r} = f(0)$$

and

$$\int \delta(\mathbf{r})\,d\mathbf{r} = 1,$$

where $d\mathbf{r} = dx\,dy\,dz$ is the volume element and the integration is carried out over the whole three-dimensional space. The inverse Fourier transform of the three-dimensional δ-function is given by

$$\delta(\mathbf{r}) = \frac{1}{(2\pi)^3}\int e^{i\mathbf{kr}}\,d\mathbf{k}. \tag{A.42}$$

A.2.2 Sokhotsky Integral Formula

It is often necessary, particularly dealing with the Green functions, to calculate an expression of the form

$$\lim_{\eta\to 0^+}\int_{-\infty}^{\infty}\frac{f(x)}{x\pm i\eta}\,dx \equiv \int_{-\infty}^{\infty}\frac{f(x)}{x\pm i0^+}\,dx\,.$$

For a well-behaved function $f(x)$ (in particular, continuous at $x = 0$) the *Sokhotsky formula* holds:

$$\lim_{\eta\to 0^+}\int_{-\infty}^{\infty}\frac{f(x)}{x\pm i\eta}\,dx = \mathcal{P}\int_{-\infty}^{\infty}\frac{f(x)}{x}\,dx \mp i\pi f(0), \tag{A.43}$$

where $\mathcal{P}$ denotes the *principal value* of the integral (for proof see, e.g. [11, Appendix III]). Clearly, for any fixed x_0, we have

$$\lim_{\eta\to 0^+}\int_{-\infty}^{\infty}\frac{f(x)}{x-x_0\pm i\eta}\,dx = \mathcal{P}\int_{-\infty}^{\infty}\frac{f(x)}{x-x_0}\,dx \mp i\pi f(x_0). \tag{A.44}$$

The latter is often expressed in the compact form

$$\frac{1}{x-x_0\pm i0^+} = \mathcal{P}\frac{1}{x-x_0} \mp i\pi\delta(x-x_0).$$

A.2.3 Smoothing by Convolution

Consider the convolution of the functions $f(x)$ and $g(x)$ (see, e.g. [4]):

$$h(y) = \int_{-\infty}^{\infty} f(x)g(y-x)\,dx.$$

If $g(x)$ is a smooth function, then so is $h(y)$, even if $f(x)$ is not. Indeed,

$$h^{(n)}(y) = \int_{-\infty}^{\infty} f(x)g^{(n)}(y-x)\,dx.$$

We take a smooth function $\rho(x)$ that approximates the δ-function such as the Gaussian density function (A.39) with a small σ or Lorentzian function (A.40) with a small γ. Then the convolution of a function $f(x)$ with the function $\rho(x)$ is a smooth function. Moreover, if $\rho(x)$ is even, then, recalling formula (A.36), we have

$$\int_{-\infty}^{\infty} f(x)\rho(y-x)\,dx \approx f(y).$$

That means the convolution with the function $\rho(x)$ close to the δ-function gives a smooth approximation of the original function $f(x)$.

Let us consider the following example. Nonphysical sharp peaks of the electrons density of states (DOS) $\nu(\varepsilon)$, which appear in energy-band calculations (see Sect. 10.5), can be smoothed by the formula [12]

$$\nu_s(\varepsilon) = \frac{1}{\pi}\mathrm{Im}\int\frac{\nu(\varepsilon')}{\varepsilon-\varepsilon'-i\Gamma}\,d\varepsilon'. \tag{A.45}$$

Indeed, from the Sokhotsky formula (A.44), we have

$$\nu(\varepsilon) = \lim_{\Gamma \to 0^+} \frac{1}{\pi} \mathrm{Im} \int \frac{\nu(\varepsilon')}{\varepsilon - \varepsilon' - \mathrm{i}\Gamma} \, \mathrm{d}\varepsilon'.$$

Hence at small $\Gamma > 0$ the smoothed DOS $\nu_s(\varepsilon)$ is close to the original one $\nu(\varepsilon)$.

The right-hand side of (A.45) can be written as the convolution. To see that we calculate the imaginary part of the integrand and recall that $\nu(\varepsilon)$ is a real function. Then

$$\nu_s(\varepsilon) = \int \nu(\varepsilon')\rho(\varepsilon - \varepsilon') \, \mathrm{d}\varepsilon',$$

where

$$\rho(\varepsilon) = \frac{1}{\pi} \, \mathrm{Im} \frac{1}{\varepsilon - \mathrm{i}\Gamma} = \frac{1}{\pi} \frac{\Gamma}{\varepsilon^2 + \Gamma^2}$$

is the Lorentzian function (A.40) of the halfwidth Γ.

A.2.4 Singularities and Contour Integrals

Integrals of functions of a real variable appearing in the Green function theory can be evaluated conveniently by regarding them as part of a contour integral in the complex z-plane (see, e.g. [11], Appendix VI; [9], Chaps. V and VI).

Recall that a function $f(z)$ is called *analytic* in a circle centred at $z = a$ if it can be expanded in the series of the form

$$f(z) = A_0 + A_1(z - a) + A_2(z - a)^2 + \cdots, \tag{A.46}$$

where A_i are constants. This is just the Taylor series for the function near $z = a$. A point at which a function fails to be analytic is called a *singular point* or *singularity* for short.

An important tool for investigating a function $f(z)$ near a singularity $z = a$ is the *Laurent series*

$$f(z) = A_0 + A_1(z - a) + A_2(z - a)^2 + \cdots + \frac{B_1}{z - a} + \frac{B_2}{(z - a)^2} + \cdots,$$

where A_i and B_i are constants. If the series of reciprocal powers is infinite, the point $z = a$ is called an *essential singularity*. If the series takes the form

$$f(z) = A_0 + A_1(z - a) + A_2(z - a)^2 + \cdots + \frac{B_1}{z - a} + \frac{B_2}{(z - a)^2} + \cdots + \frac{B_n}{(z - a)^n}, \tag{A.47}$$

where the number n is finite but nonzero, $f(z)$ is said to have a *pole of order* n at $z = a$. If $n = 1$, the singular point is called a *simple pole*. For example, a rational function

$$f(z) = \frac{p(z)}{q(z)},$$

where the polynomials $p(z)$ and $q(z)$ have no common factor, has poles at each zero of the polynomial $q(z) = Q(z - z_1)^{n_1} \ldots (z - z_k)^{n_k}$ ($Q \neq 0$ is a scalar), and the order of the pole z_i is equal to n_i.

Whatever the value $n \geq 1$ is, the coefficient B_1 of $(z - a)^{-1}$ is called the *residue* at $z = a$ and is denoted $\mathrm{res}_{z=a} f(z)$. This applies to both poles and essential singularities. If $z = a$ is a simple pole of $f(z)$, the residue can be calculated by the formula

$$B_1 = \lim_{z \to a} (z - a) f(z), \tag{A.48}$$

which immediately follows from Eq. (A.47) with $n = 1$.

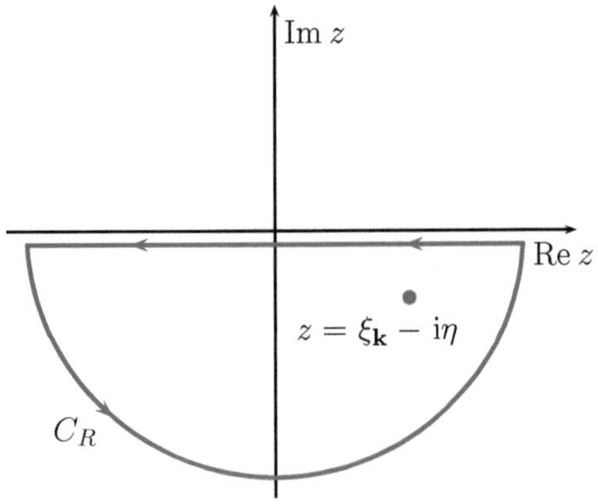

Fig. A.2 The integration contour and pole of the integrand on the left-hand side of (A.52)

The most important fact about contour integrals, the *residue theorem*, is now formulated as follows. *Let C be a closed contour without intersections, and $f(z)$ be analytic inside and on C except for a finite number of singularities z_i inside the contour, then*

$$\frac{1}{2\pi \mathrm{i}} \oint_C f(z)\,\mathrm{d}z = \sum_i \mathrm{res}_{z=z_i} f(z), \tag{A.49}$$

where the contour integral is anticlockwise. If $f(z)$ has no singularities inside or on C, the right-hand side of (A.49) reduces to zero. In other words, if the function $f(z)$ is analytic everywhere inside and on C, the contour integral is zero. This fact is known as *Cauchy's theorem.*

As an example, we apply the residue theorem to obtain expression (6.9) for the retarded real-time Green function of noninteracting electrons. Taking the inverse Fourier transformation of expression (6.8):

$$G^{\mathrm{r}0}_{\mathbf{k}}(\omega) = \lim_{\eta\to 0^+} \frac{1}{\omega - \xi_{\mathbf{k}} + \mathrm{i}\eta}, \tag{A.50}$$

we have

$$G^{\mathrm{r}0}_{\mathbf{k}}(t) = \lim_{\eta\to 0^+} \frac{1}{2\pi} \int \frac{\mathrm{e}^{-\mathrm{i}\omega t}}{\omega - \xi_{\mathbf{k}} + \mathrm{i}\eta}\,\mathrm{d}\omega, \tag{A.51}$$

where $\xi_{\mathbf{k}} = \varepsilon_{\mathbf{k}} - \mu$ (for brevity, we put $\hbar = 1$). The integrand of (A.51) is analytic everywhere except for the isolated singularity $z = \xi_{\mathbf{k}} - \mathrm{i}\eta$, where it has a simple pole. Using the residue theorem (A.49), we obtain

$$\frac{1}{2\pi \mathrm{i}} \lim_{R\to\infty} \oint_{C_R} \frac{\mathrm{e}^{-\mathrm{i}zt}}{z - \xi_{\mathbf{k}} + \mathrm{i}\eta}\,\mathrm{d}z = \mathrm{res}_{z=\xi_{\mathbf{k}}-\mathrm{i}\eta}\left(\frac{\mathrm{e}^{-\mathrm{i}zt}}{z - \xi_{\mathbf{k}} + \mathrm{i}\eta}\right), \tag{A.52}$$

where the contour C_R is shown in Fig. A.2. As $R \to \infty$ the integral over the arc of the circle tends to zero, because the integrand

$$\frac{\mathrm{e}^{-\mathrm{i}zt}}{z - \xi_{\mathbf{k}} + \mathrm{i}\eta} = \frac{\mathrm{e}^{-\mathrm{i}t\mathrm{Re}z}\,\mathrm{e}^{t\mathrm{Im}z}}{z - \xi_{\mathbf{k}} + \mathrm{i}\eta}, \qquad \mathrm{Im}z < 0,$$

vanishes exponentially for $t > 0$. Therefore,

$$\lim_{R\to\infty} \oint_{C_R} \frac{\mathrm{e}^{-\mathrm{i}zt}}{z - \xi_{\mathbf{k}} + \mathrm{i}\eta}\,\mathrm{d}z = -\int_{-\infty}^{\infty} \frac{\mathrm{e}^{-\mathrm{i}\omega t}}{\omega - \xi_{\mathbf{k}} + \mathrm{i}\eta}\,\mathrm{d}\omega, \tag{A.53}$$

when $t > 0$. Here the negative sign results from the counterclockwise integration. Calculating the residual by formula (A.48), we have

$$\mathrm{res}_{z=\xi_{\mathbf{k}}-\mathrm{i}\eta}\left(\frac{\mathrm{e}^{-\mathrm{i}zt}}{z - \xi_{\mathbf{k}} + \mathrm{i}\eta}\right) = \mathrm{e}^{-\mathrm{i}(\xi_{\mathbf{k}}-\mathrm{i}\eta)t}.$$

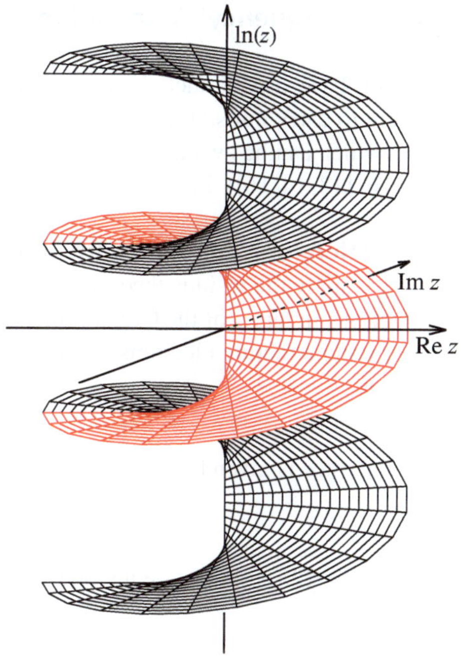

Fig. A.3 Graph of the multivalued function $\ln z = \ln|z| + \mathrm{i}\theta$ of the complex variable $z = |z|\,\mathrm{e}^{\mathrm{i}\theta}$ for $-3\pi < \theta < 3\pi$. The single-valued branch of $\ln z$ corresponding to $-\pi < \theta < \pi$ is shown by colour

Substituting this in (A.52) and taking (A.53) into account, we obtain

$$\int_{-\infty}^{\infty} \frac{\mathrm{e}^{-\mathrm{i}\omega t}}{\omega - \xi_{\mathbf{k}} + \mathrm{i}\eta}\,\mathrm{d}\omega = -2\pi\mathrm{i}\,\mathrm{e}^{-\mathrm{i}(\xi_{\mathbf{k}}-\mathrm{i}\eta)t}, \qquad t > 0.$$

When $t < 0$, we need to consider the contour integral over the semicircle in the upper half-plane. Since this contour does not contain any singularities, using the same argument, we obtain

$$\int_{-\infty}^{\infty} \frac{\mathrm{e}^{-\mathrm{i}\omega t}}{\omega - \xi_{\mathbf{k}} + \mathrm{i}\eta}\,\mathrm{d}\omega = 0, \qquad t < 0.$$

Thus, the Green function (A.51) is given by

$$G_{\mathbf{k}}^{\mathrm{r}0}(t) = -\mathrm{i}\mathrm{e}^{-\mathrm{i}\xi_{\mathbf{k}}t}\theta(t).$$

Note that trigonometric, hyperbolic, integer power and exponential functions are all single-valued. However, their inverses are multi-valued. For example, if we express z in polar form $z = |z|\,\mathrm{e}^{\mathrm{i}\theta}$, then $\ln z = \ln|z| + \mathrm{i}\theta$. The polar angle θ is defined up to $2\pi n$. Different n give the same z but different values of $\ln z$. So, at any point z, we have different branches of the logarithm (Fig. A.3). To define single-valued branches of $\ln z$ we make a *branch cut* from zero out to infinity, e.g. along the negative real axis. If we choose $-\pi < \theta < \pi$, then clearly the one-sided limits at the points of the branch cut will be different: $\ln z \to \ln|z| \pm \mathrm{i}\pi$ as $\theta \to \pm\pi$ (see Fig. A.3). A similar situation happens when we consider the integral

$$F(z) = \int_{-\infty}^{\infty} \frac{f(x)}{x - z}\,\mathrm{d}x.$$

The function $F(z)$ is analytic outside the real axis and has different one-sided limits at a point $z = x_0$ on the real axis where $f(x_0)$ is nonzero. Indeed, the Sokhotsky formula (A.44) gives

$$F(x_0 \pm \mathrm{i}0^+) = \mathcal{P}\int_{-\infty}^{\infty} \frac{f(x)}{x - x_0}\,\mathrm{d}x \mp \mathrm{i}\pi f(x_0).$$

A.2.5 "Frequency" Summation

In Chap. 2 we used Cauchy's theorem to derive the Kramers-Kronig relations (2.19)–(2.20). Here we apply the residue theorem to carry out summation over the thermodynamic "frequencies". We start with sums of a function with only *simple poles*, which can be evaluated explicitly (see also [13–15]). Then we derive summation rules for a function with *branch cuts* (see also [13, 14, 16]). Finally, we apply these summation rules to the DSFT.

Summation for a Function with Simple Poles

First, we apply the residue theorem to show that, in the case of *noninteracting* electrons, formula (6.33) for the total number of electrons in terms of the Green functions reduces to a familiar expression, which follows from the Fermi statistics (3.52).

For noninteracting electrons, we must replace the Green function G by the zeroth Green function G^0 and write formula (6.33) as

$$N_e = T\,\mathrm{Tr}G^0. \tag{A.54}$$

Here the trace is given by

$$\mathrm{Tr}G^0 = \lim_{\tau\to 0^+} \sum_{\mathbf{k}n} G^0_{\mathbf{k}}(\mathrm{i}\omega_n)\,\mathrm{e}^{\mathrm{i}\omega_n\tau}, \tag{A.55}$$

where $\omega_n = (2n+1)\pi T$ are the odd "frequencies" and

$$G^0_{\mathbf{k}}(z) = \frac{1}{z-\xi_{\mathbf{k}}} \tag{A.56}$$

is the zeroth thermodynamic Green function.

The sum over the "frequencies"

$$\sum_n G^0_{\mathbf{k}}(\mathrm{i}\omega_n)\,\mathrm{e}^{\mathrm{i}\omega_n\tau} \equiv \sum_n h(\mathrm{i}\omega_n) \tag{A.57}$$

is calculated using the Fermi function as follows. The function $h(z) = G^0_{\mathbf{k}}(z)\,\mathrm{e}^{z\tau}$ is analytic everywhere except for the isolated singularity $z=\xi_{\mathbf{k}}$, where it has a simple pole. The Fermi function $f(z) = (\exp(z/T)+1)^{-1}$ has simple poles at $\mathrm{i}\omega_n$.[2] Indeed, using formula (A.48) and L'Hopital's rule, we have

$$\mathrm{res}_{z=\mathrm{i}\omega_n} f(z) = \lim_{z\to\mathrm{i}\omega_n} \frac{z-\mathrm{i}\omega_n}{\mathrm{e}^{z/T}+1} = \lim_{z\to\mathrm{i}\omega_n} \frac{1}{(1/T)\mathrm{e}^{z/T}} = -T. \tag{A.58}$$

Therefore, using the residue theorem, we can write

$$\lim_{R\to\infty} \frac{1}{2\pi\mathrm{i}} \oint_{C_R} f(z)h(z)\,\mathrm{d}z = \mathrm{res}_{z=\xi_{\mathbf{k}}}\big(f(z)h(z)\big) + \sum_n \mathrm{res}_{z=\mathrm{i}\omega_n}\big(f(z)h(z)\big), \tag{A.59}$$

where the contour C_R is the circle of radius R centred at the origin. Since

$$f(z)\,\mathrm{e}^{z\tau} = \frac{\mathrm{e}^{z\tau}}{\mathrm{e}^{z/T}+1} \propto \begin{cases} \mathrm{e}^{\tau \mathrm{Re} z}, & \mathrm{Re} z < 0, \\ \mathrm{e}^{(\tau-1/T)\mathrm{Re} z}, & \mathrm{Re} z > 0, \end{cases}$$

and $0<\tau<1/T$, the function $f(z)\,\mathrm{e}^{z\tau}$ vanishes exponentially as $|z|\to\infty$. Therefore, the left-hand side of (A.59) tends to zero as $R\to\infty$. Similar to (A.58) we calculate

$$\mathrm{res}_{z=\xi_{\mathbf{k}}}\big(f(z)h(z)\big) = f(\xi_{\mathbf{k}})\mathrm{e}^{\xi_{\mathbf{k}}\tau}, \qquad \mathrm{res}_{z=\mathrm{i}\omega_n}\big(f(z)h(z)\big) = -Th(\mathrm{i}\omega_n).$$

Then Eq. (A.59) reduces to

$$f(\xi_{\mathbf{k}})\mathrm{e}^{\xi_{\mathbf{k}}\tau} - T\sum_n h(\mathrm{i}\omega_n) = 0.$$

[2] Note the difference in the definition of the Fermi function $f(z)$ in this subsection from the one in the main text (3.52). This definition of the Fermi function simplifies intermediate steps.

Thus, we obtain the sum over the "frequencies" (A.57):

$$T \sum_n G^0_{\mathbf{k}}(\mathrm{i}\omega_n)\, \mathrm{e}^{\mathrm{i}\omega_n \tau} = f(\xi_{\mathbf{k}}) \mathrm{e}^{\xi_{\mathbf{k}}\tau}.$$

Finally, taking the limit $\tau \to 0^+$ and summing over the wavevectors $\mathbf{k}$, we write formula (A.55) as

$$T \mathrm{Tr} G^0 = \sum_{\mathbf{k}} f(\xi_{\mathbf{k}}).$$

The latter is the total number of noninteracting electrons.

As the next application of the residue theorem, we calculate the thermodynamic susceptibility of noninteracting electrons. By formula (6.71), we have

$$\chi^0_{zz}(\mathbf{q}, \mathrm{i}\omega_m) = -g^2 \mu_{\mathrm{B}}^2 \frac{1}{2} T \sum_{\mathbf{k}n} G^0_{\mathbf{k}}(\mathrm{i}\omega_n) G^0_{\mathbf{k}+\mathbf{q}}(\mathrm{i}\omega_n + \mathrm{i}\omega_m), \tag{A.60}$$

where $G^0_{\mathbf{k}}(z)$ is given by formula (A.56) and $\omega_m = 2\pi m T$ are the even "frequencies". We need to calculate the sum over the odd "frequencies"

$$\sum_n G^0_{\mathbf{k}}(\mathrm{i}\omega_n) G^0_{\mathbf{k}+\mathbf{q}}(\mathrm{i}\omega_n + \mathrm{i}\omega_m) \equiv \sum_n h(\mathrm{i}\omega_n). \tag{A.61}$$

The function

$$h(z) = G^0_{\mathbf{k}}(z) G^0_{\mathbf{k}+\mathbf{q}}(z + \mathrm{i}\omega_m)$$

is analytic everywhere but at two isolated singularities $z_1 = \xi_{\mathbf{k}}$ and $z_2 = \xi_{\mathbf{k}+\mathbf{q}} - \mathrm{i}\omega_m$, where it has simple poles. Using the residue theorem as above, we have

$$\begin{aligned} &\lim_{R\to\infty} \frac{1}{2\pi \mathrm{i}} \oint_{C_R} f(z) h(z)\, \mathrm{d}z \\ &= \mathrm{res}_{z=z_1}\big(f(z)h(z)\big) + \mathrm{res}_{z=z_2}\big(f(z)h(z)\big) + \sum_n \mathrm{res}_{z=\mathrm{i}\omega_n}\big(f(z)h(z)\big), \end{aligned} \tag{A.62}$$

where the contour C_R is the circle of radius R centred at the origin. Since $h(z)$ is proportional to $1/z^2$ as $|z| \to \infty$, the left-hand side tends to zero as $R \to \infty$. By formula (A.48) we calculate

$$\mathrm{res}_{z=z_1}\big(f(z)h(z)\big) = \frac{f(\xi_{\mathbf{k}})}{\xi_{\mathbf{k}} - \xi_{\mathbf{k}+\mathbf{q}} + \mathrm{i}\omega_m}, \qquad \mathrm{res}_{z=z_2}\big(f(z)h(z)\big) = \frac{f(\xi_{\mathbf{k}+\mathbf{q}} - \mathrm{i}\omega_m)}{\xi_{\mathbf{k}+\mathbf{q}} - \mathrm{i}\omega_m - \xi_{\mathbf{k}}}.$$

Using $f(\xi_{\mathbf{k}+\mathbf{q}} - \mathrm{i}\omega_m) = f(\xi_{\mathbf{k}+\mathbf{q}})$, we write (A.62) as

$$-\frac{f(\xi_{\mathbf{k}}) - f(\xi_{\mathbf{k}+\mathbf{q}})}{\xi_{\mathbf{k}+\mathbf{q}} - \xi_{\mathbf{k}} - \mathrm{i}\omega_m} - T \sum_n h(\mathrm{i}\omega_n) = 0.$$

Taking (A.61) into account, we finally write the susceptibility (A.60) as

$$\chi^0_{zz}(\mathbf{q}, \mathrm{i}\omega_m) = g^2 \mu_{\mathrm{B}}^2 \frac{1}{2} \sum_{\mathbf{k}} \frac{f(\xi_{\mathbf{k}}) - f(\xi_{\mathbf{k}+\mathbf{q}})}{\xi_{\mathbf{k}+\mathbf{q}} - \xi_{\mathbf{k}} - \mathrm{i}\omega_m}. \tag{A.63}$$

If we recall the difference in definition of the Fermi function here and in Chaps. 2–6 (shift of energies by μ), we see that formulae (A.63) and (6.72) coincide.

Summation for a Function with Branch Cuts

For the *interacting* electrons system, the zeroth Green function $G^0_{\mathbf{k}}(z)$ in formula for the thermodynamic susceptibility (A.60) must be replaced by the interacting electrons Green function $\bar{G}_{\mathbf{k}}(z)$ (see formula (9.34)). Since the spectrum of $\bar{G}(z)$ cannot be written down explicitly, another summation method should be used, which we describe below.

We begin with the summation rule over the even "frequencies" $\omega_m = 2\pi m T$. The Bose function $B(z) = (\exp(z/T)-1)^{-1}$ has simple poles at $\mathrm{i}\omega_m$. Using formula (A.48) and L'Hopital's rule, we obtain

$$\mathrm{res}_{z=\mathrm{i}\omega_m} B(z) = \lim_{z\to \mathrm{i}\omega_m} \frac{z-\mathrm{i}\omega_m}{\mathrm{e}^{z/T}-1} = \lim_{z\to \mathrm{i}\omega_m} \frac{1}{(1/T)\mathrm{e}^{z/T}} = T. \tag{A.64}$$

Let $h(z)$ be analytic at the origin, analytic outside the real axis and have the one-sided limits at the real axis $h(\omega \pm \mathrm{i}0^+)$. Then, applying the residue theorem, we have

$$\lim_{R\to\infty} \frac{1}{2\pi\mathrm{i}} \oint_{C_R} B(z)h(z)\,\mathrm{d}z = \sum_m \mathrm{res}_{z=\mathrm{i}\omega_m}\big(B(z)h(z)\big), \tag{A.65}$$

where C_R is the contour in the complex plane that consists of the circle centred at zero of radius R with cuts along the real axis (Fig. A.4). Since function $h(z)$ is analytic at points $\mathrm{i}\omega_m$, similar to (A.64), we obtain

$$\mathrm{res}_{z=\mathrm{i}\omega_m}\big(B(z)h(z)\big) = Th(\mathrm{i}\omega_m). \tag{A.66}$$

If the contributions from the contour integrals over arcs can be neglected, we come to

$$T\sum_m h(\mathrm{i}\omega_m) = \frac{1}{2\pi\mathrm{i}} \int B(\omega)\big(h(\omega+\mathrm{i}0^+) - h(\omega-\mathrm{i}0^+)\big)\,\mathrm{d}\omega. \tag{A.67}$$

If, additionally, $h(z)$ is such that $h^*(z) = h(z^*)$, we have

$$T\sum_m h(\mathrm{i}\omega_m) = -\frac{1}{\pi} \int B(\omega)\mathrm{Im}h(\omega+\mathrm{i}0^+)\,\mathrm{d}\omega. \tag{A.68}$$

Similarly, we consider sums over the odd "frequencies" $\omega_n = (2n+1)\pi T$. Let $h(z)$ be as above. Then, using the residue theorem, we have

$$\lim_{R\to\infty} \frac{1}{2\pi\mathrm{i}} \oint_{C_R} f(z)h(z)\,\mathrm{d}z = \sum_n \mathrm{res}_{z=\mathrm{i}\omega_n}\big(f(z)h(z)\big), \tag{A.69}$$

where C_R is the same contour in the complex plane (Fig. A.4). If the contributions from the contour integrals over arcs can be neglected, we come to

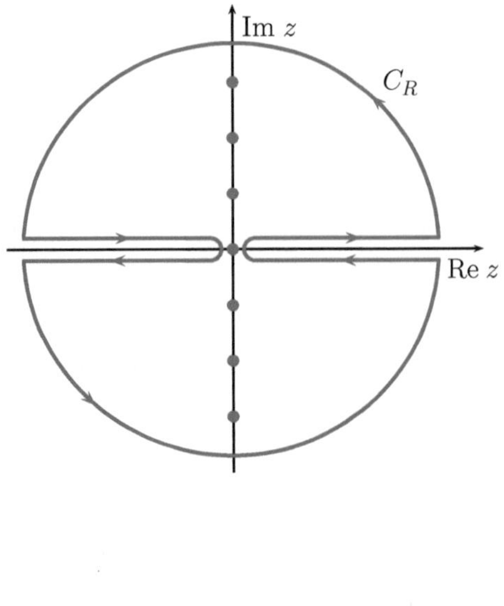

Fig. A.4 The integration contour and poles of the Bose function

$$-T\sum_n h(\mathrm{i}\omega_n) = \frac{1}{2\pi\mathrm{i}}\int f(\omega)\left(h(\omega+\mathrm{i}0^+) - h(\omega-\mathrm{i}0^+)\right)\mathrm{d}\omega.$$

If additionally $h^*(z) = h(z^*)$, then the following formula holds

$$T\sum_n h(\mathrm{i}\omega_n) = \frac{1}{\pi}\int f(\omega)\mathrm{Im}h(\omega-\mathrm{i}0^+)\,\mathrm{d}\omega. \tag{A.70}$$

"Frequency" Summation in the DSFT

First, we use the summation rule (A.68) in expression (10.16):

$$\langle\Delta V_\alpha^2\rangle' = \sum_{\mathbf{q}m}\frac{\tilde{u}T}{2}\frac{\tilde{u}\chi^{0\alpha}_{\mathbf{q}m}}{1-\tilde{u}\chi^{0\alpha}_{\mathbf{q}m}}, \tag{A.71}$$

to convert the sum with the thermodynamic susceptibility $\chi^{0\alpha}_{\mathbf{q}}(\mathrm{i}\omega_m) \equiv \chi^{0\alpha}_{\mathbf{q}m}$ over ω_m to an integral with the dynamic susceptibility $\chi^{0\alpha}_{\mathbf{q}}(\omega)$ over ω. Similar to relation (A.55) we should treat the sum over ω_m as

$$\sum_m \ldots = \lim_{\tau\to 0^+}\sum_m \ldots \mathrm{e}^{\mathrm{i}\omega_m\tau}.$$

Then by the same argument as in Appendix A.2.5, we show that the contour integrals over the arcs in formula (A.65) vanish as $R\to\infty$. Using formula (A.68) to sum over ω_m in expression (A.71), we obtain

$$\langle\Delta V_\alpha^2\rangle' = \sum_{\mathbf{q}}\frac{\tilde{u}}{2}\frac{1}{\pi}\int B(\varepsilon)\,\mathrm{Im}\,\frac{\tilde{u}\chi^{0\alpha}_{\mathbf{q}}(\varepsilon)}{1-\tilde{u}\chi^{0\alpha}_{\mathbf{q}}(\varepsilon)}\,\mathrm{d}\varepsilon, \tag{A.72}$$

where $\varepsilon = \varepsilon + \mathrm{i}0^+$. Taking $\mathrm{Re}\,\chi^{0\alpha}_{\mathbf{q}}(-\varepsilon) = \mathrm{Re}\,\chi^{0\alpha}_{\mathbf{q}}(\varepsilon)$ and $\mathrm{Im}\,\chi^{0\alpha}_{\mathbf{q}}(-\varepsilon) = -\mathrm{Im}\,\chi^{0\alpha}_{\mathbf{q}}(\varepsilon)$ into account, we see that the integrand of (A.72) is an odd function:

$$\mathrm{Im}\,\frac{\tilde{u}\chi^{0\alpha}_{\mathbf{q}}(\varepsilon)}{1-\tilde{u}\chi^{0\alpha}_{\mathbf{q}}(\varepsilon)} = \mathrm{Im}\frac{1}{1-\tilde{u}\chi^{0\alpha}_{\mathbf{q}}(\varepsilon)} = \frac{\tilde{u}\mathrm{Im}\,\chi^{0\alpha}_{\mathbf{q}}(\varepsilon)}{(1-\tilde{u}\mathrm{Re}\,\chi^{0\alpha}_{\mathbf{q}}(\varepsilon))^2 + (\tilde{u}\mathrm{Im}\,\chi^{0\alpha}_{\mathbf{q}}(\varepsilon))^2}.$$

Using the relation $B(\varepsilon) - B(-\varepsilon) = 2B(\varepsilon) + 1$, we write (A.72) in the form

$$\langle\Delta V_\alpha^2\rangle' = \sum_{\mathbf{q}}\frac{\tilde{u}}{\pi}\int_0^\infty\left(B(\varepsilon)+\frac{1}{2}\right)\mathrm{Im}\,\frac{1}{1-\tilde{u}\chi^{0\alpha}_{\mathbf{q}}(\varepsilon)}\,\mathrm{d}\varepsilon.$$

This proves formula (10.17).

Next, we use a summation rule over the odd "frequencies" slightly more complicated than (A.70) to obtain an expression for the local susceptibility:

$$\chi^{0\alpha}_{\mathrm{L}}(z) = -\frac{N_{\mathrm{d}}}{2\pi}\sum_{\gamma_1\gamma_2}\int \mathrm{Im}\Big(g^{\gamma_1}(\varepsilon)\big(g^{\gamma_2}(\varepsilon-z)+g^{\gamma_2}(\varepsilon+z)\big)$$
$$\times\,\mathrm{Sp}\big(\sigma^{\gamma_1}\sigma^\alpha\sigma^{\gamma_2}\sigma^\alpha\big)\Big)f(\varepsilon)\,\mathrm{d}\varepsilon, \tag{A.73}$$

where $g(\varepsilon)$ is the mean single-site Green function (in the grand canonical ensemble) and $f(\varepsilon) = (\exp(\varepsilon/T)+1)^{-1}$ is the Fermi function. The result is derived as follows. In expression (10.27):

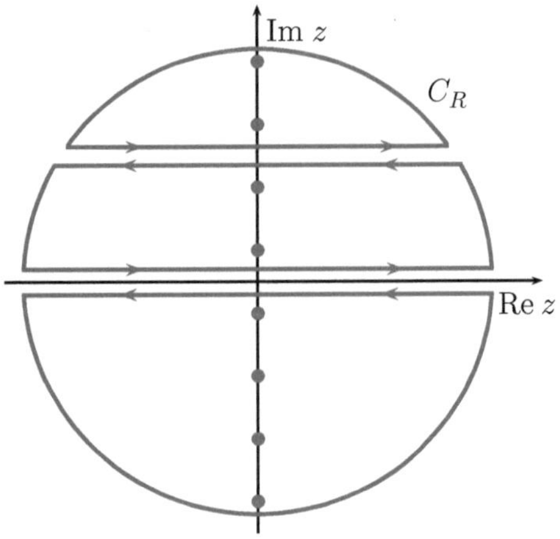

Fig. A.5 The integration contour and poles of the Fermi function

$$\chi_{\mathrm{L}}^{0\alpha}(\mathrm{i}\omega_m) = -\frac{N_\mathrm{d}}{2} T \sum_n \sum_{\gamma_1\gamma_2} g^{\gamma_1}(\mathrm{i}\omega_n) g^{\gamma_2}(\mathrm{i}\omega_n - \mathrm{i}\omega_m) \mathrm{Sp}\big(\sigma^{\gamma_1}\sigma^{\alpha}\sigma^{\gamma_2}\sigma^{\alpha}\big),$$

we need to calculate the sum

$$T \sum_n \sum_{\gamma_1\gamma_2} g^{\gamma_1}(\mathrm{i}\omega_n) g^{\gamma_2}(\mathrm{i}\omega_n - \mathrm{i}\omega_m) \mathrm{Sp}\big(\sigma^{\gamma_1}\sigma^{\alpha}\sigma^{\gamma_2}\sigma^{\alpha}\big) \equiv T \sum_n h(\mathrm{i}\omega_n). \tag{A.74}$$

Here $g^\gamma(z)$, $\gamma = 0, z$, are the components of the single-site Green function

$$g_\sigma(z) = \int \frac{\nu_\sigma(\varepsilon)}{z - \varepsilon}\,\mathrm{d}\varepsilon, \qquad \sigma = \uparrow, \downarrow, \tag{A.75}$$

where $\nu_\sigma(\varepsilon)$ is the polarized electron DOS. As we explained in the last section of Appendix A.2.4, the Cauchy integral (A.75) has a branch cut along the real axis. Therefore, the function

$$h(z) = \sum_{\gamma_1\gamma_2} g^{\gamma_1}(z) g^{\gamma_2}(z - \mathrm{i}\omega_m) \mathrm{Sp}\big(\sigma^{\gamma_1}\sigma^{\alpha}\sigma^{\gamma_2}\sigma^{\alpha}\big) \tag{A.76}$$

is analytic outside the horizontal lines $\mathrm{Im} z = 0$ and $\mathrm{Im}(z - \mathrm{i}\omega_m) = 0$, where it has one-sided limits. Therefore, applying the residue theorem, we have

$$\lim_{R\to\infty} \frac{1}{2\pi\mathrm{i}} \oint_{C_R} f(z)h(z)\,\mathrm{d}z = \sum_n \mathrm{res}_{z=\mathrm{i}\omega_n}\big(f(z)h(z)\big), \tag{A.77}$$

where C_R is the contour in the complex plane that consists of the circle centred at zero of radius R with cuts along $\mathrm{Im} z = 0$ and $\mathrm{Im}(z - \mathrm{i}\omega_m) = 0$ (Fig. A.5). Calculating the residues, we obtain

$$\sum_n \mathrm{res}_{z=\mathrm{i}\omega_n}\big(f(z)h(z)\big) = -T \sum_n h(\mathrm{i}\omega_n). \tag{A.78}$$

The functions $g^0(z)$ and $g^z(z)$ are given by

$$g^0(z) = \frac{1}{2}(g_\uparrow(z) + g_\downarrow(z)), \qquad g^z(z) = \frac{1}{2}(g_\uparrow(z) - g_\downarrow(z)).$$

From expression (A.75) we see that $g_\sigma(z)$ is proportional to $1/z$ as $|z| \to \infty$. Therefore, the product $g^{\gamma_1}(z)g^{\gamma_2}(z - \mathrm{i}\omega_m)$ is proportional to $1/z^2$ as $|z| \to \infty$, and the contour integral over the arcs in formula (A.69) can be discarded. Therefore, the left-hand side of relation (A.77) reduces to integrals over the branch cuts. Taking (A.78) into account, we obtain

$$\frac{1}{2\pi i}\bigg(\int f(\omega)\left(h(\omega+i0^+)-h(\omega-i0^+)\right)d\omega$$
$$+\int f(\omega+i\omega_m)\left(h(\omega+i\omega_m+i0^+)-h(\omega+i\omega_m-i0^+)\right)d\omega\bigg)=-T\sum_n h(i\omega_n). \tag{A.79}$$

Taking $f(\omega+i\omega_m)=f(\omega)$ into account, we have

$$\frac{1}{2\pi i}\int f(\omega)\left[h(\omega+i0^+)-h(\omega-i0^+)\right.$$
$$\left.+\,h(\omega+i\omega_m+i0^+)-h(\omega+i\omega_m-i0^+)\right]d\omega=-T\sum_n h(i\omega_n). \tag{A.80}$$

Changing the order of functions and using the cyclic property of trace in (A.76), we obtain

$$h(\omega+i0^+)=\sum_{\gamma_1\gamma_2} g^{\gamma_2}(\omega-i\omega_m+i0^+)g^{\gamma_1}(\omega+i0^+)\mathrm{Sp}\left(\sigma^{\gamma_2}\sigma^{\alpha}\sigma^{\gamma_1}\sigma^{\alpha}\right).$$

Swapping the indices γ_1 and γ_2, we have

$$\begin{aligned}h(\omega+i0^+)&=\sum_{\gamma_1\gamma_2} g^{\gamma_1}(\omega-i\omega_m+i0^+)g^{\gamma_2}(\omega+i0^+)\mathrm{Sp}\left(\sigma^{\gamma_1}\sigma^{\alpha}\sigma^{\gamma_2}\sigma^{\alpha}\right)\\&=(h(\omega+i\omega_m-i0^+))^*.\end{aligned}$$

Similarly, we prove

$$h(\omega+i\omega_m+i0^+)=(h(\omega-i0^+))^*.$$

Therefore, we can write (A.80) as

$$-\frac{1}{\pi}\int f(\omega)\mathrm{Im}\left[h(\omega-i0^+)+h(\omega+i\omega_m-i0^+)\right]d\omega=-T\sum_n h(i\omega_n). \tag{A.81}$$

Using the cyclic property of trace once again, we have

$$\begin{aligned}&\mathrm{Im}\left[h(\omega-i0^+)+h(\omega+i\omega_m-i0^+)\right]\\&\quad=\sum_{\gamma_1\gamma_2}\mathrm{Im}\left[g^{\gamma_1}(\omega-i0^+)\left(g^{\gamma_2}(\omega+i\omega_m-i0^+)+g^{\gamma_2}(\omega-i\omega_m-i0^+)\right)\right]\\&\quad\times\mathrm{Sp}\left(\sigma^{\gamma_1}\sigma^{\alpha}\sigma^{\gamma_2}\sigma^{\alpha}\right).\end{aligned} \tag{A.82}$$

Substitution of (A.82) in (A.81) yields (A.73). Shifting the integration variable by μ in (A.73), we obtain formula (10.28).

A.3 Random Variables and Fluctuating Fields

Here we summarize necessary facts about random variables and fluctuating fields (for a more detailed discussion see, e.g. [10]) and consider a number of examples.

A.3.1 Continuous Random Variables

Let x be a random variable with the probability density function $p(x)$. The *mean* of the random variable is given by the integral

$$\langle x \rangle = \int_{-\infty}^{\infty} x p(x)\, \mathrm{d}x. \tag{A.83}$$

For brevity, this number is often denoted by $\bar{x}$. Similarly, for any function $f(x)$, we have

$$\langle f(x) \rangle = \int_{-\infty}^{\infty} f(x) p(x)\, \mathrm{d}x, \tag{A.84}$$

or briefly, $\bar{f}(x)$. In particular,

$$\langle (\Delta x)^2 \rangle = \langle (x - \langle x \rangle)^2 \rangle = \langle x^2 \rangle - \langle x \rangle^2$$

is the *variance*, or the mean-square deviation.

The generalization of formulae (A.83) and (A.84) to the multidimensional case is straightforward. The random vector $\mathbf{x} = (x_1, \ldots, x_n)$ is determined by the probability density function $p(\mathbf{x})$. The *mean* $\langle \mathbf{x} \rangle$ is a vector, where $\langle x_i \rangle$ is given by the multiple integral

$$\langle x_i \rangle = \int x_i p(\mathbf{x})\, \mathrm{d}\mathbf{x}. \tag{A.85}$$

The random variables x_i are called independent if $p(\mathbf{x}) = p_1(x_1) \ldots p_n(x_n)$. The *covariance* is by definition the number

$$C_{ij} = \langle \Delta x_i\, \Delta x_j \rangle = \langle x_i x_j \rangle - \langle x_i \rangle \langle x_j \rangle,$$

where $\Delta x_i = x_i - \langle x_i \rangle$, and C is called the *covariance matrix*. For the *complex* random vector $\mathbf{z} = (z_1, \ldots, z_n)$, the *covariance* is defined as

$$C_{ij} = \langle \Delta z_i\, \Delta z_j^* \rangle,$$

where $\Delta z_i = z_i - \langle z_i \rangle$.

The most important example for us is the Gaussian random variable. In one dimension the Gaussian probability density function is given by

$$p(x) = \frac{1}{\sigma\sqrt{2\pi}}\, \mathrm{e}^{-\dfrac{(x-\mu)^2}{2\sigma^2}}, \tag{A.86}$$

where the real numbers μ and $\sigma > 0$ are the parameters of the Gaussian distribution. It is easy to check that μ is the mean: $\langle x \rangle = \mu$, and σ^2 is the variance: $\langle (\Delta x)^2 \rangle = \sigma^2$.

In the multidimensional case, the Gaussian probability density function can be written as

$$p(\mathbf{x}) = \frac{1}{\sqrt{\det \Sigma}\,(2\pi)^{n/2}} \exp\left(-\frac{1}{2} \sum_{ij} (x_i - \mu_i) \Sigma_{ij}^{-1} (x_j - \mu_j) \right), \tag{A.87}$$

where $\boldsymbol{\mu} = (\mu_1, \ldots, \mu_n)$ is a vector, $\Sigma = (\Sigma_{ij})$ is an $n \times n$ symmetric positive definite matrix and Σ^{-1} its inverse. Then $\langle x_i \rangle = \mu_i$ is the mean and *covariance* is given by

$$\langle \Delta x_i\, \Delta x_j \rangle = \Sigma_{ij}, \tag{A.88}$$

where $\Delta x_i = x_i - \mu_i$. In the following paragraph we give an intuitive proof of this result (for alternative proofs, see [10,17]).

To derive formula (A.88), we make a linear transform that yields n independent Gaussian random variables, calculate the covariance matrix (a trivial task in this case) and transform back to the original variables. The sum in the exponent of (A.87) can be written as the inner product:

$$\sum_{ij} (x_i - \mu_i) \Sigma_{ij}^{-1} (x_j - \mu_j) \equiv \big((\mathbf{x} - \boldsymbol{\mu}), \Lambda(\mathbf{x} - \boldsymbol{\mu}) \big),$$

where $\Lambda = \Sigma^{-1}$. For the symmetric matrix Λ there exists an orthogonal matrix O ($O O^{\mathrm{T}} = 1$) such that

$$\Lambda = ODO^{\mathrm{T}} \quad \text{or} \quad D = O^{\mathrm{T}}\Lambda O, \tag{A.89}$$

where D is a diagonal matrix (see, e.g. [3]). Then, applying the orthogonal transformation $\mathbf{x} = O\tilde{\mathbf{x}}$, we have

$$(\Delta\mathbf{x}, \Lambda\Delta\mathbf{x}) = (\Delta\tilde{\mathbf{x}}, O^{\mathrm{T}}\Lambda O\Delta\tilde{\mathbf{x}}) = (\Delta\tilde{\mathbf{x}}, D\Delta\tilde{\mathbf{x}}).$$

The matrix D has the eigenvalues of the matrix $\Lambda = \Sigma^{-1}$ at the main diagonal; we denote them by $1/\sigma_i^2$. Therefore, in the new variables $\tilde{\mathbf{x}}$, the density function (A.87) becomes

$$p(\tilde{\mathbf{x}}) = \frac{1}{\sigma_1 \dots \sigma_n (2\pi)^{n/2}} \exp\left(-\sum_i \frac{(\tilde{x}_i - \tilde{\mu}_i)^2}{2\sigma_i^2}\right)$$
$$= \prod_{i=1}^{n} \frac{1}{\sigma_i\sqrt{2\pi}} \exp\left(-\frac{(\tilde{x}_i - \tilde{\mu}_i)^2}{2\sigma_i^2}\right).$$

So the random variables $\tilde{x}_i$ are independent, and each $\tilde{x}_i$ is a Gaussian random variable with the density function

$$p_i(\tilde{x}_i) = \frac{1}{\sigma_i\sqrt{2\pi}} \exp\left(-\frac{(\tilde{x}_i - \tilde{\mu}_i)^2}{2\sigma_i^2}\right).$$

The covariance is given by

$$\tilde{C}_{ij} = \langle\Delta\tilde{x}_i\, \Delta\tilde{x}_j\rangle = \sigma_i^2\delta_{ij},$$

or, in the matrix form,

$$\tilde{C} = D^{-1}. \tag{A.90}$$

Transforming back to the original variable $\mathbf{x} = O\tilde{\mathbf{x}}$, we have

$$C_{ij} = \langle\Delta x_i \Delta x_j\rangle = \sum_{i'j'} O_{ii'}O_{jj'}\langle\Delta\tilde{x}_{i'}\Delta\tilde{x}_{j'}\rangle = \sum_{i'j'} O_{ii'}\tilde{C}_{i'j'}O^{\mathrm{T}}_{j'j},$$

or, in the matrix form,

$$C = O\tilde{C}O^{\mathrm{T}}.$$

Substituting (A.90), we have

$$C = OD^{-1}O^{\mathrm{T}}.$$

Inverting the relation $\Lambda = ODO^{\mathrm{T}}$ and taking $O^{-1} = O^{\mathrm{T}}$ into account, we obtain

$$C = \Lambda^{-1} = \Sigma,$$

which is the matrix form of the necessary relation (A.88).

In the multidimensional *complex* case, the Gaussian probability density function can be written as

$$p(\mathbf{z}) = \frac{1}{\det\Gamma\pi^n} \exp\left(-\sum_{ij}(z_i - \mu_i)^*\Gamma_{ij}^{-1}(z_j - \mu_j)\right), \tag{A.91}$$

where $\boldsymbol{\mu} = (\mu_1, \dots, \mu_n)$ is a complex vector and $\Gamma = (\Gamma_{ij})$ is an $n \times n$ Hermitian positive definite matrix. Then $\langle z_i\rangle = \mu_i$ is the mean and the covariance is given by

$$\langle\Delta z_i\, \Delta z_j^*\rangle = \Gamma_{ij}, \tag{A.92}$$

where $\Delta z_i = z_i - \mu_i$.

The proof of formula (A.92) is similar to the one for the real multidimensional Gaussian distribution. In the complex case, the sum in the exponent of (A.91) can be written as the inner product:

$$\sum_{ij}(z_i-\mu_i)^*\Gamma_{ij}^{-1}(z_j-\mu_j)\equiv\big((\mathbf{z}-\boldsymbol{\mu}),\Lambda(\mathbf{z}-\boldsymbol{\mu})\big),$$

where $\Lambda=\Gamma^{-1}$. For the Hermitian matrix Λ there exists a unitary matrix U ($UU^\dagger=1$) such that

$$\Lambda=UDU^\dagger \quad \text{or} \quad D=U^\dagger\Lambda U, \tag{A.93}$$

where D is the diagonal matrix (see, e.g. [3]). Then, applying the unitary transformation $\mathbf{z}=U\tilde{\mathbf{z}}$, we have

$$(\Delta\mathbf{z},\Lambda\Delta\mathbf{z})=(\Delta\tilde{\mathbf{z}},U^\dagger\Lambda U\Delta\tilde{\mathbf{z}})=(\Delta\tilde{\mathbf{z}},D\Delta\tilde{\mathbf{z}}).$$

The matrix D has the eigenvalues $1/\sigma_i^2$ of the matrix $\Lambda=\Gamma^{-1}$ at the main diagonal. Therefore, in the new variables $\tilde{\mathbf{z}}$, the density function (A.91) becomes

$$\begin{aligned}p(\tilde{\mathbf{z}})&=\frac{1}{\sigma_1^2\ldots\sigma_n^2\pi^n}\exp\left(-\sum_i\frac{|\tilde{z}_i-\tilde{\mu}_i|^2}{\sigma_i^2}\right)\\&=\prod_{i=1}^n\frac{1}{\sigma_i^2\pi}\exp\left(-\frac{|\tilde{z}_i-\tilde{\mu}_i|^2}{\sigma_i^2}\right).\end{aligned}$$

So random variables $\tilde{z}_i$ are independent, and each $\tilde{z}_i$ is a complex random variable with the Gaussian density function[3]

$$p_i(\tilde{z}_i)=\frac{1}{\sigma_i^2\pi}\exp\left(-\frac{|\tilde{z}_i-\tilde{\mu}_i|^2}{\sigma_i^2}\right).$$

The covariance is given by $\tilde{C}_{ij}=\langle\Delta\tilde{z}_i\,\Delta\tilde{z}_j^*\rangle=\sigma_i^2\delta_{ij}$, or, in the matrix form, $\tilde{C}=D^{-1}$. Transforming back to the original variable $\mathbf{z}=U\tilde{\mathbf{z}}$, we have

$$C_{ij}=\langle\Delta z_i\Delta z_j^*\rangle=\sum_{i'j'}U_{ii'}U_{jj'}^*\langle\Delta\tilde{z}_{i'}\Delta\tilde{z}_{j'}^*\rangle=\sum_{i'j'}U_{ii'}\tilde{C}_{i'j'}U_{j'j}^\dagger,$$

or, in the matrix form,

$$C=U\tilde{C}U^\dagger=UD^{-1}U^\dagger.$$

Inverting $\Lambda=UDU^\dagger$ and taking $U^{-1}=U^\dagger$ into account, we obtain

$$C=\Lambda^{-1}=\Gamma,$$

which is the matrix form of the necessary relation (A.92).

[3]Note the normalizing coefficient appearing from the complex Gaussian integral

$$\int \mathrm{e}^{-|z|^2}\,\mathrm{d}z=\iint \mathrm{e}^{-(x^2+y^2)}\,\mathrm{d}x\,\mathrm{d}y=2\pi\int \mathrm{e}^{-r^2}r\,\mathrm{d}r=\pi.$$

A.3.2 Discrete Random Variables

For a discrete random variable, the integral in (A.83) can be written as a sum. Indeed, suppose that x takes the value ξ_i with the probability p_i. In this case,

$$p(x) = \sum_i p_i \delta(x - \xi_i).$$

Inserting into (A.83) and using the identity (A.36), we obtain

$$\langle x \rangle = \sum_i p_i \xi_i .$$

Similarly, formula (A.84) becomes

$$\langle f(x) \rangle = \sum_i p_i f(\xi_i).$$

As an example of the use of a discrete random variable, we consider a simple case of *one* nondegenerate band with the mean number of electrons per site equal to n. In the paramagnetic state, the probabilities of a spin-up or spin-down electron occupying a site are equal, $p_\uparrow = p_\downarrow = p$. Then the mean number of electrons per site is given by

$$n = p_1 + 2p_2,$$

where

$$p_1 = 2p(1 - p) \tag{A.94}$$

is the probability that there is only one electron at a site and

$$p_2 = p^2$$

is the probability that there are exactly two electrons at a site (other possibilities are prohibited by the Pauli exclusion principle). From this we readily obtain $p = n/2$. Then, for the square of the local spin we have

$$\langle \mathbf{s}^2 \rangle = \frac{1}{2}\left(\frac{1}{2} + 1\right) p_1$$

(if the site is occupied by two electrons with the opposite spins, $\langle \mathbf{s}^2 \rangle$ is zero). Taking (A.94) into account, we obtain the final result (13.10):

$$\langle \mathbf{s}^2 \rangle = \frac{3}{4} 2\frac{n}{2}\left(1 - \frac{n}{2}\right) = \frac{3}{8} n(2 - n).$$

Analogously, in the ferromagnetic state with the magnetic moment $s = \frac{1}{2}(p_\uparrow - p_\downarrow)$, we obtain that the square of the local spin becomes larger by $\frac{3}{2}s^2$.

Let us consider a complete set of outcomes A for a discrete random variable. Let A' be an outcome of another discrete random variable. Then the *law of total probability* says that

$$P(A') = \sum_A P(A'|A)P(A), \tag{A.95}$$

where $P(A'|A)$ is the conditional probability of A' given the outcome A.

As an example, we apply the law of total probability to the *scattering process*. Let A label the states before the scattering and A' the states after. Then the conditional probability $P(A'|A)$ is just the *transition probability* $P_{A\to A'}$, and (A.95) becomes

$$P_{A'} = \sum_A P_A P_{A\to A'} .$$

A.3.3 Fluctuating Fields

By a *fluctuating exchange field* V we mean a family of random variables $V_j(\tau)$ depending on the parameters j and τ. Here $V_j(\tau) = \sum_\alpha \sigma^\alpha V_j^\alpha(\tau)$, where σ^α are the Pauli matrices ($\alpha = x, y, z$) and $(V_j^x(\tau), V_j^y(\tau), V_j^z(\tau))$ is a random vector for each j and τ. The covariance is defined by

$$\langle \Delta V_j^\alpha(\tau) \Delta V_{j'}^{\alpha'}(\tau') \rangle, \tag{A.96}$$

where $\Delta V_j^\alpha(\tau) = V_j^\alpha(\tau) - \langle V_j^\alpha(\tau) \rangle$. Here the average is given by the functional integral:

$$\langle \dots \rangle = \int \dots p(V) \mathrm{D}V,$$

where $p(V)$ is the probability density function. The Fourier transform $V_{\mathbf{q}m}^\alpha$ is a *complex* random variable. In this case, the covariance is defined by

$$\langle \Delta V_{\mathbf{q}m}^\alpha (\Delta V_{\mathbf{q}'m'}^{\alpha'})^* \rangle, \tag{A.97}$$

where $\Delta V_{\mathbf{q}m}^\alpha = V_{\mathbf{q}m}^\alpha - \langle V_{\mathbf{q}m}^\alpha \rangle$. In the main text, we refer to (A.96) and (A.97) as the field correlator and to

$$\langle \Delta V_{\mathbf{q}m}^\alpha (\Delta V_{\mathbf{q}m}^\alpha)^* \rangle = \langle |\Delta V_{\mathbf{q}m}^\alpha|^2 \rangle$$

as the fluctuation.

In this section, we calculate the correlator of the Gaussian fluctuating field V with the probability density $p^{(2)}(V) \propto \mathrm{e}^{-F^{(2)}(V)/T}$ given by the quadratic form

$$F^{(2)}(V) = \int_0^{1/T}\int_0^{1/T} \sum_{jj'\alpha\beta} \Delta V_j^\alpha(\tau)\, A_{jj'}^{\alpha\beta}(\tau, \tau')\, \Delta V_{j'}^\beta(\tau')\, \mathrm{d}\tau\, \mathrm{d}\tau'. \tag{A.98}$$

Due to the "time" and space translational invariance, we have

$$A_{jj'}^{\alpha\beta}(\tau, \tau') = A_{j-j'}^{\alpha\beta}(\tau - \tau').$$

Using the formulae for the inverse Fourier transformation (C.19) and (C.29), in the momentum-"frequency" representation we have

$$F^{(2)}(V) = \sum_{\mathbf{q}m\alpha\beta} \Delta V_{\mathbf{q}m}^\alpha A_{\mathbf{q}m}^{\alpha\beta} \Delta V_{-\mathbf{q}-m}^\beta, \tag{A.99}$$

where

$$A_{\mathbf{q}m}^{\alpha\beta} = \frac{N}{T}\int_0^{1/T} \sum_j A_j^{\alpha\beta}(\tau)\, \mathrm{e}^{\mathrm{i}(\mathbf{q}\mathbf{R}_j - \omega_m \tau)}\, \mathrm{d}\tau$$

(a detailed calculation similar to this one is given in Appendix C.4).

We now calculate the correlator (A.97). First, we write the quadratic form (A.99) as the inner product. Changing the signs of the summation indices $\mathbf{q}$ and m in (A.99), we have

$$F^{(2)}(V) = \sum_{\mathbf{q}m\alpha\beta} \Delta V_{-\mathbf{q}-m}^\alpha A_{-\mathbf{q}-m}^{\alpha\beta} \Delta V_{\mathbf{q}m}^\beta.$$

Since $V_j^\alpha(\tau)$ is a real function, its Fourier transform satisfies $\Delta V_{-\mathbf{q}-m}^\alpha = (\Delta V_{\mathbf{q}m}^\alpha)^*$. Similarly, $A_{-\mathbf{q}-m}^{\alpha\beta} = (A_{\mathbf{q}m}^{\alpha\beta})^*$. Hence

$$F^{(2)}(V) = \sum_{\mathbf{q}m\alpha\beta} (\Delta V_{\mathbf{q}m}^\alpha)^* (A_{\mathbf{q}m}^{\alpha\beta})^* \Delta V_{\mathbf{q}m}^\beta \equiv (\Delta V, A^* \Delta V), \tag{A.100}$$

where the elements of the matrix A^* are the complex conjugates of the ones of A. The matrix A is block-diagonal with respect to momenta and "frequency", but the 3×3 blocks $A_{\mathbf{q}m}$ are not diagonal. Applying formula (A.92), we calculate the covariance of the three-dimensional random vectors $V_{\mathbf{q}m}$:

$$\langle \Delta V^{\alpha}_{\mathbf{q}m} (\Delta V^{\beta}_{\mathbf{q}m})^* \rangle = \frac{T}{2} \left((A^*_{\mathbf{q}m})^{-1} \right)^{\alpha\beta}, \tag{A.101}$$

where $(A^*_{\mathbf{q}m})^{-1}$ is the inverse of the 3×3 matrix $A^*_{\mathbf{q}m}$. Recalling $\Delta V^{\alpha}_{-\mathbf{q}-m} = (\Delta V^{\alpha}_{\mathbf{q}m})^*$ and $A^{\alpha\beta}_{-\mathbf{q}-m} = (A^{\alpha\beta}_{\mathbf{q}m})^*$, we write formula (A.101) as

$$\langle \Delta V^{\alpha}_{\mathbf{q}m} \Delta V^{\beta}_{-\mathbf{q}-m} \rangle = \frac{T}{2} \left(A^{-1}_{-\mathbf{q}-m} \right)^{\alpha\beta}.$$

Changing the signs of $\mathbf{q}$ and m and swapping the indices α and β, we finally obtain

$$\langle \Delta V^{\alpha}_{\mathbf{q}m} \Delta V^{\beta}_{-\mathbf{q}-m} \rangle = \frac{T}{2} \left(A^{-1}_{\mathbf{q}m} \right)^{\beta\alpha}. \tag{A.102}$$

References

1. I.V. Savelyev, *Fundamentals of Theoretical Physics (vol. 2). Quantum Mechanics* (Mir Publ., Moscow, 1982)
2. C. Cohen-Tannoudji, B. Diu, F. Laloe, *Quantum Mechanics* (Wiley, New York, 1991)
3. E.P. Wigner, *Group Theory. Expanded and Improved Edition* (Academic, New York, 1959)
4. G.A. Korn, T.M. Korn, *Mathematical Handbook for Scientists and Engineers*, 2nd edn. (McGraw-Hill, New York, 1968)
5. V.V. Voevodin, Y.A. Kuznetsov, *Matrices and Calculations* (Nauka, Moscow, 1984) [in Russian]
6. R.D. Mattuck, *A Guide to Feynman Diagrams in the Many-Body Problem*, 2nd edn. (McGraw-Hill, New York, 1976)
7. G.L. Squires, *Introduction to the Theory of Thermal Neutron Scattering*, 2nd edn. (Dover, New York, 1996)
8. D.A. Varshalovich, A.N. Moskalev, V.K. Khersonskii, *Quantum Theory of Angular Momemtum* (World Scientific, Singapore, 1988)
9. E.T. Whittaker, G.N. Watson, *A Course of Modern Analysis*, 4th edn. (Cambridge University Press, Cambridge, 1962)
10. A. Papoulis, *Probability, Random Variables and Stochastic Processes*, 3rd edn. (McGraw-Hill, New York, 1991)
11. S. Raimes, *Many-Electron Theory* (North-Holland, Amsterdam, 1972)
12. B.I. Reser, V.I. Grebennikov, Phys. Met. Metallogr. **83**, 127 (1997)
13. D.J. Kim, *New Perspectives in Magnetism of Metals* (Kluwer/Plenum, New York, 1999)
14. H. Bruus, K. Flensberg, *Many-Body Quantum Theory in Condensed Matter Physics*, 3rd edn. (Oxford University Press, Oxford, 2004)
15. N.B. Melnikov, B.I. Reser, Theor. Math. Phys. **181**(2), 1435 (2014)
16. N.B. Melnikov, B.I. Reser, Procs. Steklov Inst. Math. **271**, 149 (2010)
17. L.D. Landau, E.M. Lifshitz, *Statistical Physics, Course on Theoretical Physics*, vol. 5, 3rd edn. (Pergamon, Oxford, 1985)

Concepts Related to Functional Integral

B

B.1 "Time"-Ordered Exponential

In the interacting electrons Hamiltonian $\mathcal{H} = \mathcal{H}_0' + \mathcal{H}_\mathrm{I}$ the operators $\mathcal{H}_0'$ and $\mathcal{H}_\mathrm{I}$ do not commute

$$[\mathcal{H}_0', \mathcal{H}_\mathrm{I}] \neq 0,$$

therefore

$$\mathrm{e}^{-(\mathcal{H}_0'+\mathcal{H}_\mathrm{I})/T} \neq \mathrm{e}^{-\mathcal{H}_0'/T}\,\mathrm{e}^{-\mathcal{H}_\mathrm{I}/T}$$

(see Appendix A.1.3). The correct formula using the *"time"-ordered exponential* is as follows

$$\mathrm{e}^{-(\mathcal{H}_0'+\mathcal{H}_\mathrm{I})/T} = \mathrm{e}^{-\mathcal{H}_0'/T}\, T_\tau \exp\left(-\int_0^{1/T} \mathcal{H}_\mathrm{I}(\tau)\,\mathrm{d}\tau\right), \tag{B.1}$$

where

$$\mathcal{H}_\mathrm{I}(\tau) = \mathrm{e}^{\mathcal{H}_0'\tau}\,\mathcal{H}_\mathrm{I}\,\mathrm{e}^{-\mathcal{H}_0'\tau}$$

is the "interaction" representation of the operator $\mathcal{H}_\mathrm{I}$. For the sake of completeness, we give a simple derivation of this well-known formula (see, e.g. [1, 2]).

It is convenient to introduce the new variable $\beta = 1/T$. From the operator function $\exp\big(-\beta(\mathcal{H}_0' + \mathcal{H}_\mathrm{I})\big)$ we extract the multiplier $\exp\big(-\beta\mathcal{H}_0'\big)$, and denote the rest by $S(\beta)$:

$$\mathrm{e}^{-\beta(\mathcal{H}_0'+\mathcal{H}_\mathrm{I})} = \mathrm{e}^{-\beta\mathcal{H}_0'}\,S(\beta). \tag{B.2}$$

Let us obtain an expression for $S(\beta)$. Differentiating (B.2) with respect to β, we have

$$-(\mathcal{H}_0' + \mathcal{H}_\mathrm{I})\,\mathrm{e}^{-\beta(\mathcal{H}_0'+\mathcal{H}_\mathrm{I})} = -\mathcal{H}_0'\,\mathrm{e}^{-\beta\mathcal{H}_0'}\,S(\beta) + \mathrm{e}^{-\beta\mathcal{H}_0'}\frac{\mathrm{d}S(\beta)}{\mathrm{d}\beta}. \tag{B.3}$$

Substituting (B.2) in the left-hand side of (B.3), we obtain

$$-(\mathcal{H}_0' + \mathcal{H}_\mathrm{I})\,\mathrm{e}^{-\beta\mathcal{H}_0'}\,S(\beta) = -\mathcal{H}_0'\,\mathrm{e}^{-\beta\mathcal{H}_0'}\,S(\beta) + \mathrm{e}^{-\beta\mathcal{H}_0'}\frac{\mathrm{d}S(\beta)}{\mathrm{d}\beta}.$$

Hence

$$-\mathcal{H}_\mathrm{I}\,\mathrm{e}^{-\beta\mathcal{H}_0'}\,S(\beta) = \mathrm{e}^{-\beta\mathcal{H}_0'}\frac{\mathrm{d}S(\beta)}{\mathrm{d}\beta}. \tag{B.4}$$

N. B. Melnikov, B. I. Reser, *Dynamic Spin Fluctuation Theory of Metallic Magnetism*,
https://doi.org/10.1007/978-3-319-92974-3

Multiplying (B.4) from the left by $\exp(\beta\mathcal{H}_0')$, we come to

$$\frac{\mathrm{d}S(\beta)}{\mathrm{d}\beta} = -\mathrm{e}^{\beta\mathcal{H}_0'}\,\mathcal{H}_\mathrm{I}\,\mathrm{e}^{-\beta\mathcal{H}_0'}\,S(\beta) \equiv -\mathcal{H}_\mathrm{I}(\beta)S(\beta). \tag{B.5}$$

Integrating both sides of (B.5) we have

$$S(\beta) = S(0) - \int_0^\beta \mathcal{H}_\mathrm{I}(\tau_1)\,S(\tau_1)\,\mathrm{d}\tau_1\,.$$

From (B.2) it follows that $S(0) = 1$. Therefore,

$$S(\beta) = 1 - \int_0^\beta \mathcal{H}_\mathrm{I}(\tau_1)\,S(\tau_1)\,\mathrm{d}\tau_1\,. \tag{B.6}$$

One can solve this equation by iterations. In the first step, substituting

$$S(\tau_1) = 1 - \int_0^{\tau_1} \mathcal{H}_\mathrm{I}(\tau_2)\,S(\tau_2)\,\mathrm{d}\tau_2\,,$$

we obtain

$$S(\beta) = 1 - \int_0^\beta \mathcal{H}_\mathrm{I}(\tau_1)\,\mathrm{d}\tau_1 + \int_0^\beta \mathcal{H}_\mathrm{I}(\tau_1)\,\mathrm{d}\tau_1 \int_0^{\tau_1} \mathcal{H}_\mathrm{I}(\tau_2)\,S(\tau_2)\,\mathrm{d}\tau_2\,.$$

Continuing the process, we come to

$$S(\beta) = \sum_{k=0}^{\infty}(-1)^k \int_0^\beta \mathcal{H}_\mathrm{I}(\tau_1)\,\mathrm{d}\tau_1 \int_0^{\tau_1} \mathcal{H}_\mathrm{I}(\tau_2)\,\mathrm{d}\tau_2 \ldots \int_0^{\tau_{k-1}} \mathcal{H}_\mathrm{I}(\tau_k)\,\mathrm{d}\tau_k\,. \tag{B.7}$$

Here it is convenient to introduce the "time"-ordering operator T_τ, so that we do not have to take care about the order of the operators $\mathcal{H}_\mathrm{I}(\tau_i)$ at different "times". It is easy to convert expression (B.7) to the form[1]

$$S(\beta) = T_\tau \sum_{k=0}^{\infty} \frac{(-1)^k}{k!} \int_0^\beta \mathcal{H}_\mathrm{I}(\tau_1)\,\mathrm{d}\tau_1 \int_0^\beta \mathcal{H}_\mathrm{I}(\tau_2)\,\mathrm{d}\tau_2 \ldots \int_0^\beta \mathcal{H}_\mathrm{I}(\tau_k)\,\mathrm{d}\tau_k\,. \tag{B.8}$$

The right-hand side of (B.8) is called the *"time"-ordered exponential*, and is denoted as

$$S(\beta) = T_\tau \exp\left(-\int_0^\beta \mathcal{H}_\mathrm{I}(\tau)\,\mathrm{d}\tau\right). \tag{B.9}$$

Substitution of (B.9) in (B.2) gives formula (B.1).

B.2 Functional Derivative

The concept of *functional derivative*, which is familiar not to everybody, can be explained in a simple way (see [3]). The value of the functional $F[x(t)]$ is defined for any function $x(t)$. One can ask the question: how does this value change if we change the argument function $x(t)$? In other words, how large will be the difference $F[x(t) + \eta(t)] - F[x(t)]$ if $\eta(t)$ is small? In the first approximation in η this difference is linear in η expression of the kind $\int K(s)\,\eta(s)\,\mathrm{d}s$. Thus defined

[1]The number of permutations of k operators is equal to $k!$ Therefore, if we replace the integration limits 0 to τ_k by 0 to β, the integral will increase by $k!$ times. Hence we need to divide by $k!$ to get the correct result.

function $K(s)$ is called the *functional derivative* of the functional F with respect to the function $x(t)$ at the point s and is denoted by $\delta F/\delta x(s)$. Therefore, up to linear terms, one can write

$$F[x(t)+\eta(t)] = F[x(t)] + \int \frac{\delta F}{\delta x(s)} \eta(s)\,\mathrm{d}s + \cdots$$

Clearly, the derivative $\delta F/\delta x(s)$ depends on the function $x(t)$ as well as on the value s, i.e. it is a functional of $x(t)$ and a function of s. For practical purposes, it is convenient to use the following equivalent definition (see [4, p. 173]):

$$\frac{\delta F}{\delta x(s)} = \lim_{\varepsilon\to 0} \frac{F[x(t)+\varepsilon\delta(t-s)] - F[x(t)]}{\varepsilon}, \tag{B.10}$$

where $\delta(t)$ is the Dirac delta function.

As an illustration, we calculate the functional derivative of the integrand in (8.26) with respect to the charge field $V_j^0(\tau)$:

$$\frac{\delta}{\delta V_j^0(\tau)} \mathrm{e}^{-(F_0(V)+\Omega_1(V))/T} = -\frac{1}{T}\left(\frac{\delta F_0(V)}{\delta V_j^0(\tau)} + \frac{\delta \Omega_1(V)}{\delta V_j^0(\tau)}\right) \mathrm{e}^{-(F_0(V)+\Omega_1(V))/T}. \tag{B.11}$$

Using formula (8.27):

$$F_0(V) = T\int_0^{1/T} \sum_{j'} \left(\frac{1}{u}\mathbf{V}_{j'}^2(\tau') - \frac{1}{u_0}(V_{j'}^0(\tau'))^2\right) \mathrm{d}\tau',$$

by the definition (B.10) we have

$$\frac{\delta F_0(V)}{\delta V_j^0(\tau)} = -\frac{2T}{u_0} V_j^0(\tau). \tag{B.12}$$

Similarly, using formula (8.28):

$$\Omega_1(V) = -T \ln \mathrm{Tr}\left[T_\tau \exp\left(-\int_0^{1/T} \mathcal{H}'(V)\,\mathrm{d}\tau'\right)\right],$$

we calculate

$$\frac{\delta \Omega_1(V)}{\delta V_j^0(\tau)} = -T \frac{\mathrm{Tr}\left[T_\tau \frac{\delta}{\delta V_j^0(\tau)}\left(-\int_0^{1/T} \mathcal{H}'(V)\,\mathrm{d}\tau'\right) \exp\left(-\int_0^{1/T} \mathcal{H}'(V)\,\mathrm{d}\tau'\right)\right]}{\mathrm{Tr}\left[T_\tau \exp\left(-\int_0^{1/T} \mathcal{H}'(V)\,\mathrm{d}\tau'\right)\right]}. \tag{B.13}$$

By formula (8.29):

$$\mathcal{H}'(V) = \mathcal{H}_0' + 2\sum_{j'}\left(\mathbf{V}_{j'}(\tau')\boldsymbol{\rho}_{j'}(\tau') + V_{j'}^0(\tau')\rho_{j'}^0(\tau')\right),$$

we obtain

$$\frac{\delta}{\delta V_j^0(\tau)}\left(-\int_0^{1/T} \mathcal{H}'(V)\,\mathrm{d}\tau'\right) = -2\rho_j^0(\tau).$$

Substituting this to expression (B.13), we have

$$\frac{\delta \Omega_1(V)}{\delta V_j^0(\tau)} = 2T\rho_j^0(V), \tag{B.14}$$

where

$$\rho_j^0(V) = \frac{\mathrm{Tr}\left[T_\tau\, \rho_j^0(\tau) \exp\left(-\int_0^{1/T} \mathcal{H}'(V)\, \mathrm{d}\tau'\right)\right]}{\mathrm{Tr}\left[T_\tau \exp\left(-\int_0^{1/T} \mathcal{H}'(V)\, \mathrm{d}\tau'\right)\right]}. \tag{B.15}$$

Substitution of (B.12) and (B.14) in (B.11) gives the result (8.31):

$$\frac{\delta}{\delta V_j^0(\tau)} \mathrm{e}^{-(F_0(V)+\Omega_1(V))/T} = \left(\frac{2V_j^0(\tau)}{u_0} - 2\rho_j^0(V)\right) \mathrm{e}^{-(F_0(V)+\Omega_1(V))/T}.$$

Expression (B.15) can be written in the compact form

$$\rho_j^0(V) = \mathrm{Tr}\left[T_\tau\, \rho_j^0(\tau) \exp \int_0^{1/T} \left(\Omega_1(V) - \mathcal{H}'(V)\right) \mathrm{d}\tau'\right].$$

B.3 Stratonovich-Hubbard Transformation

B.3.1 Ising Model

To calculate the partition function of the interacting spins system, we use the Stratonovich-Hubbard transformation [5, 6]. This method consists in replacing the pair interaction of spins with the interaction of spins with a fluctuating field. Here we give the necessary mathematical details.

The key element of the Stratonovich-Hubbard transformation is the identity

$$\exp\left(\frac{A^2}{a}\right) = \sqrt{\frac{a}{\pi}} \int \exp(-ax^2 + 2Ax)\, \mathrm{d}x, \tag{B.16}$$

which is valid for any *real* A and $a > 0$. This identity is obtained by shifting the dummy variable in the improper integral over the real line:

$$\int \mathrm{e}^{-ax^2}\, \mathrm{d}x = \int \exp\left(-a\left(x - \frac{A}{a}\right)^2\right) \mathrm{d}x.$$

Replacing the integral on the left-hand side by its value $\sqrt{\pi/a}$ and rearranging, we come to

$$\sqrt{\frac{\pi}{a}} = \int \exp(-ax^2 + 2Ax - A^2/a)\, \mathrm{d}x.$$

From this we immediately obtain identity (B.16). The main feature of formula (B.16) is that the exponent on the left-hand side is the square of A and the one on the right-hand side is linear in A.

Similarly, in the *complex* case, we use the integral $\int \mathrm{e}^{-a(x^2+y^2)}\, \mathrm{d}x\, \mathrm{d}y = \int \mathrm{e}^{-a|z|^2}\, \mathrm{d}z$, where $z = x + \mathrm{i}y$ and $\mathrm{d}z = \mathrm{d}x\, \mathrm{d}y$. Rearranging the equality

$$\frac{\pi}{a} = \int \mathrm{e}^{-a|z|^2}\, \mathrm{d}z = \int \exp\left(-a\left|z - \frac{A}{a}\right|^2\right) \mathrm{d}z,$$

we obtain the complex analogue of relation (B.16):

$$\mathrm{e}^{|A|^2/a} = \frac{a}{\pi} \int \exp(-a|z|^2 + z^* A + zA^*)\, \mathrm{d}z, \qquad a > 0. \tag{B.17}$$

In the momentum representation the Hamiltonian of the Ising model $\mathcal{H}$ is given by a sum of squares (7.2):

$$\mathcal{H} = -\frac{1}{2N} \sum_{\mathbf{q}} J_{\mathbf{q}} |S_{\mathbf{q}}|^2.$$

Therefore, the partition function $Z = \mathrm{Tr} e^{-\mathcal{H}/T}$ can be written as

$$Z = \mathrm{Tr} \prod_{\mathbf{q}} \exp\left(\frac{J_{\mathbf{q}}}{2NT} |S_{\mathbf{q}}|^2\right).$$

Using the real identity (B.16) for $q = 0$ with $a = 1/2$ and $A = \sqrt{J_0/NT}\, S_0/2$ and complex identity (B.17) for $\mathbf{q} \neq 0$ with $a = 1/2$ and $A = \sqrt{J_{\mathbf{q}}/NT}\, S_{\mathbf{q}}/2$, we obtain

$$\begin{aligned} Z = & \left(\int \exp\left(-\frac{x_0^2}{2}\right) \mathrm{d}x_0\right)^{-1} \left(\prod_{\mathbf{q}\neq 0} \int \exp\left(-\frac{|z_{\mathbf{q}}|^2}{2}\right) \mathrm{d}x_{\mathbf{q}} \mathrm{d}y_{\mathbf{q}}\right)^{-1} \\ & \times \mathrm{Tr}\left(\int \exp\left(-\frac{x_0^2}{2} + \sqrt{\frac{J_0}{NT}}\, S_0 x_0\right) \mathrm{d}x_0 \right. \\ & \left. \times \prod_{\mathbf{q}\neq 0} \int \exp\left(-\frac{|z_{\mathbf{q}}|^2}{2} + \sqrt{\frac{J_{\mathbf{q}}}{NT}}\, S_{\mathbf{q}} z_{-\mathbf{q}}\right) \mathrm{d}x_{\mathbf{q}} \mathrm{d}y_{\mathbf{q}}\right). \end{aligned}$$

Simplifying, we come to formula (7.6):

$$Z = \left[\int \exp\left(-\sum_{\mathbf{q}} \frac{|z_{\mathbf{q}}|^2}{2}\right) \mathrm{d}\mathbf{z}\right]^{-1} \mathrm{Tr}\left[\int \exp\left(-\sum_{\mathbf{q}} \frac{|z_{\mathbf{q}}|^2}{2} + \sqrt{\frac{J_{\mathbf{q}}}{NT}} \sum_{\mathbf{q}} S_{\mathbf{q}} z_{-\mathbf{q}}\right) \mathrm{d}\mathbf{z}\right],$$

where $\mathrm{d}\mathbf{z} = \mathrm{d}x_0 \prod_{\mathbf{q}\neq 0} \mathrm{d}x_{\mathbf{q}} \mathrm{d}y_{\mathbf{q}}$ (for details, see [7]).

B.3.2 Functional Integrals

In the itinerant electron system, the Stratonovich-Hubbard transformation [5, 6] is based on the following identity:

$$\exp\left(\frac{\mathcal{A}^2}{a}\right) = \sqrt{\frac{a}{\pi}} \int \exp(-ax^2 + 2\mathcal{A}x)\, \mathrm{d}x, \qquad a > 0,$$

where $\mathcal{A}$ is an *operator*. Applying this identity to the "time"-ordered exponential of the operator $\mathcal{A}^2$,

$$T_\tau \exp\left(\int_0^{1/T} \mathcal{A}^2(\tau)\, \mathrm{d}\tau\right) = \lim_{N\to\infty} \exp\big(\mathcal{A}^2(\tau_N)\Delta\tau\big) \ldots \exp\big(\mathcal{A}^2(\tau_1)\Delta\tau\big),$$

we have

$$\begin{aligned} T_\tau \exp\left(\int_0^{1/T} \mathcal{A}^2(\tau)\, \mathrm{d}\tau\right) &= \lim_{N\to\infty} T_\tau \prod_{n=1}^{N} \exp\left(\mathcal{A}^2(\tau_n)\Delta\tau\right) \\ &= \lim_{N\to\infty} T_\tau \prod_{n=1}^{N} \left\{\frac{1}{\sqrt{\pi\Delta\tau}} \int \exp\left(-\frac{x^2(\tau_n)}{\Delta\tau} + 2\mathcal{A}(\tau_n)x(\tau_n)\right) \mathrm{d}x(\tau_n)\right\} \\ &= \lim_{N\to\infty} T_\tau \int \cdots \int \exp\left(\sum_{n=1}^{N}\left(-\frac{x^2(\tau_n)}{\Delta\tau} + 2\mathcal{A}(\tau_n)x(\tau_n)\right)\right) \left(\prod_{n=1}^{N} \frac{\mathrm{d}x(\tau_n)}{\sqrt{\pi\Delta\tau}}\right). \end{aligned}$$

Replacing the dimensionless variable $x(\tau_n)$ by the new variable $v(\tau_n) = -x(\tau_n)/\Delta\tau$, which has dimensions of energy, we come to

$$T_\tau \exp\left(\int_0^{1/T} \mathcal{A}^2(\tau)\,\mathrm{d}\tau\right) \tag{B.18}$$

$$= \lim_{N\to\infty} T_\tau \int \cdots \int \exp\left(-\sum_{n=1}^{N}\left(v^2(\tau_n) + 2\mathcal{A}(\tau_n)v(\tau_n)\right)\Delta\tau\right)\left((-1)^N \prod_{n=1}^{N}\sqrt{\frac{\Delta\tau}{\pi}}\,\mathrm{d}v(\tau_n)\right).$$

In particular, for $\mathcal{A} = 0$ we have

$$1 = \lim_{N\to\infty} T_\tau \int \cdots \int \exp\left(-\sum_{n=1}^{N} v^2(\tau_n)\Delta\tau\right)\left((-1)^N \prod_{n=1}^{N}\sqrt{\frac{\Delta\tau}{\pi}}\,\mathrm{d}v(\tau_n)\right). \tag{B.19}$$

Dividing the right-hand side of (B.18) by the one of (B.19), we obtain

$$T_\tau \exp\left(\int_0^{1/T} \mathcal{A}^2(\tau)\,\mathrm{d}\tau\right) = \frac{T_\tau \int \exp\left(-\int_0^{1/T}\left(v^2(\tau) + 2\mathcal{A}(\tau)v(\tau)\right)\mathrm{d}\tau\right)\mathrm{D}v(\tau)}{T_\tau \int \exp\left(-\int_0^{1/T} v^2(\tau)\,\mathrm{d}\tau\right)\mathrm{D}v(\tau)}, \tag{B.20}$$

where the notation

$$\mathrm{D}v(\tau) = \lim_{N\to\infty}\left(\prod_{n=1}^{N}\mathrm{d}v(\tau_n)\right)$$

indicates the functional integration over $v(\tau)$ on the "time" interval $[0, 1/T]$.[2] Since the multiband Hubbard Hamiltonian is transformed to a sum of squares in the site representation, in the DSFT we need to carry out the Stratonovich-Hubbard transformation only in the real space.

B.4 Optimal Gaussian Approximation

B.4.1 Ising Model

First, we consider the Ising model, where the functional integral reduces to a multidimensional integral in the Euclidean space.

We obtain the parameters of the optimal Gaussian approximation using the free energy minimum principle (9.25). In the class of quadratic "modelling" functions $F^{(2)}(\mathbf{V})$, inequality (9.25) is written as

$$\mathcal{F} \le \mathcal{F}^{(2)} + \left\langle F(\mathbf{V}) - F^{(2)}(\mathbf{V})\right\rangle_{(2)}. \tag{B.21}$$

The minimum of the right-hand side is determined by the system of equations

$$\frac{\partial}{\partial \bar{V}_{\mathbf{q}}}\left[\mathcal{F}^{(2)} + \left\langle F(\mathbf{V}) - F^{(2)}(\mathbf{V})\right\rangle_{(2)}\right] = 0, \tag{B.22}$$

$$\frac{\partial}{\partial A_{\mathbf{q}}}\left[\mathcal{F}^{(2)} + \left\langle F(\mathbf{V}) - F^{(2)}(\mathbf{V})\right\rangle_{(2)}\right] = 0. \tag{B.23}$$

Here

$$\mathcal{F}^{(2)} = -T \ln Z^{(2)} = -T \ln \int \mathrm{e}^{-F^{(2)}(\mathbf{V})/T}\,\mathrm{d}\mathbf{V} \tag{B.24}$$

[2]Functional integrals are also known under the name *path integrals* (see, e.g. [3, 8]).

is the total free energy and the Gaussian average is determined by

$$\langle \dots \rangle_{(2)} = \frac{1}{Z^{(2)}} \int \dots \mathrm{e}^{-F^{(2)}(\mathbf{V})/T} \, \mathrm{d}\mathbf{V}.$$

The translational invariance of the crystal allows to consider quadratic functions of the form

$$F^{(2)}(\mathbf{V}) = \sum_{\mathbf{q}} A_{\mathbf{q}} |\Delta V_{\mathbf{q}}|^2 = \sum_{\mathbf{q}} \Delta V_{\mathbf{q}} A_{\mathbf{q}} \Delta V_{-\mathbf{q}}, \tag{B.25}$$

where $\Delta V_{\mathbf{q}} = V_{\mathbf{q}} - \bar{V}_{\mathbf{q}}$ and $A_{\mathbf{q}} > 0$. Then

$$p^{(2)}(\mathbf{V}) = \left(\int \mathrm{e}^{-F^{(2)}(\mathbf{V})/T} \, \mathrm{d}\mathbf{V} \right)^{-1} \mathrm{e}^{-F^{(2)}(\mathbf{V})/T} \tag{B.26}$$

is the probability density of the Gaussian fluctuating field. The mean values $\langle V_{\mathbf{q}} \rangle = \bar{V}_{\mathbf{q}}$ and correlators

$$\langle \Delta V_{\mathbf{q}} \Delta V_{-\mathbf{q}'} \rangle = \frac{T}{2A_{\mathbf{q}}} \delta_{\mathbf{q},\mathbf{q}'}$$

completely describe the Gaussian fluctuating field (for details, see Appendix A.3.3).

We start with calculating the left-hand side of Eq. (B.22). Using formula (B.25), we evaluate the Gaussian partition function

$$\begin{aligned} Z^{(2)} &= \int \exp\Big(-\frac{1}{T} \sum_{\mathbf{q}} A_{\mathbf{q}} |\Delta V_{\mathbf{q}}|^2\Big) \, \mathrm{d}\mathbf{V} \\ &= \int \exp\Big(-\frac{1}{T} A_0 \Delta V_0^2\Big) \, \mathrm{d}V_0 \prod_{\mathbf{q} \neq 0} \int \exp\Big(-\frac{1}{T} A_{\mathbf{q}} |\Delta V_{\mathbf{q}}|^2\Big) \, \mathrm{d}V_{\mathbf{q}} \\ &= \sqrt{\frac{\pi T}{A_0}} \prod_{\mathbf{q} \neq 0} \frac{\pi T}{A_{\mathbf{q}}} \end{aligned} \tag{B.27}$$

Since $Z^{(2)}$ and hence $\mathcal{F}^{(2)}$ are independent of $\bar{V}_{\mathbf{q}}$, we obtain

$$\begin{aligned} \frac{\partial}{\partial \bar{V}_{\mathbf{q}}} \Big[\mathcal{F}^{(2)} + \big\langle F(\mathbf{V}) - F^{(2)}(\mathbf{V}) \big\rangle_{(2)} \Big] &= \frac{\partial}{\partial \bar{V}_{\mathbf{q}}} \langle F(\mathbf{V}) - F^{(2)}(\mathbf{V}) \rangle_{(2)} \\ &= \frac{1}{Z^{(2)}} \frac{\partial}{\partial \bar{V}_{\mathbf{q}}} \int \big(F(\mathbf{V}) - F^{(2)}(\mathbf{V})\big) \exp\Big(-\frac{1}{T} F^{(2)}(\mathbf{V})\Big) \, \mathrm{d}\mathbf{V}. \end{aligned}$$

Changing the dummy variable to $\mathbf{X} = \mathbf{V} - \bar{\mathbf{V}}$, we have

$$\begin{aligned} &\frac{\partial}{\partial \bar{V}_{\mathbf{q}}} \big\langle F(\mathbf{V}) - F^{(2)}(\mathbf{V}) \big\rangle_{(2)} \\ &= \frac{1}{Z^{(2)}} \frac{\partial}{\partial \bar{V}_{\mathbf{q}}} \int \Big(F(\mathbf{X} + \bar{\mathbf{V}}) - \sum_{\mathbf{q}'} A_{\mathbf{q}'} |X_{\mathbf{q}'}|^2 \Big) \exp\Big(-\frac{1}{T} \sum_{\mathbf{q}'} A_{\mathbf{q}'} |X_{\mathbf{q}'}|^2\Big) \, \mathrm{d}\mathbf{X} \\ &= \frac{1}{Z^{(2)}} \int \frac{\partial F(\mathbf{X} + \bar{\mathbf{V}})}{\partial \bar{V}_{\mathbf{q}}} \exp\Big(-\frac{1}{T} \sum_{\mathbf{q}'} A_{\mathbf{q}'} |X_{\mathbf{q}'}|^2\Big) \, \mathrm{d}\mathbf{X}. \end{aligned}$$

Reverting to the original variable $\mathbf{V}$, we obtain

$$\frac{\partial}{\partial \bar{V}_{\mathbf{q}}}\langle F(\mathbf{V}) - F^{(2)}(\mathbf{V})\rangle_{(2)} \\ = \frac{1}{Z^{(2)}}\int \frac{\partial F(\mathbf{V})}{\partial V_{\mathbf{q}}} \exp\Big(-\frac{1}{T}F^{(2)}(\mathbf{V})\Big)\,\mathrm{d}\mathbf{V} = \left\langle \frac{\partial F(\mathbf{V})}{\partial V_{\mathbf{q}}}\right\rangle_{(2)}. \tag{B.28}$$

Thus, Eq. (B.22) is written as

$$\left\langle \frac{\partial F(\mathbf{V})}{\partial V_{\mathbf{q}}}\right\rangle_{(2)} = 0. \tag{B.29}$$

To calculate the left-hand side of Eq. (B.23), we transform the expression

$$\langle F(\mathbf{V}) - F^{(2)}(\mathbf{V})\rangle_{(2)} = \frac{1}{Z^{(2)}}\int \Big[F(\mathbf{V}) - F^{(2)}(\mathbf{V})\Big] \exp\Big(-\frac{1}{T}F^{(2)}(\mathbf{V})\Big)\,\mathrm{d}\mathbf{V}.$$

Making the change of variables $V_{\mathbf{q}} = Y_{\mathbf{q}}/\sqrt{A_{\mathbf{q}}} + \bar{V}_{\mathbf{q}}$ such that $|Y_{\mathbf{q}}|^2 = A_{\mathbf{q}}|\Delta V_{\mathbf{q}}|^2$, we have

$$\langle F(\mathbf{V}) - F^{(2)}(\mathbf{V})\rangle_{(2)} = \int \Big[F\Big(\frac{\mathbf{Y}}{\sqrt{\mathbf{A}}} + \bar{\mathbf{V}}\Big) - \sum_{\mathbf{q}'}|Y_{\mathbf{q}'}|^2\Big] \exp\Big(-\frac{1}{T}\sum_{\mathbf{q}'}|Y_{\mathbf{q}'}|^2\Big)\,\mathrm{d}\mathbf{Y}.$$

Here we used the brief notation $\mathbf{V} = \mathbf{Y}/\sqrt{\mathbf{A}} + \bar{\mathbf{V}}$ and

$$\mathrm{d}\mathbf{Y} = \Big(\frac{\mathrm{d}Y_0}{\sqrt{\pi T}}\Big)\prod_{\mathbf{q}'\neq 0}\Big(\frac{\mathrm{d}Y_{\mathbf{q}'}}{\pi T}\Big).$$

Differentiating, for $\mathbf{q} \neq 0$ we obtain

$$\frac{\partial}{\partial A_{\mathbf{q}}}\langle F(\mathbf{V}) - F^{(2)}(\mathbf{V})\rangle_{(2)} \\ = -\frac{1}{2A_{\mathbf{q}}^{3/2}}\int \frac{\partial F}{\partial(Y_{\mathbf{q}}/\sqrt{A_{\mathbf{q}}})}\Big(\frac{\mathbf{Y}}{\sqrt{\mathbf{A}}} + \bar{\mathbf{V}}\Big)\, Y_{\mathbf{q}} \exp\Big(-\frac{1}{T}\sum_{\mathbf{q}'}|Y_{\mathbf{q}'}|^2\Big)\,\mathrm{d}\mathbf{Y}.$$

Integrating by parts and reverting to the original variable $\mathbf{V}$, we have

$$\frac{\partial}{\partial A_{\mathbf{q}}}\langle F(\mathbf{V}) - F^{(2)}(\mathbf{V})\rangle_{(2)} = -\frac{T}{2A_{\mathbf{q}}^2}\frac{1}{Z^{(2)}}\int \frac{\partial^2 F(\mathbf{V})}{\partial V_{\mathbf{q}}\,\partial V_{-\mathbf{q}}} \exp\Big(-\frac{1}{T}F^{(2)}(\mathbf{V})\Big)\,\mathrm{d}\mathbf{V}.$$

Taking (B.27) into account, for $\mathbf{q} \neq 0$ we write the right-hand side of (B.23) as

$$\frac{\partial}{\partial A_{\mathbf{q}}}\Big[\mathcal{F}^{(2)} + \langle F(\mathbf{V}) - F^{(2)}(\mathbf{V})\rangle_{(2)}\Big] = \frac{T}{A_{\mathbf{q}}} - \frac{T}{2A_{\mathbf{q}}^2}\left\langle \frac{\partial^2 F(\mathbf{V})}{\partial V_{\mathbf{q}}\,\partial V_{-\mathbf{q}}}\right\rangle_{(2)}.$$

Similarly, for $\mathbf{q} = 0$ we obtain one half of this expression. Hence Eq. (B.23) becomes

$$A_{\mathbf{q}} = \frac{1}{2}\left\langle \frac{\partial^2 F(\mathbf{V})}{\partial V_{\mathbf{q}}\,\partial V_{-\mathbf{q}}}\right\rangle_{(2)}. \tag{B.30}$$

B.4.2 Functional Integrals

In the itinerant-electrons system, it is necessary to consider a more general class of quadratic "modelling" functions in the inequality (B.21). As we showed, the quadratic approximation of the free energy in the DSFT is given by (A.99). The difference from the previous subsection is that the matrix of the quadratic form is nondiagonal.

To obtain the parameters of the optimal Gaussian approximation, we first write the quadratic form as the inner product (similar to Appendix A.3.3):

$$F^{(2)}(V) = (\Delta V, A^* \Delta V),$$

where the elements of the matrix A^* are the complex conjugates of the ones of A. Next, we carry out the unitary transformation $V = U\tilde{V}$ ($UU^\dagger = 1$) that diagonalizes the Hermitian matrix A^*, i.e.

$$A^* = UDU^\dagger, \tag{B.31}$$

where D is a real diagonal matrix. Then

$$(\Delta V, A^* \Delta V) = (\Delta \tilde{V}, U^\dagger A^* U \Delta \tilde{V}) = (\Delta \tilde{V}, D \Delta \tilde{V}),$$

and the quadratic form in the new coordinates $\tilde{F}^{(2)}(\tilde{V}) \equiv F^{(2)}(U\tilde{V})$ is written as

$$\tilde{F}^{(2)}(\tilde{V}) = \sum_i \Delta \tilde{V}_i D_i \Delta \tilde{V}_i^*,$$

where i is a multi-index (not a lattice site). Now, following the same line as in the previous subsection, we obtain the equations for the parameters of the optimal Gaussian approximation in the new variables:

$$\left\langle \frac{\partial \tilde{F}(\tilde{V})}{\partial \tilde{V}_i} \right\rangle_{(2)} = 0, \qquad D_i = \frac{1}{2} \left\langle \frac{\partial^2 \tilde{F}(\tilde{V})}{\partial \tilde{V}_i \, \partial \tilde{V}_i^*} \right\rangle_{(2)}, \tag{B.32}$$

where $\tilde{F}(\tilde{V}) \equiv F(U\tilde{V}) = F(V)$. Reverting to the original variables by the formula

$$\tilde{V}_j = \sum_{j'} U_{jj'}^\dagger V_{j'},$$

we have

$$\left\langle \frac{\partial F(V)}{\partial V_i} \right\rangle_{(2)} = \left\langle \sum_j \frac{\partial \tilde{F}(\tilde{V})}{\partial \tilde{V}_j} \frac{\partial \tilde{V}_j}{\partial V_i} \right\rangle_{(2)} = \sum_j \left\langle \frac{\partial \tilde{F}(\tilde{V})}{\partial \tilde{V}_j} \right\rangle_{(2)} U_{ji}^\dagger.$$

Taking into account the first equation in (B.32), we obtain

$$\left\langle \frac{\partial F(V)}{\partial V_i} \right\rangle_{(2)} = 0.$$

Similarly, we carry out the coordinate transformation $V = U\tilde{V}$ of the second derivative

$$D_i = \frac{1}{2} \left\langle \frac{\partial^2 \tilde{F}(\tilde{V})}{\partial \tilde{V}_i \, \partial \tilde{V}_i^*} \right\rangle_{(2)} = \frac{1}{2} \left\langle \sum_{jj'} \frac{\partial V_j}{\partial \tilde{V}_i} \frac{\partial^2 F(V)}{\partial V_j \, \partial V_{j'}^*} \frac{\partial V_{j'}^*}{\partial \tilde{V}_i^*} \right\rangle_{(2)}.$$

Using the relation

$$V_j = \sum_{j'} U_{jj'} \tilde{V}_{j'},$$

we obtain

$$D_i = \frac{1}{2}\sum_{jj'} U_{ji}\left\langle \frac{\partial^2 F(V)}{\partial V_j\, \partial V^*_{j'}}\right\rangle_{(2)} (U_{j'i})^* = \frac{1}{2}\sum_{jj'} U_{ji}\left\langle \frac{\partial^2 F(V)}{\partial V_j\, \partial V^*_{j'}}\right\rangle_{(2)} U^\dagger_{ij'}.$$

Swapping the indices j and j', we have

$$D_i = \frac{1}{2}\sum_{jj'} U^\dagger_{ij}\left\langle \frac{\partial^2 F(V)}{\partial V_{j'}\, \partial V^*_{j}}\right\rangle_{(2)} U_{j'i}.$$

On the other hand, from relation (B.31), we obtain $D = U^\dagger A^* U$, or in coordinates

$$D_i = \sum_{jj'} U^\dagger_{ij} A^*_{jj'} U_{j'i}.$$

Comparing the last two equations, we finally obtain

$$A_{jj'} = \frac{1}{2}\left\langle \frac{\partial^2 F(V)}{\partial V_j\, \partial V^*_{j'}}\right\rangle_{(2)}.$$

Specifically, in the momentum-"frequency" representation, we get the result [9]

$$\left\langle \frac{\partial F(V)}{\partial V^\alpha_{\mathbf{q}m}}\right\rangle_{(2)} = 0, \qquad A^{\alpha\beta}_{\mathbf{q}m} = \frac{1}{2}\left\langle \frac{\partial^2 F(V)}{\partial V^\alpha_{\mathbf{q}m}\, \partial V^\beta_{-\mathbf{q}-m}}\right\rangle_{(2)}. \tag{B.33}$$

References

1. A.A. Abrikosov, L.P. Gorkov, I.E. Dzyaloshinski, *Methods of Quantum Field Theory in Statistical Physics* (Prentice-Hall, Englewood Cliffs, 1963)
2. A.L. Fetter, J.D. Walecka, *Quantum Theory of Many-Particle Systems* (McGraw-Hill, New York, 1971)
3. R.P. Feynman, R.P. Hibbs, *Quantum Mechanics and Path Integrals* (McGraw-Hill, New York, 1965)
4. L.H. Ryder, *Quantum Field Theory*, 2nd edn. (Cambridge University Press, Cambridge, 1996)
5. R.L. Stratonovich, Sov. Phys. Dokl. **2**, 416 (1958)
6. J. Hubbard, Phys. Rev. Lett. **3**(2), 77 (1959)
7. N.B. Melnikov, Yu.A. Romanenko, Vychisl. Metody Programm. **13**(3), 452 (2012) [in Russian]
8. J. Zinn-Justin, *Path Integrals in Quantum Mechanics* (Oxford University Press, Oxford, 2004)
9. N.B. Melnikov, B.I. Reser, Procs. Steklov Inst. Math. **271**, 149 (2010)

Fourier Transformations

C

C.1 Translationally Invariant Systems

Translational invariance of a homogeneous media favours use of the Fourier transformation. This fact is quite general and has been exploited many times in this book. Here we illustrate the idea in the example of the Ising model (see Chap. 7).

Since we are interested in bulk properties of the crystal rather than surface effects, we can use the boundary conditions that are mathematically most convenient. As is customary in solid state physics, we apply the cyclic boundary conditions. That means the spin $S(\mathbf{R})$ as a function of the lattice site $\mathbf{R}$ is assumed to be periodic with the period $N_i a_i$ in the direction of the ith primitive vector $\mathbf{a}_i$:

$$S(\mathbf{R}) = S(\mathbf{R} + N_i\mathbf{a}_i), \qquad i = 1, 2, 3.$$

The periodicity allows us to introduce the Fourier transformation of spins in the usual way:

$$S_{\mathbf{q}} = \sum_j S_j\, \mathrm{e}^{-\mathrm{i}\mathbf{q}\mathbf{R}_j}, \qquad S_j = \frac{1}{N}\sum_{\mathbf{q}} S_{\mathbf{q}}\mathrm{e}^{\mathrm{i}\mathbf{q}\mathbf{R}_j}, \tag{C.1}$$

where $N = N_1N_2N_3$ is the number of sites, $\mathbf{q}$ is a vector of the reciprocal lattice and the sum is carried out over N sites of the Brillouin zone (see, e.g. [1]). Similarly, we define the Fourier transform and its inverse for the interaction coefficients $J_{jj'}$ by the formulae

$$J_{\mathbf{q}\mathbf{q}'} = \frac{1}{N}\sum_{jj'} J_{jj'}\mathrm{e}^{-\mathrm{i}\mathbf{q}\mathbf{R}_j}\mathrm{e}^{\mathrm{i}\mathbf{q}'\mathbf{R}_{j'}}, \qquad J_{jj'} = \frac{1}{N}\sum_{\mathbf{q}\mathbf{q}'} J_{\mathbf{q}\mathbf{q}'}\mathrm{e}^{\mathrm{i}\mathbf{q}\mathbf{R}_j}\mathrm{e}^{-\mathrm{i}\mathbf{q}'\mathbf{R}_{j'}}. \tag{C.2}$$

The interaction coefficients $J_{jj'} = J_{j-j'}$ in the Hamiltonian (7.1) depend only on the distance between the sites but not on their position in the crystal. Then the Fourier coefficients $J_{\mathbf{q}\mathbf{q}'}$ will not change if we shift the crystal as a whole by a lattice vector $\mathbf{R}_l$:

$$\begin{aligned} J_{\mathbf{q}\mathbf{q}'} &= \frac{1}{N}\sum_{jj'} J_{j-j'}\mathrm{e}^{-\mathrm{i}\mathbf{q}(\mathbf{R}_j+\mathbf{R}_l)}\mathrm{e}^{\mathrm{i}\mathbf{q}'(\mathbf{R}_{j'}+\mathbf{R}_l)} \\ &= \frac{1}{N}\sum_{jj'} J_{j-j'}\mathrm{e}^{-\mathrm{i}\mathbf{q}\mathbf{R}_j}\mathrm{e}^{\mathrm{i}\mathbf{q}'\mathbf{R}_{j'}}\mathrm{e}^{-\mathrm{i}(\mathbf{q}-\mathbf{q}')\mathbf{R}_l}. \end{aligned}$$

Averaging over all lattice sites $\mathbf{R}_l$, we come to

$$J_{\mathbf{q}\mathbf{q}'} = \frac{1}{N}\sum_{jj'} J_{j-j'}\mathrm{e}^{-\mathrm{i}\mathbf{q}\mathbf{R}_j}\mathrm{e}^{\mathrm{i}\mathbf{q}'\mathbf{R}_{j'}}\frac{1}{N}\sum_l \mathrm{e}^{-\mathrm{i}(\mathbf{q}-\mathbf{q}')\mathbf{R}_l}.$$

It is easy to verify the formula

$$\sum_l \mathrm{e}^{\mathrm{i}\mathbf{k}\mathbf{R}_l} = N\delta_{\mathbf{k}0} \tag{C.3}$$

N. B. Melnikov, B. I. Reser, *Dynamic Spin Fluctuation Theory of Metallic Magnetism*,
https://doi.org/10.1007/978-3-319-92974-3

(calculate the geometric series over each coordinate). Using the latter, we obtain

$$J_{\mathbf{q}\mathbf{q}'} = \frac{1}{N}\sum_{jj'} J_{j-j'} \mathrm{e}^{-\mathrm{i}\mathbf{q}\mathbf{R}_{j-j'}} \delta_{\mathbf{q}\mathbf{q}'}.$$

Thus, the Fourier coefficients form a diagonal matrix: $J_{\mathbf{q}\mathbf{q}'} = J_{\mathbf{q}}\delta_{\mathbf{q}\mathbf{q}'}$, where

$$J_{\mathbf{q}} = \sum_j J_j \, \mathrm{e}^{-\mathrm{i}\mathbf{q}\mathbf{R}_j}.$$

As we already mentioned the matrix $J_{jj'} = J_{j-j'}$ possesses the *circulant* property: $J_{jj'} = J_{j+l,j'+l}$, i.e. the matrix has the same elements at each diagonal parallel to the principal one. Moreover, using the cyclic boundary conditions, we see that there are only N different elements (e.g. $J_{-1} = J_{N-1}$). The Fourier coefficients $J_{\mathbf{q}}$ of those N elements J_j form the principal diagonal of the matrix $J_{\mathbf{q}\mathbf{q}'}$:

$$\begin{bmatrix} J_0 & J_1 & J_2 & \cdots & J_{N-1} \\ J_{N-1} & J_0 & J_1 & \cdots & J_{N-2} \\ J_{N-2} & J_{N-1} & J_0 & \cdots & J_{N-3} \\ \vdots & \vdots & & \ddots & \vdots \\ J_1 & J_2 & J_3 & \cdots & J_0 \end{bmatrix} \longleftrightarrow \begin{bmatrix} J_{\mathbf{q}_1} & 0 & 0 & \cdots & 0 \\ 0 & J_{\mathbf{q}_2} & 0 & \cdots & 0 \\ 0 & 0 & J_{\mathbf{q}_3} & \cdots & 0 \\ \vdots & \vdots & & \ddots & \vdots \\ 0 & 0 & 0 & \cdots & J_{\mathbf{q}_N} \end{bmatrix}$$

Thus, the Fourier transformation diagonalizes a circulant matrix, and the inverse Fourier transformation converts a diagonal matrix into a circulant matrix.[1] The formalism is similar in the continuous case, which we consider in the next section.

C.2 Exchange Field and Green Functions

As we explained in the previous section, working with translationally invariant systems, it is often useful to switch between the real space and momentum space using the Fourier transformation and its inverse. To do this in a quantum-mechanical system we need to make basis changes in the second quantization, which is demonstrated below.

The external field operator was defined in the Wannier representation as

$$\hat{V} = \sum_{jj'} V_{jj'}(\tau, \tau') a_j^\dagger(\tau) a_{j'}(\tau'), \tag{C.4}$$

where we introduced the matrix

$$V_{jj'}(\tau, \tau') = V_j(\tau)\delta_{jj'}\delta(\tau - \tau') \tag{C.5}$$

(for brevity we omit the spin indices here) and $a_j(\tau) = \mathrm{e}^{\mathcal{H}'\tau} a_j \, \mathrm{e}^{-\mathcal{H}'\tau}$ is the "Heisenberg" representation.

We begin with the spatial Fourier transform. The creation-annihilation operators in the Bloch and Wannier representations (3.6) are related by the formulae

$$a_j^\dagger = \frac{1}{\sqrt{N}}\sum_{\mathbf{k}} \mathrm{e}^{-\mathrm{i}\mathbf{k}\mathbf{R}_j} a_{\mathbf{k}}^\dagger, \qquad a_j = \frac{1}{\sqrt{N}}\sum_{\mathbf{k}} \mathrm{e}^{\mathrm{i}\mathbf{k}\mathbf{R}_j} a_{\mathbf{k}}. \tag{C.6}$$

Substituting (C.6) in (C.4), we obtain

$$\hat{V} = \sum_{\mathbf{k}\mathbf{k}'} V_{\mathbf{k}\mathbf{k}'}(\tau, \tau') a_{\mathbf{k}}^\dagger(\tau) a_{\mathbf{k}'}(\tau),$$

where

[1] For more on circulant matrices and their use in physics see, e.g. [2].

$$V_{\mathbf{kk}'}(\tau,\tau') = \frac{1}{N}\sum_{jj'} V_{jj'}(\tau,\tau')\mathrm{e}^{-\mathrm{i}\mathbf{k}\mathbf{R}_j}\mathrm{e}^{\mathrm{i}\mathbf{k}'\mathbf{R}_{j'}} \tag{C.7}$$

is the spatial Fourier transform of (C.5). Since the initial matrix (C.5) is site-diagonal, its Fourier transform (C.7) is translationally invariant in the momentum space:

$$V_{\mathbf{kk}'}(\tau,\tau') = V_{\mathbf{k}-\mathbf{k}'}(\tau)\delta(\tau-\tau'), \qquad V_{\mathbf{k}}(\tau) = \frac{1}{N}\sum_j V_j(\tau)\mathrm{e}^{-\mathrm{i}\mathbf{k}\mathbf{R}_j}.$$

On the opposite, the mean Green function

$$\bar{G}_{jj'}(\tau,\tau') = -\langle T_\tau\, a_j(\tau)a^\dagger_{j'}(\tau')\rangle$$

(for details, see Sect. 8.3.1) is translationally invariant in space and "time":

$$\bar{G}_{jj'}(\tau,\tau') = \bar{G}_{j-j'}(\tau-\tau'),$$

because the system is homogeneous and the Hamiltonian is "time" independent. Hence the spatial Fourier transform of $\bar{G}$ is a $\mathbf{k}$-diagonal matrix:

$$\bar{G}_{\mathbf{kk}'}(\tau-\tau') = \bar{G}_{\mathbf{k}}(\tau-\tau')\delta_{\mathbf{kk}'}, \qquad \bar{G}_{\mathbf{k}}(\tau-\tau') = \sum_j \bar{G}_j(\tau-\tau')\mathrm{e}^{-\mathrm{i}\mathbf{k}\mathbf{R}_j}.$$

We now proceed to the Fourier transforms in "frequencies" corresponding to the "time" variable. As a function of the "time" difference $\bar{G}(\tau-\tau')$ is defined on the interval $[-1/T, 1/T]$. For the function $\bar{G}(\tau)$ the following relation holds

$$\bar{G}_{j-j'}(\tau-1/T) = -\bar{G}_{j-j'}(\tau), \qquad \tau > 0. \tag{C.8}$$

Indeed, using the definition of the mean Green function and cyclic property of trace, we have

$$\begin{aligned}
\bar{G}_{j-j'}(\tau-1/T) &= \langle a^\dagger_{j'}a_j(\tau-1/T)\rangle\\
&= \frac{1}{\Xi}\mathrm{Tr}\left[a^\dagger_{j'}\mathrm{e}^{\mathcal{H}'(\tau-1/T)}a_j\mathrm{e}^{-\mathcal{H}'(\tau-1/T)}\,\mathrm{e}^{-\mathcal{H}'/T}\right]\\
&= \frac{1}{\Xi}\mathrm{Tr}\left[a^\dagger_{j'}\mathrm{e}^{\mathcal{H}'\tau}\mathrm{e}^{-\mathcal{H}'/T}a_j\mathrm{e}^{-\mathcal{H}'\tau}\right]\\
&= \frac{1}{\Xi}\mathrm{Tr}\left[a_j\mathrm{e}^{-\mathcal{H}'\tau}a^\dagger_{j'}\mathrm{e}^{\mathcal{H}'\tau}\,\mathrm{e}^{-\mathcal{H}'/T}\right]\\
&= \frac{1}{\Xi}\mathrm{Tr}\left[\mathrm{e}^{\mathcal{H}'\tau}a_j\mathrm{e}^{-\mathcal{H}'\tau}a^\dagger_{j'}\,\mathrm{e}^{-\mathcal{H}'/T}\right]\\
&= \langle a_j(\tau)a^\dagger_{j'}\rangle = -\bar{G}_{j-j'}(\tau),
\end{aligned}$$

where $\Xi = \mathrm{Tr}\mathrm{e}^{-\mathcal{H}'/T}$ is the grand canonical partition function. Similarly, we verify that

$$\bar{G}_{j-j'}(\tau+1/T) = -\bar{G}_{j-j'}(\tau), \qquad \tau < 0. \tag{C.9}$$

We introduce the Fourier series by the formula

$$\bar{G}_{j-j'}(\tau-\tau') = T\sum_{s=-\infty}^{\infty}\bar{G}_{j-j'}(\omega_s)\mathrm{e}^{-\mathrm{i}\omega_s(\tau-\tau')},$$

where

$$\bar{G}_{j-j'}(\omega_s) = \frac{1}{2}\int_{-1/T}^{1/T}\bar{G}_{j-j'}(\tau)\mathrm{e}^{\mathrm{i}\omega_s\tau}\,\mathrm{d}\tau$$

is the Fourier coefficient and $\omega_s = s\pi T$ is the "frequency". By splitting the integral into two

$$\bar{G}_{j-j'}(\omega_s) = \frac{1}{2}\int_{-1/T}^{0} \bar{G}_{j-j'}(\tau)e^{i\omega_s\tau}\,d\tau + \frac{1}{2}\int_{0}^{1/T} \bar{G}_{j-j'}(\tau)e^{i\omega_s\tau}\,d\tau$$

and making use of (C.9), we obtain

$$\bar{G}_{j-j'}(\omega_s) = -\frac{1}{2}\int_{-1/T}^{0} \bar{G}_{j-j'}(\tau + 1/T)e^{i\omega_s\tau}\,d\tau + \frac{1}{2}\int_{0}^{1/T} \bar{G}_{j-j'}(\tau)e^{i\omega_s\tau}\,d\tau$$
$$= \frac{1}{2}(1 - e^{-i\omega_s/T})\int_{0}^{1/T} \bar{G}_{j-j'}(\tau)e^{i\omega_s\tau}\,d\tau.$$

The latter is zero for even s. Hence the mean Green function is expanded in odd "frequencies" $\omega_n = (2n+1)\pi T$, $n = 0, \pm 1, \pm 2, \ldots$:

$$\bar{G}_{j-j'}(\tau - \tau') = T\sum_{n} \bar{G}_{j-j'}(\omega_n)e^{-i\omega_n(\tau-\tau')},$$

where the Fourier coefficient is calculated as

$$\bar{G}_{j-j'}(\omega_n) = \int_{0}^{1/T} \bar{G}_{j-j'}(\tau)e^{i\omega_n\tau}\,d\tau.$$

To define the "temporal" Fourier transform of the fluctuating field, we extend the function $V_j(\tau)$ to the whole interval $[-1/T, 1/T]$ in such a way that the equality

$$V_j(\tau - 1/T) = -V_j(\tau)$$

holds for $\tau > 0$, just as (C.8) for the Green function $\bar{G}$. Then the function

$$V_{jj'}(\tau, \tau') = V_j(\tau)\delta_{jj'}\delta(\tau - \tau')$$

is expanded in the Fourier series in the odd "frequencies":

$$V_{jj'}(\tau, \tau') = T\sum_{nn'} V_{jj'}(\omega_n, \omega_{n'})e^{-i\omega_n\tau + i\omega_{n'}\tau'}. \tag{C.10}$$

Since $V_{jj'}(\tau, \tau')$ is "diagonal" with respect to "time", the Fourier coefficient is a function of the "frequency" difference:

$$V_{jj'}(\omega_n, \omega_{n'}) = V_j(\omega_n - \omega_{n'})\delta_{jj'} \equiv V_j(\omega_{n-n'})\delta_{jj'}.$$

Thus, we come to the Fourier expansion

$$V_j(\tau) = \sum_{m} V_j(\omega_m)e^{-i\omega_m\tau}, \qquad V_j(\omega_m) = T\int_{0}^{1/T} V_j(\tau)e^{i\omega_m\tau}\,d\tau, \tag{C.11}$$

in the even "frequencies" $\omega_m = 2m\pi T$, $m = 0, \pm 1, \pm 2, \ldots$

To sum up, in the site-"time" representation, the matrices of the mean Green function $\bar{G}$ and external field V have the form

$$\bar{G}_{jj'}(\tau, \tau') = \bar{G}_{j-j'}(\tau - \tau'), \qquad V_{jj'}(\tau, \tau') = V_j(\tau)\delta_{jj'}\delta(\tau - \tau'),$$

and, in the momentum-"frequency" (Fourier) representation, they have the form

$$\bar{G}_{\mathbf{k}\mathbf{k}'nn'} = \bar{G}_{\mathbf{k}n}\delta_{\mathbf{k}\mathbf{k}'}\delta_{nn'}, \qquad V_{\mathbf{k}\mathbf{k}'nn'} = V_{\mathbf{k}-\mathbf{k}',n-n'}, \tag{C.12}$$

where we used the shorthand notation

$$\bar{G}_{\mathbf{k}n} \equiv \bar{G}_{\mathbf{k}}(\omega_n), \qquad V_{\mathbf{k}m} \equiv V_{\mathbf{k}}(\omega_m),$$

and $\omega_n = (2n+1)\pi T$ are odd and $\omega_m = 2m\pi T$ are even "frequencies".

Note that the zeroth Green function G^0 is also translationally invariant in the site space and depends on the "time" difference. Hence

$$G^0_{jj'}(\tau,\tau') = G^0_{j-j'}(\tau-\tau'), \qquad G^0_{\mathbf{k}\mathbf{k}'nn'} = G^0_{\mathbf{k}n}\delta_{\mathbf{k}\mathbf{k}'}\delta_{nn'}.$$

As an example, we calculate the RPA susceptibility by the functional integral method. The second-order expansion of the free energy $F_1(V) = T\mathrm{Tr}\ln G(V)$ around the mean field $\bar{V}$ is given by (see Appendix A.1.2)

$$F_1(V) = T\mathrm{Tr}(G(\bar{V})\Delta V) + \frac{1}{2}T\mathrm{Tr}(G(\bar{V})\Delta V G(\bar{V})\Delta V) + \cdots \tag{C.13}$$

By the choice of the mean field $\bar{V}$, the linear term in the expansion vanishes (see Sect. 9.2). To calculate the susceptibility, we write the quadratic term in the momentum-"frequency" representation. Using formulae (C.12), we have

$$\mathrm{Tr}\big(G(\bar{V})\Delta V G(\bar{V})\Delta V\big) = N_\mathrm{d}\sum_{\mathbf{k}\mathbf{k}'nn'}\mathrm{Sp}\big(G_{\mathbf{k}n}(\bar{V})\Delta V_{\mathbf{k}-\mathbf{k}',n-n'}G_{\mathbf{k}'n'}(V)\Delta V_{\mathbf{k}'-\mathbf{k},n'-n}\big).$$

Introducing the new summation indices $\mathbf{q} = \mathbf{k}-\mathbf{k}'$ and $m = n-n'$, we write

$$\mathrm{Tr}\big(G(\bar{V})\Delta V G(\bar{V})\Delta V\big) = N_\mathrm{d}\sum_{\mathbf{q}\mathbf{k}mn}\mathrm{Sp}\big(G_{\mathbf{k}n}(\bar{V})\Delta V_{\mathbf{q}m}G_{\mathbf{k}-\mathbf{q},\,n-m}(\bar{V})\Delta V_{-\mathbf{q}-m}\big). \tag{C.14}$$

The unenhanced susceptibility (in units of $g^2\mu_\mathrm{B}^2/2$) is given by formula (9.21):

$$\chi^{0\alpha\beta}_{\mathbf{q}m} = -\frac{1}{2}\frac{\partial^2 F_1(\bar{V})}{\partial V^\alpha_{\mathbf{q}m}\,\partial V^\beta_{-\mathbf{q}-m}}. \tag{C.15}$$

Substituting (C.13) and (C.14) in (C.15) and using the cyclic property of trace, we obtain

$$\chi^{0\alpha\beta}_{\mathbf{q}m} = -\frac{N_\mathrm{d}}{2}T\sum_{\mathbf{k}n}\mathrm{Sp}\big(G_{\mathbf{k}n}(\bar{V})\sigma^\alpha G_{\mathbf{k}-\mathbf{q},n-m}(\bar{V})\sigma^\beta\big).$$

C.3 Summary of the Fourier Transformations

In this section we summarize the formulae for the Fourier transformation of different physical quantities that are used in the book.

C.3.1 Site and Momentum Representations

In the DSFT, the choice of the normalization factor in the *spatial* Fourier transformation depends on the type of physical quantity. We adopt the following convention (see also [3], pp. 21 and 47). The Fourier transformations of quantities that do not depend on the number of sites N (such as the magnetic h_j and exchange V_j fields) are defined with the factor $1/N$, whereas the transformations of quantities depending on N (such as the Green functions G^0_j and $\bar{G}_j$ or single-site spin s_j and charge n_j operators) are defined without it. This way we obtain convenient expressions (without powers of $1/N$) in formulae that contain products of V and G such as (C.14).

The Fourier transformation and its inverse of the single-site spin operator s_j (similarly, charge operator n_j) are given by

$$s^\alpha_\mathbf{q} = \sum_j s^\alpha_j\,\mathrm{e}^{-\mathrm{i}\mathbf{q}\mathbf{R}_j}, \tag{C.16}$$

$$s^\alpha_j = \frac{1}{N}\sum_\mathbf{q} s^\alpha_\mathbf{q}\,\mathrm{e}^{\mathrm{i}\mathbf{q}\mathbf{R}_j}, \tag{C.17}$$

where $\alpha = x, y, z$. For the exchange field V_j (similarly, for magnetic field h_j) we have

$$V_{\mathbf{q}}^{\alpha} = \frac{1}{N} \sum_{j} V_j^{\alpha} \, \mathrm{e}^{-\mathrm{i}\mathbf{q}\mathbf{R}_j}, \tag{C.18}$$

$$V_j^{\alpha} = \sum_{\mathbf{q}} V_{\mathbf{q}}^{\alpha} \, \mathrm{e}^{\mathrm{i}\mathbf{q}\mathbf{R}_j} \tag{C.19}$$

(in this subsection the time argument is omitted for brevity). The mean Green function $\bar{G}_{j-j'}$ (similarly, zeroth Green function $G^0_{j-j'}$) is Fourier transformed as

$$\bar{G}_{\mathbf{k}}^{\gamma} = \sum_{j} \bar{G}_j^{\gamma} \, \mathrm{e}^{-\mathrm{i}\mathbf{k}\mathbf{R}_j}, \tag{C.20}$$

$$\bar{G}_j^{\gamma} = \frac{1}{N} \sum_{\mathbf{k}} \bar{G}_{\mathbf{k}}^{\gamma} \, \mathrm{e}^{\mathrm{i}\mathbf{k}\mathbf{R}_j}, \tag{C.21}$$

where $\gamma = 0, z$. The unity function in the site space is Fourier transformed by the formulae

$$\delta_{\mathbf{q}0} = \frac{1}{N} \sum_{j} \mathrm{e}^{-\mathrm{i}\mathbf{q}\mathbf{R}_j}, \tag{C.22}$$

$$1 = \sum_{\mathbf{q}} \delta_{\mathbf{q}0} \, \mathrm{e}^{\mathrm{i}\mathbf{q}\mathbf{R}_j}. \tag{C.23}$$

whereas, the Kronecker delta δ_{j0} in the site space is Fourier transformed by the formulae

$$1 = \sum_{j} \delta_{j0} \, \mathrm{e}^{-\mathrm{i}\mathbf{q}\mathbf{R}_j}, \tag{C.24}$$

$$\delta_{j0} = \frac{1}{N} \sum_{\mathbf{q}} \mathrm{e}^{\mathrm{i}\mathbf{q}\mathbf{R}_j}. \tag{C.25}$$

C.3.2 "Time" and "Frequency" Representations

The normalization factor in the Fourier transformation with respect to "time" depends on the quantity's dimensions. The Fourier transform of the spin $s_j(\tau)$ and charge $n_j(\tau)$ operators in the "Heisenberg" representation, just as the exchange field $V_j(\tau)$, are defined with the factor T so that their Fourier transforms have the same dimension, whereas the Fourier transform of the Green functions $G_j^0(\tau)$ and $\bar{G}_j(\tau)$ are defined without the factor T. As a result, we obtain expressions without powers of T in formulae that contain products of V and G such as (C.14).

The spin density in the "Heisenberg" representation $s_{\mathbf{q}}^{\alpha}(\tau) = \mathrm{e}^{\mathcal{H}\tau} s_{\mathbf{q}}^{\alpha} \, \mathrm{e}^{-\mathcal{H}\tau}$ is Fourier transformed as

$$s_{\mathbf{q}m}^{\alpha} = T \int_0^{1/T} s_{\mathbf{q}}^{\alpha}(\tau) \, \mathrm{e}^{\mathrm{i}\omega_m \tau} \, \mathrm{d}\tau, \tag{C.26}$$

$$s_{\mathbf{q}}^{\alpha}(\tau) = \sum_{m} s_{\mathbf{q}m}^{\alpha} \, \mathrm{e}^{-\mathrm{i}\omega_m \tau}, \tag{C.27}$$

where $\omega_m = 2m\pi T$ are even thermodynamic "frequencies". Both $s_{\mathbf{q}m}^{\alpha}$ and $s_{\mathbf{q}}^{\alpha}(\tau)$ are dimensionless.

The exchange field $V_{\mathbf{q}}(\tau)$ is transformed as

$$V_{\mathbf{q}m}^{\alpha} = T \int_0^{1/T} V_{\mathbf{q}}^{\alpha}(\tau) \, \mathrm{e}^{\mathrm{i}\omega_m \tau} \, \mathrm{d}\tau, \tag{C.28}$$

$$V_{\mathbf{q}}^{\alpha}(\tau) = \sum_{m} V_{\mathbf{q}m}^{\alpha} \, \mathrm{e}^{-\mathrm{i}\omega_m \tau}, \tag{C.29}$$

so that $V_{\mathbf{q}}^{\alpha}(\tau)$ and $V_{\mathbf{q}m}^{\alpha}$ have dimensions of energy.

The mean Green function $\bar{G}_{\mathbf{k}}(\tau)$ (similarly, zeroth Green function $G_{\mathbf{k}}^0(\tau)$) is transformed as

$$\bar{G}^{\gamma}_{\mathbf{k}n} = \int_0^{1/T} \bar{G}^{\gamma}_{\mathbf{k}}(\tau)\,\mathrm{e}^{\mathrm{i}\omega_n\tau}\,\mathrm{d}\tau, \tag{C.30}$$

$$\bar{G}^{\gamma}_{\mathbf{k}}(\tau) = T\sum_n \bar{G}^{\gamma}_{\mathbf{k}n}\,\mathrm{e}^{-i\omega_n\tau}, \tag{C.31}$$

where $\omega_n = (2n+1)\pi T$ are odd thermodynamic "frequencies". The Green function $\bar{G}^{\gamma}_{\mathbf{k}}(\tau)$ is dimensionless, while its Fourier transform $\bar{G}^{\gamma}_{\mathbf{k}n}$ has dimensions of inverse energy. As a result, the product of the Fourier transforms such as $V^{\alpha}_{\mathbf{q}m}\bar{G}^{\gamma}_{\mathbf{k}n}$ is dimensionless.

The "time" unity function is transformed by the formulae

$$\delta_{m0} = T\int_0^{1/T} \mathrm{e}^{\mathrm{i}\omega_m\tau}\,\mathrm{d}\tau, \tag{C.32}$$

$$1 = \sum_m \delta_{m0}\,\mathrm{e}^{-i\omega_m\tau}, \tag{C.33}$$

where both functions are dimensionless. The "time" delta function is transformed by the formulae

$$1 = \int_0^{1/T} \delta(\tau)\,\mathrm{e}^{\mathrm{i}\omega_m\tau}\,\mathrm{d}\tau, \tag{C.34}$$

$$\delta(\tau) = T\sum_m \mathrm{e}^{-i\omega_m\tau}. \tag{C.35}$$

The delta function $\delta(\tau)$ has dimensions of energy.

C.4 Relation Between Spin and Field Correlators

Here we prove the relation

$$\left\langle \Delta s^{\alpha}_j(\tau)\Delta s^{\beta}_{j'}(\tau')\right\rangle = \frac{1}{u^2}\left\langle \Delta V^{\alpha}_j(\tau)\Delta V^{\beta}_{j'}(\tau')\right\rangle - \frac{1}{2u}\delta_{jj'}\delta(\tau-\tau')\delta_{\alpha\beta}. \tag{C.36}$$

First, we carry out the Fourier transformation of this formula, then derive the expression in the momentum-"frequency" representation.

We start with the Fourier transformation of relation (C.36) in sites, temporarily omitting the argument τ. From the translational invariance it follows that the spin correlator does not change if we shift both spins by a lattice vector $\mathbf{R}_l$: $\langle \Delta s^{\alpha}_j \Delta s^{\beta}_{j'}\rangle = \langle \Delta s^{\alpha}_{j+l}\Delta s^{\beta}_{j'+l}\rangle$. Averaging the latter over all $\mathbf{R}_l$, we obtain

$$\langle \Delta s^{\alpha}_j \Delta s^{\beta}_{j'}\rangle = \frac{1}{N}\sum_l \langle \Delta s^{\alpha}_{j+l}\Delta s^{\beta}_{j'+l}\rangle. \tag{C.37}$$

Using relation (C.17), we can write

$$\langle \Delta s^{\alpha}_{j+l}\Delta s^{\beta}_{j'+l}\rangle = \frac{1}{N^2}\sum_{\mathbf{q}\mathbf{q}'}\langle \Delta s^{\alpha}_{\mathbf{q}}\Delta s^{\beta}_{\mathbf{q}'}\rangle\,\mathrm{e}^{\mathrm{i}(\mathbf{q}+\mathbf{q}')\mathbf{R}_l}\mathrm{e}^{\mathrm{i}\mathbf{q}\mathbf{R}_j}\mathrm{e}^{\mathrm{i}\mathbf{q}'\mathbf{R}_{j'}}. \tag{C.38}$$

Substitution of (C.38) in (C.37) leads to

$$\langle \Delta s^{\alpha}_j \Delta s^{\beta}_{j'}\rangle = \frac{1}{N^2}\sum_{\mathbf{q}\mathbf{q}'}\langle \Delta s^{\alpha}_{\mathbf{q}}\Delta s^{\beta}_{\mathbf{q}'}\rangle\,\mathrm{e}^{\mathrm{i}\mathbf{q}\mathbf{R}_j}\mathrm{e}^{\mathrm{i}\mathbf{q}'\mathbf{R}_{j'}}\frac{1}{N}\sum_l \mathrm{e}^{\mathrm{i}(\mathbf{q}+\mathbf{q}')\mathbf{R}_l}.$$

Taking the identity (C.22) into account, we obtain

$$\langle \Delta s_j^\alpha \Delta s_{j'}^\beta \rangle = \frac{1}{N^2} \sum_{\mathbf{q}} \langle \Delta s_{\mathbf{q}}^\alpha \Delta s_{-\mathbf{q}}^\beta \rangle \, \mathrm{e}^{\mathrm{i}\mathbf{q}(\mathbf{R}_j - \mathbf{R}_{j'})}. \tag{C.39}$$

Similarly, using the inverse Fourier transformation (C.19) of the fluctuating field V, we have

$$\langle \Delta V_j^\alpha \Delta V_{j'}^\beta \rangle = \sum_{\mathbf{q}} \langle \Delta V_{\mathbf{q}}^\alpha \Delta V_{-\mathbf{q}}^\beta \rangle \, \mathrm{e}^{\mathrm{i}\mathbf{q}(\mathbf{R}_j - \mathbf{R}_{j'})}. \tag{C.40}$$

Substituting (C.39) and (C.40) in (C.36) and taking (C.25) into account, we come to

$$\begin{aligned} \frac{1}{N^2} \sum_{\mathbf{q}} \langle \Delta s_{\mathbf{q}}^\alpha \Delta s_{-\mathbf{q}}^\beta \rangle \, \mathrm{e}^{\mathrm{i}\mathbf{q}(\mathbf{R}_j - \mathbf{R}_{j'})} &= \frac{1}{u^2} \sum_{\mathbf{q}} \langle \Delta V_{\mathbf{q}}^\alpha \Delta V_{-\mathbf{q}}^\beta \rangle \, \mathrm{e}^{\mathrm{i}\mathbf{q}(\mathbf{R}_j - \mathbf{R}_{j'})} \\ &\quad - \frac{1}{2u} \delta_{\alpha\beta} \frac{1}{N} \sum_{\mathbf{q}} \mathrm{e}^{\mathrm{i}\mathbf{q}(\mathbf{R}_j - \mathbf{R}_{j'})}. \end{aligned}$$

Equating the Fourier coefficients, we have

$$\langle \Delta s_{\mathbf{q}}^\alpha(\tau) \Delta s_{-\mathbf{q}}^\beta(\tau') \rangle = \frac{1}{u^2} N^2 \langle \Delta V_{\mathbf{q}}^\alpha(\tau) \Delta V_{-\mathbf{q}}^\beta(\tau') \rangle - \frac{1}{2u} N \delta(\tau - \tau') \delta_{\alpha\beta}. \tag{C.41}$$

Next, we carry out the Fourier transformation of relation (C.36) with respect to the "time" τ, temporarily omitting the argument $\mathbf{q}$. Using formula (C.27), we write the spin correlator as

$$\langle \Delta s^\alpha(\tau) \Delta s^\beta(\tau') \rangle = \sum_{mm'} \langle \Delta s_m^\alpha \Delta s_{m'}^\beta \rangle \, \mathrm{e}^{-\mathrm{i}\omega_m \tau} \mathrm{e}^{-\mathrm{i}\omega_{m'} \tau'}.$$

Shifting τ and τ' by s and integrating over the interval $[0, 1/T]$, we obtain

$$\langle \Delta s^\alpha(\tau) \Delta s^\beta(\tau') \rangle = \sum_{mm'} \langle \Delta s_m^\alpha \Delta s_{m'}^\beta \rangle \, \mathrm{e}^{-\mathrm{i}\omega_m \tau} \mathrm{e}^{-\mathrm{i}\omega_{m'} \tau'} T \int_0^{1/T} \mathrm{e}^{-\mathrm{i}(\omega_m + \omega_{m'})s} \, \mathrm{d}s.$$

By (C.32) the latter reduces to

$$\langle \Delta s^\alpha(\tau) \Delta s^\beta(\tau') \rangle = \sum_{m} \langle \Delta s_m^\alpha \Delta s_{-m}^\beta \rangle \, \mathrm{e}^{-\mathrm{i}\omega_m (\tau - \tau')}. \tag{C.42}$$

Similarly, for the fluctuating field correlator we have

$$\langle \Delta V^\alpha(\tau) \Delta V^\beta(\tau') \rangle = \sum_{m} \langle \Delta V_m^\alpha \Delta V_{-m}^\beta \rangle \, \mathrm{e}^{-\mathrm{i}\omega_m (\tau - \tau')}. \tag{C.43}$$

Substituting (C.42) and (C.43) in (C.41) and taking (C.35) into account, we obtain

$$\begin{aligned} &\sum_{m} \langle \Delta s_{\mathbf{q}m}^\alpha \Delta s_{-\mathbf{q}-m}^\beta \rangle \, \mathrm{e}^{-\mathrm{i}\omega_m (\tau - \tau')} \\ &\quad = \frac{N^2}{u^2} \sum_{m} \langle \Delta V_{\mathbf{q}m}^\alpha \Delta V_{-\mathbf{q}-m}^\beta \rangle \, \mathrm{e}^{-\mathrm{i}\omega_m (\tau - \tau')} - \frac{N}{2u} \delta_{\alpha\beta} T \sum_{m} \mathrm{e}^{-\mathrm{i}\omega_m (\tau - \tau')}. \end{aligned}$$

Equating the Fourier coefficients, we finally rewrite relation (C.36) as

$$\langle \Delta s^{\alpha}_{\mathbf{q}m} \Delta s^{\beta}_{-\mathbf{q}-m} \rangle = \frac{1}{\tilde{u}^2} \langle \Delta V^{\alpha}_{\mathbf{q}m} \Delta V^{\beta}_{-\mathbf{q}-m} \rangle - \frac{T}{2\tilde{u}} \delta_{\alpha\beta}, \tag{C.44}$$

where $\tilde{u} = u/N$ is the Fourier transform of the effective interaction constant u.

To show that Eq. (C.44) holds, first we prove the identity

$$\begin{aligned} &\left\langle \frac{\partial^2 F_0(V)}{\partial V^{\alpha}_{\mathbf{q}m} \partial V^{\beta}_{-\mathbf{q}-m}} \right\rangle - \frac{1}{T} \left\langle \frac{\partial F_0(V)}{\partial V^{\alpha}_{\mathbf{q}m}} \frac{\partial F_0(V)}{\partial V^{\beta}_{-\mathbf{q}-m}} \right\rangle \\ &= \left\langle \frac{\partial^2 F_1(V)}{\partial V^{\alpha}_{\mathbf{q}m} \partial V^{\beta}_{-\mathbf{q}-m}} \right\rangle - \frac{1}{T} \left\langle \frac{\partial F_1(V)}{\partial V^{\alpha}_{\mathbf{q}m}} \frac{\partial F_1(V)}{\partial V^{\beta}_{-\mathbf{q}-m}} \right\rangle, \end{aligned} \tag{C.45}$$

which is independent of a particular form of the functions $F_0(V)$ and $F_1(V)$ defining the partition function (8.49). We start off with the average

$$\left\langle \frac{\partial^2 F_1(V)}{\partial V^{\alpha}_{\mathbf{q}m} \partial V^{\beta}_{-\mathbf{q}-m}} \right\rangle = Q^{-1} \int \frac{\partial^2 F_1(V)}{\partial V^{\alpha}_{\mathbf{q}m} \partial V^{\beta}_{-\mathbf{q}-m}} \mathrm{e}^{-(F_0(V)+F_1(V))/T} \, \mathrm{D}V,$$

where

$$Q = \int \mathrm{e}^{-(F_0(V)+F_1(V))/T} \, \mathrm{D}V$$

is the normalizing factor. Integrating by parts, we come to

$$\begin{aligned} \left\langle \frac{\partial^2 F_1(V)}{\partial V^{\alpha}_{\mathbf{q}m} \partial V^{\beta}_{-\mathbf{q}-m}} \right\rangle &= (TQ)^{-1} \int \frac{\partial F_1(V)}{\partial V^{\beta}_{-\mathbf{q}-m}} \mathrm{e}^{-(F_0(V)+F_1(V))/T} \frac{\partial F_0(V)}{\partial V^{\alpha}_{\mathbf{q}m}} \, \mathrm{D}V \\ &+ (TQ)^{-1} \int \frac{\partial F_1(V)}{\partial V^{\beta}_{-\mathbf{q}-m}} \mathrm{e}^{-(F_0(V)+F_1(V))/T} \frac{\partial F_1(V)}{\partial V^{\alpha}_{\mathbf{q}m}} \, \mathrm{D}V. \end{aligned} \tag{C.46}$$

The first term on the right-hand side can be rewritten as

$$I_1 = -Q^{-1} \int \frac{\partial}{\partial V^{\beta}_{-\mathbf{q}-m}} \left(\mathrm{e}^{-F_1(V)/T} \right) \mathrm{e}^{-F_0(V)/T} \frac{\partial F_0(V)}{\partial V^{\alpha}_{\mathbf{q}m}} \, \mathrm{D}V.$$

Integrating by parts, we have

$$\begin{aligned} I_1 &= -(TQ)^{-1} \int \mathrm{e}^{-(F_0(V)+F_1(V))/T} \frac{\partial F_0(V)}{\partial V^{\alpha}_{\mathbf{q}m}} \frac{\partial F_0(V)}{\partial V^{\beta}_{-\mathbf{q}-m}} \, \mathrm{D}V \\ &+ Q^{-1} \int \mathrm{e}^{-(F_0(V)+F_1(V))/T} \frac{\partial^2 F_0(V)}{\partial V^{\alpha}_{\mathbf{q}m} \partial V^{\beta}_{-\mathbf{q}-m}} \, \mathrm{D}V \end{aligned}$$

or, equivalently,

$$I_1 = -\frac{1}{T} \left\langle \frac{\partial F_0(V)}{\partial V^{\alpha}_{\mathbf{q}m}} \frac{\partial F_0(V)}{V^{\beta}_{-\mathbf{q}-m}} \right\rangle + \left\langle \frac{\partial^2 F_0(V)}{\partial V^{\alpha}_{\mathbf{q}m} \partial V^{\beta}_{-\mathbf{q}-m}} \right\rangle. \tag{C.47}$$

The second term on the right-hand side of (C.46) can be written as

$$I_2 = \frac{1}{T} \left\langle \frac{\partial F_1(V)}{\partial V^{\alpha}_{\mathbf{q}m}} \frac{\partial F_1(V)}{\partial V^{\beta}_{-\mathbf{q}-m}} \right\rangle. \tag{C.48}$$

Substituting (C.47) and (C.48) in (C.46), we come to the identity (C.45).

Next, we obtain explicit expressions for the derivatives in the identity (C.45). The free energy is given by (8.60):

$$F_1(V) = -T \ln \mathrm{Tr} e^{-\mathcal{H}(V)/T},$$

where

$$\mathcal{H}(V) = \mathcal{H}_0 + 2 \sum_{\mathbf{q}m\alpha} V^\alpha_{\mathbf{q}m} s^\alpha_{-\mathbf{q}-m}.$$

Differentiating, we have

$$\frac{\partial F_1(V)}{\partial V^\alpha_{\mathbf{q}m}} = 2 \frac{\mathrm{Tr}(s^\alpha_{-\mathbf{q}-m}\, e^{-\mathcal{H}(V)/T})}{\mathrm{Tr} e^{-\mathcal{H}(V)/T}} = 2 s^\alpha_{-\mathbf{q}-m}(V) \tag{C.49}$$

and

$$\begin{aligned} \frac{\partial^2 F_1(V)}{\partial V^\alpha_{\mathbf{q}m} \partial V^\beta_{-\mathbf{q}-m}} &= -\frac{4}{T} \frac{\mathrm{Tr}(s^\beta_{\mathbf{q}m} s^\alpha_{-\mathbf{q}-m}\, e^{-\mathcal{H}(V)/T})}{\mathrm{Tr} e^{-\mathcal{H}(V)/T}} \\ &+ \frac{4}{T} \frac{\mathrm{Tr}(s^\beta_{\mathbf{q}m}\, e^{-\mathcal{H}(V)/T}) \mathrm{Tr}(s^\alpha_{-\mathbf{q}-m}\, e^{-\mathcal{H}(V)/T})}{(\mathrm{Tr} e^{-\mathcal{H}(V)/T})^2} \\ &= -\frac{4}{T}\big((s^\beta_{\mathbf{q}m} s^\alpha_{-\mathbf{q}-m})(V) - s^\beta_{\mathbf{q}m}(V) s^\alpha_{-\mathbf{q}-m}(V)\big). \end{aligned}$$

Averaging the latter over V and recalling (8.51), we come to

$$\left\langle \frac{\partial^2 F_1(V)}{\partial V^\alpha_{\mathbf{q}m} \partial V^\beta_{-\mathbf{q}-m}} \right\rangle = -\frac{4}{T}\big(\langle s^\beta_{\mathbf{q}m} s^\alpha_{-\mathbf{q}-m}\rangle - \langle s^\beta_{\mathbf{q}m}(V) s^\alpha_{-\mathbf{q}-m}(V)\rangle\big). \tag{C.50}$$

Using (C.49) and (C.50), we write the right-hand side of (C.45) as

$$\left\langle \frac{\partial^2 F_1(V)}{\partial V^\alpha_{\mathbf{q}m} \partial V^\beta_{-\mathbf{q}-m}} \right\rangle - \frac{1}{T} \left\langle \frac{\partial F_1(V)}{\partial V^\alpha_{\mathbf{q}m}} \frac{\partial F_1(V)}{\partial V^\beta_{-\mathbf{q}-m}} \right\rangle = -\frac{4}{T} \langle s^\beta_{\mathbf{q}m} s^\alpha_{-\mathbf{q}-m} \rangle. \tag{C.51}$$

The energy of the field is given by (8.57):

$$F_0(V) = \frac{1}{\tilde{u}} \sum_{\mathbf{q}m\alpha} |V^\alpha_{\mathbf{q}m}|^2 = \frac{1}{\tilde{u}} \sum_{\mathbf{q}m\alpha} V^\alpha_{\mathbf{q}m} V^\alpha_{-\mathbf{q}-m}.$$

Differentiating, we have

$$\frac{\partial F_0(V)}{\partial V^\alpha_{\mathbf{q}m}} = \frac{2V^\alpha_{-\mathbf{q}-m}}{\tilde{u}}, \qquad \frac{\partial^2 F_0(V)}{\partial V^\alpha_{\mathbf{q}m} \partial V^\beta_{-\mathbf{q}-m}} = \frac{2}{\tilde{u}} \delta_{\alpha\beta}. \tag{C.52}$$

Substituting (C.51) and (C.52) in (C.45), we come to the relation

$$\langle s^\alpha_{\mathbf{q}m} s^\beta_{-\mathbf{q}-m} \rangle = \frac{1}{\tilde{u}^2} \langle V^\alpha_{\mathbf{q}m} V^\beta_{-\mathbf{q}-m} \rangle - \frac{T}{2\tilde{u}} \delta_{\alpha\beta}. \tag{C.53}$$

Finally, we pass to the spin *fluctuation* using the well-known formula $\langle \Delta a \Delta b \rangle = \langle ab \rangle - \langle a \rangle \langle b \rangle$, where $\Delta a = a - \langle a \rangle$. Writing the mean-field equation (8.54) in the Fourier representation

$$\langle s^\alpha_{\mathbf{q}m} \rangle = -\frac{1}{\tilde{u}} \langle V^\alpha_{\mathbf{q}m} \rangle,$$

we obtain the relation between the spin and field correlators (C.44).

C.5 Fourier Transformation of the Susceptibility

The Fourier transformation of the dynamic susceptibility is given by

$$\chi(\mathbf{q},\omega) = \iint \chi(\mathbf{r},t)\,\mathrm{e}^{-\mathrm{i}(\mathbf{qr}-\omega t)}\,\mathrm{d}\mathbf{r}\,\mathrm{d}t$$

and its inverse is given by

$$\chi(\mathbf{r},t) = \frac{1}{V}\sum_{\mathbf{q}}\frac{1}{2\pi}\int \chi(\mathbf{q},\omega)\,\mathrm{e}^{\mathrm{i}(\mathbf{qr}-\omega t)}\,\mathrm{d}\omega.$$

Here $\omega = \omega + \mathrm{i}0^{+}$, which ensures the convergence of the integral. Let us explain the origin of this infinitesimal shift.

An explicit expression for the dynamic susceptibility is obtained in the linear response theory in Sect. 2.2.2. Since the field is turned on adiabatically, we have to multiply the left-hand side of (2.37) by $\mathrm{e}^{\eta t}$ and the right-hand side by $\mathrm{e}^{\eta t'}$, both of them vanish at $-\infty$ for $\eta > 0$. Then, repeating the derivation, we obtain the magnetization

$$M_{\alpha}(\mathbf{r},t) = \mathrm{i}\iint_{-\infty}^{t}\sum_{\beta}\langle[\mathcal{M}_{\alpha}(\mathbf{r},t-t'),\mathcal{M}_{\beta}(\mathbf{r}')]\rangle\,H_{\beta}(\mathbf{r}',t')\,\mathrm{e}^{-\eta(t-t')}\mathrm{d}\mathbf{r}'\,\mathrm{d}t'.$$

Hence the linear response is

$$\chi_{\alpha\beta}(\mathbf{q},t) = \mathrm{i}\langle[\mathcal{M}_{\alpha}(\mathbf{q},t),\mathcal{M}_{\beta}(-\mathbf{q})]\rangle\theta(t)\,\mathrm{e}^{-\eta t}.$$

Making the Fourier transform, we obtain

$$\chi_{\alpha\beta}(\mathbf{q},\omega) = \mathrm{i}\int_{0}^{\infty}\langle[\mathcal{M}_{\alpha}(\mathbf{q},t),\mathcal{M}_{\beta}(-\mathbf{q})]\rangle\,\mathrm{e}^{\mathrm{i}\omega t-\eta t}\,\mathrm{d}t.$$

The magnetic field is turned on slowly, so η is a small positive number. Taking the limit, we have

$$\begin{aligned}\chi_{\alpha\beta}(\mathbf{q},\omega) &= \lim_{\eta\to 0^{+}}\mathrm{i}\int_{0}^{\infty}\langle[\mathcal{M}_{\alpha}(\mathbf{q},t),\mathcal{M}_{\beta}(-\mathbf{q})]\rangle\,\mathrm{e}^{\mathrm{i}\omega t-\eta t}\,\mathrm{d}t\\ &\equiv \mathrm{i}\int_{0}^{\infty}\langle[\mathcal{M}_{\alpha}(\mathbf{q},t),\mathcal{M}_{\beta}(-\mathbf{q})]\rangle\,\mathrm{e}^{\mathrm{i}\omega t}\,\mathrm{d}t,\end{aligned}$$

where $\omega = \omega + \mathrm{i}0^{+}$.

References

1. C. Kittel, *Introduction to Solid State Physics*, 8th edn. (Wiley, New York, 2005)
2. R. Aldrovandi, *Special Matrices of Mathematical Physics* (World Scientific, Singapore, 2001)
3. D.J. Kim, *New Perspectives in Magnetism of Metals* (Kluwer/Plenum, New York, 1999)

Dynamic Susceptibility in the RPA

D

Here we derive two commutators that appear in the random phase approximation (RPA) expressions for the dynamic susceptibility.

D.1 Longitudinal Susceptibility

Equation of motion (6.44) was used to calculate the longitudinal susceptibility. We consider the commutator $[a^{\dagger}_{\mathbf{k}\sigma} a_{\mathbf{k}+\mathbf{q},\sigma}, \mathcal{H}_{\mathrm{I}}]$ on the right-hand side of (6.44) and calculate it in the RPA. Here

$$\mathcal{H}_{\mathrm{I}} = \frac{1}{2}\tilde{U} \sum_{\mathbf{klq}\sigma\sigma'} a^{\dagger}_{\mathbf{k}\sigma} a^{\dagger}_{\mathbf{l}\sigma'} a_{\mathbf{l}+\mathbf{q},\sigma'} a_{\mathbf{k}-\mathbf{q},\sigma}$$

is the interaction part of the Hubbard Hamiltonian (4.6), where $\tilde{U} = U/N$ is the Fourier transform of the interaction constant.

The commutator and anticommutator

$$[A, B] = AB - BA, \qquad \{A, B\} = AB + BA,$$

satisfy the following relations:

$$[AB, C] = [A, C]B + A[B, C], \qquad [B, CD] = \{B, C\}D - C\{B, D\}. \tag{D.1}$$

Introducing the notation

$$\begin{aligned}\left[a^{\dagger}_{\mathbf{k}\sigma} a_{\mathbf{k}+\mathbf{q},\sigma}, \mathcal{H}_{\mathrm{I}}\right] &= \frac{1}{2}\tilde{U} \sum_{\mathbf{k}'\mathbf{k}''\mathbf{p}\sigma'\sigma''} \left[a^{\dagger}_{\mathbf{k}\sigma} a_{\mathbf{k}+\mathbf{q},\sigma}, a^{\dagger}_{\mathbf{k}'\sigma'} a^{\dagger}_{\mathbf{k}''\sigma''} a_{\mathbf{k}''+\mathbf{p},\sigma''} a_{\mathbf{k}'-\mathbf{p},\sigma'}\right] \\ &\equiv \frac{1}{2}\tilde{U} \sum [AB, CDEF],\end{aligned}$$

and using relations (D.1), we evaluate the sum on the right-hand side as

$$\begin{aligned}&\sum [AB, CDEF] \\ &= \sum \left(A[B, CDEF] + [A, CDEF]B\right) \\ &= \sum \Big(\underbrace{ACD[B, EF]}_{=0} + A[B, CD]EF + CD[A, EF]B + \underbrace{[A, CD]EFB}_{=0}\Big) \\ &= \sum \big(A\{B, C\}DEF - AC\{B, D\}EF + CD\{A, E\}FB - CDE\{A, F\}B\big).\end{aligned} \tag{D.2}$$

N. B. Melnikov, B. I. Reser, *Dynamic Spin Fluctuation Theory of Metallic Magnetism*,
https://doi.org/10.1007/978-3-319-92974-3

In the second to the last line we used the anticommutation relations for the creation-annihilation operators (3.12). Further use of the anticommutation relations (3.12), leads to expression (6.48):

$$
\begin{aligned}
\left[a^{\dagger}_{\mathbf{k}\sigma}a_{\mathbf{k+q},\sigma}, \mathcal{H}_{\mathrm{I}}\right] = \frac{1}{2}\tilde{U}\Bigg(&\sum_{\mathbf{k}''\mathbf{p}\sigma''} a^{\dagger}_{\mathbf{k}\sigma}a^{\dagger}_{\mathbf{k}''\sigma''}a_{\mathbf{k}''+\mathbf{p},\sigma''}a_{\mathbf{k+q-p},\sigma} \\
&- \sum_{\mathbf{k}'\mathbf{p}\sigma'} a^{\dagger}_{\mathbf{k}\sigma}a^{\dagger}_{\mathbf{k}'\sigma'}a_{\mathbf{k+q+p},\sigma}a_{\mathbf{k}'-\mathbf{p},\sigma'} + \sum_{\mathbf{k}'\mathbf{p}\sigma'} a^{\dagger}_{\mathbf{k}'\sigma'}a^{\dagger}_{\mathbf{k-p},\sigma}a_{\mathbf{k}'-\mathbf{p},\sigma'}a_{\mathbf{k+q},\sigma} \\
&- \sum_{\mathbf{k}''\mathbf{p}\sigma''} a^{\dagger}_{\mathbf{k+p},\sigma}a^{\dagger}_{\mathbf{k}''\sigma''}a_{\mathbf{k}''+\mathbf{p},\sigma''}a_{\mathbf{k+q},\sigma}\Bigg).
\end{aligned}
\tag{D.3}
$$

Next we apply the mean-field approximation to the right-hand side of (D.3). In the 1st term, we average the 1st operator with 4th one, then 1st operator with 3rd one, then 2nd operator with 3rd one and finally 2nd operator with 4th one to obtain

$$
\begin{aligned}
&\sum_{\mathbf{k}''\mathbf{p}\sigma''} a^{\dagger}_{\mathbf{k}\sigma}a^{\dagger}_{\mathbf{k}''\sigma''}a_{\mathbf{k}''+\mathbf{p},\sigma''}a_{\mathbf{k+q-p},\sigma} \\
&\quad = \bar{n}_{\mathbf{k}\sigma}\sum_{\mathbf{k}''\sigma''} a^{\dagger}_{\mathbf{k}''\sigma''}a_{\mathbf{k}''+\mathbf{q},\sigma''} - \bar{n}_{\mathbf{k}\sigma}\sum_{\mathbf{p}} a^{\dagger}_{\mathbf{k-p},\sigma}a_{\mathbf{k+q-p},\sigma} \\
&\quad\quad + a^{\dagger}_{\mathbf{k}\sigma}a_{\mathbf{k+q},\sigma}\sum_{\mathbf{k}''\sigma''}\bar{n}_{\mathbf{k}''\sigma''} - a^{\dagger}_{\mathbf{k}\sigma}a_{\mathbf{k+q},\sigma}\sum_{\mathbf{k}''}\bar{n}_{\mathbf{k}''\sigma},
\end{aligned}
\tag{D.4}
$$

where we used

$$
\left\langle a^{\dagger}_{\mathbf{k}\sigma}a_{\mathbf{k}'\sigma'}\right\rangle = \bar{n}_{\mathbf{k}\sigma}\delta_{\mathbf{kk}'}\delta_{\sigma\sigma'}.
$$

Applying the mean-field approximation to the 2nd term, we have

$$
\begin{aligned}
&-\sum_{\mathbf{k}'\mathbf{p}\sigma'} a^{\dagger}_{\mathbf{k}\sigma}a^{\dagger}_{\mathbf{k}'\sigma'}a_{\mathbf{k+q+p},\sigma}a_{\mathbf{k}'-\mathbf{p},\sigma'} \\
&\quad = -\bar{n}_{\mathbf{k}\sigma}\sum_{\mathbf{p}} a^{\dagger}_{\mathbf{k+p},\sigma}a_{\mathbf{k+q+p},\sigma} + \bar{n}_{\mathbf{k}\sigma}\sum_{\mathbf{k}'\sigma'} a^{\dagger}_{\mathbf{k}'\sigma'}a_{\mathbf{k}'+\mathbf{q},\sigma'} \\
&\quad\quad - a^{\dagger}_{\mathbf{k}\sigma}a_{\mathbf{k+q},\sigma}\sum_{\mathbf{k}'}\bar{n}_{\mathbf{k}'\sigma} + a^{\dagger}_{\mathbf{k}\sigma}a_{\mathbf{k+q},\sigma}\sum_{\mathbf{k}'\sigma'}\bar{n}_{\mathbf{k}'\sigma'}.
\end{aligned}
\tag{D.5}
$$

We see that (D.4) and (D.5) coincide (equal terms are marked with lines of the same type). Similarly, applying the mean-field approximation to the 3rd term, we obtain

$$
\begin{aligned}
&\sum_{\mathbf{k}'\mathbf{p}\sigma'} a^{\dagger}_{\mathbf{k}'\sigma'}a^{\dagger}_{\mathbf{k-p},\sigma}a_{\mathbf{k}'-\mathbf{p},\sigma'}a_{\mathbf{k+q},\sigma} \\
&\quad = \bar{n}_{\mathbf{k+q},\sigma}\sum_{\mathbf{p}} a^{\dagger}_{\mathbf{k-p},\sigma}a_{\mathbf{k+q-p},\sigma} - a^{\dagger}_{\mathbf{k},\sigma}a_{\mathbf{k+q},\sigma}\sum_{\mathbf{k}'\sigma'}\bar{n}_{\mathbf{k}'\sigma'} \\
&\quad\quad + a^{\dagger}_{\mathbf{k}\sigma}a_{\mathbf{k+q},\sigma}\sum_{\mathbf{p}}\bar{n}_{\mathbf{k-p},\sigma} - \bar{n}_{\mathbf{k+q},\sigma}\sum_{\mathbf{k}'\sigma'} a^{\dagger}_{\mathbf{k}'\sigma'}a_{\mathbf{k}'+\mathbf{q},\sigma'}.
\end{aligned}
\tag{D.6}
$$

Applying the mean-field approximation to the 4th term, we have

$$
\begin{aligned}
&-\sum_{\mathbf{k}''\mathbf{p}\sigma''} a^\dagger_{\mathbf{k}+\mathbf{p},\sigma} a^\dagger_{\mathbf{k}''\sigma''} a_{\mathbf{k}''+\mathbf{p},\sigma''} a_{\mathbf{k}+\mathbf{q},\sigma} \\
&= -\bar{n}_{\mathbf{k}+\mathbf{q},\sigma} \sum_{\mathbf{k}''\sigma''} a^\dagger_{\mathbf{k}''\sigma''} a_{\mathbf{k}''+\mathbf{q},\sigma''} + a^\dagger_{\mathbf{k}\sigma} a_{\mathbf{k}+\mathbf{q},\sigma} \sum_{\mathbf{p}} \bar{n}_{\mathbf{k}+\mathbf{p},\sigma} \\
&\quad - a^\dagger_{\mathbf{k},\sigma} a_{\mathbf{k}+\mathbf{q},\sigma} \sum_{\mathbf{k}''\sigma''} \bar{n}_{\mathbf{k}''\sigma''} + \bar{n}_{\mathbf{k}+\mathbf{q},\sigma} \sum_{\mathbf{p}} a^\dagger_{\mathbf{k}+\mathbf{p},\sigma} a_{\mathbf{k}+\mathbf{q}+\mathbf{p},\sigma}.
\end{aligned} \tag{D.7}
$$

Expressions (D.6) and (D.7) coincide (equal terms are marked with lines of the same type). Making use of

$$\sum_{\mathbf{p}} a^\dagger_{\mathbf{k}+\mathbf{p},\sigma} a_{\mathbf{k}+\mathbf{p}+\mathbf{q},\sigma'} = \sum_{\mathbf{k}} a^\dagger_{\mathbf{k}\sigma} a_{\mathbf{k}+\mathbf{q},\sigma'}, \tag{D.8}$$

we see that (D.4) and (D.5) can be written as

$$\sum_{\mathbf{k}''\mathbf{p}\sigma''} a^\dagger_{\mathbf{k}\sigma} a^\dagger_{\mathbf{k}''\sigma''} a_{\mathbf{k}''+\mathbf{p},\sigma''} a_{\mathbf{k}+\mathbf{q}-\mathbf{p},\sigma} = \bar{n}_{\mathbf{k}\sigma} \sum_{\mathbf{k}'} a^\dagger_{\mathbf{k}'\bar{\sigma}} a_{\mathbf{k}'+\mathbf{q},\bar{\sigma}} + N_{\bar{\sigma}} a^\dagger_{\mathbf{k}\sigma} a_{\mathbf{k}+\mathbf{q},\sigma}.$$

Similarly, (D.6) and (D.7) can be written as

$$\sum_{\mathbf{k}'\mathbf{p}\sigma'} a^\dagger_{\mathbf{k}'\sigma'} a^\dagger_{\mathbf{k}-\mathbf{p},\sigma} a_{\mathbf{k}'-\mathbf{p},\sigma'} a_{\mathbf{k}+\mathbf{q},\sigma} = -N_{\bar{\sigma}} a^\dagger_{\mathbf{k}\sigma} a_{\mathbf{k}+\mathbf{q},\sigma} - \bar{n}_{\mathbf{k}+\mathbf{q},\sigma} \sum_{\mathbf{k}'} a^\dagger_{\mathbf{k}'\bar{\sigma}} a_{\mathbf{k}'+\mathbf{q},\bar{\sigma}}.$$

Thus, we are left with only two terms in the commutator (D.3) and come to expression (6.49):

$$\left[a^\dagger_{\mathbf{k}\sigma} a_{\mathbf{k}+\mathbf{q},\sigma}, \mathcal{H}_{\mathrm{I}}\right] = \tilde{U}(\bar{n}_{\mathbf{k}\sigma} - \bar{n}_{\mathbf{k}+\mathbf{q},\sigma}) \sum_{\mathbf{k}'} a^\dagger_{\mathbf{k}'\bar{\sigma}} a_{\mathbf{k}'+\mathbf{q},\bar{\sigma}}.$$

D.2 Transverse Susceptibility

The equation of motion (6.53) was used to calculate the transverse susceptibility. We consider the commutator $[a^\dagger_{\mathbf{k}\downarrow} a_{\mathbf{k}+\mathbf{q},\uparrow}, \mathcal{H}_{\mathrm{I}}]$ on the right-hand side of (6.53) and calculate it in the RPA. Using (D.1) just as in (D.2), we obtain

$$
\begin{aligned}
\left[a^\dagger_{\mathbf{k}\downarrow} a_{\mathbf{k}+\mathbf{q},\uparrow}, \mathcal{H}_{\mathrm{I}}\right] &= \frac{1}{2}\tilde{U}\Bigg(\sum_{\mathbf{k}''\mathbf{p}\sigma''} a^\dagger_{\mathbf{k}\downarrow} a^\dagger_{\mathbf{k}''\sigma''} a_{\mathbf{k}''+\mathbf{p},\sigma''} a_{\mathbf{k}+\mathbf{q}-\mathbf{p},\uparrow} \\
&\quad - \sum_{\mathbf{k}'\mathbf{p}\sigma'} a^\dagger_{\mathbf{k}\downarrow} a^\dagger_{\mathbf{k}'\sigma'} a_{\mathbf{k}+\mathbf{q}+\mathbf{p},\uparrow} a_{\mathbf{k}'-\mathbf{p},\sigma'} + \sum_{\mathbf{k}'\mathbf{p}\sigma'} a^\dagger_{\mathbf{k}'\sigma'} a^\dagger_{\mathbf{k}-\mathbf{p},\downarrow} a_{\mathbf{k}'-\mathbf{p},\sigma'} a_{\mathbf{k}+\mathbf{q},\uparrow} \\
&\quad - \sum_{\mathbf{k}''\mathbf{p}\sigma''} a^\dagger_{\mathbf{k}+\mathbf{p},\downarrow} a^\dagger_{\mathbf{k}''\sigma''} a_{\mathbf{k}''+\mathbf{p},\sigma''} a_{\mathbf{k}+\mathbf{q},\uparrow} \Bigg).
\end{aligned} \tag{D.9}
$$

Next we apply the mean-field approximation to the right-hand side of (D.9). In the 1st term, we average the 1st operator with 4th one, then 1st operator with 3rd one, then 2nd operator with 3rd one and then 2nd operator with 4th one to obtain

$$\sum_{\mathbf{k}''\mathbf{p}\sigma''} a^{\dagger}_{\mathbf{k}\downarrow} a^{\dagger}_{\mathbf{k}''\sigma''} a_{\mathbf{k}''+\mathbf{p},\sigma''} a_{\mathbf{k}+\mathbf{q}-\mathbf{p},\uparrow}$$

$$= -\bar{n}_{\mathbf{k}\downarrow} \sum_{\mathbf{p}} a^{\dagger}_{\mathbf{k}-\mathbf{p},\downarrow} a_{\mathbf{k}+\mathbf{q}-\mathbf{p},\uparrow} + a^{\dagger}_{\mathbf{k}\downarrow} a_{\mathbf{k}+\mathbf{q},\uparrow} \sum_{\mathbf{k}''\sigma''} \bar{n}_{\mathbf{k}''\sigma''} - a^{\dagger}_{\mathbf{k}\downarrow} a_{\mathbf{k}+\mathbf{q},\uparrow} \sum_{\mathbf{k}''} \bar{n}_{\mathbf{k}''\uparrow}. \tag{D.10}$$

Applying the mean-field approximation to the 2nd term, we have

$$-\sum_{\mathbf{k}'\mathbf{p}\sigma'} a^{\dagger}_{\mathbf{k}\downarrow} a^{\dagger}_{\mathbf{k}'\sigma'} a_{\mathbf{k}+\mathbf{q}+\mathbf{p},\uparrow} a_{\mathbf{k}'-\mathbf{p},\sigma'}$$

$$= -\bar{n}_{\mathbf{k}\downarrow} \sum_{\mathbf{p}} a^{\dagger}_{\mathbf{k}+\mathbf{p},\downarrow} a_{\mathbf{k}+\mathbf{q}+\mathbf{p},\uparrow} - a^{\dagger}_{\mathbf{k}\downarrow} a_{\mathbf{k}+\mathbf{q},\uparrow} \sum_{\mathbf{k}'} \bar{n}_{\mathbf{k}'\uparrow} + a^{\dagger}_{\mathbf{k}\downarrow} a_{\mathbf{k}+\mathbf{q},\uparrow} \sum_{\mathbf{k}'\sigma'} \bar{n}_{\mathbf{k}'\sigma'}. \tag{D.11}$$

Applying the mean-field approximation to the 3rd term, we have

$$\sum_{\mathbf{k}'\mathbf{p}\sigma'} a^{\dagger}_{\mathbf{k}'\sigma'} a^{\dagger}_{\mathbf{k}-\mathbf{p},\downarrow} a_{\mathbf{k}'-\mathbf{p},\sigma'} a_{\mathbf{k}+\mathbf{q},\uparrow}$$

$$= \bar{n}_{\mathbf{k}+\mathbf{q},\uparrow} \sum_{\mathbf{p}} a^{\dagger}_{\mathbf{k}-\mathbf{p},\downarrow} a_{\mathbf{k}+\mathbf{q}-\mathbf{p},\uparrow} - a^{\dagger}_{\mathbf{k}\downarrow} a_{\mathbf{k}+\mathbf{q},\uparrow} \sum_{\mathbf{k}'\sigma'} \bar{n}_{\mathbf{k}'\sigma'} + a^{\dagger}_{\mathbf{k}\downarrow} a_{\mathbf{k}+\mathbf{q},\uparrow} \sum_{\mathbf{p}} \bar{n}_{\mathbf{k}-\mathbf{p},\downarrow}. \tag{D.12}$$

Applying the mean-field approximation to the 4th term, we have

$$-\sum_{\mathbf{k}''\mathbf{p}\sigma''} a^{\dagger}_{\mathbf{k}+\mathbf{p},\downarrow} a^{\dagger}_{\mathbf{k}''\sigma''} a_{\mathbf{k}''+\mathbf{p},\sigma''} a_{\mathbf{k}+\mathbf{q},\uparrow}$$

$$= a^{\dagger}_{\mathbf{k}\downarrow} a_{\mathbf{k}+\mathbf{q},\uparrow} \sum_{\mathbf{p}} \bar{n}_{\mathbf{k}+\mathbf{p},\downarrow} - a^{\dagger}_{\mathbf{k}\downarrow} a_{\mathbf{k}+\mathbf{q},\uparrow} \sum_{\mathbf{k}''\sigma''} \bar{n}_{\mathbf{k}''\sigma''} + \bar{n}_{\mathbf{k}+\mathbf{q},\uparrow} \sum_{\mathbf{p}} a^{\dagger}_{\mathbf{k}+\mathbf{p},\downarrow} a_{\mathbf{k}+\mathbf{q}+\mathbf{p},\uparrow}. \tag{D.13}$$

Using relation (D.8), we see that both (D.10) and (D.11) are equal to

$$\sum_{\mathbf{k}''\mathbf{p}\sigma''} a^{\dagger}_{\mathbf{k}\downarrow} a^{\dagger}_{\mathbf{k}''\sigma''} a_{\mathbf{k}''+\mathbf{p},\sigma''} a_{\mathbf{k}+\mathbf{q}-\mathbf{p},\uparrow} = -\bar{n}_{\mathbf{k}\downarrow} \sum_{\mathbf{k}'} a^{\dagger}_{\mathbf{k}'\downarrow} a_{\mathbf{k}'+\mathbf{q},\uparrow} + N_{\downarrow} a^{\dagger}_{\mathbf{k}\downarrow} a_{\mathbf{k}+\mathbf{q},\uparrow}.$$

Similarly, (D.12) and (D.13) are equal to

$$\sum_{\mathbf{k}'\mathbf{p}\sigma'} a^{\dagger}_{\mathbf{k}'\sigma'} a^{\dagger}_{\mathbf{k}-\mathbf{p},\downarrow} a_{\mathbf{k}'-\mathbf{p},\sigma'} a_{\mathbf{k}+\mathbf{q},\uparrow} = \bar{n}_{\mathbf{k}+\mathbf{q},\uparrow} \sum_{\mathbf{k}'} a^{\dagger}_{\mathbf{k}'\downarrow} a_{\mathbf{k}'+\mathbf{q},\uparrow} - N_{\uparrow} a^{\dagger}_{\mathbf{k}\downarrow} a_{\mathbf{k}+\mathbf{q},\uparrow}.$$

Thus, we write (D.9) in the form

$$\left[a^{\dagger}_{\mathbf{k}\downarrow} a_{\mathbf{k}+\mathbf{q},\uparrow}, \mathcal{H}_{\mathrm{I}}\right] = -\tilde{U}(\bar{n}_{\mathbf{k}\downarrow} - \bar{n}_{\mathbf{k}+\mathbf{q},\uparrow}) \sum_{\mathbf{k}'} a^{\dagger}_{\mathbf{k}'\downarrow} a_{\mathbf{k}'+\mathbf{q},\uparrow} + \tilde{U}(N_{\downarrow} - N_{\uparrow}) a^{\dagger}_{\mathbf{k}\downarrow} a_{\mathbf{k}+\mathbf{q},\uparrow},$$

which coincides with expression (6.57).

Proofs of Four Results in the DSFT

E

E.1 From Grand Canonical to Canonical Ensemble

We typically use the *grand* canonical ensemble to derive formulae in the many-body theory, and then switch to canonical ensemble in calculations for real metals. In the grand canonical ensemble all the characteristics are determined by the Hamiltonian $\mathcal{H}' = \mathcal{H} - \mu\mathcal{N}$, where $\mathcal{H}$ is the Hamiltonian in the canonical ensemble, μ is the chemical potential and $\mathcal{N}$ is the number of electrons operator. In particular, the one-electron part of the Hamiltonian $\mathcal{H}_0' = \mathcal{H}_0 - \mu\mathcal{N}$ can be written as

$$\mathcal{H}_0' = \sum_{\nu\mathbf{k}\sigma}(\varepsilon_{\mathbf{k}} - \mu)n_{\nu\mathbf{k}\sigma} \equiv \sum_{\nu\mathbf{k}\sigma} \xi_{\mathbf{k}} n_{\nu\mathbf{k}\sigma}.$$

Clearly, the energy spectrum $\xi_{\mathbf{k}}$ of $\mathcal{H}_0'$ is obtained by shifting the spectrum $\varepsilon_{\mathbf{k}}$ of $\mathcal{H}_0$ to the left by the chemical potential μ. Therefore, to obtain the electron density of states in the canonical ensemble $\nu(\varepsilon)$ from the electron density of states in the *grand* canonical ensemble $\nu'(\varepsilon)$, we need to shift the energy scale by μ to the right: $\nu_\sigma(\varepsilon) = \nu_\sigma'(\varepsilon - \mu)$. The same rule works for the mean single-site Green function and coherent potential: $g_\sigma(\varepsilon) = g_\sigma'(\varepsilon - \mu)$ and $\Sigma_\sigma(\varepsilon) = \Sigma_\sigma'(\varepsilon - \mu)$.

The mean single-site Green function in the *grand* canonical ensemble is given by formula (10.10):

$$g_\sigma'(\varepsilon) = \int \frac{\nu'(\varepsilon')}{\varepsilon - \varepsilon' - \Sigma_\sigma'(\varepsilon)}\, d\varepsilon'. \tag{E.1}$$

Shifting the energy variable ε by μ, we have

$$g_\sigma'(\varepsilon - \mu) = \int \frac{\nu'(\varepsilon')}{\varepsilon - \mu - \varepsilon' - \Sigma_\sigma'(\varepsilon - \mu)}\, d\varepsilon'.$$

Changing the dummy variable to $\tilde{\varepsilon} = \varepsilon' + \mu$, we obtain

$$g_\sigma'(\varepsilon - \mu) = \int \frac{\nu'(\tilde{\varepsilon} - \mu)}{\varepsilon - \tilde{\varepsilon} - \Sigma_\sigma'(\varepsilon - \mu)}\, d\tilde{\varepsilon}.$$

Recalling that $g_\sigma(\varepsilon) = g_\sigma'(\varepsilon - \mu)$, $\nu_\sigma(\varepsilon) = \nu_\sigma'(\varepsilon - \mu)$ and $\Sigma_\sigma(\varepsilon) = \Sigma_\sigma'(\varepsilon - \mu)$, we come to the expression in the canonical ensemble

$$g_\sigma(\varepsilon) = \int \frac{\nu(\tilde{\varepsilon})}{\varepsilon - \tilde{\varepsilon} - \Sigma_\sigma(\varepsilon)}\, d\tilde{\varepsilon}.$$

Thus, the relation between the mean single-site Green function, the electron density of states and coherent potential in the canonical ensemble is the same as the one in the *grand* canonical ensemble.

N. B. Melnikov, B. I. Reser, *Dynamic Spin Fluctuation Theory of Metallic Magnetism*,
https://doi.org/10.1007/978-3-319-92974-3

E.2 Single-Site Coherent Potential

In Chap. 10 we presented a method of calculating the Green function based on the single-site quasistatic approximation. Here we derive an expression for the corresponding coherent potential following the paper [1].

The single-site approximation of Eq. (9.44) gives the coherent-potential equation (10.36):

$$\Delta\Sigma(\varepsilon) = \langle [1 + (\Sigma(\varepsilon) - V)g(\varepsilon)]^{-1}\Delta V\rangle, \tag{E.2}$$

where $\Delta\Sigma(\varepsilon) = \Sigma(\varepsilon) - \bar{V}$. Calculating the inverse 2×2-matrix, we write (E.2) as

$$\Delta\Sigma = \left\langle \frac{1}{\det D + \det N}(\tilde{D} - N)(\Delta V^{\mathrm{D}} + \Delta V^{\mathrm{N}})\right\rangle, \tag{E.3}$$

where

$$D = \begin{pmatrix} 1 + (\Sigma_1 - V_1)g_1 & 0 \\ 0 & 1 + (\Sigma_2 - V_2)g_2 \end{pmatrix}, \qquad N = \begin{pmatrix} 0 & -V_- g_2 \\ -V_+ g_1 & 0 \end{pmatrix},$$
$$\Delta V^{\mathrm{D}} = \begin{pmatrix} \Delta V_1 & 0 \\ 0 & \Delta V_2 \end{pmatrix}, \qquad \Delta V^{\mathrm{N}} = \begin{pmatrix} 0 & V_- \\ V_+ & 0 \end{pmatrix}, \tag{E.4}$$

matrix $\tilde{D}$ is obtained from the matrix D by swapping its diagonal elements, $V_1 = V_z$, $V_2 = -V_z$ and $V_\pm = V_x \pm \mathrm{i}V_y$. In the second-order perturbation theory with respect to ΔV, we can neglect the term $\det N = -V_+V_-g_1g_2$ in the denominator of (E.3). Then

$$\Delta\Sigma = \left\langle \frac{1}{\det D}\tilde{D}\Delta V^{\mathrm{D}}\right\rangle - \left\langle \frac{1}{\det D}\right\rangle \langle N\Delta V^{\mathrm{N}}\rangle. \tag{E.5}$$

Consider the first term, i.e. the contribution from the fluctuation of V_z. Taking (E.4) into account, we have

$$\Delta\Sigma^z = \left\langle \frac{1}{\det D}\tilde{D}\Delta V^{\mathrm{D}}\right\rangle = \langle D^{-1}\Delta V^{\mathrm{D}}\rangle$$
$$= \left\langle \begin{pmatrix} [1 + (\Sigma_1 - V_1)g_1]^{-1}\Delta V_1 & 0 \\ 0 & [1 + (\Sigma_2 - V_2)g_2]^{-1}\Delta V_2 \end{pmatrix}\right\rangle$$

or, equivalently,

$$\Delta\Sigma_i^z = \left\langle \frac{\Delta V_i}{1 + (\Sigma_i - V_i)g_i}\right\rangle$$
$$= \left\langle \frac{[1 + (\Sigma_i + V_i)g_i]\Delta V_i}{[1 + (\Sigma_i - V_i)g_i][1 + (\Sigma_i + V_i)g_i]}\right\rangle, \qquad i = 1, 2. \tag{E.6}$$

Since the denominator is an even function of V_i, we split (E.6) by the formula [2]

$$\langle V_z^{2n+1}\rangle \simeq \langle V_z^{2n}\rangle\langle V_z\rangle \simeq \langle V_z^2\rangle^n\langle V_z\rangle, \qquad n = 0, 1, 2, \ldots \tag{E.7}$$

and keep the terms up to the second order with respect to ΔV_i. Then, taking $\langle\Delta V_i\rangle = 0$ into account, we obtain

$$\Delta\Sigma_i^z = \frac{\langle [1 + (2\langle V_i\rangle + \Delta\Sigma_i + \Delta V_i)g_i]\Delta V_i\rangle}{\langle [1 + (\Delta\Sigma_i - \Delta V_i)g_i][1 + (2\langle V_i\rangle + \Delta\Sigma_i + \Delta V_i)g_i]\rangle}$$
$$= \frac{\langle \Delta V_i g_i \Delta V_i\rangle}{1 + 2\langle V_i\rangle g_i} + o(\langle\Delta V_z^2\rangle)$$

or

$$\Delta\Sigma_\sigma^z(\varepsilon) = \frac{g_\sigma(\varepsilon)\langle\Delta V_z^2\rangle}{1 + 2\sigma\langle V_z\rangle g_\sigma(\varepsilon)}, \qquad \sigma = \pm 1. \tag{E.8}$$

Now we consider the second term in (E.5), i.e. the transverse fluctuation $\Delta V_\perp^2 = \Delta V_x^2 + \Delta V_y^2$ contribution. Taking (E.4) into account, we have

$$\Delta\Sigma_\sigma^\perp(\varepsilon) = \left\langle\frac{1}{\det D}\right\rangle\langle V_+V_-g_{\bar\sigma}\rangle = \left\langle\frac{1}{\det D}\right\rangle g_{\bar\sigma}(\varepsilon)\langle\Delta V_\perp^2\rangle, \qquad \bar\sigma \equiv -\sigma. \tag{E.9}$$

In $\langle 1/\det D\rangle$ we separate the zero-order term with respect to ΔV. To do that, we transform it so that the denominator becomes an even function of V_z, just as we did with $\Delta\Sigma_i^z$:

$$\left\langle\frac{1}{\det D}\right\rangle = \left\langle\prod_i \frac{1}{1 + (\Sigma_i - V_i)g_i}\right\rangle$$
$$= \left\langle\prod_i \frac{1 + (\Sigma_i + V_i)g_i}{[1 + (\Sigma_i - V_i)g_i][1 + (\Sigma_i + V_i)g_i]}\right\rangle,$$

and apply the splitting (E.7) to obtain

$$\left\langle\frac{1}{\det D}\right\rangle = \frac{\left\langle\prod_i[1 + (2\langle V_i\rangle + \Delta\Sigma_i + \Delta V_i)g_i]\right\rangle}{\left\langle\prod_i[1 + (\Delta\Sigma_i - \Delta V_i)g_i][1 + (2\langle V_i\rangle + \Delta\Sigma_i + \Delta V_i)g_i]\right\rangle}$$
$$= \frac{\prod_i(1 + 2\langle V_i\rangle g_i)}{\prod_i(1 + 2\langle V_i\rangle g_i)} + o(\langle\Delta V_z^2\rangle) = 1 + o(\langle\Delta V_z^2\rangle).$$

Using the latter, we write (E.9) as

$$\Delta\Sigma_\sigma^\perp(\varepsilon) = g_{\bar\sigma}(\varepsilon)\langle\Delta V_\perp^2\rangle. \tag{E.10}$$

Substituting (E.8) and (E.10) in (E.5), we finally obtain

$$\Delta\Sigma_\sigma(\varepsilon) = \frac{g_\sigma(\varepsilon)\langle\Delta V_z^2\rangle}{1 + 2\sigma\langle V_z\rangle g_\sigma(\varepsilon)} + g_{\bar\sigma}(\varepsilon)\langle\Delta V_\perp^2\rangle.$$

E.3 Higher-Order Correction Coefficient

We consider the third-order term $\mathrm{Tr}\big(G\underbrace{\Delta V G\Delta V}G\Delta V\big)$ in the renormalized Gaussian approximation of the DSFT. Here the underbrace means the Gaussian average

$$(\cdots\underbrace{\Delta V\cdots\Delta V}\cdots) = \int(\cdots\Delta V\cdots\Delta V\cdots)p^{(2)}(V)\,DV, \tag{E.11}$$

where $p^{(2)}(V) \propto \exp(-F^{(2)}(V)/T)$ the probability density. Our goal is to derive the splitting formula[1]

[1] A simpler, and probably sufficient, splitting formula is

$$\mathrm{Tr}\big(\bar G\underbrace{\Delta V\bar G\Delta V}\bar G\Delta V\big) = (\mathrm{Tr}1)^{-1}\mathrm{Tr}\big(\bar G\underbrace{\Delta V\bar G\Delta V}\big)\mathrm{Tr}\big(\bar G\Delta V\big),$$

$$\mathrm{Tr}\Big(\bar{G}\underbrace{\Delta V\bar{G}\Delta V}\bar{G}\Delta V\Big) \approx \frac{\pi T}{W}\frac{1}{2N_{\mathrm{d}}N}\mathrm{Tr}\Big(\bar{G}\underbrace{\Delta V\bar{G}\Delta V}\Big)\mathrm{Tr}(\bar{G}\Delta V) \equiv \eta\mathrm{Tr}(\bar{G}\Delta V).$$

First, we separate the spin-dependent part. Using

$$\bar{G} = \sum_{\gamma=0,z}\bar{G}^{\gamma}\sigma^{\gamma}, \qquad V = \sum_{\alpha=x,y,z}V^{\alpha}\sigma^{\alpha},$$

we write

$$\mathrm{Tr}\Big(\bar{G}\underbrace{\Delta V\bar{G}\Delta V}\bar{G}\Delta V\Big) = \sum_{\substack{\gamma_1\gamma_2\gamma_3\\ \alpha_1\alpha_2\alpha_3}}\mathrm{Tr}\big(\bar{G}^{\gamma_1}\underbrace{\Delta V^{\alpha_1}\bar{G}^{\gamma_2}\Delta V^{\alpha_2}}\bar{G}^{\gamma_3}\Delta V^{\alpha_3}\big)$$
$$\times\,\mathrm{Sp}\big(\sigma^{\gamma_1}\sigma^{\alpha_1}\sigma^{\gamma_2}\sigma^{\alpha_2}\sigma^{\gamma_3}\sigma^{\alpha_3}\big). \tag{E.12}$$

Now we transform the first product term in (E.12). In the momentum-"frequency" representation, we have

$$\bar{G}^{\gamma}_{\mathbf{kk}'nn'} = \bar{G}^{\gamma}_{\mathbf{k}n}\delta_{\mathbf{kk}'}\delta_{nn'}, \qquad V^{\alpha}_{\mathbf{kk}'nn'} = V^{\alpha}_{\mathbf{k}-\mathbf{k}',\,n-n'}.$$

Hence

$$\mathrm{Tr}\big(\bar{G}^{\gamma_1}\Delta V^{\alpha_1}\bar{G}^{\gamma_2}\Delta V^{\alpha_2}\bar{G}^{\gamma_3}\Delta V^{\alpha_3}\big)$$
$$= N_{\mathrm{d}}\sum_{\substack{\mathbf{kk}_1\mathbf{k}_2\\ nn_1n_2}}\bar{G}^{\gamma_1}_{\mathbf{k}n}\Delta V^{\alpha_1}_{\substack{\mathbf{k}-\mathbf{k}_1\\ n-n_1}}\bar{G}^{\gamma_2}_{\mathbf{k}_1n_1}\Delta V^{\alpha_2}_{\substack{\mathbf{k}_1-\mathbf{k}_2\\ n_1-n_2}}\bar{G}^{\gamma_3}_{\mathbf{k}_2n_2}\Delta V^{\alpha_3}_{\substack{\mathbf{k}_2-\mathbf{k}\\ n_2-n}}. \tag{E.13}$$

Introducing the new summation indices

$$\mathbf{q}_1 = \mathbf{k}-\mathbf{k}_1, \qquad \mathbf{q}_2 = \mathbf{k}_1-\mathbf{k}_2, \qquad \mathbf{q}_3 = \mathbf{k}_2-\mathbf{k},$$

$$m_1 = n-n_1, \qquad m_2 = n_1-n_2, \qquad m_3 = n_2-n,$$

we write

$$\mathrm{Tr}\big(\bar{G}^{\gamma_1}\underbrace{\Delta V^{\alpha_1}\bar{G}^{\gamma_2}}\Delta V^{\alpha_2}\bar{G}^{\gamma_3}\Delta V^{\alpha_3}\big)$$
$$= N_{\mathrm{d}}\sum_{\substack{\mathbf{kq}_1\mathbf{q}_2\mathbf{q}_3\\ nm_1m_2m_3}}\bar{G}^{\gamma_1}_{\mathbf{k}n}\underbrace{\Delta V^{\alpha_1}_{\mathbf{q}_1m_1}\bar{G}^{\gamma_2}_{\substack{\mathbf{k}-\mathbf{q}_1\\ n-m_1}}}\Delta V^{\alpha_2}_{\mathbf{q}_2m_2}\bar{G}^{\gamma_3}_{\substack{\mathbf{k}-\mathbf{q}_1-\mathbf{q}_2\\ n-m_1-m_2}}\Delta V^{\alpha_3}_{\mathbf{q}_3m_3}$$
$$= N_{\mathrm{d}}\sum_{\substack{\mathbf{kq}_1\mathbf{q}_2\mathbf{q}_3\\ nm_1m_2m_3}}\bar{G}^{\gamma_1}_{\mathbf{k}n}\bar{G}^{\gamma_2}_{\substack{\mathbf{k}-\mathbf{q}_1\\ n-m_1}}\langle\Delta V^{\alpha_1}_{\mathbf{q}_1m_1}\Delta V^{\alpha_2}_{\mathbf{q}_2m_2}\rangle\bar{G}^{\gamma_3}_{\substack{\mathbf{k}-\mathbf{q}_1-\mathbf{q}_2\\ n-m_1-m_2}}\Delta V^{\alpha_3}_{\mathbf{q}_3m_3}, \tag{E.14}$$

where $\mathbf{q}_1+\mathbf{q}_2+\mathbf{q}_3=0$ and $m_1+m_2+m_3=0$. Making use of

$$\langle\Delta V^{\alpha_1}_{\mathbf{q}_1m_1}\Delta V^{\alpha_2}_{\mathbf{q}_2m_2}\rangle = \langle|\Delta V^{\alpha_1}_{\mathbf{q}_1m_1}|^2\rangle\delta_{\mathbf{q}_1,-\mathbf{q}_2}\delta_{m_1,-m_2}\delta_{\alpha_1\alpha_2},$$

---

where

$$\mathrm{Tr}1 = N_{\mathrm{d}}\int_0^{1/T}\sum_j\mathrm{Sp}\sigma^0\mathrm{d}\tau = 2N_{\mathrm{d}}N/T.$$

we write (E.14) as

$$\mathrm{Tr}\big(\bar{G}^{\gamma_1}\underbrace{\Delta V^{\alpha_1}\bar{G}^{\gamma_2}\Delta V^{\alpha_2}}\bar{G}^{\gamma_3}\Delta V^{\alpha_3}\big) = N_{\mathrm{d}} \sum_{\mathbf{kq}nm} \bar{G}^{\gamma_1}_{\mathbf{k}n}\bar{G}^{\gamma_2}_{\substack{\mathbf{k-q}\\ n-m}} \langle|\Delta V^{\alpha_1}_{\mathbf{q}m}|^2\rangle \bar{G}^{\gamma_3}_{\mathbf{k}n}\Delta V^{\alpha_3}_{00}. \tag{E.15}$$

Replacing the mean Green functions by their single-site approximation,

$$\bar{G}^{\gamma_1}_{\mathbf{k}n} \approx g^{\gamma_1}(\omega_n), \qquad \bar{G}^{\gamma_2}_{\mathbf{k-q},n-m} \approx g^{\gamma_2}(\omega_n - \omega_m),$$

we obtain

$$\mathrm{Tr}\big(\bar{G}^{\gamma_1}\underbrace{\Delta V^{\alpha_1}\bar{G}^{\gamma_2}\Delta V^{\alpha_2}}\bar{G}^{\gamma_3}\Delta V^{\alpha_3}\big) = N_{\mathrm{d}}N \sum_{\mathbf{q}nm} g^{\gamma_1}_n g^{\gamma_2}_{n-m}\langle|\Delta V^{\alpha_1}_{\mathbf{q}m}|^2\rangle g^{\gamma_3}_n \Delta V^{\alpha_3}_{00} \tag{E.16}$$

This completes the splitting in the momenta.

To split (E.16) in "frequencies", first we replace the summation in n by the integration over the energy variable with the Fermi function $f(\varepsilon)$:

$$\begin{aligned}\sum_n g^{\gamma_1}_n g^{\gamma_2}_{n-m} g^{\gamma_3}_n &= \frac{1}{\pi T}\int f(\varepsilon)\mathrm{Im}(g^{\gamma_1}(\varepsilon)g^{\gamma_2}(\varepsilon - \mathrm{i}\omega_m)g^{\gamma_3}(\varepsilon))\,\mathrm{d}\varepsilon \\ &= \frac{1}{\pi T}\int_0^W f(\varepsilon)\mathrm{Im}(g^{\gamma_1}(\varepsilon)g^{\gamma_2}(\varepsilon - \mathrm{i}\omega_m)g^{\gamma_3}(\varepsilon))\,\mathrm{d}\varepsilon,\end{aligned} \tag{E.17}$$

where W is the band-width. Replacing the first two product terms in the integrand $g^{\gamma_1}(\varepsilon)g^{\gamma_2}(\varepsilon - \mathrm{i}\omega_m)$ by the average

$$\{g^{\gamma_1}(\varepsilon)g^{\gamma_2}(\varepsilon - \mathrm{i}\omega_m)\} \equiv \frac{1}{W}\int_0^W f(\varepsilon)\mathrm{Im}(g^{\gamma_1}(\varepsilon)g^{\gamma_2}(\varepsilon - \mathrm{i}\omega_m))\,\mathrm{d}\varepsilon,$$

we write (E.17) as

$$\begin{aligned}\sum_n g^{\gamma_1}_n g^{\gamma_2}_{n-m} g^{\gamma_3}_n &= \frac{1}{\pi T}\int_0^W f(\varepsilon)\{g^{\gamma_1}(\varepsilon)g^{\gamma_2}(\varepsilon - \mathrm{i}\omega_m)\}\mathrm{Im}g^{\gamma_3}(\varepsilon)\,\mathrm{d}\varepsilon \\ &= \left(\frac{1}{W}\int_0^W f(\varepsilon)\mathrm{Im}(g^{\gamma_1}(\varepsilon)g^{\gamma_2}(\varepsilon - \mathrm{i}\omega_m))\,\mathrm{d}\varepsilon\right)\frac{1}{\pi T}\int_0^W f(\varepsilon)\mathrm{Im}g^{\gamma_3}(\varepsilon)\,\mathrm{d}\varepsilon.\end{aligned}$$

Returning to the sum over "frequencies", we obtain

$$\sum_n g^{\gamma_1}_n g^{\gamma_2}_{n-m} g^{\gamma_3}_n = \frac{\pi T}{W}\sum_n g^{\gamma_1}_n g^{\gamma_2}_{n-m}\sum_{n'} g^{\gamma_3}_{n'}. \tag{E.18}$$

Finally, substituting the latter in (E.16), we have

$$\begin{aligned}&\mathrm{Tr}\big(\bar{G}^{\gamma_1}\underbrace{\Delta V^{\alpha_1}\bar{G}^{\gamma_2}\Delta V^{\alpha_2}}\bar{G}^{\gamma_3}\Delta V^{\alpha_3}\big) \\ &= \frac{\pi T}{W}\sum_{\mathbf{q}nm} g^{\gamma_1}_n g^{\gamma_2}_{n-m}\langle|\Delta V^{\alpha_1}_{\mathbf{q}m}|^2\rangle\delta_{\alpha_1\alpha_2}\left(N_{\mathrm{d}}N\sum_{n'} g^{\gamma_3}_{n'}\Delta V^{\alpha_3}_{00}\right) \\ &= \frac{\pi T}{W}\frac{1}{N_{\mathrm{d}}N}\mathrm{Tr}\big(\bar{G}^{\gamma_1}\underbrace{\Delta V^{\alpha_1}\bar{G}^{\gamma_2}\Delta V^{\alpha_2}}\big)\mathrm{Tr}\big(\bar{G}^{\gamma_3}\Delta V^{\alpha_3}\big).\end{aligned} \tag{E.19}$$

Now we carry out the splitting over spin in (E.12), using the approximation

$$\mathrm{Sp}\big(\sigma^{\gamma_1}\sigma^{\alpha_1}\sigma^{\gamma_2}\sigma^{\alpha_2}\sigma^{\gamma_3}\sigma^{\alpha_3}\big) \approx \frac{1}{2}\mathrm{Sp}\big(\sigma^{\gamma_1}\sigma^{\alpha_1}\sigma^{\gamma_2}\sigma^{\alpha_2}\big)\mathrm{Sp}\big(\sigma^{\gamma_3}\sigma^{\alpha_3}\big), \tag{E.20}$$

which correlates with the approximations made above. Then (E.12) becomes

$$
\begin{aligned}
&\mathrm{Tr}\Big(\bar{G}\Delta \underbracket{V\bar{G}\Delta V}\bar{G}\Delta V\Big) \\
&= \frac{\pi T}{W}\frac{1}{2N_\mathrm{d}N}\sum_{\substack{\gamma_1\gamma_2\\ \alpha_1\alpha_2}} \mathrm{Tr}\big(\bar{G}^{\gamma_1}\Delta \underbracket{V^{\alpha_1}\bar{G}^{\gamma_2}\Delta V^{\alpha_2}}\big)\mathrm{Sp}\big(\sigma^{\gamma_1}\sigma^{\alpha_1}\sigma^{\gamma_2}\sigma^{\alpha_2}\big) \\
&\quad\times \sum_{\gamma_3\alpha_3} \mathrm{Tr}\big(\bar{G}^{\gamma_3}\Delta V^{\alpha_3}\big)\mathrm{Sp}\big(\sigma^{\gamma_3}\sigma^{\alpha_3}\big) \\
&= \frac{\pi T}{W}\frac{1}{2N_\mathrm{d}N}\mathrm{Tr}\Big(\bar{G}\Delta \underbracket{V\bar{G}\Delta V}\Big)\mathrm{Tr}\Big(\bar{G}\Delta V\Big).
\end{aligned}
\tag{E.21}
$$

The correction coefficient

$$
\eta \equiv \frac{\pi T}{W}\frac{1}{2N_\mathrm{d}N}\mathrm{Tr}\Big(\bar{G}\Delta \underbracket{V\bar{G}\Delta V}\Big)
$$

is calculated by the formula [3]

$$
\eta = -\frac{\pi}{WN_\mathrm{d}}\sum_{\alpha} \grave{\chi}_\mathrm{L}^{\alpha}(0)\langle|\Delta V_\alpha|^2\rangle',
$$

where

$$
\grave{\chi}_\mathrm{L}^{\alpha}(0) = -\frac{N_\mathrm{d}}{\pi}\sum_{\gamma_1\gamma_2}\int \mathrm{Im}\big(g^{\gamma_1}(\varepsilon)g^{\gamma_2}(\varepsilon)\big)\mathrm{Sp}\big(\sigma^{\gamma_1}\sigma^{\alpha}\sigma^{\gamma_2}\sigma^{\alpha}\big)f(\varepsilon)\,\mathrm{d}\varepsilon
$$

is the static local susceptibility in the Gaussian approximation.

E.4 Uniform Spin Fluctuations

In Sect. 11.2.2 we improved the effect of nonlocality in the site-diagonal mean Green function by taking into account uniform fluctuations. Following [3,4], here we derive the approximation of the field $\tilde{V}$ by a field that takes just two values $\pm\bar{V}_z$, which is used to obtain formula (11.18).

To explain the main idea of the interpolation, let first V be a scalar fluctuating field with the probability density $p(V)$, so that the average of a function $f(V)$ is $\langle f(V)\rangle = \int f(V)p(V)\,\mathrm{d}V$. Our goal is to approximate the initial continuous distribution by a discrete one that takes two symmetric values $\pm v$ with the probabilities $P(\pm v)$, so that the integral could be performed as $\langle f(V)\rangle \approx P(-v)f(-v) + P(v)f(v)$.

We use the splitting $\langle V^{2n}\rangle \approx \langle V^2\rangle^n$ and $\langle V^{2n+1}\rangle \approx \langle V^2\rangle^n\langle V\rangle$ [2], so that the power series expansion of the function $f(V)$ about the origin can be written as

$$
\langle f(V)\rangle \approx f(0) + f'(0)\langle V\rangle + \frac{1}{2!}f''(0)\langle V^2\rangle + \frac{1}{3!}f'''(0)\langle V^2\rangle\langle V\rangle + \frac{1}{4!}f^{(4)}(0)\langle V^2\rangle^2 + \cdots \tag{E.22}
$$

Hence, expressing $f(V)$ as the sum of its even and odd parts, we have

$$
\begin{aligned}
\langle f(V)\rangle &= \left\langle \frac{f(V)+f(-V)}{2}\right\rangle + \left\langle \frac{f(V)-f(-V)}{2V}V\right\rangle \\
&\approx \frac{f(v)+f(-v)}{2} + \frac{f(v)-f(-v)}{2v}\langle V\rangle,
\end{aligned}
$$

where $v = \sqrt{\langle V^2\rangle}$. Rearranging the expression on the right-hand side, we come to

$$
\langle f(V)\rangle \approx \sum_{\sigma'}\frac{1}{2}\left(1+\frac{\langle V\rangle}{\sigma' v}\right)f(\sigma' v). \tag{E.23}
$$

From (E.22) we see that the splitting affects only higher-order terms of the function $f(V)$ keeping the first moment $\langle V \rangle$ and second moment $v^2 = \langle V^2 \rangle$ of the initial field unchanged. Since $v^2 = \langle V \rangle^2 + \langle \Delta V^2 \rangle$, the approximation (E.23) is completely described by the mean $\langle V \rangle$ and variance $\langle \Delta V^2 \rangle$ of the initial field.

Similar to (E.23), we approximate the average

$$\tilde{g}(\varepsilon) \equiv \big\langle g^0(\varepsilon - \tilde{V} - \Delta\Sigma(\varepsilon, \tilde{V})) \big\rangle,$$

where $g^0(\varepsilon)$ is the single-site zeroth Green function, $\tilde{V} = \sum_\alpha \tilde{V}_\alpha \sigma^\alpha$ is the uniform fluctuating field and $\Delta\Sigma(\varepsilon, \tilde{V})$ is the fluctuational contribution to the self-energy. Recalling that $\bar{V}_\sigma = \sigma \bar{V}_z$, we write

$$\tilde{g}_\sigma(\varepsilon) \approx \sum_{\sigma'} P_{\sigma\sigma'} g^0(\varepsilon - \sigma' v - \Delta\Sigma_\sigma(\varepsilon, \sigma' v)), \tag{E.24}$$

where $P_{\sigma\sigma'} = \frac{1}{2}(1 + (\sigma \bar{V}_z)/(\sigma' v))$ and $v = (\bar{V}_z^2 + \langle \Delta V_x^2 \rangle + \langle \Delta V_y^2 \rangle + \langle \Delta V_z^2 \rangle)^{1/2}$.

Note that, for large V, the exact distribution $p(V) \propto \exp(-F(V)/T)$ is spherically symmetric and does not depend on the electronic structure. The RGA distribution $p^{(2)}(V)$ is shifted by $\bar{V}$ (in the ferromagnetic case). The discrete approximation (E.24) of the field V is constructed so that it restores the symmetry with respect to the origin.

References

1. B.I. Reser, V.I. Grebennikov, Phys. Met. Metallogr. **83**, 127 (1997)
2. H. Hasegawa, J. Phys. Soc. Jpn. **46**, 1504 (1979)
3. N.B. Melnikov, B.I. Reser, V.I. Grebennikov, J. Phys. Condens. Matter **23**, 276003 (2011)
4. V.I. Grebennikov, Phys. Solid State **40**, 79 (1998)

Basic Approximations in Scattering Theory

F

In this chapter we present two important applications of the first-order perturbation theory to neutron scattering in a crystal. For a more general treatment of the perturbation theory and scattering theory, see, e.g. [1–6].

F.1 Fermi's Golden Rule

The scattering process is described by the time-dependent Hamiltonian

$$\mathcal{H}(t) = \mathcal{H}_0 + V(t),$$

where $\mathcal{H}_0$ is the Hamiltonian of the neutron and crystal without interaction between them, and $V(t)$ is the interaction potential. The potential $V(t) = \eta W(t)$ is considered as a perturbation, where $\eta \ll 1$ is a small parameter. For $t < 0$ we have $V(t) \equiv 0$ and at $t = 0$ the potential begins to act on the system.

We assume that, for $t < 0$, the system is in one of the stationary states ψ_m of the unperturbed Hamiltonian:

$$\mathcal{H}_0 \psi_m = E_m \psi_m,$$

where m stands for $(\mathbf{k}, \sigma, \lambda)$. When $t > 0$ the system evolves according to the time-dependent Shrödinger equation

$$\mathrm{i}\hbar \frac{\partial \psi(t)}{\partial t} = (\mathcal{H}_0 + V(t))\psi(t) \tag{F.1}$$

with the initial condition $\psi(0) = \psi_m$. The wave function $\psi(t)$ can be represented as a superposition

$$\psi(t) = \sum_n c_n(t)\psi_n, \tag{F.2}$$

where the initial values of coefficients $c_n(0)$ are all zero except for $c_m(0) = 1$. The probability $P_{m\to m'}(t)$ of finding the system in a stationary state $\psi_{m'}$ at a time $t > 0$ is equal to $|c_{m'}(t)|^2$. Then the transition probability per unit time is given by

$$w_{m\to m'} = \frac{\mathrm{d}}{\mathrm{d}t} P_{m\to m'} = \frac{\mathrm{d}}{\mathrm{d}t} |c_{m'}(t)|^2. \tag{F.3}$$

First we make use of the *nonstationary* perturbation theory to calculate the coefficient $c_{m'}(t)$ to the first order in $V(t)$, and then we obtain an asymptotic expression for $w_{m\to m'}$ at large t that corresponds to the *stationary* scattering process.

We substitute expansion (F.2) to Eq. (F.1) and take the inner product of both sides with $\psi_{m'}$. Making use of orthogonality, we come to

$$\mathrm{i}\hbar \frac{\mathrm{d}}{\mathrm{d}t} c_{m'}(t) = E_{m'} c_{m'}(t) + \sum_n V_{m'n}(t) c_n(t), \tag{F.4}$$

N. B. Melnikov, B. I. Reser, *Dynamic Spin Fluctuation Theory of Metallic Magnetism*,
https://doi.org/10.1007/978-3-319-92974-3

where $V_{m'n}(t) = (\psi_{m'}, V(t)\psi_n)$ is the matrix element. We seek the solution of equation (F.4) as a series in the small parameter η:

$$c_{m'}(t) = c_{m'}^{(0)}(t) + \eta c_{m'}^{(1)}(t) + \cdots \tag{F.5}$$

Substituting the latter in (F.4) and collecting the terms of the same order in η, we obtain

$$\mathrm{i}\hbar \frac{\mathrm{d}}{\mathrm{d}t} c_{m'}^{(0)}(t) = E_{m'} c_{m'}^{(0)}(t) \tag{F.6}$$

and

$$\mathrm{i}\hbar \frac{\mathrm{d}}{\mathrm{d}t} c_{m'}^{(s)}(t) = E_{m'} c_{m'}^{(s)}(t) + \sum_n W_{m'n}(t) c_n^{(s-1)}(t) \tag{F.7}$$

for $s = 1, 2, \ldots$ Integrating equation (F.6) and taking the initial condition $\psi(0) = \psi_m$ into account, we have

$$c_{m'}^{(0)}(t) = \delta_{m'm}\, \mathrm{e}^{-\mathrm{i}E_{m'}t/\hbar}. \tag{F.8}$$

Substitution of (F.8) in (F.7) at $s = 1$ leads to

$$\mathrm{i}\hbar \frac{\mathrm{d}}{\mathrm{d}t} c_{m'}^{(1)}(t) = E_{m'} c_{m'}^{(1)}(t) + W_{m'm}(t)\, \mathrm{e}^{-\mathrm{i}E_m t/\hbar}. \tag{F.9}$$

Integrating this equation, we obtain

$$c_{m'}^{(1)}(t) = \left(-\frac{\mathrm{i}}{\hbar} \int_0^t W_{m'm}(\tau)\, \mathrm{e}^{\mathrm{i}\omega_{m'm}\tau}\, \mathrm{d}\tau \right) \mathrm{e}^{-\mathrm{i}E_{m'}t/\hbar}, \tag{F.10}$$

where $\hbar\omega_{m'm} = E_{m'} - E_m$. Substituting (F.8) and (F.10) in (F.5) and keeping only the terms up to the first order in $V(t) = \eta W(t)$, we write

$$c_{m'}(t) = \left(\delta_{m'm} - \frac{\mathrm{i}}{\hbar} \int_0^t V_{m'm}(\tau)\, \mathrm{e}^{\mathrm{i}\omega_{m'm}\tau}\, \mathrm{d}\tau \right) \mathrm{e}^{-\mathrm{i}E_{m'}t/\hbar}. \tag{F.11}$$

It is often convenient to consider scattering as a stationary process, i.e. as a continuous flux of incoming particles that is transformed into a flux of scattered particles. The density of the particles in the flux must be sufficiently low for the interaction between the incident particles to be negligibly small. In the stationary treatment, the scattering problem consists in calculation of the flux of scattered particles at an infinitely great distance from the scattering system. In this case, we can assume that the scattering potential is time-independent, $V(t) \equiv V$, which considerably simplifies calculating the transition probability $P_{m \to m'}(t) = |c_{m'}(t)|^2$. Using expression (F.11), for $m' \neq m$ we obtain

$$P_{m \to m'}(t) = \frac{1}{\hbar^2} |V_{m'm}|^2 \left| \int_0^t \mathrm{e}^{\mathrm{i}\omega_{m'm}\tau}\, \mathrm{d}\tau \right|^2 = \frac{2\pi t}{\hbar^2} |V_{m'm}|^2 f(\omega_{m'm}, t), \tag{F.12}$$

where

$$f(\omega, t) = \frac{1}{2\pi t} \left(\frac{\sin(\omega t/2)}{\omega/2} \right)^2. \tag{F.13}$$

Figure F.1 shows the variation of the function $f(\omega, t)$ with respect to ω, when t is fixed. Using $\sin x \approx x$ for small $x = \omega t/2$, at $\omega = 0$ we have

$$f(0, t) = \frac{t}{2\pi} \to \infty, \qquad t \to \infty,$$

and at $\omega \neq 0$ we have

$$f(\omega, t) \to 0, \qquad t \to \infty.$$

Calculating the integral of $f(\omega, t)$, we obtain

$$\int f(\omega, t)\, \mathrm{d}\omega = \frac{1}{2\pi t} \int \left(\frac{\sin(\omega t/2)}{\omega/2} \right)^2 \mathrm{d}\omega = \frac{1}{\pi} \int \frac{\sin^2 x}{x^2}\, \mathrm{d}x = 1.$$

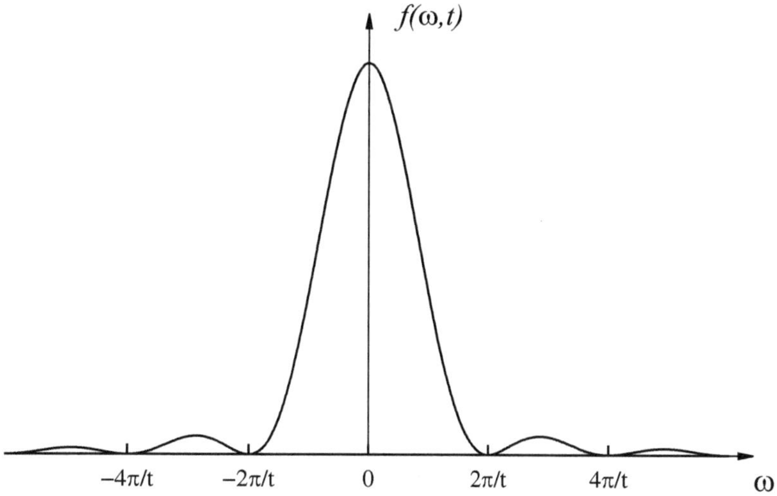

Fig. F.1 The graph of the function (F.13) for a fixed t

Thus, for large t the function $f(\omega, t)$ can be approximated by the δ-function. Using formula (F.12), we come to the asymptotic formula for large t:

$$P_{m\to m'}(t) = \frac{2\pi t}{\hbar^2}\,|V_{m'm}|^2\,\delta(\omega_{m'm}) = \frac{2\pi t}{\hbar}\,|V_{m'm}|^2\,\delta(E_{m'} - E_m).$$

Taking formula (F.3) into account, we write the transition probability per unit time as

$$w_{m\to m'} = \frac{2\pi}{\hbar}|V_{m'm}|^2\delta(E_{m'} - E_m). \tag{F.14}$$

This relation is known as *Fermi's golden rule* (see, e.g. [3, 5, 6]).

F.2 The Born Approximation

Calculation of the matrix element $V_{m'm}$ in Fermi's golden rule requires further approximations. In this section we show that, in the first-order perturbation theory, it is sufficient to use plane waves as the neutron wave functions.

The neutron wave function $\psi_{\mathbf{k}}(\mathbf{r})$ satisfies the stationary Schrödinger equation

$$(\nabla^2 + k^2)\psi_{\mathbf{k}}(\mathbf{r}) = \frac{2m}{\hbar^2}\,V(\mathbf{r})\psi_{\mathbf{k}}(\mathbf{r}), \qquad k^2 = \frac{2mE}{\hbar^2}, \quad E > 0, \tag{F.15}$$

where m is the mass of a neutron. Assuming that the interaction range of the potential $V(\mathbf{r})$ is bounded ($r \leq d$), we wish to obtain the asymptotic form of the solution of equation (F.15) at large distances ($r \gg d$).

First, we convert the Schödinger equation (F.15) to an integral equation. Introducing the Green function $G(\mathbf{r}, \mathbf{r}')$ as the solution to

$$(\nabla^2 + k^2)G(\mathbf{r}, \mathbf{r}') = \delta(\mathbf{r} - \mathbf{r}'), \tag{F.16}$$

we write Eq. (F.15) as[1]

$$\psi_{\mathbf{k}}(\mathbf{r}) = \psi_{\mathbf{k}}^{(0)}(\mathbf{r}) + \frac{2m}{\hbar^2}\int G(\mathbf{r}, \mathbf{r}')V(\mathbf{r}')\psi_{\mathbf{k}}(\mathbf{r}')\,d\mathbf{r}', \tag{F.17}$$

[1]Applying the operator $\nabla^2 + k^2$ to both sides of the integral equation (F.17) and taking relation (F.16) into account, we come to the original equation (F.15).

where $\psi_{\mathbf{k}}^{(0)}(\mathbf{r}) = \mathrm{e}^{\mathrm{i}\mathbf{kr}}$ is the wave function of the free neutron. Treating the potential $V(\mathbf{r})$ as a small perturbation, we can write

$$\psi_{\mathbf{k}}(\mathbf{r}) = \psi_{\mathbf{k}}^{(0)}(\mathbf{r}) + \psi_{\mathbf{k}}^{(1)}(\mathbf{r}) + \cdots, \tag{F.18}$$

where $\psi_{\mathbf{k}}^{(1)}(\mathbf{r})$ is the first-order term in $V(\mathbf{r})$. Substituting this in (F.17) and omitting the second-order terms, we obtain

$$\psi_{\mathbf{k}}^{(1)}(\mathbf{r}) = \frac{2m}{\hbar^2}\int G(\mathbf{r},\mathbf{r}')V(\mathbf{r}')\psi_{\mathbf{k}}^{(0)}(\mathbf{r}')\,\mathrm{d}\mathbf{r}'. \tag{F.19}$$

For the elastic scattering (momenta of the incident and scattered neutron have equal modulus: $k = k'$), asymptotic expression for the wave function $\psi_{\mathbf{k}}(\mathbf{r})$ is obtained as follows. The Green function is given by (see, e.g. [4, Appendix VI])

$$G(\mathbf{r},\mathbf{r}') = -\frac{1}{4\pi|\mathbf{r}-\mathbf{r}'|}\mathrm{e}^{\mathrm{i}k|\mathbf{r}-\mathbf{r}'|}. \tag{F.20}$$

At large distances ($r \gg d$), we can write $k|\mathbf{r}-\mathbf{r}'| \approx kr\sqrt{1-2\mathbf{rr}'/r^2} \approx kr - \mathbf{k}'\mathbf{r}'$, where $\mathbf{k}' = k\mathbf{r}'/r'$. In the denominator of (F.20) we assume $|\mathbf{r}-\mathbf{r}'| \approx r$. Then the function (F.19) becomes

$$\begin{aligned}\psi_{\mathbf{k}}^{(1)}(\mathbf{r}) &= -\frac{2m}{\hbar^2}\frac{\mathrm{e}^{\mathrm{i}kr}}{r}\int \mathrm{e}^{-\mathrm{i}\mathbf{k}'\mathbf{r}'}\,V(\mathbf{r}')\,\mathrm{e}^{\mathrm{i}\mathbf{kr}'}\,\mathrm{d}\mathbf{r}'\\ &= -\frac{2m}{\hbar^2}\frac{\mathrm{e}^{\mathrm{i}kr}}{r}V_{\boldsymbol{\kappa}},\end{aligned}$$

where

$$V_{\boldsymbol{\kappa}} = \int V(\mathbf{r})\,\mathrm{e}^{-\mathrm{i}\boldsymbol{\kappa}\mathbf{r}}\,\mathrm{d}\mathbf{r}$$

is the Fourier transform of the potential and $\boldsymbol{\kappa} = \mathbf{k}-\mathbf{k}'$ is the scattering vector. The first-order approximation of the wave function (F.18) is finally given by

$$\psi_{\mathbf{k}}(\mathbf{r}) = \mathrm{e}^{\mathrm{i}\mathbf{kr}} + A(\Omega')\,\frac{\mathrm{e}^{\mathrm{i}kr}}{r}, \tag{F.21}$$

where

$$A(\Omega') = -\frac{m}{2\pi\hbar^2}V_{-\boldsymbol{\kappa}} \tag{F.22}$$

is the scattering amplitude and Ω' is the scattering angle. The first term in (F.21) describes the neutron before the scattering and the second term describes the scattered neutron at large distances. Expression (F.22) is called the *Born approximation* for the scattering amplitude.

Finally, we calculate the first-order approximation, or the Born approximation, of the matrix element over neutron states in formula (F.14):

$$\langle\mathbf{k}'|\,V\,|\mathbf{k}\rangle = \int \psi_{\mathbf{k}'}^{*}(\mathbf{r})V(\mathbf{r})\psi_{\mathbf{k}}(\mathbf{r})\,\mathrm{d}\mathbf{r}.$$

Since the second term in (F.18) is of first order in V, it is sufficient to keep only the first term, i.e. the plain wave (normalized to the volume of the crystal $\mathcal{V}$). Then

$$\langle\mathbf{k}'|\,V\,|\mathbf{k}\rangle = \frac{1}{\mathcal{V}}\int \mathrm{e}^{-\mathrm{i}\mathbf{k}'\mathbf{r}}\,V(\mathbf{r})\,\mathrm{e}^{\mathrm{i}\mathbf{kr}}\,\mathrm{d}\mathbf{r} = \frac{1}{\mathcal{V}}V_{-\boldsymbol{\kappa}}. \tag{F.23}$$

In other words, the first-order approximation of the matrix element $\langle\mathbf{k}'|\,V\,|\mathbf{k}\rangle$ is proportional to the Fourier transform of the potential $V_{-\boldsymbol{\kappa}}$.

References

1. E. Fermi, *Notes on Quantum Mechanics* (University Chicago Press, Chicago, 1961)
2. A. Messiah, *Quantum Mechanics* (North-Holland, Amsterdam, 1961)
3. L. Schiff, *Quantum Mechanics*, 3rd edn. (McGraw-Hill, New York, 1968)
4. I.V. Savelyev, *Fundamentals of Theoretical Physics (vol. 2). Quantum Mechanics* (Mir Publ., Moscow, 1982)
5. C. Cohen-Tannoudji, B. Diu, F. Laloe, *Quantum Mechanics* (Wiley, New York, 1991)
6. E. Mertzbacher, *Quantum Mechanics*, 3rd edn. (Wiley, New York, 1997)

Lattice Vibrations in the Harmonic Approximation

G

Here we present necessary facts about the quantum lattice vibrations (phonons) in the harmonic approximation [1]. A more detailed presentation of phonons and their applications can be found in [2–6]. First, we obtain an expression for displacement of an atom in a crystal lattice from its equilibrium position. We start with the classical mechanics treatment of the normal modes and then show how they are quantized. Next, we prove that normal modes have the Gaussian probability distribution. Finally, we present the Debye model and apply it to calculate sums over the normal modes. As an illustration, we calculate the Debye-Waller factor (DWF) in the Debye model.

G.1 Normal Modes and Their Quantization

In the harmonic approximation, the *classical* Hamiltonian of a three-dimensional crystal lattice is (see, e.g. [6])

$$H = \sum_j \frac{\mathbf{p}_j^2}{2M} + \frac{1}{2}\sum_{jj'} \mathbf{u}_j \cdot D_{jj'}\mathbf{u}_{j'}, \tag{G.1}$$

where M is the mass of an atom, $\mathbf{p}_j$ is the momentum of the jth atom, $\mathbf{u}_j$ is the displacement of the jth atom from its equilibrium position and $D_{jj'}$ is the 3×3-matrix of the form

$$D_{jj'} = \left.\frac{\partial^2 U}{\partial \mathbf{u}_j \partial \mathbf{u}_{j'}}\right|_{\mathbf{u}=0}. \tag{G.2}$$

Here $U = U(\mathbf{u}_1, \ldots, \mathbf{u}_N)$ is the potential energy, which attains its minimum at the equilibrium $\mathbf{u} = 0$. From the Hamiltonian system,

$$\frac{\mathrm{d}\mathbf{p}_j}{\mathrm{d}t} = -\frac{\partial H}{\partial \mathbf{u}_j}, \qquad \frac{\mathrm{d}\mathbf{u}_j}{\mathrm{d}t} = \frac{\partial H}{\partial \mathbf{p}_j},$$

we obtain the equation of motion

$$M\frac{\mathrm{d}^2\mathbf{u}_j}{\mathrm{d}t^2} = -\sum_{j'} D_{jj'}\mathbf{u}_{j'}. \tag{G.3}$$

We seek a particular solution in the form

$$\mathbf{u}_{j\mathbf{q}}(t) = Q_\mathbf{q}\mathbf{e}_\mathbf{q}\mathrm{e}^{\mathrm{i}\mathbf{q}\mathbf{R}_j - \mathrm{i}\omega_\mathbf{q}t}, \tag{G.4}$$

where $\mathbf{e}_\mathbf{q}$ is the polarization vector, $Q_\mathbf{q}$ is the (complex) amplitude and $\omega_\mathbf{q}$ is the frequency of the oscillation. The wave determined by (G.4) is called *the normal mode*. Substituting expression (G.4) in (G.3), we obtain

$$M\omega_\mathbf{q}^2\mathbf{e}_\mathbf{q} = \sum_{j'} D_{jj'}\mathbf{e}_\mathbf{q}\mathrm{e}^{-\mathrm{i}\mathbf{q}\mathbf{R}_{j-j'}}. \tag{G.5}$$

N. B. Melnikov, B. I. Reser, *Dynamic Spin Fluctuation Theory of Metallic Magnetism*,
https://doi.org/10.1007/978-3-319-92974-3

Due to homogeneity of the crystal, the matrix $D_{jj'}$ depends only on the distance between the sites: $D_{jj'} = D_{j-j'}$. Hence its Fourier transform is a diagonal matrix with the elements (for details, see Appendix C.1)

$$D_{\mathbf{q}} = \sum_{j'} D_{jj'} \mathrm{e}^{-\mathrm{i}\mathbf{q}\mathbf{R}_{j-j'}} = \sum_{j'} D_{j-j'} \mathrm{e}^{-\mathrm{i}\mathbf{q}\mathbf{R}_{j-j'}} \tag{G.6}$$

at the diagonal. Substituting (G.6) in (G.5), we see that $\mathbf{e_q}$ is an eigenvector of $D_{\mathbf{q}}$:

$$D_{\mathbf{q}} \mathbf{e_q} = M\omega_{\mathbf{q}}^2 \mathbf{e_q}.$$

Since the matrix (G.2) is symmetric, its eigenvalues are real and its eigenvectors $\mathbf{e}_{\mathbf{q}i}$, $i = 1, 2, 3$, can be chosen such that they form an orthonormal basis in the three-dimensional space,

$$D_{\mathbf{q}} \mathbf{e}_{\mathbf{q}i} = M\omega_{\mathbf{q}i}^2 \mathbf{e}_{\mathbf{q}i}, \qquad i = 1, 2, 3. \tag{G.7}$$

We can also choose $\mathbf{e}_{\mathbf{q}i}$ such that $\mathbf{e}_{-\mathbf{q}i} = \mathbf{e}_{\mathbf{q}i}$. Each of the vectors $\mathbf{e}_{\mathbf{q}i}$ determines the direction of the normal mode oscillation with the frequency $\omega_{\mathbf{q}i}$. The general solution to Eq. (G.3) is given by the superposition

$$\mathbf{u}_j(t) = N^{-1/2} \sum_{\mathbf{q}i} Q_{\mathbf{q}i} \mathbf{e}_{\mathbf{q}i} \mathrm{e}^{\mathrm{i}\mathbf{q}\mathbf{R}_j - \mathrm{i}\omega_{\mathbf{q}i} t}, \tag{G.8}$$

where the summation over $\mathbf{q}$ is carried out over the Brillouin zone. Then the momentum is written as

$$\mathbf{p}_j(t) = N^{-1/2} \sum_{\mathbf{q}i} P_{\mathbf{q}i} \mathbf{e}_{\mathbf{q}i} \mathrm{e}^{\mathrm{i}\mathbf{q}\mathbf{R}_j - \mathrm{i}\omega_{\mathbf{q}i} t}. \tag{G.9}$$

Since the displacement (G.8) and momentum (G.9) are real quantities, the Fourier coefficients $P_{\mathbf{q}i}$ and $Q_{\mathbf{q}i}$ satisfy $P_{-\mathbf{q}i} = P_{\mathbf{q}i}^*$ and $Q_{-\mathbf{q}i} = Q_{\mathbf{q}i}^*$.

We use $P_{\mathbf{q}i}$ and $Q_{\mathbf{q}i}$ as new coordinates, in which the system becomes an ensemble of *independent* oscillators. Substituting expressions (G.8) and (G.9) at $t = 0$ in the Hamiltonian (G.1) and using the identity

$$\sum_j \mathrm{e}^{\mathrm{i}(\mathbf{q}+\mathbf{q}')\mathbf{R}_j} = N\delta_{\mathbf{q}',-\mathbf{q}}, \tag{G.10}$$

we write the first term of (G.1) as

$$H_0 = \frac{1}{2M} \sum_{\mathbf{q}ii'} P_{\mathbf{q}i} P_{-\mathbf{q}i'} \mathbf{e}_{\mathbf{q}i} \cdot \mathbf{e}_{-\mathbf{q}i'}. \tag{G.11}$$

Similarly, using (G.6) and (G.10), we develop the second term of (G.1) to

$$H_\mathrm{I} = \frac{1}{2} \sum_{\mathbf{q}ii'} Q_{\mathbf{q}i} Q_{-\mathbf{q}i'} \mathbf{e}_{\mathbf{q}i} \cdot D_{\mathbf{q}} \mathbf{e}_{-\mathbf{q}i'}.$$

Recalling that $\mathbf{e}_{\mathbf{q}i}$ is an eigenvector of the matrix $D_{\mathbf{q}}$ and applying Eq. (G.7), we obtain

$$H_\mathrm{I} = \frac{1}{2} \sum_{\mathbf{q}ii'} M\omega_{\mathbf{q}i}^2 Q_{\mathbf{q}i} Q_{-\mathbf{q}i'} \mathbf{e}_{\mathbf{q}i} \cdot \mathbf{e}_{-\mathbf{q}i'}. \tag{G.12}$$

Finally, applying $\mathbf{e}_{\mathbf{q}i} \cdot \mathbf{e}_{-\mathbf{q}i'} = \mathbf{e}_{\mathbf{q}i} \cdot \mathbf{e}_{\mathbf{q}i'} = \delta_{ii'}$ to (G.11) and (G.12), we write the Hamiltonian (G.1) as

$$H = H_0 + H_\mathrm{I} = \sum_{\mathbf{q}i} \left(\frac{1}{2M} |P_{\mathbf{q}i}|^2 + \frac{1}{2} M\omega_{\mathbf{q}i}^2 |Q_{\mathbf{q}i}|^2 \right). \tag{G.13}$$

We now replace the classical variables $P_{\mathbf{q}i}$ and $Q_{\mathbf{q}i}$ by the operators $\hat{P}_{\mathbf{q}i}$ and $\hat{Q}_{\mathbf{q}i}$ to obtain the quantum-mechanical form of the classical Hamiltonian (G.13):

$$\mathcal{H} = \sum_{\mathbf{q}i} \left(\frac{1}{2M} |\hat{P}_{\mathbf{q}i}|^2 + \frac{1}{2} M\omega_{\mathbf{q}i}^2 |\hat{Q}_{\mathbf{q}i}|^2 \right), \tag{G.14}$$

where $|\hat{P}_{\mathbf{q}i}|^2 \equiv \hat{P}_{\mathbf{q}i}\hat{P}_{\mathbf{q}i}^\dagger = \hat{P}_{\mathbf{q}i}\hat{P}_{-\mathbf{q}i}$ and $|\hat{Q}_{\mathbf{q}i}|^2 \equiv \hat{Q}_{\mathbf{q}i}\hat{Q}_{\mathbf{q}i}^\dagger = \hat{Q}_{\mathbf{q}i}\hat{Q}_{-\mathbf{q}i}$. The creation and annihilation operators for the $\mathbf{q}i$-th mode are defined by

$$\begin{aligned} b_{\mathbf{q}i}^\dagger &= (2M\hbar\omega_{\mathbf{q}i})^{-1/2} \left(M\omega_{\mathbf{q}i}\hat{Q}_{-\mathbf{q}i} - \mathrm{i}\hat{P}_{\mathbf{q}i} \right), \\ b_{\mathbf{q}i} &= (2M\hbar\omega_{\mathbf{q}i})^{-1/2} \left(M\omega_{\mathbf{q}i}\hat{Q}_{\mathbf{q}i} + \mathrm{i}\hat{P}_{-\mathbf{q}i} \right). \end{aligned} \tag{G.15}$$

These quantized normal modes represent the quasi-particles called *phonons*. Taking into account the commutation relations

$$[\hat{Q}_{\mathbf{q}i}, \hat{Q}_{\mathbf{q}'i'}] = [\hat{P}_{\mathbf{q}i}, \hat{P}_{\mathbf{q}'i'}] = 0, \quad [\hat{P}_{\mathbf{q}i}, \hat{Q}_{\mathbf{q}'i'}] = \delta_{\mathbf{q},\mathbf{q}'}\delta_{ii'},$$

we write the Hamiltonian (G.14) in the second-quantized form:

$$\mathcal{H} = \sum_{\mathbf{q}i} \hbar\omega_{\mathbf{q}i} \left(b_{\mathbf{q}i}^\dagger b_{\mathbf{q}i} + \frac{1}{2} \right). \tag{G.16}$$

Next, we obtain the operator form of the displacement $\hat{\mathbf{u}}_j$. Expressing $\hat{Q}_{\mathbf{q}i}$ from relations (G.15),

$$\hat{Q}_{\mathbf{q}i} = \left(\frac{\hbar}{2M\omega_{\mathbf{q}i}} \right)^{1/2} (b_{\mathbf{q}i} + b_{-\mathbf{q}i}^\dagger),$$

and substituting the latter in the quantized form of (G.8), we obtain

$$\hat{\mathbf{u}}_j = \left(\frac{\hbar}{2MN} \right)^{1/2} \sum_{\mathbf{q}i} \omega_{\mathbf{q}i}^{-1/2} \mathbf{e}_{\mathbf{q}i} (b_{\mathbf{q}i} + b_{-\mathbf{q}i}^\dagger) \mathrm{e}^{\mathrm{i}\mathbf{q}\mathbf{R}_j}.$$

Making use of

$$\sum_{\mathbf{q}i} \omega_{\mathbf{q}i}^{-1/2} \mathbf{e}_{\mathbf{q}i} b_{-\mathbf{q}i}^\dagger \mathrm{e}^{\mathrm{i}\mathbf{q}\mathbf{R}_j} = \sum_{\mathbf{q}i} \omega_{\mathbf{q}i}^{-1/2} \mathbf{e}_{\mathbf{q}i} b_{\mathbf{q}i}^\dagger \mathrm{e}^{-\mathrm{i}\mathbf{q}\mathbf{R}_j},$$

we write the displacement operator in the second-quantized form:

$$\hat{\mathbf{u}}_j = \left(\frac{\hbar}{2MN} \right)^{1/2} \sum_s \omega_s^{-1/2} \mathbf{e}_s \left(b_s \mathrm{e}^{\mathrm{i}\mathbf{q}\mathbf{R}_j} + b_s^\dagger \mathrm{e}^{-\mathrm{i}\mathbf{q}\mathbf{R}_j} \right), \tag{G.17}$$

where $s = (\mathbf{q}, i)$. Since we consider noninteracting phonons, the creation and annihilation operators in the Heisenberg representation satisfy

$$b_s^\dagger(t) = b_s^\dagger \, \mathrm{e}^{\mathrm{i}\omega_s t}, \qquad b_s(t) = b_s \, \mathrm{e}^{-\mathrm{i}\omega_s t}$$

(the proof is similar to the corresponding result (3.44) for noninteracting electrons). Then the Heisenberg representation of the displacement operator (G.17) is given by

$$\hat{\mathbf{u}}_j(t) = \left(\frac{\hbar}{2MN} \right)^{1/2} \sum_s \omega_s^{-1/2} \mathbf{e}_s \left(b_s \mathrm{e}^{\mathrm{i}(\mathbf{q}\mathbf{R}_j - \omega_s t)} + b_s^\dagger \mathrm{e}^{-\mathrm{i}(\mathbf{q}\mathbf{R}_j - \omega_s t)} \right). \tag{G.18}$$

G.2 Gaussian Distribution of the Normal Modes

First of all we define the probability density function. We start with the canonical average

$$\langle f(\hat{u}_j^\alpha)\rangle = \frac{1}{Z}\mathrm{Tr}\left(f(\hat{u}_j^\alpha)\,\mathrm{e}^{-\mathcal{H}/T}\right), \tag{G.19}$$

where $f(\hat{u}_j^\alpha)$ is a function of the displacement $\hat{u}_j^\alpha$ (hereafter the indices are omitted for brevity), $Z = \mathrm{Tr}\mathrm{e}^{-\mathcal{H}/T}$ is the partition function and $\mathcal{H}$ is the Hamiltonian of the crystal lattice. Assuming E_λ and ψ_λ to be the eigenvalues and eigenfunctions of the Hamiltonian:

$$\mathcal{H}\psi_\lambda = E_\lambda\psi_\lambda, \qquad \mathrm{e}^{-\mathcal{H}/T}\psi_\lambda = \mathrm{e}^{-E_\lambda/T}\psi_\lambda,$$

we transform the trace in (G.19) as

$$\langle f(\hat{u})\rangle = \frac{1}{Z}\sum_\lambda \mathrm{e}^{-E_\lambda/T}\int f(u)|\psi_\lambda(u)|^2\,\mathrm{d}u.$$

Introducing the probability density function

$$p(u) = \frac{1}{Z}\sum_\lambda |\psi_\lambda(u)|^2\,\mathrm{e}^{-E_\lambda/T},$$

we obtain the integral representation for the canonical average (G.19):

$$\langle f(\hat{u})\rangle = \int f(u)p(u)\,\mathrm{d}u. \tag{G.20}$$

We show that, for *noninteracting* phonons, $p(u)$ is the *Gaussian* probability density. This result was first derived by Bloch [7]. We prove it by calculating the characteristic function[1]:

$$\varphi(x) = \int p(u)\mathrm{e}^{\mathrm{i}xu}\,\mathrm{d}u, \tag{G.21}$$

which is the Fourier transform of $p(u)$. Hence the probability density function $p(u)$ is obtained by the inverse Fourier transform of the characteristic function:

$$p(u) = \frac{1}{2\pi}\int \varphi(x)\mathrm{e}^{-\mathrm{i}xu}\,\mathrm{d}x. \tag{G.22}$$

We calculate the characteristic function as follows. Since the integral representation (G.20) is equivalent to the canonical average (G.19), we rewrite (G.21) as

$$\varphi(x) = \frac{1}{Z}\mathrm{Tr}\left(\mathrm{e}^{\mathrm{i}x\hat{u}}\,\mathrm{e}^{-\mathcal{H}/T}\right) = \langle \mathrm{e}^{\mathrm{i}x\hat{u}}\rangle. \tag{G.23}$$

Taking expression (G.17) into account, we write the displacement operator $\hat{u}$ as

$$\hat{u} = \sum_s (\gamma_s b_s + \gamma_s^* b_s^\dagger), \tag{G.24}$$

where we introduced the shorthand notation

$$\gamma_s = \left(\frac{\hbar}{2MN}\right)^{1/2}\frac{\mathbf{e}_s}{\sqrt{\omega_s}}\,\mathrm{e}^{\mathrm{i}\mathbf{q}\mathbf{R}_j}. \tag{G.25}$$

[1] In [2] the calculation of the characteristic function is rather tedious. The formula proved by Mermin [8] allows to derive the characteristic function almost in one line [1].

Using expression (G.24) and formula for the canonical average of exponentials of operators linear in b_s and $b_s^\dagger$ (14.36), we write the characteristic function (G.23) in the form

$$\varphi(x) = \mathrm{e}^{-\frac{1}{2}x^2\sigma^2}, \tag{G.26}$$

where

$$\sigma^2 = \sum_s |\gamma_s|^2 \coth\left(\frac{\hbar\omega_s}{2T}\right). \tag{G.27}$$

Substituting (G.26) in (G.22), we obtain

$$p(u) = \frac{1}{2\pi}\int \mathrm{e}^{-\mathrm{i}xu}\mathrm{e}^{-\frac{1}{2}x^2\sigma^2}\,\mathrm{d}x.$$

After completing the square, we have

$$p(u) = \frac{1}{2\pi}\,\mathrm{e}^{-\frac{1}{2}u^2/\sigma^2}\int \mathrm{e}^{-\frac{1}{2}(x\sigma - \mathrm{i}u/\sigma)^2}\,\mathrm{d}x = \frac{1}{\sigma\sqrt{2\pi}}\,\mathrm{e}^{-\frac{1}{2}u^2/\sigma^2}.$$

The latter is the Gaussian probability density function with the zero mean and mean-square displacement $\langle u^2\rangle = \sigma^2$ (see Appendix A.3).

Now we can apply the above result to the DWF $\mathrm{e}^{-2W(\kappa)}$. Substituting (G.25) in (G.27), we have

$$\sigma^2 = \frac{\hbar}{2MN}\sum_s \frac{1}{\omega_s}\coth\left(\frac{\hbar\omega_s}{2T}\right). \tag{G.28}$$

Comparing the latter with the expression (14.41) for the DWF exponent:

$$2W(\kappa) = \frac{\hbar\kappa^2}{6MN}\sum_s \frac{1}{\omega_s}\coth\left(\frac{\hbar\omega_s}{2T}\right), \tag{G.29}$$

and taking $\langle u^2\rangle = \sigma^2$ into account, we obtain

$$2W(\kappa) = \frac{1}{3}\kappa^2\langle u^2\rangle.$$

G.3 The Debye Model

The DWF can be expressed in terms of the phonon density of states $n(\omega)$. By definition, the phonon density of states is given by

$$n(\omega) = \frac{\mathrm{d}N(\omega)}{\mathrm{d}\omega}, \tag{G.30}$$

where $N(\omega)$ is the number of normal modes with phonon frequencies less than or equal to ω. First, we calculate the number of normal modes $N(q)$ with magnitude of the phonon wavevectors less than or equal to q. Since the wavevector $\mathbf{q}$ takes N discrete values in the Brillouin zone, we have one $\mathbf{q}$ value per $(2\pi)^3/\mathcal{V}$ volume. Then $N(q)$ is obtained by dividing the volume of the sphere of radius q by the volume $(2\pi)^3/\mathcal{V}$ and multiplying by three (the number of polarization modes i):

$$N(q) = \frac{4\pi q^3\mathcal{V}}{(2\pi)^3}. \tag{G.31}$$

In the *Debye model* it is assumed that $\omega = vq$, where v is the velocity of sound. Substituting $q = \omega/v$ in (G.31), we have

$$N(\omega) = \frac{\omega^3\mathcal{V}}{2\pi^2 v^3}.$$

Differentiating the latter, we write (G.30) as

$$n(\omega) = \frac{3\omega^2 \mathcal{V}}{2\pi^2 v^3}. \tag{G.32}$$

Replacing the Brillouin zone by the equal-volume sphere of the Debye radius q_D, we have $N(q_D) = 3N$. Hence from (G.31) we obtain $q_D^3 = 6\pi^2 N/\mathcal{V}$. Substitution of the latter in $v^3 = q_D^3/\omega_D^3$ gives

$$v^3 = \frac{\mathcal{V}\omega_D^3}{6\pi^2 N}, \tag{G.33}$$

where the Debye frequency ω_D is the maximum frequency of the normal modes. Substituting (G.33) in (G.32), we obtain the final expression for the phonon density of states in the Debye model:

$$n(\omega) = \frac{9N\omega^2}{\omega_D^3}, \qquad 0 \le \omega \le \omega_D. \tag{G.34}$$

Summation over frequencies of the normal modes ω_s is converted to the integration over the frequencies with the phonon density of states $n(\omega)$ by the formula

$$\sum_s f(\omega_s) = \int f(\omega) n(\omega)\, d\omega, \tag{G.35}$$

where $f(\omega)$ is an arbitrary function. Using formula (G.35), we write the exponent of the Debye-Waller factor (G.29) in the form

$$2W(\kappa) = \frac{\hbar\kappa^2}{6MN} \int \frac{1}{\omega} \coth\left(\frac{\hbar\omega}{2T}\right) n(\omega)\, d\omega. \tag{G.36}$$

Substituting the Debye density of states (G.34) in (G.36), we have

$$2W(\kappa) = \frac{3\hbar\kappa^2}{2M\omega_D^3} \int_0^{\omega_D} \coth\left(\frac{\hbar\omega}{2T}\right) \omega\, d\omega. \tag{G.37}$$

At high temperatures, using $\coth x \approx 1/x$ for $x \ll 1$, we finally obtain

$$2W(\kappa) = \frac{3\hbar^2\kappa^2}{M} \frac{T}{\Theta_D^2}, \qquad T \gg \Theta_D, \tag{G.38}$$

where $\Theta_D = \hbar\omega_D$ is the Debye temperature.

References

1. G.V. Paradezhenko, N.B. Melnikov, B.I. Reser, arXiv:1711.04302v1 [cond-mat.mtrl-sci] (2017)
2. A.A. Maradudin, E.W. Montroll, G.H. Weiss, *Theory of Lattice Dynamics in the Harmonic Approximation* (Academic, New York, 1963)
3. J.A. Reissland, *Physics of Phonons* (Wiley, London, 1973)
4. G.L. Squires, *Introduction to the Theory of Thermal Neutron Scattering*, 2nd edn. (Dover, New York, 1996)
5. C. Kittel, *Introduction to Solid State Physics*, 8th edn. (Wiley, New York, 2005)
6. J. Quinn, K. Yi, *Solid State Physics. Principles and Modern Applications* (Springer, Berlin, 2009)
7. F. Bloch, Z. Physik **74**, 295 (1932)
8. N.D. Mermin, J. Math. Phys. **7**, 1038 (1966)

H Numerical Integral Transformations

H.1 Hilbert Transformation

As we showed in Chap. 2, the susceptibility is fully reconstructed from its imaginary part by applying the *Hilbert transformation* (see, e.g. [1, 2])

$$\chi(z) = -\frac{1}{\pi}\int \frac{\mathrm{Im}\chi(\omega')}{z-\omega'}\,\mathrm{d}\omega', \tag{H.1}$$

where z is a complex number from the upper half-plane and $\omega' = \omega' - \mathrm{i}0^+$. Indeed, the right-hand side of (H.1) can be written as

$$-\frac{1}{\pi}\int \frac{\mathrm{Im}\chi(\omega')}{z-\omega'+\mathrm{i}0^+}\,\mathrm{d}\omega' = \frac{1}{\pi}\int \frac{\mathrm{Im}\chi(\omega')}{\omega'-z-\mathrm{i}0^+}\,\mathrm{d}\omega'.$$

Using the Sokhotsky formula (A.44), we have

$$\frac{1}{\pi}\int \frac{\mathrm{Im}\chi(\omega')}{\omega'-z-\mathrm{i}0^+}\,\mathrm{d}\omega' = \frac{1}{\pi}\mathcal{P}\int \frac{\mathrm{Im}\chi(\omega')}{\omega'-z}\,\mathrm{d}\omega' + \mathrm{i}\mathrm{Im}\chi(z).$$

From the Kramers-Kronig relation (2.19), we finally obtain

$$\frac{1}{\pi}\int \frac{\mathrm{Im}\chi(\omega')}{\omega'-z-\mathrm{i}0^+}\,\mathrm{d}\omega' = \mathrm{Re}\chi(z) + \mathrm{i}\mathrm{Im}\chi(z) = \chi(z).$$

Direct calculation by formula (H.1) is problematic. Following [3], we obtain a formula that is applicable for numerical calculations.

First we convert the tabular function $\mathrm{Im}\chi(\omega')$ into a piecewise-linear function:

$$\mathrm{Im}\chi(\omega') = \begin{cases} a_i(\omega'-\omega_i)+b_i, & \omega_i \le \omega' \le \omega_{i+1}, \\ 0, & \omega' < \omega_1 \text{ and } \omega' > \omega_{n+1}, \end{cases} \tag{H.2}$$

where $a_i = (b_{i+1}-b_i)/(\omega_{i+1}-\omega_i)$, $b_i = -\pi^{-1}\mathrm{Im}\chi(\omega_i)$, $i = 1, \ldots, n+1$, and n is the number of intervals. Substitution of (H.2) into (H.1) gives

$$\chi(z) = \sum_{i=1}^{n}\int_{\omega_i}^{\omega_{i+1}} \frac{a_i(\omega'-\omega_i)+b_i}{z-\omega'}\,\mathrm{d}\omega'.$$

After simple calculations with $b_1 = b_{n+1} = 0$ we obtain

$$\chi(z) = \sum_{i=1}^{n+1} A_i(z-\omega_i)\ln(z-\omega_i) \equiv \sum_{i=1}^{n+1} A_i B_i(z), \tag{H.3}$$

where $A_1 = a_1$, $A_i = a_i - a_{i-1}$, $i = 2, \ldots, n$ and $A_{n+1} = -a_n$.

N. B. Melnikov, B. I. Reser, *Dynamic Spin Fluctuation Theory of Metallic Magnetism*,
https://doi.org/10.1007/978-3-319-92974-3

For complex numbers $z_j = \omega_j + \mathrm{i}\delta$, $\delta > 0$, $j = 1, \ldots, n$, given with a *uniform* step in ω, the calculation is significantly simplified. First, we store all the values A_i, which are independent of z. Then we compute and store $B_i(z_1)$, $i = 1, \ldots, n+1$. Since $z_j - \omega_i = z_{j-1} - \omega_{i-1}$, we have

$$B_i(z_j) = B_{i-1}(z_{j-1}),$$

where $B_i(z_j) = (z_j - \omega_i)\ln(z_j - \omega_i)$. Therefore, for each $z_j = \omega_j + \mathrm{i}\delta$ in the loop over $j = 2, \ldots, n+1$, we already know all $B_i(z_j)$, $i = 2, \ldots, n+1$ from the previous step, and so we need to calculate just one missing value $B_1(z_j)$. Thus, explicit calculation of *all* $B_i(z)$, $i = 1, \ldots, n+1$, is carried out only for $z = z_1$.

Similarly, we obtain a formula for numerical calculations of the single-site Green function $g(z)$. Just as the susceptibility, $g(z)$ is given by the Hilbert transform (10.6):

$$g(z) = \int \frac{\nu(\varepsilon')}{z - \varepsilon'}\,\mathrm{d}\varepsilon', \tag{H.4}$$

where $\nu(\varepsilon') = \pi^{-1}\mathrm{Im}g(\varepsilon')$ is the DOS and $\varepsilon' = \varepsilon' - \mathrm{i}0^+$ (the sign is different from the one in (H.1), because we use the *advanced* Green function). Using linear interpolation, we transform the tabular function $\nu(\varepsilon')$ into a piecewise-linear function

$$\nu(\varepsilon') = \begin{cases} a_i(\varepsilon' - \varepsilon_i) + b_i, & \varepsilon_i \le \varepsilon' \le \varepsilon_{i+1}, \\ 0, & \varepsilon' < \varepsilon_1 \text{ and } \varepsilon' > \varepsilon_{n+1}, \end{cases} \tag{H.5}$$

where $a_i = (b_{i+1} - b_i)/(\varepsilon_{i+1} - \varepsilon_i)$, $b_i = \nu(\varepsilon_i)$, $i = 1, \ldots, n+1$ and n is the number of intervals. Then $g(z)$ is given by [3]

$$g(z) = \sum_{i=1}^{n+1} A_i(z - \varepsilon_i)\ln(z - \varepsilon_i), \tag{H.6}$$

where $A_1 = a_1$, $A_i = a_i - a_{i-1}$, $i = 2, \ldots, n$, $A_{n+1} = -a_n$, just as in (H.3).

H.2 Integrals with the Fermi Function

Calculation of many quantities in solid state physics, such as the electron DOS and total energy, local magnetic moments and susceptibilities, reduces to the computation of the so-called *Fermi integrals*

$$I = \int_{-\infty}^{\infty} g(\varepsilon) f(\varepsilon)\,\mathrm{d}\varepsilon, \tag{H.7}$$

where $g(\varepsilon)$ is an arbitrary function vanishing fast enough as $\varepsilon \to \pm\infty$, and $f(\varepsilon) = [\exp((\varepsilon - \mu)/T) + 1]^{-1}$ is the Fermi function. In practice it is necessary to compute repeatedly various Fermi integrals for tabular functions $g(\varepsilon)$. We present a general numerical method for calculation of this kind of integrals developed by Reser [4].

The method works as follows. The tabular function $g(\varepsilon)$ is linearly interpolated by the formula

$$g(\varepsilon) = \sum_{i=1}^{N} \Delta g_i'(\varepsilon - \varepsilon_i)\theta(\varepsilon - \varepsilon_i), \tag{H.8}$$

where

$$\begin{aligned} &\Delta g_1' = g_1', \qquad \Delta g_i' = g_i' - g_{i-1}', \quad i = 2, 3, \ldots, N-1, \qquad \Delta g_N' = -g_{N-1}'; \\ &g_i' = (g_{i+1} - g_i)/(\varepsilon_{i+1} - \varepsilon_i), \qquad g_i \equiv g(\varepsilon_i), \quad i = 1, 2, \ldots, N-1, \end{aligned} \tag{H.9}$$

N is the number of interpolation points, and $\theta(x)$ is the step function equal to zero for $x < 0$ and unity for $x \ge 0$. Substitution of (H.8) into (H.7) gives

$$I = \sum_{i=1}^{N} \Delta g_i' \int_0^\infty f(\varepsilon, \xi_i, T)\varepsilon\, d\varepsilon = -\frac{1}{2}\sum_{i=1}^{N} \Delta g_i' \int_0^\infty \varepsilon^2 \frac{\partial f(\varepsilon, \xi_i, T)}{\partial \varepsilon}\, d\varepsilon, \tag{H.10}$$

where $\xi_i = \mu - \varepsilon_i$. The integral

$$J_1(\xi, T) \equiv \int_0^\infty \varepsilon^2 \frac{\partial f(\varepsilon, \xi, T)}{\partial \varepsilon}\, d\varepsilon \tag{H.11}$$

is represented as

$$J_1(\xi, T) = -2T^2 F_1(-\xi/T), \tag{H.12}$$

where

$$F_1(x) = \int_x^\infty \ln(1 + e^{-x_1})\, dx_1. \tag{H.13}$$

Taking (H.11) and (H.12) into account, we can write the integral (H.10) in the form

$$I(T) = T^2 \sum_{i=1}^{N} \Delta g_i' F_1(-\xi_i/T).$$

Dividing the sum over i into two sums over $x_i \equiv -\xi_i/T = -(\mu - \varepsilon_i)/T < 0$ and $x_i \geq 0$, and taking into account that [4]

$$F_1(-|x|) = \frac{x^2}{2} + \zeta(2) - F_1(|x|), \qquad x < 0, \tag{H.14}$$

we have

$$I(T) = T^2 \left\{ \sum_{x_i<0} \Delta g_i' \left[\frac{x_i^2}{2} + \zeta(2) - F_1(|x_i|) \right] + \sum_{x_i \geq 0} \Delta g_i' F_1(x_i) \right\}, \tag{H.15}$$

where $\zeta(2) = \pi^2/6$ is the Riemann zeta function. From formula (H.13) we see that $F_1(x_i) \to 0$ as $x_i \to 0^+$. Since $x_i = -(\mu - \varepsilon_i)/T$, the second term in (H.15) vanishes as $T \to 0$. The first term in (H.15) reduces to

$$I(0) = \frac{1}{2} \sum_{\xi_i>0} \Delta g_i' \xi_i^2.$$

From expression (H.15) we see that calculation of the Fermi integrals (H.7) by using the linear interpolation (H.8) is reduced to calculation of the integral $F_1(x)$ only for $x \geq 0$. The integral (H.13) cannot be calculated analytically. Substituting the power series of $\ln(1 + e^{-x_1})$ into (H.13) and integrating term by term, which is valid because of uniform convergence of the series at $x \geq 0$, we obtain

$$F_1(x) = \sum_{k=1}^{\infty} (-1)^{k+1} \frac{\exp(-kx)}{k^2}. \tag{H.16}$$

Hence, at $x \to \infty$ the function $F_1(x)$ decreases as $\exp(-x)$. Taking into account the asymptotic value at infinity, we have the following simple approximate formula for calculating $F_1(x)$:

$$F_1(x) \simeq F_1^{\mathrm{R}}(x) = \left[\exp(x) + q_1\right]^{-1}, \tag{H.17}$$

where the parameter q_1 is chosen so that $F_1(x)$ and $F_1^{\mathrm{R}}(x)$ coincide at $x = 0$. Using (H.14), we obtain

$$q_1 = [F_1(0)]^{-1} - 1 = \frac{2}{\zeta(2)} - 1. \tag{H.18}$$

Numerical calculations show that the error $\Delta F_1^{\mathrm{R}}(x) = |F_1(x) - F_1^{\mathrm{R}}(x)|$ in calculation of $F_1(x)$ by (H.17) over the whole interval $[0, \infty)$ does not exceed 3.1×10^{-3}. For many problems using the Fermi integrals, this accuracy of the $F_1(x)$ calculation is quite sufficient. For exact values of $F_1(x)$ we used partial sums of the expansion (H.16) by which $F_1(x)$ can be computed to any required accuracy. Since the series (H.16) is alternating and its elements monotonically decrease in absolute value, the absolute value of the remainder is less than its first term. Clearly, for practical calculation of $F_1(x)$ at small x the series (H.16) is unsuitable; for calculation of $F_1(x)$, say with the accuracy 10^{-8}, it is necessary to add 10^4 terms together.

The following method is suitable for effective and practical exact calculation of the integrals $F_1(x)$. By substituting $t = \exp(-x_1)$ the integral $F_1(x)$ is reduced to a dilogarithm:

$$F_1(x) = L[\exp(-x)], \qquad x \geq 0,$$

$$L(y) = \int_0^y \frac{\ln(1+t)}{t}\,\mathrm{d}t, \qquad 0 \leq y \leq 1.$$

Using the expansion of a dilogarithm in the series of Chebyshev polynomials

$$L(y) = \sum_{k=1}^{\infty} a_k T_k^*(y), \qquad 0 \leq y \leq 1,$$

where a_k are the Chebyshev coefficients, and

$$T_k^*(y) = \sum_{j=0}^{k} b_{kj} y^{k-j}$$

are shifted Chebyshev polynomials of the first kind, we obtain

$$F_1(x) = \sum_{k=0}^{\infty} a_k T_k^*[\exp(-x)] = \sum_{k=0}^{\infty} a_k \sum_{j=0}^{k} b_{kj} \exp[-(k-j)x], \qquad x \geq 0. \tag{H.19}$$

Since these series converge very rapidly and the coefficients a_k and b_{kj} are known up to large k-values (see, e.g. [5, pp. 74 and 494]), the integral $F_1(x)$ can be calculated with a high accuracy from (H.19). The mth partial sum of Chebyshev polynomials series

$$F_{1m}^{\mathrm{C}}(x) = \sum_{k=0}^{m} a_k \sum_{j=0}^{k} b_{kj} \exp[-(k-j)x], \qquad x \geq 0, \tag{H.20}$$

gives an approximation of the integral $F_1(x)$ with the maximum error decreasing very quickly with increase in m; already at $m = 7$ the error becomes equal to 4.5×10^{-8}, i.e. five orders smaller than in approximation by (H.17) (see Table 1 in [4]). However, unlike $\Delta F_1^{\mathrm{R}}(x)$ that vanishes with increase in x, the error $\Delta F_{1m}^{\mathrm{C}}(x)$ for large values of x remains small but finite:

$$F_{1m}^{\mathrm{C}}(\infty) = \lim_{x\to\infty} F_{1m}^{\mathrm{C}}(x) = \sum_{k=0}^{m} a_k b_{kk}.$$

Therefore, in the calculation of the integral $F_1(x)$ the function $F_{1m}^{\mathrm{C}}(x)$ is substituted by the function

$$\tilde{F}_{1m}^{\mathrm{C}}(x) \equiv F_{1m}^{\mathrm{C}}(x) - F_{1m}^{\mathrm{C}}(\infty) = \sum_{k=1}^{m} a_k \sum_{j=0}^{k-1} b_{kj} \exp[-(k-j)x], \tag{H.21}$$

which has correct asymptotic behaviour at infinity.

The subroutine for calculation of the Fermi integrals $I(T)$, named FINT and based on the method outlined above, was presented in [6]. The subroutine allows us to calculate $I(T)$ using both the simple approximate formula (H.17) and the Chebyshev approximation (H.21). By FINT the Fermi integral can be calculated, if necessary, to any required accuracy. The subroutine FINT is written in FORTRAN and is included in the illustrating program that runs a test. As the test, the program calculates the Fermi integral of the model electron DOS for iron at the Curie temperature (for details, see [4, 6]).

H.3 Integrals with the Derivative of the Fermi Function

Here we present a simple method [7] for calculation of the integrals with the derivative of the Fermi function

$$I = \int_{-\infty}^{\infty} g(\varepsilon) \left(-\frac{\partial f(\varepsilon)}{\partial \varepsilon} \right) \mathrm{d}\varepsilon \tag{H.22}$$

under the same assumptions as in the previous section. Let us represent the function $f(\varepsilon)$ as a sum of the step function

$$f_1(\varepsilon - \mu) = \begin{cases} 1, & \text{when } \varepsilon \le \mu, \\ 0, & \text{when } \varepsilon > \mu, \end{cases}$$

(it would be the Fermi function at $T = 0$ if we put ε_F instead of μ) and the remainder

$$f_2(\varepsilon) = \mathrm{sgn}(\varepsilon - \mu) \left[e^{|\varepsilon - \mu|/T} + 1 \right]^{-1} = -T \frac{\partial}{\partial \varepsilon} \ln \left(1 + e^{-|\varepsilon - \mu|/T} \right). \tag{H.23}$$

Substituting $f(\varepsilon) = f_1(\varepsilon - \mu) + f_2(\varepsilon)$ into (H.22), we split the integral I into two integrals:

$$I = \int g(\varepsilon) \left(-\frac{\partial f_1(\varepsilon - \mu)}{\partial \varepsilon} \right) \mathrm{d}\varepsilon + \int g(\varepsilon) \left(-\frac{\partial f_2(\varepsilon)}{\partial \varepsilon} \right) \mathrm{d}\varepsilon \equiv I_1 + I_2.$$

For the first integral, using the properties of the delta-function (A.36) and (A.38), we immediately obtain

$$I_1 = \int g(\varepsilon)\, \delta(\varepsilon - \mu)\, \mathrm{d}\varepsilon = g(\mu). \tag{H.24}$$

For the second one we first use integration by parts:

$$I_2 = \int \frac{\partial g(\varepsilon)}{\partial \varepsilon} f_2(\varepsilon)\, \mathrm{d}\varepsilon. \tag{H.25}$$

Using the linear interpolation (H.5) with $b_i = g(\varepsilon_i)$, we transform the tabular function $g(\varepsilon)$ into a piecewise-linear function. Then substitution of this piecewise-linear function $g(\varepsilon)$ and (H.23) into (H.25) yields

$$I_2 = \sum_{i=1}^{n} a_i \int_{\varepsilon_i}^{\varepsilon_{i+1}} f_2(\varepsilon)\, \mathrm{d}\varepsilon = T \sum_{i=1}^{n} a_i \left[\ln \left(1 + \mathrm{e}^{-|\varepsilon_i - \mu|/T} \right) - \ln \left(1 + \mathrm{e}^{-|\varepsilon_{i+1} - \mu|/T} \right) \right]. \tag{H.26}$$

Taking into account (H.24) and (H.26), for the initial integral (H.22) we obtain

$$I = g(\mu) + T {\sum_i}' a_i \left[\ln \left(1 + \mathrm{e}^{-|\varepsilon_i - \mu|/T} \right) - \ln \left(1 + \mathrm{e}^{-|\varepsilon_{i+1} - \mu|/T} \right) \right]. \tag{H.27}$$

The prime indicates that the sum includes only the terms for which $|\varepsilon_i - \mu| \le t \ln \beta T$, where β is the base and t precision of the floating-point system in a specific computer (see, e.g. [8]).

H.4 Integrals over the Brillouin Zone

Calculation of the mean-square fluctuations in the DSFT requires evaluation of the integral over the wavevector modulus (11.6):

$$I = \int_0^1 \frac{1}{a^2 + b^2 k^2} \frac{2}{\pi} \arctan \frac{c}{a^2 + b^2 k^2} 3k^2\, \mathrm{d}k, \qquad 0 < a^2 < 1, \tag{H.28}$$

where a, b and c are given by (10.25) (we omit the index α). To obtain an approximate analytic formula for the integral (H.28), we replace the integration variable k by $x = bk$ and approximate the function $(2/\pi)\arctan(c/(a^2+x^2))$ by two asymptotic expansions matched (together with their derivatives) at the point $c/(a^2+x^2) = 1$. This leads to

$$I = \frac{3}{b^3}\int_0^b \mathrm{d}x \frac{x^2}{a^2+x^2}\begin{cases} \dfrac{1}{2}\dfrac{c}{a^2+x^2}, & \dfrac{c}{a^2+x^2} \le 1, \\ 1 - \dfrac{1}{2}\dfrac{a^2+x^2}{c}, & \dfrac{c}{a^2+x^2} > 1. \end{cases}$$

Then, introducing the notation

$$\begin{aligned} F_1(x) &= \frac{c}{2}\int \frac{x^2}{(a^2+x^2)^2}\,\mathrm{d}x = \frac{c}{2}\left(-\frac{x}{2(a^2+x^2)} + \frac{1}{2a}\arctan\frac{x}{a}\right), \\ F_2(x) &= \int \frac{x^2}{a^2+x^2}\,\mathrm{d}x - \frac{1}{2c}\int x^2\,\mathrm{d}x = x - a\arctan\frac{x}{a} - \frac{x^3}{6c}, \end{aligned} \tag{H.29}$$

we finally obtain

$$I = \frac{3}{b^3}\begin{cases} F_1(b), & x_0^2 \le 0, \\ F_2(x_0) + F_1(b) - F_1(x_0), & 0 < x_0 < b, \\ F_2(b), & x_0 \ge b, \end{cases} \tag{H.30}$$

where $x_0^2 = c - a^2$.

H.5 Inverse Cosine Transformation

For calculating the oscillating part $A(t)$ of the temporal correlation function $F(t)$ (see relation (13.8)), we need an approximate formula for the Fourier cosine transformation (for details see, e.g. [9])

$$A(t) = \frac{1}{\pi}\int_0^\infty A(\omega)\cos(\omega t)\,\mathrm{d}\omega. \tag{H.31}$$

In the DFT calculations [10], the inverse Fourier cosine transform of the function $A(\omega)$ given at a uniform grid with the step h was computed by interpolating $A(\omega)$ in each segment of the length $2h$ with a second-degree polynomial and then integrating it analytically by parts (see Sect. 13.3.2). In the DSFT, the function $A(\omega)$ is specified at a fairly fine mesh; therefore linear interpolation of the form (H.2) with $b_i = A(\omega_i)$ proves to be sufficient. Substituting the linear interpolation in (H.31), after simple calculations with $\omega_1 = 0$, similar to (H.3) we obtain [3]

$$A(t) = \frac{1}{\pi t}\left[b_{n+1}\sin(\omega_{n+1}t) - \frac{2}{t}\sin\left(\frac{h}{2}t\right)\sum_{i=1}^{n} a_i \sin\left(\frac{\omega_{i+1}+\omega_i}{2}t\right)\right]. \tag{H.32}$$

References

1. V.A. Ditkin, A.P. Prudnikov, *Integral Transforms and Operational Calculus* (Pergamon, New York, 1965)
2. F.W. King, *Hilbert Transforms* (Cambridge University Press, New York, 2009)
3. B.I. Reser, Phys. Met. Metallogr. **77**, 451 (1994)
4. B.I. Reser, J. Phys. Condens. Matter **8**, 3151 (1996)
5. Y.L. Luke, *Mathematical Functions and their Approximations* (Academic, New York, 1975)
6. B.I. Reser, V.B. Lyskov, Calculation of the Fermi integrals on a computer (1995) [in Russian]. Dep. VINITI 1267-B95
7. B.I. Reser, J. Phys. Condens. Matter **11**, 4871 (1999)

8. G.E. Forsythe, M.A. Malcolm, C.B. Moler, *Computer Methods for Mathematical Computations* (Prentice-Hall, Englewood Cliffs, 1977)
9. V.I. Krylov, N.S. Skoblya, *A Handbook of Methods of Approximate Fourier Transformation and Inversion of the Laplace Transformation* (Mir Publ., Moscow, 1977)
10. B.I. Reser, E.V. Rosenfeld, E.V. Shipitsyn, Phys. Met. Metallogr. **69**(6), 48 (1990)

DSFT Solution Methods and Software

I

We obtain the temperature dependence of magnetic characteristics using the parameter continuation (Chap. 10). As we discussed in Chap. 11, it can become unstable at high temperatures. In this case, solution of the DSFT system of nonlinear equations becomes extremely sensitive to the choice of the initial condition, and a smaller step size in temperature is required. For the same reason, we avoid numerical methods that use derivatives, because their difference approximations can destroy the stability. Instead, we use the coordinate bisection method and coordinate descent methods. Here we describe these solution methods [1, 2] and give a brief overview of the MAGPROP software suite [3][1].

I.1 Solution Methods

We consider a system of nonlinear equations in the n-dimensional Euclidian space:

$$\begin{aligned} \varphi_1(x_1, x_2, \dots, x_n) &= 0, \\ \varphi_2(x_1, x_2, \dots, x_n) &= 0, \\ &\dots\dots\dots\dots\dots \\ \varphi_n(x_1, x_2, \dots, x_n) &= 0. \end{aligned} \tag{I.1}$$

Let the system (I.1) have at least one solution such that its ith component belongs to the interval $[a_i, b_i]$, $i = 1, \dots, n$.

I.1.1 Coordinate Bisection Method

One-dimensional bisection method is well known, but we have not come across any generalizations of the bisection method to *systems* of equations. The *coordinate bisection method* was developed and used for calculations of magnetic properties in the paper [4]. To a general system of equations of the form (I.1) the method is applied as follows [1]. We fix all variables but the first one: $x_2 = a_2, \dots, x_n = a_n$. Then $\varphi_1(x_1, a_2, \dots, a_n)$ is a function of one variable x_1. In the interval $[a_1, b_1]$, we find the root $x_1^{(1)}(a_2, a_3, \dots, a_n)$ of the first equation of system (I.1) by the (one-dimensional) bisection method. The value of the function φ_2 at the left endpoint of the interval for x_2 will be $\varphi_2(x_1^{(1)}, a_2, \dots, a_n)$. Next, we set $x_2 = b_2$ and find the root of the first equation of the system $x_1^{(2)}(b_2, a_3, \dots, a_n)$ by the bisection method. With the new x_1, the value of the function φ_2 at the right endpoint of the interval for x_2 will be $\varphi_2(x_1^{(2)}, b_2, a_3, \dots, a_n)$. Continuing the bisection of the second equation, we find the root $x_2^{(1)}(a_3, a_4, \dots, a_n)$, with the root x_1 being updated to a certain $x_1^{(k_1)}(x_2^{(1)}, a_3, \dots, a_n)$. Now the value of the third function at the left endpoint $x_3 = a_3$ is equal to $\varphi_3(x_1^{(k_1)}, x_2^{(1)}, a_3, a_4, \dots, a_n)$. Similarly, at $x_3 = b_3$ we find the root of the second equation $x_2^{(2)}(b_3, a_4, \dots, a_n)$ by the bisection method, with the root of the first equation updated to a certain $x_1^{(k_1+k_2)}$. Then the value of the third function at the right endpoint for x_3 will be $\varphi_3(x_1^{(k_1+k_2)}, x_2^{(2)}, b_3, a_4, \dots, a_n)$.

[1] Upgraded version of the program is B.I. Reser, G.V. Paradezhenko, N.B. Melnikov, Program suite MAGPROP 2.0. Federal Service for Intellectual Property (ROSPATENT), RU 2018617208, 2018.

N. B. Melnikov, B. I. Reser, *Dynamic Spin Fluctuation Theory of Metallic Magnetism*,
https://doi.org/10.1007/978-3-319-92974-3

Continuing the bisection of the third equation, we obtain the root of the third equation $x_3^{(1)}(a_4, \ldots, a_n)$, with x_2 and x_1 updated to $x_2^{(l_1)}(x_3^{(1)}, a_4, \ldots, a_n)$ and $x_1^{(k_1+k_2+\cdots+k_{l_1})}(x_2^{(l_1)}, x_3^{(1)}, a_4, \ldots, a_n)$, respectively. Analogous calculations for the equations $\varphi_4, \ldots, \varphi_n$ give the roots $x_4, \ldots, x_n$. Thus, the solution of the system of nonlinear equations (I.1) reduces to a multiple solution of one-dimensional problems by the bisection method.

An advantage of the coordinate bisection method is that it does not use either derivatives or their approximations as opposed to quasi-Newton methods (see the next subsection). A large number of steps in the coordinate bisection method is not a serious weakness for the solution of the DSFT system (11.1)–(11.4):

$$\varphi_1(\mu, \bar{s}^z, \zeta^x, \zeta^z) \equiv n_\uparrow + n_\downarrow - n_e = 0, \tag{I.2}$$

$$\varphi_2(\mu, \bar{s}^z, \zeta^x, \zeta^z) \equiv (n_\uparrow - n_\downarrow)/2 - \bar{s}^z = 0, \tag{I.3}$$

$$\varphi_3(\mu, \bar{s}^z, \zeta^x, \zeta^z) \equiv uT/(2\lambda_L^x)I^x - \zeta^x = 0, \tag{I.4}$$

$$\varphi_4(\mu, \bar{s}^z, \zeta^x, \zeta^z) \equiv uT/(2\lambda_L^z)I^z - \zeta^z = 0, \tag{I.5}$$

because its dimension is not high and the computation time for the functions φ_i is small. To secure convergence of the method it is necessary to specify the search domain with precision (for details, see [2]). Indeed, if the intervals $[a_i, b_i]$ are too large, the solution cannot be found (in contrast to the one-dimensional bisection, where a root is always found once there is a sign change of the function). The specific character of the problem under consideration allows us to choose the initial approximation at zero temperature with great precision, and the small step size over the temperature ensures the proximity of the initial approximation in the successive calculations.

Since the spin fluctuations ζ^x and ζ^z at each temperature are of the same order, one can use the *mean* fluctuation in the numerical calculations. This idea was successfully used in the coordinate bisection method. The modification of the method is as follows. Instead of the spin fluctuations ζ^x and ζ^z, we introduce the mean value $\bar{\zeta} = (2\zeta^x + \zeta^z)/3$ and the difference $\Delta\zeta = \zeta^x - \zeta^z$ (recall that $\zeta^y = \zeta^x$). Substituting

$$\zeta^x = \bar{\zeta} + \Delta\zeta/3, \qquad \zeta^z = \bar{\zeta} - 2\Delta\zeta/3, \tag{I.6}$$

we transform the initial system (I.2)–(I.5) to an equivalent one

$$\tilde{\varphi}_1(\mu, \bar{s}^z, \bar{\zeta}, \Delta\zeta) = 0, \tag{I.7}$$

$$\tilde{\varphi}_2(\mu, \bar{s}^z, \bar{\zeta}, \Delta\zeta) = 0, \tag{I.8}$$

$$\tilde{\varphi}_3(\mu, \bar{s}^z, \bar{\zeta}, \Delta\zeta) \equiv (2\tilde{g}_3(\mu, \bar{s}^z, \bar{\zeta}, \Delta\zeta) + \tilde{g}_4(\mu, \bar{s}^z, \bar{\zeta}, \Delta\zeta))/3 - \bar{\zeta} = 0, \tag{I.9}$$

$$\tilde{\varphi}_4(\mu, \bar{s}^z, \bar{\zeta}, \Delta\zeta) \equiv \tilde{g}_3(\mu, \bar{s}^z, \bar{\zeta}, \Delta\zeta) - \tilde{g}_4(\mu, \bar{s}^z, \bar{\zeta}, \Delta\zeta) - \Delta\zeta = 0, \tag{I.10}$$

where

$$g_3(\mu, \bar{s}^z, \zeta^x, \zeta^z) = uT/(2\lambda_L^x)I^x, \qquad g_4(\mu, \bar{s}^z, \zeta^x, \zeta^z) = uT/(2\lambda_L^z)I^z,$$

and tilde stands for the result of the substitution (I.6). Now, instead of system (I.7)–(I.10), we solve the system of the first three equations (I.7)–(I.9), as if $\Delta\zeta$ were fixed, while at each step $\Delta\zeta$ is refined using the fixed-point iterations of the last equation:

$$\Delta\zeta_{k+1} = \tilde{g}_3(\mu_k, \bar{s}_k^z, \bar{\zeta}_k, \Delta\zeta_k) - \tilde{g}_4(\mu_k, \bar{s}_k^z, \bar{\zeta}_k, \Delta\zeta_k),$$

where $k = 0, 1, 2, \ldots$ is the step number of the solution algorithm for the system (I.7)–(I.9), and $\bar{\zeta}_0$ and $\Delta\zeta_0$ are taken from the calculation with the preceding value of the temperature. At each step, the values of μ_k, $\bar{s}_k^z$ and $\bar{\zeta}_k$ are determined by the coordinate bisection method with a small step size in $\bar{\zeta}$. That means the search interval for $\bar{\zeta}$ is additionally divided into N subinterval of equal length. If a sign change of $\tilde{\varphi}_3$ occurs at the first subinterval, it is further used for more accurate calculation of $\bar{\zeta}$. Otherwise, we check for a sign change at the second subinterval, etc.

To control the accuracy of the numerical solution at a fixed temperature, we explicitly calculate the residual by substituting the coordinate bisection results μ, $\bar{s}^z$, ζ^x and ζ^z into the initial system of equations (I.2)–(I.5).

I.1.2 Multidimensional Minimization

Solution of the system of nonlinear equations (I.1) can be reduced to minimization of the function

$$\Phi(x_1, x_2, \ldots, x_n) = \sum_{i=1}^{n} \varphi_i^2(x_1, x_2, \ldots, x_n). \tag{I.11}$$

The function (I.11) is nonnegative and vanishes if and only if all the equations $\varphi_i(x) = 0$ are satisfied. Often φ_i have different orders of magnitude. Therefore, the function that has the smallest order of magnitude might be practically ignored in the minimization process. This is usually prevented by introducing new functions that are properly scaled: $f_i = c_i\varphi_i$, where c_i are the scale factor coefficients chosen according to a specific character of the problem. Then the system of equations (I.1) is reduced to the minimization problem

$$F(x_1, x_2, \ldots, x_n) = \sum_{i=1}^{n} f_i^2(x_1, x_2, \ldots, x_n). \tag{I.12}$$

To solve this problem one can use an iterative method (see, e.g. [5, 6]). In the case of the fixed-point iteration, from the explicit form of the functions φ_2, φ_3 and φ_4 in (I.2)–(I.5) we see that their values are exactly equal to the differences of the respective variables from the current and preceding steps:

$$\varphi_2(\mu_k, \bar{s}_k^z, \zeta_k^x, \zeta_k^z) = \bar{s}_k^z - \bar{s}_{k-1}^z, \tag{I.13}$$

$$\varphi_3(\mu_k, \bar{s}_k^z, \zeta_k^x, \zeta_k^z) = \zeta_k^x - \zeta_{k-1}^x, \tag{I.14}$$

$$\varphi_4(\mu_k, \bar{s}_k^z, \zeta_k^x, \zeta_k^z) = \zeta_k^z - \zeta_{k-1}^z. \tag{I.15}$$

If another iterative method is used, it is natural to assume that these relations are satisfied approximately for the iterations close enough to the minimum point. Therefore, the scale factor coefficients c_2, c_3 and c_4 are chosen in such a way that the magnitudes of f_2, f_3 and f_4 are approximately equal to the relative errors:

$$c_1 = 1, \qquad c_2 = 1/\bar{s}_z(0), \qquad c_3 = c_4 = 1/(\bar{V}_z(0))^2 = 1/(u\bar{s}_z(0))^2.$$

Quasi-Newton Minimization Methods

The classical Newton's method for a function minimization operates as follows (see, e.g. [7, 8]). At the kth step, to refine the approximation $x^{(k)}$ of the local minimum x^* of the multivariable function $F(x)$ one constructs its quadratic Taylor approximation

$$F(x) \approx F(x^{(k)}) + (x - x^{(k)})^{\mathrm{T}}\nabla F(x^{(k)}) + \frac{1}{2}(x - x^{(k)})^{\mathrm{T}}\nabla^2 F(x^{(k)})(x - x^{(k)}), \tag{I.16}$$

where $(x - x^{(k)})^{\mathrm{T}}$ is the row vector, $\nabla F(x)$ is the gradient of the function F at the point x and $\nabla^2 F(x)$ is the *Hessian matrix* of the second partial derivatives of the function F at the point x. The minimum of the function (I.16) is attained at the point

$$x^{(k+1)} = x^{(k)} - [\nabla^2 F(x^{(k)})]^{-1}\nabla F(x^{(k)}), \tag{I.17}$$

which is taken as the next approximation. Then

$$F(x^{(k+1)}) \approx F(x^{(k)}) - \frac{1}{2}(x^{(k+1)} - x^{(k)})^{\mathrm{T}}\nabla^2 F(x^{(k)})(x^{(k+1)} - x^{(k)}).$$

Hence the value of the function at the kth step decreases if the matrix $\nabla^2 F(x^{(k)})$ is positively defined. It is not difficult to show that Newton's method converges in a sufficiently small neighbourhood of a local minimum if the function has continuous first and second derivatives. Moreover, the rate of convergence is quadratic near the solution.

Newton's method is mostly used as a part of a more general algorithm. The reason is that far from the minimum point the matrix $\nabla^2 F(x^{(k)})$ may be not positively defined, and thus the shift by the vector

$$p = -[\nabla^2 F(x^{(k)})]^{-1} \nabla F(x^{(k)})$$

in formula (I.17) may not lead to decrease of the function. In this case the matrix $\nabla^2 F(x^{(k)})$ is replaced by a close symmetric positively defined matrix $B^{(k)}$. Positive definiteness of the matrix $B^{(k)}$ ensures that the function decreases: $F(x^{(k+1)}) < F(x^{(k)})$. The step direction p is now obtained as the solution of the system $B^{(k)} p = -\nabla F(x^{(k)})$ and the next iteration is chosen along this direction, but not necessarily all the way: $x^{(k+1)} = x^{(k)} + \alpha p$. The length of the step $\alpha > 0$ in the direction p can be obtained by one-dimensional optimization method.

The methods of this kind are called *quasi-Newton methods*. In practice, the vector $\nabla F(x^{(k)})$ is usually replaced by a finite-difference approximation. The matrix $B^{(k)}$ is typically updated by a simple low-rank matrix.

To apply quasi-Newton method, we used the unconstrained minimization double precision (UNCMND) Fortran routine [6], which seeks a minimum of a function with line search. Results of the calculations did not give satisfactory agreement with the ones of the coordinate bisection method even in the temperature interval $T \leq 0.68\, T_C^{\exp}$. Instead of the solution to the absolute minimum of the function (I.12), which is the solution of the system of nonlinear equations (I.1), the routine converges to a local minimum far off the absolute one. It is necessary to note that the routine UNCMND does not give a user full control over the low level parameters (such as maximal step size). The routine also assumes that the function values are obtained accurately (to an accuracy comparable to the precision of the computer arithmetic). Since we aimed at implementing a reliable method that would work in a wide range of temperatures, we rejected the idea of tuning the quasi-Newton method. Instead, we implemented coordinate descent method, which is more slow but easier to control.

Coordinate Descent Method

Multidimensional minimization of the function $F(x_1, x_2, \ldots, x_n)$ can be performed by the slower but more reliable coordinate descent method. The gist of the method is as follows. As an initial approximation, we choose a point M_0 with coordinates $(x_1^{(0)}, x_2^{(0)}, \ldots, x_n^{(0)})$. We fix all coordinates but the first one. Then $F(x_1, x_2^{(0)}, \ldots, x_n^{(0)})$ is a function of one variable, x_1. Solving the one-dimensional optimization problem for this function, we replace the point M_0 by the point $M_1 = (x_1^{(1)}, x_2^{(0)}, \ldots, x_n^{(0)})$, where the function F takes on the minimal value with respect to x_1 with other coordinates being fixed. Now we fix all coordinates except for x_2, and consider $F(x_1^{(1)}, x_2, x_3^{(0)}, \ldots, x_n^{(0)})$ as a function of this coordinate. Once again solving the one-dimensional optimization problem, we find its minimum point $x_2 = x_2^{(1)}$, which gives us the next point $M_2 = (x_1^{(1)}, x_2^{(1)}, x_3^{(0)}, \ldots, x_n^{(0)})$. Similarly, we perform the descent over the coordinates $x_3, x_4, \ldots, x_n$, then start a new cycle from x_1 to x_n and so on. Finally, there is a sequence of the points $M_0, M_1, \ldots$ such that the values of the function F at these points form a nonincreasing sequence $F(M_0) \geq F(M_1) \geq \ldots$. The process stops when we reach either accuracy of the function, or tolerance of the arguments, or the maximum number of iterations (cycles).

For smooth functions, given a good initial approximation, coordinate descent converges to the minimum. Among the advantages of the coordinate descent is the possibility to use simple algorithms of one-dimensional optimization. For one-dimensional minimization, we use the method that combines the golden-section search with successive parabolic interpolation [5,6]. The method is characterized by the quadratic rate of convergence in a neighbourhood of the minimum of a smooth function and guaranteed linear rate of convergence in the case of nonsmooth functions.

I.2 MAGPROP Software Suite

We calculate the temperature dependence of magnetic characteristics in transition metals and alloys in the DSFT using the software suite MAGPROP [3], which is written in Fortran.

The program MAGPROP allows the user to solve the DSFT system of nonlinear equations using one of the two numerical methods: coordinate bisection method (subroutine TMAGB) or coordinate descent method (subroutine TMAGD). The initial physical parameters of the calculation are the nonmagnetic DOS of a particular metal and its value of the mean magnetic moment at absolute zero. All quantities that have dimensions of energy and temperature (in energy units) are used in units of the energy bandwidth W.

MAGPROP has three possible models for the spin fluctuations:

- only *transverse* fluctuations in the plane perpendicular to the magnetization;
- only *longitudinal* fluctuations along the magnetization axis;
- *all* fluctuations in the three-dimensional space.

One can also run the program with the *mean* fluctuation and without fluctuations (Stoner mean-field theory). Moreover, the program allows to use four different approximations in the SFT: static local, static nonlocal, dynamic local and dynamic nonlocal.

In the dynamic nonlocal approximation of the SFT there is an option to reduce the spin fluctuations by

- changing the effective interaction constant in the expression for the mean-square fluctuation;
- adding a correction term to the denominator of the enhanced susceptibility;
- explicitly taking higher-order terms of the free energy into account.

In the last case it is possible to use either a simplified formula with free parameters or the expression obtained by the partial Gaussian averaging. The latter is the most consistent variant from the theoretical point of view.

The user interface allows to assign the temperature interval $[T_1, T_2]$, where the temperature dependence is investigated, and the step size in temperature ΔT. There is a possibility to go in the positive temperature direction (from T_1 to T_2) as well as in the negative direction (from T_2 to T_1) with a temperature step ΔT. When $\Delta T > 0$, there are two possible modes: with automatic transition from ferro- to paramagnetic region (once the mean field vanishes, it is set to zero for larger temperatures) and without it (MAGPROP solves all four DSFT equations at all temperatures). When $\Delta T < 0$, there is only one variant: with no automatic transition from para- to ferromagnetic region (MAGPROP solves all four DSFT equations at all temperatures).

Besides the mean spin moment and mean-square fluctuations, one can obtain temperature dependencies for other magnetic characteristics: local and effective magnetic moments, ferro- and paramagnetic Curie temperatures, inverse paramagnetic susceptibility, nuclear spin relaxation rates, etc.

References

1. B.I. Reser, N.B. Melnikov, J. Phys. Condens. Matter **20**, 285205 (2008)
2. N.B. Melnikov, B.I. Reser, in *Proceedings of Dynamic Systems and Applications*, vol. 5 (Dynamic Publ., Atlanta, 2008), pp. 312–316
3. B.I. Reser, N.B. Melnikov, Software suite MAGPROP for calculating magnetic properties of transition metals and alloys in the dynamic spin-fluctuation theory [in Russian]. Joint Fund of Electronic Resources for Science and Education, Software certificate No. 17257, July 12 (2011)
4. B.I. Reser, Phys. Met. Metallogr. **77**, 451 (1994)
5. G.E. Forsythe, M.A. Malcolm, C.B. Moler, *Computer Methods for Mathematical Computations* (Prentice-Hall, Englewood Cliffs, 1977)
6. D. Kahaner, C. Moler, S. Nash, *Numerical Methods and Software* (Prentice-Hall, Englewood Cliffs, 1989)
7. P.E. Gill, W. Murray, M.H. Wright, *Practical Optimization* (Academic, London, 1981)
8. J.E. Dennis Jr., R.B. Schnabel, *Numerical Methods for Unconstrained Optimization and Nonlinear Equations* (Prentice-Hall, Englewood Cliffs, 1983)

Index

N. B. Melnikov, B. I. Reser, *Dynamic Spin Fluctuation Theory of Metallic Magnetism*,
https://doi.org/10.1007/978-3-319-92974-3

MIX
Papier aus verantwortungsvollen Quellen
Paper from responsible sources
FSC® C105338

If you have any concerns about our products,
you can contact us on
ProductSafety@springernature.com

In case Publisher is established outside the EU,
the EU authorized representative is:
Springer Nature Customer Service Center GmbH
Europaplatz 3, 69115 Heidelberg, Germany

Printed by Libri Plureos GmbH
in Hamburg, Germany